Pacific Forest

Pacific Forest

A History of Resource Control and Contest
in Solomon Islands, c. 1800–1997

Judith A. Bennett

The White Horse Press

Copyright © Judith A. Bennett, 2000

First published 2000 by

The White Horse Press, 10 High Street, Knapwell, Cambridge CB3 8NR, UK
and
Brill Academic Publishers, PO Box 9000, 2300 PA Leiden, The Netherlands

Paperback edition published 2022 by The White Horse Press

www.whpress.co.uk

Set in 10.5 point Palatino

All rights reserved. Except for the quotation of short passages for the purpose of criticism or review, no part of this book may be reprinted or reproduced or utilised in any form or by any electronic, mechanical or other means, including photocopying or recording, or in any information storage or retrieval system, without permission from the publishers.

British Library Cataloguing in Publication Data
A catalogue record for this book is available from the British Library

ISBN   1-874267-43-X (White Horse Press)
           90-04-11960-4 (Brill Academic Publishers)

           978-1-912186-54-9 (Paperback)

Front cover: *On the Fringe of a Primeval Forest, Simbo, Solomon Islands*, painted by Norman Hardy. From E. Way Elkington, *The Savage South Seas* (London, A. and C. Black, 1907). Reproduced by permission of the publishers.

Back cover: 'The cut', Tulagi, 1920 (Metcalfe collection); Lever's Pacific Timbers operations, 1970 (*Australian Timber Journal*).

To the memory of the late Ombudsman Isaac Qoloni, OBE, guardian of the people and their forests, and to the women of Solomon Islands who work hard in the forests and waters of Solomons to support their families.

Contents

Acknowledgments xiv

Introduction 1

Chapter 1. Solomons Forests 4

Chapter 2. To Open the Forest 18

Chapter 3. A Great Coconut Estate 36

Chapter 4. From the Shortlands to Santa Cruz:
  Beginning the Timber Industry 62

Chapter 5. The Vanikoro Kauri Timber Company 91

Chapter 6. World War Two: Focus on Forests 115

Chapter 7. Domestic Needs and Overseas Markets, c. 1950–1963 142

Chapter 8. The Government is but a Stranger … 163

Chapter 9. The Forestry Department and the Control of Species,
  c. 1960–1980 186

Chapter 10. Contest for the Forests, c. 1963–1985 209

Chapter 11. Attempts at Control? c. 1980–1990 236

Chapter 12. Aiding the Forests 254

Chapter 13. Provincial Ambitions, Community Aspirations:
  The Western Islands, 1985–1995 277

Chapter 14. Provincial Ambitions, Community Aspirations:
  Central and Eastern Islands, 1985–1995 298

Chapter 15. Profits and Loss, c. 1982–1996 319
Chapter 16. Contested State: Contest of States, 1993–1997 339
Conclusion. The Forest, Contested and Uncontested 361
Afterword 379
Appendices 384
Notes 407
Bibliography 469
Index 489

Tables

Table 1. Major forest types found in Solomon Islands.	9
Table 2. Number and origin of Vanikolo employees.	99
Table 3. Annual replanting by the Forestry Department.	193
Table 4. Rates of duty on round logs 1988–1994.	322
Table 5. Log exports, key financial indicators.	347

Maps

Map 1. Solomon Islands, c. 1978.	5
Map 2. Cyclone tracks 1960–1990.	15
Map 3. Shortland Islands.	70
Map 4. Vanikolo.	105
Map 5. Savo and Florida, with some wartime bases.	124
Map 6. Guadalcanal.	152
Map 7. Timber tracts on public land of Solomon Islands, 1968, exclusive of declared forest areas.	168
Map 8. Forest areas declared in 1968, exclusive of land purchased by the government.	173
Map 9. Eastern New Georgia Islands.	221
Map 10. West New Georgia Islands.	283
Map 11. Choiseul.	291
Map 12. Santa Isabel.	295
Map 13. Malaita.	301
Map 14. Makira.	308
Map 15. Solomon Islands Forest Estate, 1987, on land owned by the government.	353

Other Illustrations

Fig. 1. Profile of the regional forest communities of Solomon Islands.	8
Fig. 2. Forest on the north coast of Kolombangara.	12
Fig. 3. Forest on the west coast of Kolombangara.	12
Fig. 4. Simplified diagram of inorganic nutrient cycling in a Melanesian rainforest.	16
Fig. 5. Schematic representation of expansion of lineage land and customary tenure.	26
Fig. 6. Scrub-clearing.	39
Fig. 7. Cleared land at Bingera, Queensland.	40
Fig. 8. Acres cultivated for plantations, 1898–1929.	49
Fig. 9. Australian timber for Methodist mission buildings at Roviana lagoon.	51
Fig. 10. 'Letting in the sunshine', Munda.	51
Fig. 11. Methodist mission land cleared for coconut planting, Munda.	52
Fig. 12. The Goldie residence on Kokenggolo hill.	53
Fig. 13. Typical land clearance in early 1900s.	52–3
Fig. 14. Clearance and path construction, Penjuku.	55
Fig. 15. Sketch map of Tulagi in 1930s.	57
Figs 16–19. Tulagi in 1920s–1930s.	58–9
Fig. 20. Eric Monckton's logging team.	67
Fig. 21. Loading log sections into steamer.	68
Fig. 22. The first pre-fabricated house brought to the Solomons by the Seventh Day Adventists.	73
Fig. 23. On the verandah of the mission house at Wainoni Bay.	74
Fig. 24. Roman Catholic Cathedral, 1930.	74
Fig. 25. The saw pit team, 1920s.	75
Figs 26–33. Seventh Day Adventists gathering logs for the first Mbatuna saw-mill.	77–80
Fig. 34. The refurbished saw-mill, Mbatuna.	81
Fig. 35. Bringing up sawn timber for construction, Mbatuna.	81
Fig. 36. Making oars from local timber.	82
Fig. 37. Pivu cutting a new stern.	82
Fig. 38. Furniture made by Marovo pupils.	83
Fig. 39. Paeu from the sea.	93
Fig. 40. Clearing at Paeu for company headquarters.	93
Fig. 41. Football team at Paeu.	96
Fig. 42. The reason they were there.	100
Fig. 43. Axeman felling the kauri.	101
Fig. 44. Hauling the log.	101

Fig. 45. Europeans and Solomon Islanders hard at work.	102
Fig. 46. Shay locomotive and logs at landing.	103
Fig. 47. Log 'rafts' ready to be hoisted aboard.	103
Fig. 48. Log exports, 1922–1942.	107
Figs 49 and 50. New Zealand engineers at work, Vella Lavella, 1943.	117–18
Fig. 51. American soldiers at work, 1943.	119
Fig. 52. Captured Japanese tractor.	120
Fig. 53. Loading logs cut from north coast of Guadalcanal.	121
Fig. 54. Logs arriving at one of RNZAF's mills.	122
Fig. 55. Sawing logs at RNZAF mill, Guadalcanal.	123
Fig. 56. Sketch map of Hombu Hombu, c. 1944.	126
Fig. 57. War's ways of clearing the land. Tunnel entrance to Kokenggolo hill in the centre of Munda's wartime airfield.	130
Figs 58 and 59. Two views of Munda's wartime airfield.	130–1
Fig. 60. American Seabees repairing Munda airfield.	131
Fig. 61. Lambete landing east of Munda.	132
Fig. 62. Total consumption of sawn timber, 1953–1970.	148
Fig. 63. Sawn timber, local production, 1956–1967.	149
Figs 64 and 65. What upset the Brisbane plywood millers.	159–60
Fig. 66. Exports by logging companies, 1951–1966.	171
Fig. 67. Yarder for high-lead log extraction.	179
Fig. 68. Log extraction by crawler tractor in late 1960s.	188
Fig. 69. Keith Trenaman and assistant.	189
Fig. 70. Forestry workers and others at Vanikolo, 1954.	192
Fig. 71. Experimental plot of pencil cedar.	194
Fig. 72. Experimental plot of *Eucalyptus deglupta*.	195
Fig. 73. The impact of cyclone Ida, Santa Isabel.	211
Fig. 74. Part of the road built by Allardyce Lumber Company.	211
Figs 75–78. Lever's Pacific Timbers. Cat D7 tractors, Kenwoth three-axle truck and dump loading by spar tree.	214–15
Figs 79–82. Ringgi Cove, fallers and equipment.	216–17
Fig. 83. Logging camp at Ringgi Cove.	222
Fig. 84. Children of workers at Ringgi Cove.	223
Fig. 85. Australian and Fijian nurses at clinic.	223
Fig. 86. People at Kaonasughu – royalty payment.	231
Fig. 87. Government land reforested, 1966–1990.	255
Fig. 88. Poster publicity for replanting.	264
Fig. 89. Cover of *Link* magazine.	270
Fig. 90. Former Prime Minister Solomon Mamaloni and John Cunning.	272

Fig. 91. *Greased Dinner* by J. Makini.	296
Fig. 92. Tan Sri Dato Chee Yioun of Berjaya, 1994.	309
Fig. 93. Sustainable agriculture compared to unsustainable agriculture.	328
Fig. 94. Three trees felled within the boundary of Honiara botanical gardens.	334
Fig. 95. Former Prime Minister Francis Billy Hilly, 1994.	343
Fig. 96. Log exports and sustainable rate of logging.	351

Measurements

Units of weight, distance, area and money are given in the original form used in the sources, except where comparisons across different systems of measurement are made. In Solomon Islands, the imperial system prevailed. Currencies used before World War Two were pound sterling and the Australian pound, which retained parity until the Great Depression. In 1932, a pound sterling was equal to A£1.25, or US$4.00. In a pound (£1), there were 20 shillings; in a shilling (s), there were 12 pence (d). In 1966, Australia converted to the decimal system and the Solomon Islands followed. At this time, A£1 converted to A$2.00 (16 shillings sterling or US$2.25). In 1978, when the archipelago achieved independence, Solomon Islands introduced its own decimal currency. Measures of weight, distance, and area were converted to the metric system in 1975.

In relation to measurements of timber, one superfoot of sawn timber is a piece of wood, one foot long, one foot wide, and one inch thick. Volume of logs standing in a forest or felled was calculated in superfeet Hoppus measure, though the term 'Hoppus' was not always stated. The volume of sawn timber produced by a saw-mill was calculated in superfeet true measure. All metric units are in true measure. For logs, 100 superfeet Hoppus = 0.301 cubic metres; for sawn timber, 100 superfeet true measure = 0.236 cubic metres.

Abbreviations

ADAB Australian Development Assistance Bureau
Ag.   Agriculture
AIDAB Australian International Development Assistance Bureau
AR   Annual Report
ARLD Annual Report Labour Department
AUSAID   Australian Agency for International Development

BPA	Burns Philp Archives
BPSS	Burns Philp (South Seas)
BSIP	British Solomon Islands Protectorate
C	Commissioner
CBAR	Central Bank of Solomon Islands Annual Report
CDC	Commonwealth Development Corporation
CFO	Chief Forestry Officer
CO	Colonial Office
COF	Commissioner of Forests
DC	District Commissioner
DNA	Development and Native Affairs
DO	District Officer
DP	Deland Papers
EC	European Community
EU	European Union
FAO	Food and Agriculture Organisation
FD	Forestry Department/Division
FDAR	Forestry Department Annual Report
FIB	Foreign Investment Board
FRIS	Forest Resources Information System
FRP	Fairley Rigby Papers
JICA	Japan International Co-operation Agency
JOAA	Japanese Overseas Afforestation Association
KFPL	Kolombangara Forest Products Pty Ltd.
LTO	Land Titles Office
MCNZ	Methodist Church of New Zealand
MOF	Ministry of Finance
MOM	Methodist Overseas Mission
NCP	National Coalition Party
NGOs	Non-governmental Organisations
ODA	Overseas Development Administration
OF	Old Files
OPM	Office of the Prime Minister
PIM	Pacific Islands Monthly
PMB	Pacific Manuscripts Bureau
PNG	Papua New Guinea
PRO	Public Record Office
QRSC	Quarterly Reports Santa Cruz
QRSI	Quarterly Reports Shortland Islands
RCo	Register of Companies
RO	Report of Ombudsman

SDA	Seventh Day Advestist
SI	Solomon Islands
SICHE	Solomon Islands College of Higher Education
SID	Solomon Island Development Company
SIDT	Solomon Islands Development Trust
SIFIA	Solomon Islands Forest Industries Association
SIG	Solomon Islands Government
SINFRI	*Solomon Islands National Forest Resources Inventory*
SINURP	Solomon Islands National Unity, Reconciliation and Progressive Party
SS	Solomon Star
SV	Solomons Voice
TFAP	Tropical Forest Action Plan
VKTCo	Vanikoro Kauri Timber Company
VR	Vanikoro Records
VTC	Vanikoro Timber Company
WP	Woodford Papers
WPHC	Western Pacific High Commission

Acknowledgments

This book was meant to be a history of the Dominican nuns in the South Island, New Zealand. In discussing that project, one of the Dominicans in Dunedin, Sandra Winton, mentioned that the Dominicans in Solomons were concerned with problems arising among the people from commercial logging there. Sandra, knowing my long-standing interest in these islands, thought I might be interested. So, for the second time in my life, I was diverted to Solomons when I had planned to do other things. So thanks, Sandra, for that. The Domincans in New Zealand still await a new history, alas!

I owe thanks to so many supportive, generous people and institutions both in Solomons and beyond. In Honiara, Rex Horoi, former principal of the Solomon Islands College of Higher Education (SICHE) and the Council of the College afforded me an opportunity to work again in Solomons for several months, through a Senior Fellowship in 1992. Murray Chapman, teacher, friend and colleague of many years, one foot in Solomons and the other in the University of Hawaii, gave his usual unstinting bounty of enthusiasm and advice.

The archivists, librarians, and record clerks in the National Archives, the National Library, the University of the South Pacific, the National Parliament, and the Ministry of Natural Resources, Solomon Islands, were all helpful and I thank them.

Sam Gaviro of the Forestry Division was generous with his time and access to his staff. Rosemary Kinne of SICHE was a source of advice and knowledge, as was John Roughan.

I also thank the following who assisted in a variety of ways while I was in Solomons: all the people interviewed, listed in the Bibliography, and the staff of SICHE, Ezekial Alebua, Abraham Baeanisia, John Beverley, Walter Bila, Sylvia Bluck, Spencer Brook, Jennifer Boseto, Ella Bugotu, Jean and Bob Butman, Alan Cameron, Joanna Daiwo, the Dominican nuns at Mole,

Geoff Dennis, Jo Eagle, Dorothy and Ken Ferris, Steven Foster, Ian Frazer, Eric Havea, Jan Henderson, the late Captain Hone of *Baruku*, representatives of Hyundai on Vella Lavella, Enele Kwainarara, Clement Kengava, Bill Kranenberg, John Lamani, John Lapli, Patrick Lavery, Bishop Lazarus, William Leliti, Philip Leobea, Ruth Liligula, Michael Lomiri, Jully Makini, Joan Morris, Lesley Moseley, Peter Noli, Claire O' Brien, Mary O'Dea, Elliot Cortez Pade, Henry Paroi, Ruth Pierce, Chris Ramoli, N. Sala, Kabini Sanga, John Schenk, George Scott, Myknee Qusa Sirokolo, Hylton Taylor, Tim Thorpe, Alan Taro, Ben Tua and the people of Buma, Victor Tutu, Esau Tuza, John Vadaka, Wayne Woofe, and Daniel Zamae. Sir Lesley Boseto and his family were a great help to me on my visit to Choiseul, as were Bishop Philemon Riti and his wife Nancy on New Georgia. On Vella Lavella, my friends Sir Lloyd Maepeza Gina and Lady Olive Gina and their family were marvellous hosts.

Others outside the islands have been of great help, particularly those who work or have worked in forestry. Tony Fearnside was tremendously generous in several ways; his interest in forestry is both broad and deep. John Groome, Ross Cassells, Ken Marten, Devon Minchin, Ian Rogers, Eric Kes, and John Dargavel have been very helpful. I am particularly grateful for the detailed information the late Keith Trenaman sent me from England, even though he was gravely ill. I also thank his widow, Marion for the photograph she gave me. Michael Hadley also kindly supplied a photograph of his late father, Chris.

Particular thanks go to those who helpfully commented upon parts of the draft manuscript: Tim Bayliss-Smith, Nola Devoe, Rosemary Kinne and John Roughan for their comments on the entire draft; Paul Dalziel for Chapters 15 and 16; Jennifer Corrin for parts of Chapter 13; Tony Hughes for Chapter 8; and Ken Marten for parts of Chapter 9. And to Lin Chapman for general advice, I offer my aroha. Any errors and omissions, alas, are all my own work!

My thanks too go to the staff of the Public Record Office, Kew, and the Commonwealth and Foreign Office, Milton Keynes. Elizabeth Blott and Joan MacPherson were of great assistance to me there. In Hanslope, Nigel and Sarah Stacey assisted in several practical ways, as did Lorna Cady in London. The archivist, Michael Rigby and others at Ellen G. White Research Centre, Seventh Day Adventist Church, Avondale were most co-operative and enthusiastic. I would also like to thank Frank Strahan and the staff of the archives, Melbourne University. My special thanks go to Leigh Swancott who ferreted out material for me even when I was back in New Zealand. In Australia, I thank the Division of Pacific and Asian History for hospitality at the close of 1992. Donald Denoon and Mary were simply wonderful. In

New Zealand, the staff of the National Archives and the National Library, Wellington and David McDonald of the Hocken Library, Dunedin have been, as always, helpful and efficient.

For her wise words and encouragement while we both worked on our books, I thank Dorothy Shineberg, a gifted Pacific historian and a loyal friend. A special thanks to Dorothy Page, who as Head of the History Department at Otago strongly supported this work.

My warmest thanks and aroha go to the Macmillan Brown Centre for Pacific Studies for granting me a fellowship to commence the writing of this book. The former Director Malama Meleisea, Garth Cant, Bill Wilmott, Ann Parsonson, and Kate Scott gave generously of their support as did another Fellow, Bill Clarke. Without the Macmillan Brown Centre's help, this book would still be in progress. My own University of Otago has been supportive with reseach grants for copying of photographs and map-making as well as research leave. I thank Bill Mooney of the Geography Department for his cartography and Paul D'Arcy, Lisa Early, Aaron Fox, and Ian and Lala Frazer for their help.

For permission to reproduce illustrations, I thank the National Archives of New Zealand for World War Two material; John Lamani, Solomon Star Company Ltd for photographs; A. and C. Black, publishers, for the reproduction of Norman Hardy's painting used for chapter headings; Oxford University Press for material from T. C. Whitmore's, *An Introduction to Tropical Rainforests*, 1990; the John Oxley Library, Brisbane for photographs from their Kanaka collection; the Mitchell Library, Sydney for photographs from the George Brown collection; the Methodist Church of New Zealand for photographs from their Solomons collection; the Foreign and Commonwealth Office for several photographs; the Melbourne University Archives for photographs from the Vanikoro Papers; the Ellen G. White Research Centre for several photographs; Graeme Golden for photographs; the National Archives and Records Administration of the United States of America for photographs relating to World War Two; Peter Sheehan of the Forestry Division for permission to use photographs from departmental reports and for the forest types table based on illustrations in *Solomon Islands National Forest Resources Inventory*, vol. 1, 1995, and an illustration of regional forest communities and cyclone paths based on John Schenk's *Forest Ecology and Biogeography in the Solomons*, 1994. Jully Makini kindly gave permission for reproduction of her poem, *Greased Dinner*.

I thank Andrew Johnson at the White Horse Press for his patience and encouragement.

To Khyla Russell for her unwavering support on land and, more especially, sea where I am useless, I offer my thanks and love.

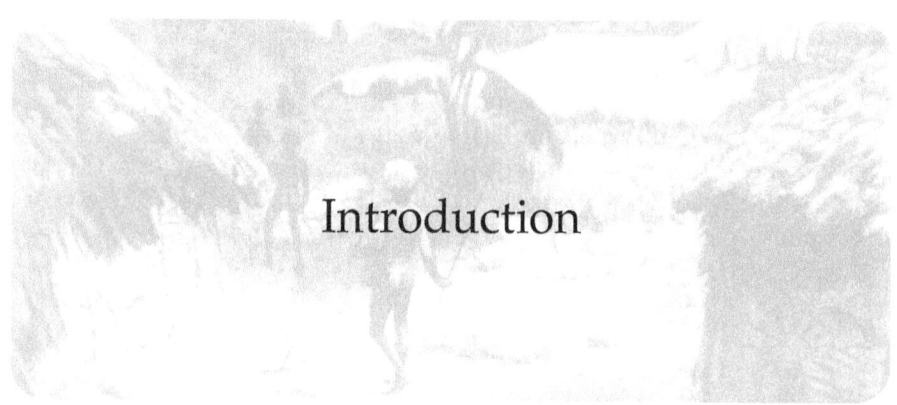

Introduction

This book presents a view of Solomon Islands forest history. As an aspect of environmental history, it attempts to address three levels: the natural environment of the past, particularly the forest environment; the interaction between the human productive technology and that environment; and human perceptions and beliefs about the forest and its use over time. These are the three foci that Donald Worster has described as interesting most environmental historians.[1] Though environmental history is an established sub-discipline of history in the United States, Europe, Australasia, and increasingly in India and Africa, it has not been a concern of Pacific historians, though some Pacific anthropologists and several prehistorians have been much more aware of the interaction of human beings with their environment.

This history is also a case study of a much larger process. On the global scale, many countries have severely depleted their natural forests. Particularly in the New World, early European settlers saw the forests as endless and literally mined them until national shortages threatened human livelihood. In North America, Australia and New Zealand conservation and reforestation arose mainly out of necessity, but not without conflict. More recently, in the tropical forests of the Philippines, east Malaysia, and Indonesia, various overseas and local companies have logged rapaciously mainly to feed the great conglomerates of Japan, the *sogo shosha*, as Peter Dauverge has so well demonstrated. As log stocks fell off, they expanded into Papua New Guinea, the Solomon Islands, and Vanuatu and now range into tropical South America and Africa seeking forests to consume.[2] The process continues with vast environmental and social implications.

Commercial logging has a relatively short history in Solomons, but the use of the forests by human beings goes back to the dawn of human settlement, perhaps 28,000 years ago. This is where the account starts with the ancestral Melanesians and their gardening within the forest. The demands on the forest accelerated with the incoming metal-based technology of the West in the nineteenth century and with the imperatives of the plantation economy under the British Protectorate. European settlers or sojourners were at first more concerned to cut down the lowland forest for plantations, than they were to fell it for timber, but by the 1920s the special qualities of certain tropical timbers were valued in Australia and a small log export industry began, though with little profit to Solomon Islanders or indeed to the loggers. The Melanesian tropics, at least until World War Two, were hard on men and material, so logging was a minor industry.

Military forces during World War Two utilised local timbers briefly and extended knowledge of their potential. As the war drew to a close there was a heightening of interest by outsiders in the forests, but this subsided to mainly domestic demand. Anticipating future timber markets, the colonial government invoked a model of forest use common to many British and other imperial powers, the establishment of sustainable plantation forestry on government land. About the same time, overseas markets along with more powerful technology of chain-saws, tractors and bulldozers drew foreign loggers to the forests in the 1960s. The growing international impetus to decolonisation meant that ways and means had to be found for the Solomons to finance its own government and future, so the extension of government plantations seemed a perfect solution in a sparsely-settled tropical archipelago. The colonial government was determined to control not simply access to log supplies, but also the location of logging, a policy opposed and finally rescinded by Solomon Islanders who wanted to deal directly with loggers and recoup more localised benefit from their own forests. Coinciding to some extent with Independence in 1978 was the beginning of a decline in sources of tropical timbers in South-east Asia, resulting in interest and then dominance by Asian loggers and contractors of the Solomons industry in the 1980s and 1990s. As the contribution of logging to national income increased in the 1980s and early 1990s, coupled with flawed economic decisions in government, logging became more politicised and unsustainable, with extraction rates far exceeding forest increment.

This history also addresses the uniqueness of the Solomons situation: the complex interplay among timber-rights holders, their governments and the Forestry Department/Division, foreign companies and aid-donor governments during the late colonial period and the years since Independence. Within this matrix too are some churches and nongovernment organisations which straddled the divide between local and foreign with greater ease than governments. Few of the groups concerned are homogeneous, particularly the Solomon Islanders, and the study attempts to reveal the levels of social and political interest groups trying to control the disposition of forest resources. As the sub-title implies the forests were highly contested by those

Introduction

who wanted solely their timbers and those who saw other values therein.

Underlying all this is the interplay of different perceptions, attitudes and beliefs about the forest. The early Melanesians saw their forests and seas as the means to life, and themselves as part of that environment. The European planters perceived the forest as simply an obstacle to development – their development, primarily. Missionaries saw the forest as the source of timber to facilitate literally their church-building work and, particularly the Methodists, as potential crop-bearing land to improve the material lives of their followers. Post-war log demand alerted Solomon Islanders to the cash value of their forests which became a seeming unlimited and untouched resource for their social development and further participation in the modern world. Increasingly, since at least 1990, that perception has faded with the realisation of what has been lost in the loggers' wake.

A word regarding sources: from Chapter 10 on, I have adopted the conventions of the journalist, rather than the historian, in relation to confidentiality of sources, since several people who entrusted information to me might suffer repercussions if I detailed informants and sources. I can supply further reference details to bona fide inquirers. Moreover, some of what I say is based purely on my own observations while in Solomon Islands. Solomon Islands is a small country. Where the media struggle for the freedom to inform and comment, 'leaks' – often from the most unlikely of people, at all levels of the government, administration, church and business – are common, and in keeping with long-established methods of social commentary. It was a case of 'Ask and you shall receive, knock and it shall be opened unto you.' Because of my training, I am loath to withhold information and apologise to the purist historians, but I consider peoples' lives more important than locating a source.

— 1 —
Solomons Forests

> The rain forest canopy is in a state of continual flux and different tree species have different recruitment and death rates. It is difficult to know whether a climax forest maintains the same species composition over several forest growth cycles....
>
> T. C. Whitmore, *An Introduction to Tropical Rain Forests* (Oxford, 1990), 132.

Tropical rainforests occur in three areas where land lies near the Equator – in central America, central Africa and Indo-Malaya where the mean temperature of the coldest month is 18 degrees Centigrade or more.[1] Unlike the monsoon-dominated forests of tropical areas, these forests are permanently wet, with a monthly rainfall of 100 mm or more and only very short dry spells of a few days or weeks. It is self-evident that there can be no forest without land. For much of the tropical world within which Solomon Islands is located, the reverse almost holds true, if the land, as always, is judged by Imperial Adam in terms of its utility to his own presently highly successful species. So it is to the land we first turn and to the processes that give it its character.

The modern Solomon Islands consist of a double chain of islands ranging across 1400 kilometres, along a north-west–south-east axis, from 5 degrees to 12 degrees south of the equator. In geological time, the islands are 'young' and unstable, say, in comparison with ancient, arid Australia. They are still growing out of the sea. However, there is localised subsidence and erosion as well, since their soils are subject to instability, and all the more so under the constant, heavy rainfall. With a land area of around 28,000 square kilometres, the islands are within the Pacific's 'rim of fire' and have both active and dormant volcanoes that over millennia have created

Solomons Forests

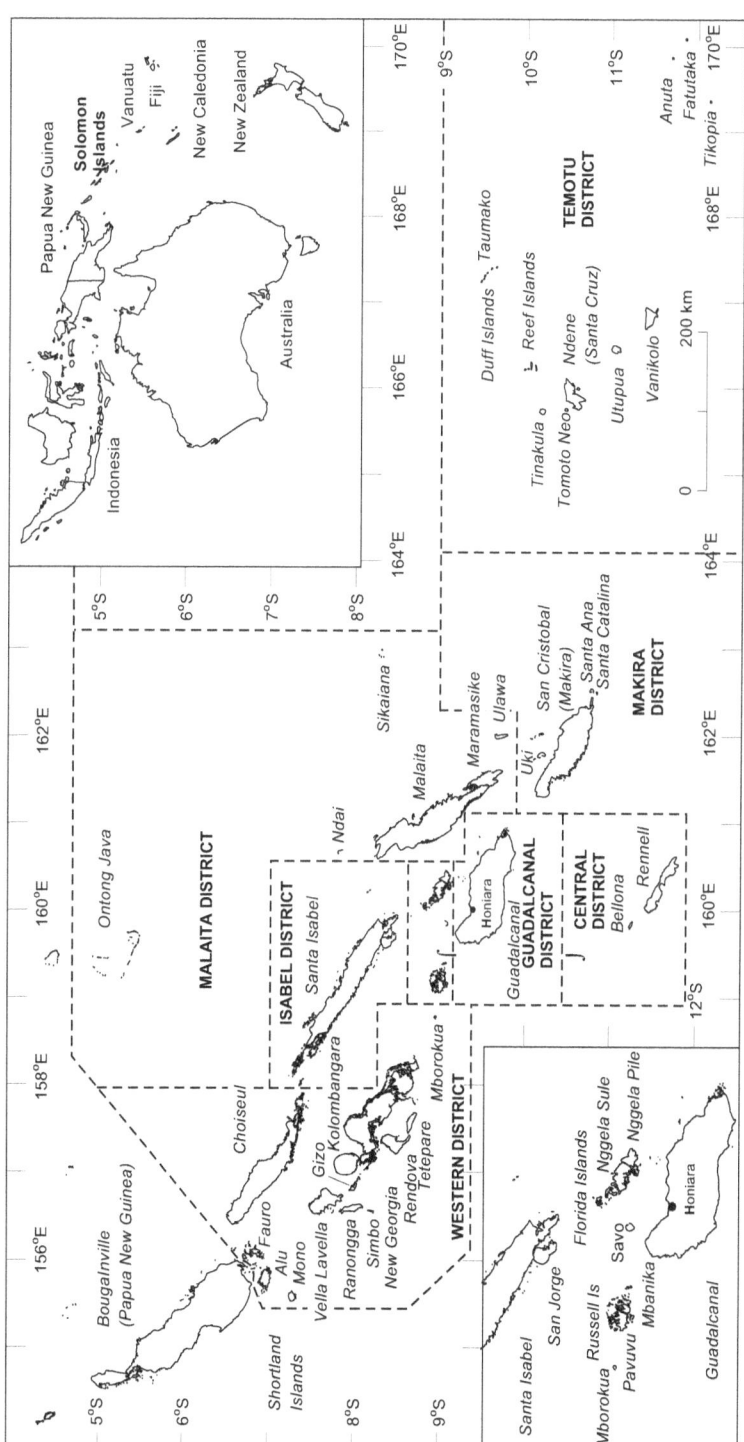

Map 1. Solomon Islands, c. 1978.

new land. Moreover, the group is subject to earthquakes, which also assist in the erosive process.

The islands' geology results from massive tectonic forces at work in the Earth's crust for millions of years. On the macro-scale, the islands are located on a convergence zone where the great Indo-Australian plate moving north and the Pacific plate moving west, ran into each other about 65 million years ago, resulting in the Pacific plate's slow subduction under the Indo-Australian plate. On a more localised scale, about 10 million years ago a portion of the Indo-Australian plate was wrenched off and the mother plate began to push under or subduct beneath it. In the east, the Pacific Plate collided with the detached section and began to push over it. Parts of these crustal plates rose from the sea, forming islands, with the molten magma from the mantle breaking through weak points in the form of volcanoes.[2] This explains in part the presence of three major and two minor geological provinces in the modern archipelago. In the north-east of the main chain, the islands of Malaita, Ulawa, and the north-east of Santa Isabel are within the Pacific crustal plate, forming the Pacific Geological Province. Makira (San Cristobal), Guadalcanal, the Florida group, the south-west of Santa Isabel and Choiseul (Lauru), are parts of the detached section of the Indo-Australian plate and constitute the Central Geological Province. In the west lies the Volcanic Geological Province which includes the volcanic islands of New Georgia, the Russell Islands, the Shortlands, the north-west of Guadalcanal and Savo. Temotu, the most easterly political region in the Solomons archipelago, is within the Oceanic Volcano Province, while the uplifted atolls of Rennell, Bellona, and Ontong Java are within the Oceanic Atoll Province.[3]

This varied geological history probably explains much regarding the development of both flora and fauna. For many species, crossing the sea barrier unaided by human beings is rare, if not impossible. Research on flora is more developed than that on fauna, but, for example, of 163 species of land birds that breed in the archipelago, 72 or 44 per cent are endemic. Another 62 species or 38 per cent occur elsewhere, but are represented in the Solomon Islands by unique sub-species. Several mammals, rats and bats, are endemic and live in undisturbed rainforests.[4] The zoologist Jared Diamond believes that the phenomenon of faunal speciation and of population variation between islands is more common in the group than anywhere else in the world.[5]

The majority of indigenous vegetation communities of the Solomon Islands are related to those of South-east Asia, although there are some species originating from what is now Australia and from extremely ancient links with south America.[6] The Indo-Malaysian rain forest includes those of New Guinea (West Irian and Papua New Guinea). This region is known to botanists as Malesia, and is the earth's richest in plant species. On its

Solomons Forests

edges are the rainforests of northeast Australia and island Melanesia, extending into Tonga and Samoa. The region, including island New Guinea (Bismarcks), Solomon Islands, Vanuatu, New Caledonia, Fiji, Samoa and Tonga, is known as the Melanesian floristic province. There are major differences in forest types between Malesia and north-east Australia. From New Guinea across into the south-east there is a floristic continuum, marked however, by rapidly decreasing variety. With increasing distance from the source biota, the numbers of plant families on the western Pacific islands decrease as do their genera and species. The Solomons is poorer floristically in comparison to say New Guinea, and far more so in relation to South-east Asia, but richer than the islands to the south and east.[7]

In the contemporary Solomons, this pattern is found within the islands themselves. Species which are prevalent in the west of the group, for example, *Campnosperma brevipetiolata*, are not found in Guadalcanal and Makira (San Cristobal). Species commonly found on these islands, such as *Terminalia brassii*, *Vitex cofassus*, and *Dillenia* spp. are similarly absent from the distant Temotu group. Geologically part of the Vanuatu islands, Temotu has the only kauri (*Agathis macrophylla*) found in Solomon Islands,[8] but isolated Vanikolo lacks most common Solomons secondary forest species except where these have been introduced in historic times.[9]

Botanists believe that many tree species, such as those in the Dipterocarpaceae, failed to reach the Solomons from Malesia because of the sea distances between islands. Those species that are well-established may well have spread from Malesia earlier during the Quaternary Period before the end of the last Ice Age when the sea level was lower. Modern New Guinea was joined to Australia and there was probably one land mass from Bougainville to Nggela. Once the land bridges were gone or where they never existed, flora of mountainous inland areas where montane forest dominates remained particularly isolated because of the cordons of the lowland forest and the sea barrier.[10]

Many botanists, including T. C. Whitmore, suggest that in the Solomons flora lacks endemicity. However, there appears to be a pattern of intra-species variation in several species. Within the same species on different islands there are clear physical differences that may indicate the beginnings of evolutionary adaptation to different environmental conditions. As John Schenk states, 'although speciation is low, specific variation is high'. Because many typical Malesian elements are absent from the flora of the Solomons both plants and animals may be taking advantage of niche opportunities to evolve and better exploit the varying conditions on each island. Schenk suggests that, in the light of the relative geological youth of the archipelago, the evolutionary processes of species formation are in an early or dynamic phase.[11]

Pacific Forest

Although in comparison to Malesia, including New Guinea, Solomons could be classed as the poor relation in terms of flora, it ranks second only to New Caledonia among the rest of the south Pacific countries, in its floristic wealth. It also is a melting pot for flora originating mainly in South-east Asia, but also from the Pacific and South America.[12]

Forest types

The rainforest is not just trees; but, in terms of contemporary international trading, they are generally perceived as the most valuable component of the forest. Consequently Western-trained botanists have concentrated less on indigenous herbaceous species, palms, epiphytes and ferns, and many remain unidentified.[13] The vegetation communities of Solomon Islands have been variously classified, most recently by the National Forest Resources Inventory project, completed in 1994, which identifies six basic forest types, based on aerial photographic information as well as site surveys (see Figure 1 and Table 1).[14] The project's emphasis, not surprisingly, tends to be from the standpoint of commercial forest potential. Melanesians, free of science as it has developed in the West, may well have other systems of classification of the forests, as they have of the plants within.[15] The Solomon Islands Forest Resources Inventory includes forests that have been altered by Melanesians over millennia and by more recent activities associated with capitalism. The Inventory defines a 'forest type' as 'any group of tree-dominated stands which possess general similarity in composition'. The forest types include areas dominated by a single species or by mixed species with a distinctive

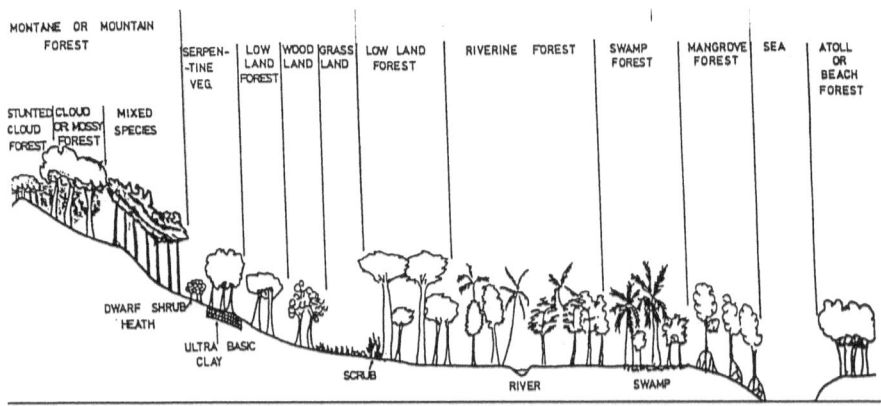

Figure 1. Profile of the regional forest communities of Solomon Islands.

(*Foreign Investment Bulletin, No. 5, 1984*)

Solomons Forests

SALINE SWAMP FOREST (S)

Saline swamp forest, mixed spp composition
Degraded forest

FRESHWATER SWAMP AND RIVERINE FOREST (F)

Casuarina dominated freshwater swamp/riverine forest
Hibiscus tilacus dominated freshwater swamp/riverine forest
Campnosperma dominated freshwater swamp/riverine forest
Logged forest
Freshwater swamp/riverine forest, mixed spp composition
Degraded forest
Pandanus dominated freshwater swamp/riverine forest
Metroxglon swamp forest
Terminalia dominated freshwater swamp/riverine forest

LOWLAND FORESTS (ON NEARLY LEVEL LANDS) (L)

Lowland beach forest
Casuarina dominated lowland rainforest
Logged lowland rainforest
Campnosperma dominated lowland rainforest
Lowland rainforest, mixed spp composition
Degraded lowland rainforest

HILL FORESTS (H)

Paraserianthes falcataria dominated hills rainforest
Casuarina papuana dominated hills rainforest
Campnosperma dominated hills rainforest
Logged hills forest
Hills forest, mixed species composition
Degraded lowland rainforest on hills
Maritime atoll hills rainforest
Agathis dominated hills rainforest

MONTANE FOREST (M)

Upland forest on hills, mixed spp composition

NON FOREST AND OTHER AREAS (N)

Herbaceous swamps, mixed spp composition
Lakes
Plantations (including forest plantations)
Braided river courses

Table 1. Major forest types found in Solomon Islands.

(Based on *Solomon Islands National Forest Resources Inventory*, vol. 1)

character. Forest types or formations, by and large, are determined by habitat, that is, by climate, elevation, soil and geology.[16]

The saline swamp forest is found in areas subject to tidal influences, such as estuaries and foreshores. The dominant species are mangroves, particularly *Bruguiera* spp. and *Rhizophora* spp. In the ecology, mangroves are important as they protect the reefs, acting as a filter or dam for eroded materials washed from the hinterland. They also act as a buffer to high seas.

Freshwater swamp and riverine forest is found in poorly draining areas. These areas frequently produce *Terminalia brassii* and the sago palm (*Metroxglon solomononesis*), but other species such as *Calophyllum* spp. and, in the western Solomons, *Campnosperma brevipetiolata* may dominate.

Lowland rainforest is found on fairly level ground, although there is a variation on better drained hillsides. Except for localised riparian and swamp vegetation within this type, there is no *Terminalia brassii* or sago.

Hill forest is complex in structure and composition. Perhaps the most notable variant of this class of forest, at least to the foresters who carried out the recent inventory, was that where *Casuarina papuana* occurred on the very alkaline ultramafic soils of south Santa Isabel and south-east Choiseul, often almost pure stands are found.

Upland rainforest on hills is montane (or 'moss' forest) forest of mixed species that occurs on higher elevation ridge tops and mountain summits, usually above 600 metres, although it can be found at lower altitudes in less hospitable conditions, such as on basalt formations.

Nonforest and other areas comprise areas with little or no tree cover, such as herbaceous swamp communities and commercial plantations, including timber-growing.[17]

Very few islands contain examples of all these forest types. Today, even fewer retain or sustain primary vegetation from sea level to the mountain peaks. However, on Kolombangara in a few places and on Vangunu a succession of forest species can be observed: from low elevations, where the big-leafed, buttressed trees grow 35–45 metres tall, with big woody climbers and big-leafed epiphytes, to the peaks, where the trees have small leaves and grow only to 6–12 metres, with few woody climbers, and the forest is draped with a mass of mosses, ferns and hepatics. Both Kolombangara and Vangunu are relatively small islands and, along with all the mountainous islands in Solomons, demonstrate the Massenerhebung effect, whereby the zonation of forest that normally correlates with varying elevation on large land masses is compressed. The montane or mossy forest that occurs at 2,100–2,400 metres on the great island of New Guinea develops as low as 700 metres on Vangunu. Why this is so remains an enigma.[18]

The dynamic forest

The European navigator Dumont d'Urville gave the name Melanesia to the area inhabited by black-skinned people, including Australia, although that island-continent is no longer considered a part of the region.[19] The name could just as easily allude to the islands of 'very dark almost blackish green rain forest mantle covering the land' with the mountain ranges frequently clothed in heavy, dark storm clouds.[20] To the casual visitor, the islands' vegetation has a solidity about it; yet it is constantly changing. As the noted forest botanist T. C. Whitmore points out, the forests are a mosaic with some parts going through a gap-phase, a building-phase, others a mature-phase and others a decay phase. This is one way of looking at the forests' continuous growth cycle. Starting arbitrarily with the gap phase, a large tree dies of old age, or more commonly is blown over or struck by lightning and, as it falls takes others, large and small, with it. The gap left will vary in size depending upon what caused the tree's demise. Plant ecologists distinguish between the kind of forest produced as a result of tree-by-tree replacement and that resulting from a large-scale blowdown, because the form of the forest produced will be different. In the case of a tree slowly decaying from age, the gap will be small; in the case of a major cyclone the gap could be measured in kilometres. The level of destruction of vegetation will vary according to the susceptibility of the site, particularly such factors as slope and nature of the geology. Once a gap occurs in the forest canopy, the area below is more exposed to both sunlight and direct rain.

The size of the gap in the forest is important because some species need more shade than others to grow. Thus in a small gap the shade tolerant seedlings already established before the gap occurred grow further, usually near their parent trees. The species that grow in this environment are called climax species and are usually slow-growing hardwoods. Among climax species in Solomon Islands are *Dillenia salomonensis*, *Maranthes corymbosa*, *Parinari salomonesis*, *Schizomeria serrata*, *Calophyllum kajewski*, *Campnosperma brevipetiolata*, and *Terminalia calamansanai*. Forests characterised by a preponderance of climax species are often described as primary forests. Where the gap in the forest is large and the amount of sunlight penetrating the canopy is plentiful, tree seeds stored in the soil or introduced by birds germinate and grow, along with other herbaceous vegetation such as climbing vines. These trees are pioneer species comparatively rapid growers and most often softwoods; many are short-lived. Common species include *Endospermum medullosum*, *Macaranga similis* and spp., *Gmelina moluccana*, and *Albizia* spp. Under them, a seedling bank of the shade-liking climax species develops, awaiting an opportunity to grow to maturity as the pioneer species mature and die or are killed by other means. When this patch of secondary forest of

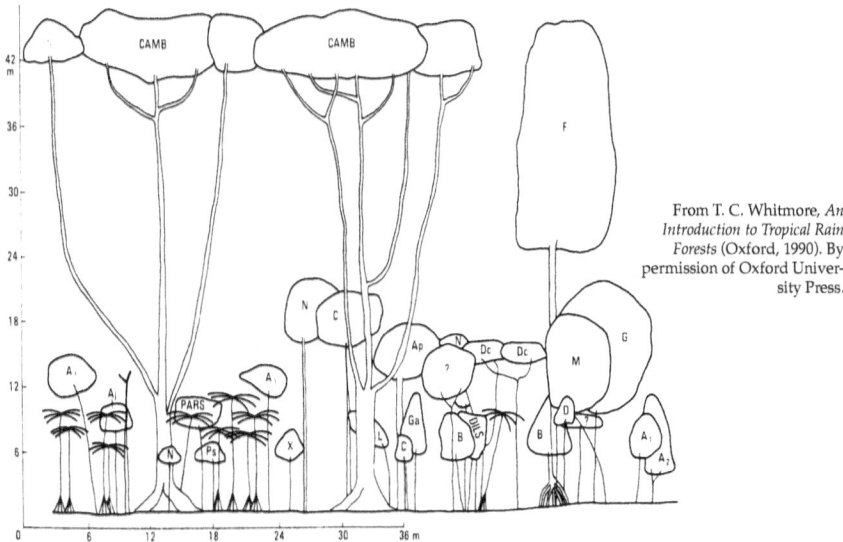

Figure 2. Forest on north coast of Kolombangara, domnated by over-mature trees of light-demanding climax species Campnosperma brevipetiolata *(CAMB) which is not regenerating itself. This particular forest resulted from massive cyclonic disturbance and, unless another were to follow, it will change its composition to resemble Figure 3, whose species are present in the lower part of the canopy.*

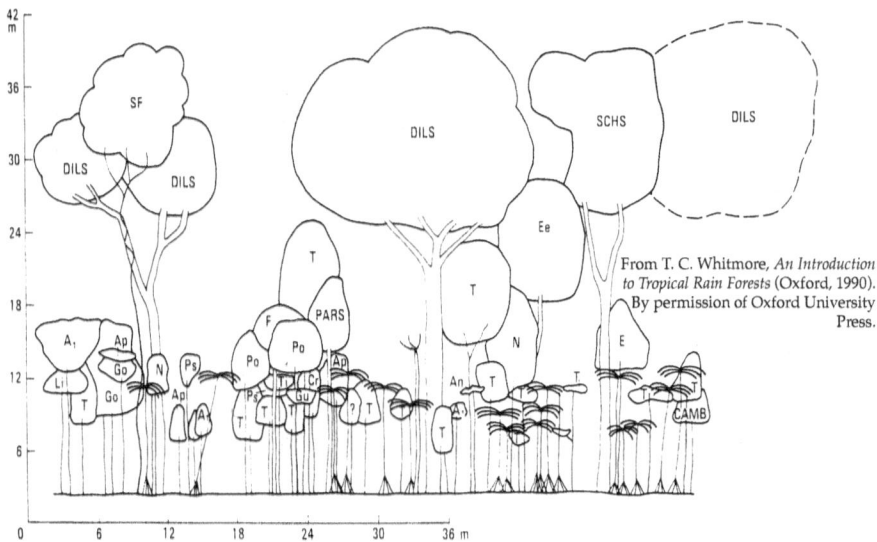

Figure 3. Forest on the west coast of Kolombangara, dominated by slow-growing, shade-tolerant climax species Dillenia salomonensis *(DILS) and* Schizomeria serrata *(SCHS), which are regenerating themselves.*

mainly pioneers gradually dies the succession of climax species re-establishes primary forest. The actual composition may vary, in part because of the size of the original gap – a very large gap resulting from major disturbance could encourage the growth of light-demanding climax species, such as *Campnosperma brevipetiolata* while a very small gap or one where pioneer species already have created much shade could encourge shade-tolerant climax species such as *Dillenia salomonensis* and *Schizomeria serrata*.

At any one time a natural rainforest demonstrates all these phases in different areas. The forest is thus a mosaic of patches of vegetation at various stages of the cycle. Because long-term records of the forest composition of Solomon Islands do not exist and palynology in the group is in its infancy, it is uncertain whether or not the overall composition of climax forest species has remained relatively constant in the recent pre-historic period.[21] It does seem likely that even in this period the impact of human beings, the ancestors of the present inhabitants, was considerable and accelerated changes in the forest's composition (see Chapter 2). In terms of composition and structure, the forests the first ancestral Solomon Islanders beheld were almost certainly not the forests seen by the first Europeans.

Cyclone adaptation.

The Solomon Islands archipelago lies within the cyclone or hurricane belt, although the central western islands appear to be less so, and Papua New Guinea virtually beyond it, except, on rare occasions, the south-east coast and off-shore islands. Generally cyclones are most common from November to April, but are not unknown at other times of the year. In the western Pacific, they form around 8 degrees south, track west, swing south and often turn again to the east.[22] The forests have had to cope with this; the tree species that could not, did not survive. Those species most adapted to cyclones or to taking advantage of the forest disturbance caused by cyclones will dominate. This is obvious, yet it was not until the 1960s, with Whitmore's work, that the significance of this began to dawn on foresters.[23] What it means is that the forests can take a very severe battering and recuperate. It also has ramifications for other disturbance regimes such as human habitation, animal activity and logging, as will be seen in later chapters. It also may mean that cyclones are necessary to maintain the natural forest of Solomon Islands, although putting this to controlled tests is beyond the realm of possibility.

Cyclones in the Solomons vary in their severity and intensity. Recent records indicate that typical cyclonic winds peak at between 80 and 110 kilometres per hour, though it is possible that in the past they have been greater. They can dramatically change the landscape as well as the composition of the forest. The impact of a typical cyclone that hit the southern Santa Cruz district (Temotu province) on 10 December 1935 was recorded by the district officer as he sailed back to the government station at Paeu, Vanikolo:[24]

> Coming up the coast the island looks as if it has been fired, the trees are just brown sticks, and leafless. [Utupua]...is completely destroyed, resembling a newly ploughed field, and from a distance, the same brown colour of newly turned earth, the tops are blown off the coconuts, and breadfruit, bananas and even pineapples are uprooted.[25]

As a result of a cyclone, trees are affected in a variety of ways ranging from minor damage through defoliation, desiccation of foliage, sometimes involving salt spray burn, loss of small branches and twigs to uprooting by wind, uprooting by landslips and washing away by floods. Any severe breakages to the tree enables disease to enter, weaken it, and eventually accelerate its death. Some locations appear more susceptible to greater damage – ridge tops and over the crest of lee slopes, where turbulence will be concentrated. The effects of cyclonic winds are exacerbated where heavy rains persist and soil becomes so saturated it literally slides off sub-strata. High winds and rain can have impacts outside the actual centre of the cyclone when seas are whipped up; all the more so when earthquakes follow, as was the case in part of south Guadalcanal in 1935:

> During December, mountainous seas occurred there for three days, with no wind. This is believed to be the result of the hurricane at Santa Cruz. Considerable damage was done to... the foreshore. On December 15th there was a severe earthquake, felt throughout the District, followed by several shocks over a period of two weeks. The shocks were most severe in the central bush region of Tatuve and on the South coast where extensive damage was done to villages, and by big landslides to gardens on steep slopes. A small tidal wave occurred at the same time on the South coast, and also on the North coast between Kaukau [Kaoka] and here [Aola]. It is feared that the damage to crops may result in a shortage of food.[26]

In a cyclone, damage to particular trees depends on their position in the forest canopy, the nature of their root system, their form and leaf characteristics as well as the type and depth of the soil. Many forest trees have a marked coppicing and sprouting ability and, of course, seedlings and seeds capitalise on the gaps created by the cyclone and the biomass providing nutrients for growth.[27] With the exception of trees and vegetative debris washed out to sea (some to float ashore elsewhere and fuel another strand ecosystem) the loss of potential litter is minimal, despite the apparent damage to established trees.

Cyclones have a substantial immediate impact, as anyone who has lived through one knows. To the novice observer the damage a cyclone causes is visually devastating. Yet the forest soon springs back to life as the

Solomons Forests

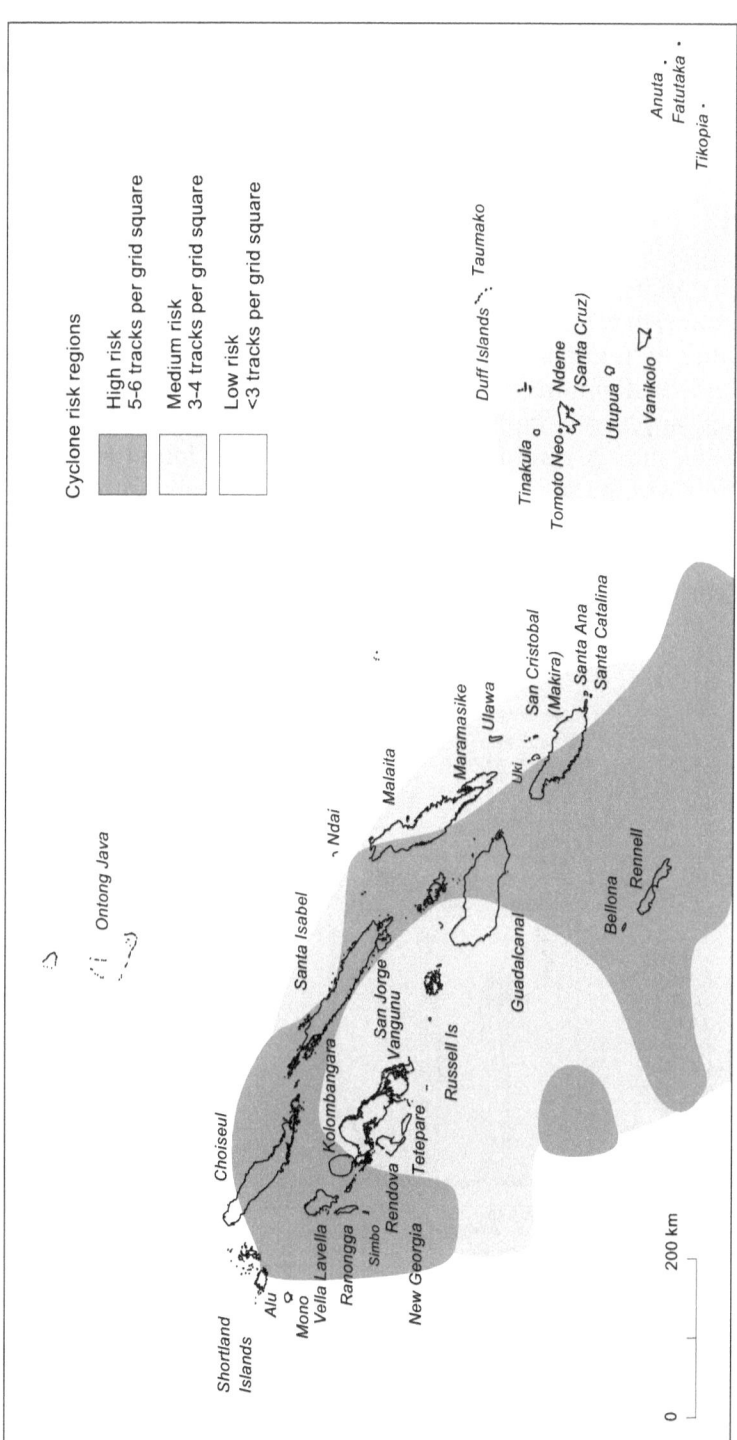

Map 2. Cyclone tracks 1960–1990.

Based on John R. Schenk, *Forest Ecology and Biogeography in the Solomons* (Canberra and Honiara, 1994). By permission, Forestry Division of Solomon Islands.

soils are undamaged and the litter very slowly fertilises all over again. The human inhabitants of such an environment lived under the impression that trees that were broken by the wind were replaced, that the canopy soon returned, and that all within a short time would be more or less the same.

Back to earth

There is much variation in soils in Solomon Islands although, with some notable exceptions, they do not greatly influence forest composition where land use regimes by humans beings are not a factor. The forest and soils are interdependent. Rainforest trees draw up nutrients from the earth, some with roots apparently tapping minerals in subsoil from parent rock at a depth greater than 30 centimetres, but most roots are found within 10 and 30 centimetres of the surface, where the topsoil is constantly enriched by the uppermost organic mat layer. As the trees and other vegetation and fauna die and decay on the forest floor, the resultant litter, with the help of termites and earthworms, eventually provides nutrients which gradually

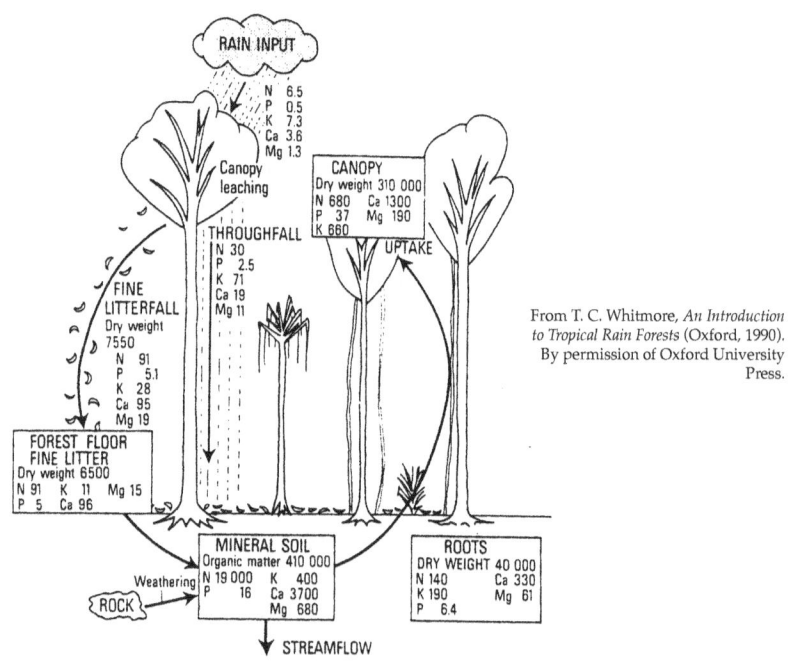

From T. C. Whitmore, *An Introduction to Tropical Rain Forests* (Oxford, 1990). By permission of Oxford University Press.

Figure 4. Simplified diagram of inorganic nutrient cycling in a Melanesian rainforest.

filter down to the different layers to feed the roots of various plants where their take-up is often assisted by mycorrhiza (fungi). But this litter layer has another important function essential for the health of the particular forest area. It acts as a kind of shock absorber for the impact of the constant rain. Like a sponge it holds and filters the water, helping to limit run-off and soil-erosion. The trees themselves assist this process as their canopy breaks the impact of rainfall, slowing its speed as it drips off the several levels of leaves as throughfall, or runs down tree trunks as stemflow. The rain is not simply composed of water. It has also dissolved chemicals, such as sulphur and nitrogen produced in fires elsewhere. Both throughfall and stemflow are augmented by chemicals leached from the leaves. Tropical soils, because of the high rainfall they receive are very subject to the leaching of minerals, hence the vital role of the nutrient cycle for the forest.[28]

The Solomons rainforest is very resilient. Its ability to bounce back from, and even capitalise on, cyclones is an outstanding example. The apparent wounds inflicted on the forests by the death of big trees, by cyclones, earthquakes, fire and floods soon heal, provided they do not occur continuously in the one area. The earth does not remain bare for long. Until human contact, the forest evolved to cope with and utilise these natural processes. The greatest challenge to its ability to adapt and co-evolve came as the first biped mammals, their fire sticks and stone tools in hand, penetrated its wild fastness, seeking a home and a living.

— 2 —
To Open the Forest

> It is remarkable indeed how precisely alike in the Solomon Islands, the Banks Islands, and the New Hebrides, the character of property in land reclaimed from the bush asserts itself to be.
>
> R. H. Codrington, *The Melanesians: Studies in their Anthropology and Folklore*, (Oxford, 1891), 61.

When the first settlers entered the Solomons archipelago, they beheld lands covered in rainforest; much like their old home in the west, but still pristine, untouched by human beings. It is possible that they had made their way down the Solomons chain when it was mostly a huge elongated island stretching from Buka to Nggela about 28,000 years ago.[1] If the pioneers did indeed make that journey they would have survived by utilising their hunter-gatherer and fishing skills. Naive local fauna including large rats, bats, frogs, and lizards and, from the sea, various fish, crabs, shellfish, turtles, dugongs and porpoises, probably fell easy prey to them, but edible starchy plants were scarce. Like most migrants, the more discerning might have brought a few 'home comforts' with them – some 'wild' taro-like aroids and varieties of the oily canarium almond (*Canarium* spp.) for example – to scatter among the natural clearings on the edge of the great forests; indeed they may have assisted their growth by burning to extend clearings here and there. Because fauna and flora decreased in variety in a continuum from the Asia-New Guinea region across to the eastern Pacific islands, perhaps too the first comers also captured a few breeding possums (possibly *Phalanger orientalis*) from New Guinea to stock the new land to give them something to eat of more meaty substance than bats, rats and birds.[2]

Whether such hunters and gatherers first began the 'domestication of the environment' and straddled the cusp of agriculture still awaits confirmation by prehistorians.[3] What prehistorians do tell us about more recent newcomers of 3,000 years ago is more certain. From South-east Asia probably via New Guinea, these were gardeners, keepers of pigs and chickens and introducers of the small, tasty rat, *Rattus exulans*. Their garden species included a range of domesticated root and tree crops. In husbanding these new biota, they created 'transported landscapes', the basis of permanent settlement and a new signature on the land.[4]

Forest farmers and burned clearings

To the ancestral Melanesians and to their more recent counterparts, the forests were a huge garden, with some parts, for a time, being more intensively used than others. Tree-clearance for gardens and hunting were predominantly men's work, but women foraged for small creatures, wild food, and firewood and utilised the non-timber products of the forest. The local forest region supplied a myriad of life's necessities: wood and leaves for house building, canoes, paddles, tools, food-bowls, combs, burial containers, as well as firewood for cooking and warmth; bark for clothing, twine, rattan-binding, sealers for caulking canoes; resin for torches; roots, bark, leaves, sap and seeds for the Melanesian pharmacopoeia and to tap magical power. Its earth supplied various stone and soils for tools, pigment and, in some places such as Choiseul and the Shortland Islands, for clay pots. Fossilised clam shell found in uplifted reefs provided the raw material for 'custom money' or valuables in several islands in the western Solomons. The south Nggatokae people had access to a particular stone used to make mortars, as they still do. In the forest trees, Solomon Islanders trapped birds for their red feathers on Santa Cruz and possums for their teeth on Malaita and other islands to make valuables for social and religious transactions.[5] Few had all these resources, but trade between different groups of 'salt-water' or coastal people and between them and their inland 'bush' neighbours provided distribution networks.[6]

Long exploited as a hunting ground for pigeons, feral pigs, land crabs, bats and possums or as a source of wild food, the forest was a place where gardens could also be made. The forest litter continuously fertilised the thin topsoil. Once the land was cleared of large trees and undergrowth and opened to the sun, the staple yam and taro thrived in the rich soil, but only till the action of heavy leaching rain and the crops' uptake of nutrients exhausted the fertility. Even as they harvested their root crops, the gardeners had moved on to another newly cleared patch in the forest and planted their crops anew, to return to the original garden intermittently to gather their tree crops, many of which produced for years amid the rapidly regenerating scrub on the fallowing land. This method of gardening is known as shifting or slash-and-burn cultivation. Some writers call it forest or bush fallow or swidden farming ('swidden' is the Old English word meaning 'burned clearing'). Worldwide, it is an ancient technique and the means of subsistence for millions of tropical forest peoples. The land is cropped for a much shorter period, say one to two years, than it is left fallow.[7] More intensive forms of cultivation were uncommon, but on the small Polynesian outlier of Anuta for example, where gardens were more or less permanently cultivated, mulching was used to maintain fertility of root crops alongside tree crops.[8] In 1568, the first Spanish explorers recorded irrigation systems on north Guadalcanal, indicative of more intensive agriculture.[9] Catoira goes on to mention villages 'surrounded with many plantations and groves of plantains' and the like, but little else.[10]

Three hundred years later, John Renton, a ship's deserter, had more to say, after longer acquaintance. In 1868, Renton had drifted to north Malaita where a chief, Kabau of Sulu Fou, befriended him. Although the Sulu Fou were archetypal 'saltwater' people, being dwellers on one of the artificial islands of Lau lagoon, they also had gardens on the adjoining mainland in the lower foothills, or *fafo asi*.[11] Renton related that there Kabau,

> had a good patch of land cleared and planted with yams and taro..... The mode of cultivating is to clear away the under brush and heap it up round the boles of the larger trees, after which it is set on fire and the tree thoroughly scorched. The natives then climb and lop off the branches, so as to admit the sun and air to the clearing, which they roughly fence in. The soil...[is] extremely rich. In planting taro, all that is done is to drill holes with a wooden crowbar in the earth, and insert a plant which, without further trouble, develops the desired root, but such industry is only possible in time of peace.[12]

With the exception of the shifting nature or forest fallow aspect, Renton's brief description highlights several key aspects of the common method of gardening in much of Melanesia – the clearing of undergrowth, the use of fire, coppicing of branches to admit sunlight, the retention of the dead or dying root system of large trees to prevent soil erosion and run-off, and the limited disturbance of the surface soil in planting root stock by the use of the wooden digging stick.

Fire was the most important tool for the gardener and had some positive attributes for the ecology of the forest garden. In most islands, the men, using a stone or shell adze, cut down bigger saplings and the branches of large trees. Women often assisted with knife-like stone or shell tools, such as the *gori-gori* or swamp mussel of eastern Makira, using these to cut vines and undergrowth. The gardeners then left the green debris to dry out, for one to two weeks and possibly up to a month. After the debris was semi-dried the gardeners piled it at the base of the larger trees and burnt it. Large trees could be felled by fire, but this was very slow. By ring-barking or girdling trees a year or two before, fire accelerated their death from four or five years to as many weeks, thus stopping them from drawing off nutrients from the soil while the roots held the earth in place. Favouring sloping sites for good drainage and the bonus of windrow felling, Melanesians achieved additional control of erosion by felling saplings across the slope, forming rough terraces. Branches and saplings were also used as barriers to keep pigs out. Some took root and soon became secure 'live fences' and, by the time the garden was exhausted, a basis for the re-establishment of the secondary forest. Often on the lower coastal areas and on the smaller islands where yams (*Dioscorea alata*) and *pana* (*Dioscorea esculenta*) were major crops, some dead or coppiced trees were left standing for the yam vines to climb. These vines require full sunlight, hence the pollarding of branches, noted by Renton. Coppiced trees, if not killed by fire, could regrow within a year or so, but meanwhile provided space for sunlight to get to the new plants. Valuable trees such as the canarium almond (*Canarium indicum, C. salomonense, C. harveyi*) were preserved. Gardeners checked the perimeter of the garden clearing to remove certain trees, such as *tui* and *megamega* of south-west Choiseul, which inhibit the growth of taro (*Colocasia esculenta*). In some places gardeners preserved certain trees on the edge of the garden because their leaf fall assisted the fertility of the soil or because their presence discouraged insects and disease.[13]

> The timing of a burn-off was important. Midday to early afternoon in the drier months from May to October are preferred times in many areas. Burning tended to be continuous, because some of the slashed vegetation was still green and, no matter the season, rain was frequent. Wherever possible, gardeners preferred to clear secondary as opposed to primary forest because the slash dries more rapidly and burns more completely.[14]

Fire does far more than burn old vegetation and reduce the labour of the gardener. As one resident of 23 years noted in 1933:

> The natives do understand the value of ashes as a manure and all rubbish caused by the preparation of gardens such as bracken, ferns, leaves, veins [vines] are burned when burning the small trees and larger scrub.[15]

In some places the gardeners left the ash where it burned, particularly in old or primary forest, but in others, such as in Kwara'ae, Malaita, when secondary forest was burned the ash was spread to all parts of the garden.[16]

Fire is essential to the health of the soil for the cultivated plants. Burning is a kind of catalyst which assists the transfer of nutrients from the vegetation to the topsoil where they are then available to the planted crops. Their composition varies. However, generally the topsoil is enriched by the addition of calcium, magnesium, potassium, and phosphorus, by spreading the ash from burned vegetation. There is loss of nitrogen, sulphur, and carbon, which go into the atmosphere as gases (but most eventually come down to the earth somewhere in rainfall). Burning, however, does not reduce the amount of these elements stored in the soil. Fire induces a 'partial sterilisation'. Competing seeds are killed and, with wetting, micro organisms increase their activity, aiding fertility. In burned areas there are fewer insect pests. Compared with other forms of land clearing, burning is the least damaging to rainforest soils.[17] It was only in areas of extremely heavy rainfall that fire was rarely used and the felled debris stacked on parallel lines against the slope.[18]

On Malaita, Renton recorded the still-common use of the 'wooden crowbar' or dibble to drill holes for the root crops with minimal soil disturbance. Depending on local custom, men and women planted out seed material. The women did the cultivation and carried home the produce as needed. Yams were annuals which took about eight to eleven months to mature and were valued because they could be stored after harvest for up to six months. Some taro gardens required greater care as the tubers were pulled when needed, tops replanted with the suckers planted out all year into adjacent newly-prepared ground. Somewhat different cultural practices were needed for wet land taro which takes two years to mature.[19]

Usually after the first year's crop of tubers, the garden was left to bananas (*Musa* spp.), betel pepper (*Piper betle*) and edible greens. In some places such as the cyclone-prone small Reef Islands (Temotu Province), many so-called 'wild' (meaning simply less palatable) varieties of root crops were often planted, but left to multiply for some years until the emergency arose. Most gardeners planted 'economic trees' for food in the garden as well as around villages. Several of these, as with the staple root crops, were aboriginal introductions from western Melanesia and beyond from South-east Asia. The breadfruit (*Artocarpus altilis*), the coconut (*Cocos nucifera*), bush spinach (*Hibiscus manihot*), two banana or plantain cultivars (*Musa* AAB and *Musa* ABB triploid), the wild apple (*Spondias dulcis*), and possibly the pomelo (*Citrus grandis*) were the main ones. Santa Cruz and Reef Islanders practised arboriculture to a greater degree than other Solomon Islanders and, through selection, have domesticated and improved feral tree species as they and other Islanders have done over hundreds of years with varieties of taro and yam. In almost all areas, canarium almonds and *to'oma* (*Terminalia solomonensis*) were among the most commonly planted as well as 'Malay' apple (*Eugenia malaccensis*) and the golden and wild apple (*Spondias cyathera* and *S. dulcis*).[20] However, the canarium (*Canarium harveyi*) did not reach Tikopia until the late eighteenth or early nineteenth century when the voyaging Tikopian chief, Pu Veterei brought it home from Vanikolo.[21] Elsewhere on the big islands, the abundance of the long-lived *Canarium* spp. in secondary forest is a common indicator of prior human settlement.[22] Canarium trees were valued and consciously planted, in some areas such as Makira, along with the prized areca or betel nut palm (*Areca catechu*) to serve as a ladder for the future harvesters.[23]

The breadfruit was planted widely in the Reef Inlands where it provided storable, dried *nambo*. However, among the Arosi people of Makira the breadfruit tree was believed to have a bad effect on adjacent garden crops. The Temotu gardeners also cultivated the Polynesian chestnut (*Inocarpus fagiferus*) and Oceanic lychee (*Pometia pinnata*). Throughout the archipelago, Solomon Islanders planted other useful trees and shrubs – betel nut (*Areca catechu*), the universal mild narcotic chewed by Melanesians and nearby Polynesians; the pandanus (*Pandanus* spp.) for fibre for making mats, baskets and sun shades; the *Cordyline terminalis* for magic, medicine and boundary marking; and, in wetter soils, the sago palm (*Metroxylon solomonense*) for leaf for buildings as well as, in the Shortlands, Choiseul, and north New Georgia, for human food, and elsewhere as emergency famine food and for pigs.[24]

Food trees lived for long periods. However, after the harvesting of the root crops the forest soon began to re-assert itself: last year's garden ground, *ali siu*, to the Lau people of Malaita, was a place where a person 'gets a bath', from the branches of the shrubs that have grown since the garden was abandoned.[25] The succession pattern is predictable, providing the surrounding area was covered in established secondary or primary forest and is the same pattern for any medium gap in the forest.[26] In its simplest form, vegetation returns in a herb, thicket, tree succession. The herbaceous phase lasts from six months to a year with grass species at first predominating. Light-demanding shrubs and seedlings exploit the gaps and soon crowd out the herbs. Woody species establish. The third stage sees competition from various small tree species. The breadfruit is common and to some extent the mango, usually planted earlier by the gardeners. Tree ferns and palms re-establish next with the gingers in the scrub layer. The large tree species also establish, usually the *Pometia pinnata* and *Vitex cofassus*, but others, too. As the ferns reach their maximum height and start to die back, the soil then is ready for yams and taro. Without further disturbance for about 20 to 30 years after the original clearance, the pioneer secondary forest establishes and, after about another 80 years, matures. Regarding cultivation, it is the state of the vegetation that the gardener reads as much as the smell and texture of the soil to decide whether or not to cultivate the site. In terms of years, the gardener could begin the new cycle of clearing about ten years after the land first was cultivated, or later, and he would find the soil would be 'fat', 'full-grown', or 'warm' (that is, fertile) once again.[27]

Shifting cultivation, under these circumstances, was an ecologically sensitive way of gardening in the rainforest. However, in drier areas usually bordering the rainshadow, the forest could be reduced and fire-farming of grasslands did long-term damage to soils by reducing nitrogen, sulphur and potassium, and generally demanded more labour input.[28] Within the rainforest, natural processes largely performed the work of soil fertilisation and preparation; erosion was minimised through the use of root systems, placing saplings across slopes, and limiting surface soil disturbance and erosion by the gentle tools of fire and dibble. The rapid, often human-assisted secondary forest succession guaranteed in time the restoration of the humic content to the soil. It was a grand productive cycle, sustaining humans and forest.

Shifting cultivation was also an extremely efficient way of producing food from the forest. 'The Melanesian does not cultivate largely, but he cultivates well', noted Bishop John Selwyn.[29] The average Melanesian did not have vast areas of land under cultivation at any one time, but shifting cultivation did require considerable areas of suitable forested land relative to population. And herein lies the key to keeping the garden-forest ecology balanced. For much of Melanesia and South-east Asia, shifting cultivation could only support 10–20 persons per square kilometre (0.1–0.2 persons per hectare) since about 10 per cent of the total suitable forested land is available for gardening at any one time, as the rest is at some stage of the forest fallow process.[30] Providing the socio-political system is relatively stable, if either the land available for potential gardens relative to a fixed population decreases or the population relative to a fixed potential garden land area increases, the system, unmodified, will become less efficient and ultimately break down, resulting in land degradation and human malnourishment. Within the limits of this gardening method and range of indigenous food plants, the equilibrium can be restored either by increasing the land area under primary or secondary forest available for gardening or by decreasing population.

In prehistoric times population pressure may have been a cause of environmental degradation. Yet recent research suggests that malaria acted as an effective check on rapid population increase, the size of settlements – rarely above 100 – and even their interaction.[31] People lived in scattered small groups with little production of huge surpluses. No significant export trade existed, as trade was barter or gift exchange. There is little suggestion of significant tributary regions, forced to produce massive quantities of goods for neighbouring rulers. However, there may have developed chiefly societies on the main islands, such as the Shortlands, with some kind of class structure.[32] This could have led to social competition and a resultant need for surplus production, which would have seen agriculture intensify. It seems that in pre-historic times larger, dense populations existed in areas such as Kolombangara, parts of New Georgia and north Guadalcanal, where stonework and terracing suggest irrigation and drainage systems, probably for wetland taro, and a form of intensified agriculture. Whether this was an outcome of increased population pressure, extensive forest clearance, the need to produce a major surplus for social competition or a combination of these factors is as yet unclear, though in the eastern Solomons (Temotu Province), water reticulated systems for growing varieties of taro, *Colocasia* and *Cyrtosperma*, appear to have been in use before the massive population decline of the early 1900s.[33]

Land may have been over-used in the past, but some or all of its inhabitants would have had to move out, die or adjust their way of using the environment. If this were ever the case, there is little memory of it in the present generation. What Melanesians saw, in their cultivation of the forest floor, were small patches of gardens, here and there and in various stages of returning to forest. They appear to have understood the ecology of change in these patchwork plots, but it was change on a localised, small scale and over hundreds of years when there were no written records to note pristine and changing states.

The slow rate of change was due largely to the limitations of gardening technology within a tropical environment. Except on the few rain-shadow savanna lands, man-made fires rarely got out of control; the heavy rains saw to that. Fire had to be encouraged rather than contained. The other means of forest clearance was stone and shell adzes. The enormous amount of time and labour required in clearing primary forest by tools, with or without fire, can be extrapolated from the process of merely ringbarking a tree on Guadalcanal:

> Large and medium-sized adzes were used for tree-cutting. The owner of the land would employ three to five men who were specialists in the binding of stone adzes. Each village would only have a few men specialising in this and they were considered skilled men. The landowner would reward them by the gift of a pig. The cutters would work together at different places round the tree. First a medium-sized adze, *kununga*, would be used to cut notches at two levels in the trunk. Then the large adze, *tiahila*, would be used to split off the bark and wood between the notches. In this way the tree would be ringbarked and left to die. Three to five men could cut a sufficient amount round a large tree in one or two days. There would be enough adzes for a continual repair service. As the binding worked loose it would be passed to the 'binding men' and a freshly bound adze taken up to continue working. At the beginning of each day's work the men would take the adze to a sharpening stone, *vatunasa*.[34]

In some areas, such as Malaita, Uki, and Ulawa, this process may have taken even more effort, given that the usual adze was made of flaked chert, a flint-like quartz.[35]

For the first gardener in a pristine forest the labour required to work it must have been enormous. Quite literally he and his family 'opened the forest' by clearing land to make a garden and, in time, discovering and utilising an array of forest products.[36] If, in living memory, this land had not been cultivated before, then it became the settler's (lineage A) and, in time, his heirs and assigns. Although there were regional differences in land holding, descent was significant, as was residence. Gradually, as population increase was probably very slow, the lineage bifurcated.[37] First, sub-lineage B, then sub-lineage C moved away from the foundation settlement. As each cleared new areas for gardens and established settlements and shrines they asserted rights more or less exclusively for their own sub-lineage within certain zones as well a large area common to A, B and C where they all could hunt and forage (see Figure 5).

The act of clearing the forest for human use was the ultimate claim for possession. Given the effort expended, it was well-founded. Moreover, given that the transient partial immunity (premunition) to malaria was to localised varieties, there was danger for people in moving into an ecologically different area, though this would have been perceived in cultural terms.[38] So, wherever possible, the Melanesian cultivator tried to maximise production while ensuring soil fertility by selecting a new garden site from regenerating secondary forest on old clan gardens (called *gano* or *ano* in much of Malaita) rather than the primary climax forests. This pattern of using land under secondary forest appears to have prevailed when the human population was either fairly stable or falling.[39] The resultant 'conservation' of the primary forest seems more an outcome of the desire not to expend unnecessary labour, rather than to sustain forest cover.

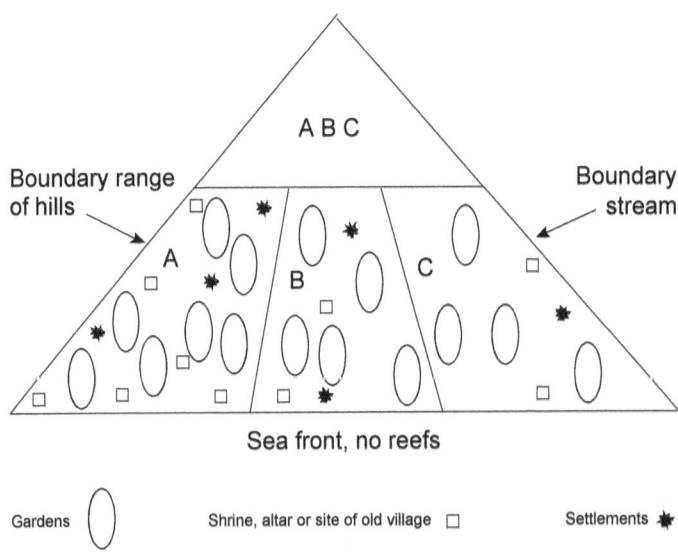

Figure 5. Schematic representation of expansion of lineage land and customary tenure.

The meaning of the forest

It is within this context of getting a living and being part of a human society that Melanesians perceived their forests and seas. Through trial and error and observation over millennia they learned much about their natural context, as is recorded in oral traditions.[40] Their botanical knowledge has impressed visitors. 'I was particularly struck with the familiar knowledge of their plants and trees which these Islanders possessed', noted naturalist Henry Guppy in 1882.[41] Solomon Islands societies developed their own plant taxonomies, revealing much regarding the way they perceived the flora. In Kwara'ae, a Malaitan language adopted by botanists for classification in the 1960s, the taxonomy 'is based on the relationship between the human community and the surrounding flora'. It consists of ordinary words describing a plant's appearance, properties or usage. Different Kwara'ae communities sometimes have different uses for the same plant, hence name variation. It is the usefulness of the plant that gives it significance. Because, for example, the many small grasses were of little use, the people rarely named them. On the other hand, for a staple food plant like the taro there were specific names for the many varieties.[42] Similar patterns of differentiation existed elsewhere: the Sa'a people of south Malaita had 120 names for taro, while the people of East Kwaio had 38.[43] On Guadalcanal, at least 108 varieties of yam (*Dioscorea alata*) and 35 varieties of pana (*Dioscorea esculenta*) are named.[44] Taxonomies are not confined to the terrestrial environment. The dwellers of Marovo lagoon, dependant on the resources of sea and lagoon, have classificatory systems of marine fauna and flora that highlight characteristics of the fish species that enable the fishers to predict behaviour and thus be more successful.[45]

Solomon Islanders ordered the plant, as well as the animal, world mainly in terms of its utility.[46] By making a thing useful they achieved a measure of control. It had a cultural value. Those things beyond real control were in a sense outside culture and were, in European terms, nature. This 'nature' was part of the natural environment, but still 'wild' or largely in the realm of wild spirits. These had to be appeased or tamed through sacrifices if one wanted access to the resources. In this way, space became place. A territory might be 'tamed' but, unless ritually or socially valued or transformed through labour, much within, such as vegetation, was not of particular worth. Land became landscape once it had human meaning. However, the domestic and wild dichotomy was very well defined in the thinking of Solomon Islanders. Hamlets and gardens with clear boundaries were culture as were those forests and seas, opened and made available for human use, with paths for their possessors. Sacred sites such as an ancestor's grave or sacrificial shrine, the founding ancestor's village, or an important ancestral battleground were all obvious signs of culture in the forest. These were the means by which clans took their purchase for land claims, as well as focal points of descent group identity. Boundaries were sometimes hazy and contested. However, within prominent topographical features like watercourses and mountain ranges or reefs and islets in lagoons as boundaries, these sites were the pivots of territory and history. They were also signs that the resident people were part of the land and, in coastal areas, the seascape.[47]

Ancestral shrines were not difficult to locate, although only relatives knew the spirit's identity. In pre-Christian times, many were kept clear of undergrowth by the descendants. In some societies the dead were cremated, cast into the sea or occasionally buried, but in most places corpses were placed in the trees.[48] Very old shrines, often simple heaped stone altars, were marked by groves of mature trees, including hardwoods, which stood out clearly from the secondary forest. These groves, in a sense, are man-made as 'no one is sacrilegious enough to put an axe to trees sacred to the ancestral dead'.[49] Depending on age, they were thus reserve areas of regenerated primary forest in stretches of secondary growth, and a potential source of seed to re-forest adjacent depleted areas, as well as a refuge for fauna. Their function was to assist in the process of balancing the cultural ecology of the descendants of human beings and, as an unintended outcome, that of the forest environment.

In many places these groves were sacrificial as well as grave sites.[50] To interfere with the trees of spirits whether ancestral or otherwise was to invite doom. The story of the ghost woman of Kirakira relates how she was captured from an enemy group in pre-Christian times and used sexually by young men who paid her 'owner'. She was eventually reduced to penury and died, her ghost abiding in a huge banyan tree nearby. She often manifested herself to young men and they soon died. When in colonial times the district officer cut the tree down she moved into the government guest house and still appears there and nearby, often to young men.[51] It has been said that when a white planter in 1919 hired several men to clear the small island of Jagi off Santa Isabel, all but one of them died because a spirit residing there was upset. Over 20 years later, a Santa Isabel man engaged six men to clear the forest and three of them died, so again the clearing ceased.[52] Elsewhere, parts of the forests were pathways of the newly dead spirits on their way to a sacred island or mountain peak. The people of Temotu Noi, in pre-Christian times treated these 'paths' in the bush with great respect.[53]

This familiar ancestral presence permeated the natural landscape, providing belonging and emotional security to Solomon Islanders on their home ground. Their ancestors, over the generations, had discovered how to manipulate and use the things of the forest to sustain and enrich human existence. The ancestors protected them from malevolent forces outside the community and guided them in their lives. Just as the forests that their ancestors had opened nurtured and empowered them, so too did the continuing ancestral presence sustain them. But ancestors did not love unconditionally. They could be offended by behaviour proscribed in the past. The ancestors also liked to receive things their descendants valued, such as pigs, betel nuts, and valuables. Solomon Islanders regularly made sacrifices of these to maintain or regain the protective, caring concern of the ancestors. Without it, Solomon Islanders were doomed – strife, poverty, illness, and sudden death were their inevitable legacy. They became cut off from their ancestors who were essential in the channelling of the inherent potential of all things in the natural world to the good of their descendants.[54]

The natural world could be unpredictable and frightening: sudden and violent cyclones could destroy gardens, sweep away housing, and strip the forests bare of vegetation and even wipe out an entire population as happened in ancient times to the people of Leli island off east Malaita.[55] Earthquakes were almost as devastating and on occasions caused islands to sink beneath the sea or conversely for reefs to rise up as atolls.[56] Volcanoes periodically disturbed some settlements, killing their inhabitants.[57] Droughts, though not frequent, did occur, causing famine.[58] The windward side of an island could have its taro crop ruined by excess rain while the lee side experienced water shortage.[59] When the forests and the sea were the domain of strangers, they possessed some of this frightening quality. Even in familiar forests and seas there dwelt various spirits. The people of Sa'a, Malaita, believed that there were wild spirits in the forest who tried to murder people.[60] Changeling ghosts might accost a person in the forest, driving away his own protecting ancestral ghost.[61] Spirits of women who died in childbirth haunted the forests and the short hairy men, the *walu masu* to the Kwara'ae, lived there, ready to spring upon the unwary.[62] The spirits of people decapitated in headhunting raids wandered the bush, reducing the spiritual potency of their descendants.[63] The *vele* man of Guadalcanal lurked there with his tiny bag of magical death seeking the lone victim,[64] as did war-bands and paid killers. No wonder the Solomon Islander dreaded to be alone in the forest depths even in the daytime, away from light, settlement and security.[65] And more so at night, when spirits – and the malarial vector, *anopheles,* – might take a deadly toll.[66] In their spiritual metaphors, the living expressed their fear of the unknown and of the forest. The forest was full of death, just as it was full of life. It was as dangerous as it was nurturing.[67] The home of ghosts and their sacred places, the forest in itself was not sacred, but the wise still trod warily there.

Through their sometimes capricious ancestors and their accumulated knowledge of their environment, Solomon Islanders had the means to work with and within the forest to win a livelihood, but never the means to control it entirely. Those who opened the forest by felling the trees and bringing into fruition the creative potential of the soil and forest became clan heroes, progenitors of families, and, in a few societies, great chiefs.[68] To have set one's seal on the land, through exerting control over a part of the forest, was both the mark and memorial of a true man.

Transformation and conservation

The first settlers had favoured the coastal fringe, being close to resources of both land and sea; but in time, various pressures, such as natural disasters, population growth, warfare and mosquito-borne malaria may have driven people inland up the valleys and out into the ranges.[69] As many oral traditions recall, over hundreds of years the distribution of population ebbed and flowed between coast and mountains; in the process local groups extended their knowledge of their environment and resources. As they did this the forest and indeed much of the natural vegetation became both product and part of culture. It seems there were few places in Solomons that were not occupied, used as hunting grounds, or cultivated sporadically at *some* time in the past.

If the activities of their neighbours who settled further south in New Hebrides and New Caledonia are any indication, ancestral Melanesians of 3,000 years ago practising their gardening began slowly to change the landscape, causing erosion on hill slopes and the consequent sedimentation on valley floors and lagoon basins. Primary forest gradually gave way to secondary growth and widening grasslands in drier areas.[70] And there may have even been a preference for open scrub and grasslands because these provided fewer habitats for *anopheles*, as the dense population seen by the Spanish in 1568 on the north Guadalcanal attests.[71]

Ancestral Polynesians, identified by Lapita pottery, settled Tikopia, part of the modern Solomons, as early as 2900 BP, probably by back migration from west Polynesia, though a Lapita fragment has recently been found in the Roviana Lagoon, New Georgia.[72] They exploited the wild, natural resources of the marine and terrestrial environment and introduced domesticated plants, pigs, dogs and fowls and the 'Polynesian' rat, *Rattus exulans*. In 800 years, they dramatically reduced the avifauna, some to extinction, as well as fish and shellfish, and had commenced to alter the landscape through forest clearance and hillside gardening. Resultant erosion deposited soil on the reef area which, coupled with organic enrichment of dunes, extended the littoral plain and entrapped a brackish lake, formerly a saltwater bay.

To Open the Forest

By historic times, land area had increased by about 40 per cent and reef area decreased by about the same, much of this transformation being due to human action on landscape.[73]

Extrapolating from this pre-historic interaction of human beings and their environment in neighbouring islands and within the Solomons group, it seems highly probable that future research in prehistory will reveal that practically all of the Solomons landscape as well as the interface of land and sea have been transformed to some extent by human action long before Europeans called there. The forest was once pristine certainly, but only as long as human beings kept out of it.

Evidence from more recent human activity indicates that Solomon Islanders were a major ecological factor in the composition and distribution of the forest and the condition of the soil. They almost certainly had a similar role in the maritime regions. In some areas, generations of shifting cultivators eliminated the primary forest and did not give it the estimated 30 years needed for it to be replaced.[74] The first British administrator, C. M. Woodford, after visiting Savo in 1898, noted that 'much of the virgin forest on Savo has been cut down'.[75] Although Savo's population had declined between 1908 to 1930,[76] by 1946, the forester Walker, confirmed that 'The forest is mainly secondary species'. The island had been so depleted of *vasa* (*Vitex cofassus*), a hard wood used for house posts, that the people had to import them from Guadalcanal.[77] Of course, other influences such as the eruption of Savo's volcano in about 1840, the island's position within a semi-rain shadow zone cast from Guadalcanal, and random cyclonic winds may have been contributing factors in preventing the re-establishment of certain tree species.[78] In the rain-shadow belt of northern Guadalcanal with its comparatively low rainfall of 2000 mm per year and its definite mid-year dry period, there are *Themeda australis* and *Imperata cylindrica* grasslands, noted by the Spaniard Mendaña in the sixteenth century. These grasslands await detailed pollen studies and could have been formed in pre-settlement times by non-human factors in extended periods of particular aridity and when the El Niño Southern Oscillation in the western Pacific was severe or frequent. Whatever, it is likely that the grasslands have been extended since settlement as an outcome of human interaction with a rain-shadow ecology, although pressure on them by Melanesians probably lessened for almost a century from about the mid-nineteenth century with the decline in the resident population due to head-hunters from the western islands, introduced diseases, and the failure of about 10–20 per cent of the young labour recruits to return from the plantations of Queensland, Fiji and Samoa. In the past, as in the present, Melanesians used fire to hunt small game, to clear paths and land around villages so enemies could not advance unnoticed, and for gardening. The fire regularly escapes and burns large areas, preventing the re-establishment of the forest, reducing the fertility of the soil.[79]

Some forest areas indicate that they were once fired as a likely preliminary to hunting. Forests of *malasulu* (*Casuarina papuana*) on ultra basic rock that have been burned a few times develop open heath communities of *Gleichenia* spp. and *Lycopodium cernum*, with bushes such as *Myrtella beccari* and ferns, as is the case on much of San Jorge Island. During the 1960s, the botanist T. C. Whitmore recorded signs of human occupation in areas then forested with high *Dillenia salomonesis* and *Calophyllum kajewski*, both climax species, on Kolombangara. This island was then almost as relatively sparsely populated as at the turn of the century when head-hunting and civil wars appear to have reduced the population dramatically. Allardyce Harbour on Santa Isabel was forested with large stands of *Campnosperma bevipetiolata*, a climax species that needs quite big gaps to establish. Whitmore believed this was the result of growth on large areas of abandoned garden sites after the Isabel people fled inland and south to escape the New Georgia headhunters from about 1840 until soon after the establishment of the Protectorate. Introduced disease also may have driven survivors away. These fugitives or their neighbours in inland Santa Isabel left signs of cultivation to within a mere couple of hundred metres of the Mt Sasari–Mt Kubunitu. Evidence of dwellings was found in then-unoccupied areas of Kolombangara and Vangunu. In central north Makira, Whitmore recorded clear paths in the largely unpeopled Wairaha river basin, probably evidence of denser settlement before the several epidemics of exotic diseases of the latter half of the nineteenth century. He also found vegetation typical of old garden sites on Mt Jonupau on Guadalcanal.[80]

Most of these changes occurred over at least decades and usually several generations. Rapid transformation of the environment except by tectonic or cyclonic activity was rare. Nonetheless, a graphic example of the interplay of natural coastal subsidence and human-induced erosion was noted on south-west Ranongga in the 1930s, an area currently considered to be highly susceptible to deterioration if forest cover is lost:[81]

> Denudation is going on rapidly. Where houses were a few years ago is now well out below water. The foreshore is all gone and the waves continue their battering. The result is that the land slopes sharply up from the beach...The hills rise sharply and are almost cliff like. Gardening is a most difficult business as the land slopes so steeply. When they clear ground they only start further denudation. They plant their food plants but when the rain comes heavily all the surface soil is carried away and with it comes the things they have planted. Then the sun bakes the soil till it is like cement.[82]

On the large islands with a relatively small population, most Melanesians were not driven to practice soil or forest conservation, other than by stabilising a current garden ground and moving on when it had become depleted of its fertility. Inhabitants of very small islands with fewer resources appear to have been more consciously conservationist because they had to be to survive.[83] On the Reef Islands and Tikopia, the people have cultivated trees, not only for food and fuel, but also to safeguard their land. The people plant *Calophyllum inophyllum, Casuarina equisetifolia, Ochrosia* spp., *Pipturus* spp. as well as *fa'ola* (*Hibiscus titiaceus*) for coastal stabilisation.[84] In the Reefs too where there is less suitable land relative to population for classic shifting horticultural methods, the people have observed that some species particularly aid the restoration of soil fertility. They deliberately plant the *fa'ola* because it is a fast grower, develops a canopy that shades the bare ground and produces substantial leaf litter. It may be that its roots play a role in nitrogen fixation; whatever the explanation, the outcome is effective. Gardeners for the same reason also encourage the growth of other trees that establish on used land, including the legume, *Schleinitzia novo-guineensis* or *karefo*.[85]

Not all practices for conserving the source of food involved the planting of trees; indeed the people of Anuta destroyed self-planted drift coconuts and any vegetable foodstuffs on small, rocky, uninhabited Fatutaka (Mitre Island). Because they periodically sailed to this islet to hunt birds, collect eggs and fish for sharks they feared that if the life-giving coconut or yams became established, settlers from elsewhere might take up residence on their hunting grounds.[86] The neighbouring Tikopians at various stages of their history have consciously eliminated pigs because of the damage these animals did to their cultivations.[87]

Fertility and protection of the food supply were sought by other means too. All societies had religious rites associated with gardening, especially planting and the first harvest. On Makira, human victims were sacrificed in a new garden and their blood poured over it or the *maniato*, the garden post or tree set up to protect the garden. Red shrubs, such as types of cordyline, to represent blood were also planted in gardens to ensure fertility.[88] Since the gardeners' ancestors had been the first to clear the forests and to demarcate clan lands, sacrifices of the first fruits were made to them.[89]

For all Solomon Islanders, gardens were the basis of their subsistence; for a few, they were the basis of socio-political power. Although large chiefdoms beyond a followership of 100–500 were not possible because of the sporadic movement of settlements following their shifting cultivation, some leaders became relatively powerful. Whether semi-hereditary chief or self-made big-man, a leader had to have at his disposal forms of wealth, pigs, and valuables – made from shell, stone, teeth or feathers. To get these or, with pigs, to raise them, he had to have access to abundant supplies of vegetables and nuts. Likewise, since feast-giving was synonymous with fame, a leader was expected to sponsor feasts marking significant events in the life cycle of his people when the ancestors were summoned to bestow their power and blessings. His control of substantial quantities of garden produce was essential.[90] With a few exceptions, such as the chiefs of Roviana who got the bulk of their vegetables and nuts from trading their locally produced valuables to other islands, this meant that a leader had to bring large areas of land under cultivation or persuade his kin to do it for him. Hence, the need for several wives and relatives to tend the gardens and feed the pigs: hence, in many places, the need for captives and adoptees to provide extra labour. In 1882, Guppy walked across Magusaia in the Shortlands, noting nearly a mile of continuous cultivation. This was the garden of chief Gorai who had a hundred wives.[91] Of course, in some societies leadership was also linked with esoteric knowledge of land-holding, warfare and religious ritual and magic. But the leaders with the greatest political power within and without their own clans and residential groups seem to have always been those who could acquire or create wealth, enabling them to give and to be patrons, to attract and keep followers. In land-based economies, these were the men who made big gardens, imposing their control over large tracts of the forest, bringing it into ordered cultivation.

In doing this, generations of Melanesian horticulturalists and hunters, whether leaders or ordinary men and women, altered the structure and composition of the forest considerably. The ancestors introduced new plant material, including trees, and selectively encouraged the growth of valued endemics. So diversification of food plants for human beings went hand in hand with simplification of the original ecology, at least within garden sites. The gardens of the ancestors had regularly occupied much of the lowland forests areas, but at times reached inland almost to the mountain tops. Manipulating the environment for subsistence induced changes. Hunting probably led to the depletion of animal life, particularly avian species, as happened elsewhere in the Pacific.[92] On south Maramasike, Malaita, the native fowl became extinct, remembered only in proverb.[93] Lowland forest ecology almost certainly altered, with increasing grasslands in some sensitive areas and the creation of heath in parts of Santa Isabel. Vast areas of primary forest were replaced by secondary growth for long periods, in areas of denser population. Solomon Islanders, during their occupancy of the archipelago, learned much about their environment and did what they had to do to win subsistence from it. In so doing, they altered the insular terrestrial and coastal maritime environment and often degraded its native ecology. Some even built new 'lands' in the lagoons – such as the artificial islands of northern Malaita. They adapted and probably modified some of their practices with experience of the degradation they had caused and seemed to have evolved virtual sustained-yield systems in horticulture, agroforestry, hunting, gathering and fishing,[94] enjoying what has been labelled optimistically 'primitive affluence', but at the price of toil and risk.[95]

To Open the Forest

However, the ancestral Solomon Islanders were not ecologically noble savages,[96] practising some kind of Green Primitivism. There was a tight restraining relationship between settlement patterns, population and the method and means of gardening. The people lived in small, relatively isolated societies, producing few surpluses, but traded regularly with close neighbours and with distant peoples across the sea to obtain materials they could not produce in their own territory. They thus made limited demands on the environment. Their environmental knowledge was deep, but almost certainly localised and utilised to benefit themselves and their families. When long-term changes occurred that simplified the ecology, life-spans were too short to record the process. Like all forest fallow gardeners, most Melanesians had no need to plan ahead more than a few years because of the relatively low labour input. Their gardening, as with much of their use of the forest, was cyclical and this meant modification of it was at any given time temporary and partial, resulting in both relative ecological and economic efficiency, but always within the constraints of population. Although there do appear to have been areas which once had higher populations than noted at European contact, the population seems never to have outrun resources entirely, except when a disaster like a drought struck.[97] In fact, as the Kwara'ae say, 'man' was 'a scarce resource'.[98] This assessment was not surprising, in societies that had an infant mortality rate of around 40 per cent, where a nuclear family could not risk more than one child-in-arms in case of raids, and where the average life expectancy in a malaria-ridden environment was short, probably about 30 years. It was far more the characteristic small, politically non-centralised socio-political and residential groups, the low population/land ratio and the limitations of stone and shell technology that circumscribed the amount of control and change Melanesians wrought on the environment, not a deep practical and ideological commitment to sustainability.[99] However, within a few hundred years, some groups, many of Polynesian stock, on small islands learned to live within the islands' carrying capacity through population controls and stewarding their resources. Following the coming of the white man, at least one of these factors was to change dramatically, as much of the old technology, the key means of production, was abandoned.

— 3 —
A Great Coconut Estate

Worst of all their bad points almost, is their incredible and incurable laziness – the heritage of all Pacific races – the result, no doubt, of the extreme fertility of a land which causes them no occasion to work in order to live.

Boyle T. Somerville, 'Ethnographical Notes on New Georgia, Solomon Islands', *Journal of the Royal Anthropological Institute* 26 (1897), 411.

To the uncultivated eye a forest appears simply as uncultivated land.

E. P. Thompson, *Whigs and Hunters: The Origins of the Black Act* (Harmondsworth, 1977), 29.

The European settlers who came to the Solomons were much more interested in the potential of the land than the forest. Although some saw a resource in the trees for timber, most perceived them simply as obstacles to the creation of their plantation estates. Settlement was made feasible by the establishment of a British presence in 1896. Colonies cost money to run, so the administration's task was to encourage use of resources to produce revenue. Thus it fostered the plantation economy. Another group who came for the long-term were the missionaries. Though having mixed attitudes to commerce, they put the interests, as they saw them, of their future converts first. Their aim was conversion of Solomon Islanders and the establishment of a church. They were vessels of Christianity, but they were inescapably vessels of their own culture and revealed similar perceptions of the landscape as their countrymen, the planters and administrators. All these newcomers were set to transform land and people to some extent, but found that process had been underway before the British flag waved in the trade winds above the islands.

Potent tools, potent weapons

The advent of metal revolutionised Solomons societies. The trickle of introduced iron and steel implements that commenced at the turn of the century with the sporadic victualling trade of the whaling ships increased in volume from about 1840 on, as white traders came seeking bêche-de-mer, tortoiseshell, pearl and trochus shell and coconut oil, superseded by the more profitable copra in the 1870s. The trickle became a flood when the labour trade reached Solomons in the 1860s–1870s. Whether kidnapped or contracted, the thousands of returners who survived their indentures in Queensland, Fiji, Samoa or New Caledonia, brought home considerable supplies of axes, adzes, knives, scissors, files, and other metal goods, including Tower muskets and Snider rifles. By the 1880s, all but a few isolated inland communities had earned, traded or extorted the durable and efficient metal tools that reduced the labour and time expended on tasks, formerly done with the stone-shell technology, by 30 per cent to 40 per cent.[1]

While they remained scarce as emblems of status, the first metal blades were treated as potent items consecrated to the ancestors.[2] The first axes into Sa'a, south Malaita, came via Wango on north Makira, gifts of the Melanesian Mission. There was no need to teach Melanesians about the power inherent in controlling the means of production: the Sa'a traders who owned the axes hired them out for 'five or ten porpoise teeth or a string of [shell] money'.[3] Clans and chiefs worked on a similar principle upon the return of the labourer: 'His father selects, with due deliberation, the best-tempered of the axes', then 'The chief's son relieves him of one of the largest knives'.[4]

In the New Georgia islands from the 1840s, the mounted axe head became a terrifying weapon, *kilakila*, in the hands of headhunters who, by the 1880s, were raiding as far away as Russell Islands, west Guadalcanal, and north Malaita, as well as Choiseul and Santa Isabel. Heads and head-hunting were both symbols of and ways to potency for the warriors and their *bangara*, chiefs.[5] With firearms added, those in possession of these weapons killed, enslaved and drove their demoralised foes into hiding.

Metal weapons had advantages especially when only one side possessed them, but metal tools had a more subtle impact. From the mid-nineteenth century, as more societies obtained the new technology its effects ramified. Where once a single house accommodated two families, two could be easily built since the supports of wood and bamboo could be readily cut with the new tools. Clearing for gardens could be both more extensive and intensive; trees could be killed, not simply coppiced, over larger areas. Yet before the advent of the colonial government, there is only occasional evidence that more land was cultivated, although its clearing was easier.⁶ From the 1860s the introduced sweet potato, the *kumara* (*Ipomoea batatas*), recorded as early as the 1840s,⁷ was becoming a hardy staple suited to poorer soils, so providing a crop for the second year of a garden and reducing the need for further clearing, though this extended cropping was very demanding on the soil. *Kumara* became a valued crop in coastal areas. Thus the technological revolution, coupled with a new horticultural crop, brought about an ecological revolution in Solomons. But the ecological revolution was to be gradual because of population changes.

Introduced continental diseases appear to have dramatically reduced the unexposed or 'virgin soil' population in some places such as Makira, Temotu (Santa Cruz district), New Georgia and the Shortlands. This would have lessened the need for expanded garden areas, just as Europeans, their plantations and their animals began to seek their own niche in the environment. The new technology meant that communities could allow their young men to work overseas or to make copra for the traders to obtain more metal goods. As well, time saved with the new tools went mainly on increased ceremony, much associated with raiding.⁸ In parts of Malaita and Choiseul, more trees were felled, not to make more gardens, but for walls of logs as high as 15 feet and thick enough to stop the bullets!⁹ However, as the twin forces of the colonial government and Christianity began to spread, raiding lessened.

Comprador capitalists or indigenous entrepreneurs?

Several returners from overseas and some deported from Queensland in 1907 after the labour trade ceased came home as Christians. Aware that their ancestor-venerating bush relatives would not welcome them, they built villages on the coasts, often where they had but tenuous claims.¹⁰ The new settlers 'proceeded to clear and work the virgin forest', using the steel axes.¹¹ This was partly from necessity, but it seems that the experience of 'scrub clearing'¹² for plantations in Queensland and Fiji turned the thinking of some in north Malaita, not only to opening the forests for subsistence, but also 'to open[ing] up our country' to commercial development as the government extended control.¹³ These Christian settlements attracted kinsfolk, some because of the anticipated development: 'Vague visions – a town, roads, a mill, perhaps, and other things caught from returned Queenslanders – float before their minds …'.¹⁴

The British had annexed the Solomons in 1893 to protect British interests – namely, strategically vulnerable Australia and New Zealand – by excluding other European powers and to further regulate the arms and

A Great Coconut Estate

labour trade in the western Pacific. However, British policies were premised on the creation of revenue to run an administration through establishing a plantation economy. As far as the British administration was concerned, the role of the Solomon Islanders was to acquiesce in the occupation of apparently 'vacant' land, to sell some of their 'occupied' land, and to provide labour for plantations.

Some in the Christian settlements of north Malaita, led by Peter Ambuaffer and Benjamin Footaboory, had wanted to deal in land. They informed the King of England in 1907 that 'the land is laying up for to sal [sell] to those who are full of money to make a Possession', but they intended it to be for the astronomical price of £8 (160 shillings) an acre for uncleared land, compared to a range of three farthings to two shillings being paid for more secure land elsewhere.[15] But Malaita, with its rugged terrain, poor soils, and its large 'bush' population uncontacted by the administration, was still a dangerous place for the white man. This dashed the 'get rich quick' real estate development schemes of the Malaitans, their petition being regarded by the Colonial Office, as 'an ingenious attempt to beg for money...'.[16] For those like the Malaitans who had only their labour to trade for manufactured goods, work on plantations seemed the only option.[17]

Kanaka Collection: John Oxley Library, Brisbane.

Figure 6. 'Scrub-clearing' in Queensland by 'kanaka' labour. Most of these workers are Gilbertese, clearing land at Farnborough, near Rockhampton, Queensland, 1895.

Kanaka Collection: John Oxley Library, Brisbane.

Figure 7. Cleared land at Bingera, Queensland. 'Kanakas' planting sugar-cane with scrub land in background.

However, Ambuaffer, Footaboory, and others tried again in 1912, after Fiji had closed, to convince Woodford to assist them in accumulating development capital through a 'collection' from labourers' wages to be set aside to finance land improvement. They stressed the need for higher wages as well as the re-opening of Queensland and Fiji. Woodford ignored the proposal. Besides his belief that the future of the Solomons would be with the white man (and, if he could get it, Asian labour), he was aware of the contentious nature of the Christian claims to coastal lands.[18]

By the early 1900s, Solomon Islanders were no longer experiencing the white man's culture second-hand through the mediation of the 30,000 returners involved in the overseas labour trade for over 40 years. Missionaries, planters, and government had appeared on the scene and soon outnumbered the 20 or so traders. Although their numbers were few, at 200–500 compared to an indigenous population of 100,000–150,000, their impact shook the land on which the Melanesians stood.

Unlimited resources?

Charles Morris Woodford, the first resident commissioner, did not have a high opinion of the Solomon Islanders' ability or desire to increase the productivity of their islands.[19] Nor did he believe in their ability to increase their numbers in the face of the incoming Europeans and local fighting, made more deadly by steel axes and firearms. No stranger to the Solomons, Woodford in 1886 had visited the group as a 'naturalist'. Being familiar with the social equivalents of Darwinian theory,[20] he predicted 'the eventual extinction of the existing native race' and expressed the hope to British readers, that the Islanders' successors 'will turn to better use the unlimited resources of these beautiful islands'.[21] In a sense, he saw the indigenous people as a part of nature, a nature that needed to be transformed by and for the capitalist enterprise; if, they as part of nature could not adapt then they would become extinct. Even in the relatively unimportant Pacific Islands, as colonial powers were expanded into seemingly near-empty lands, the horizon of opportunity appeared ever-receding. Like all would-be settlers, Woodford perceived the natural resources of this 'new' land, with the notable exception of the Islanders, as unlimited and available to those who could develop them and bring them into capitalist production.

The existing export trade was too small to provide significant revenue, so Woodford devoted much of his energy and scant funds to making the islands 'safe' and attracting large planting investors, including the Pacific Islands Company in 1897, subsequently bought out by William Lever, the British soap-manufacturing magnate.[22] Thus, the choice was made to concentrate on large-scale or mass production of raw materials, not to elaborate the small domestic production of the indigenous people. In the face of Woodford's conviction that the potential labour supply was contracting, it was a choice fraught with risk.

The government, reinforced by missionary negotiation, gradually 'pacified' the islands. Most of the western and central islands had come under control by the late 1900s, Makira by the late 1910s, the Santa Cruz group by the early 1920s and all Malaita by the late 1920s. With relative security, would-be planters and speculators, mainly from Australia, became interested in the Solomons' potential because prices for tropical products were buoyant early in the century. Although many plantation crops, such as rubber, cotton, sisal, coffee and bananas, were to be tried, coconuts were the easiest to establish, tend, and return a profit on coastal land.[23]

Until 1912, Europeans obtained most of an estimated 400,000 to 500,000 acres of land, under either the 1896 Land Regulation or the 1904 Waste Land Regulation. The waste land concept was applied to lands that the government, that is Woodford, believed to 'unowned, unoccupied and uncultivated' by Solomon Islanders, an impression created by the absence of coastal settlements in areas targeted by headhunters for decades. Woodford confused ownership with sovereignty. Any one society had a range of rights to land, as well as sea, and/or the resources thereupon in a specified location, though the exercise of some might be dormant. Within that society, be it a village, a cluster of related hamlets or several allied communities, individuals and clans had varying rights in relation to each other. These varied with time, needs, and the moral or physical force a claimant could bring to bear, but except by extinction of the particular group, were distinguishable from those of 'other' sovereign societies which had territorial rights elsewhere, though there were probably 'no-man's lands' along territorial peripheries where it was not safe to try to assert control. In Woodford's perception however, indigenous rights to land were only demonstrable by forest clearance, evidenced in the presence of hamlets and gardens; that is, land that was occupied and cultivated.[24]

Because of Woodford's views, the true status of indigenous claims on these 'waste' lands, with scant exceptions, was not thoroughly investigated at the time, although some 250,000 acres were alienated in this way on Certificates of Occupation to the new 'owners' for 999 years.[25] The biggest concession of 220,000 acres went to Lever's Pacific Plantations Limited, in the western and central Solomons.

The remaining 170,000–250,000 acres of alienated land included lands claimed by Europeans before 1896, probably about 80,000 acres, some of which was not subsequently occupied, and about 170,000 acres bought from Solomon Islanders under the 1896 Regulation which allowed direct freehold purchase. Following the abolition of direct purchase by non-Solomon Islanders in 1912 and the introduction of a system of Crown (Public) and Native leaseholds, mainly to prevent speculation, within two years another 18,110 acres were leased as Crown leasehold and 980 acres as Native leasehold, primarily to planters. Both Crown leases and Certificates of Occupation had clauses requiring the development of specified acreages within a time limit. By 1914 almost 5 per cent of the estimated land area of the archipelago was in European ownership or occupation.[26]

Missions generally purchased or leased modest acreages. Often land sold to the missions was swampy, prone to flooding or attack, or taboo, so of little use to its original claimants. The Roman Catholics obtained 4,000 acres mainly on Guadalcanal, Makira and Malaita; the Melanesian Mission bought about 2,000 acres mainly on Guadalcanal, Santa Isabel, Nggela and Makira; the Methodists acquired 2,000 acres in the western Solomons, about half of which was donated by a trader, Lars Neilsen. The South Sea Evangelical Mission had only a few small holdings on Malaita, north Makira and south Guadalcanal.[27] The Seventh Day Adventists came to the Solomons after freehold purchasing had ceased and so in the western Solomons and, later, Malaita and Guadalcanal, it negotiated Native Leases. In the 1920s these amounted to 316 acres, the largest of 200 acres being at Batuna, the mission headquarters.[28]

Developing the land

The word 'development' peppers government and commercial documents of the time. It was the ultimate rationale for the presence of Europeans and they invoked the word like a mantra.[29] Development meant using the natural resources of the Protectorate to produce economic gain for the good of the planters, the British nation, and, as some saw it, the entire world.[30] Development meant creating a political, physical and social infrastructure to facilitate the export of produce and the import of items for the establishment of the plantations. Centres for mercantile and administrative functions were to be built. Plantations had to be established. Because limited funds were available to the Protectorate, plantations often preceded government control. 'Massacres' of Europeans and their employees happened, resulting in government punitive raids that suborned the neighbouring and sometimes innocent population. Government personnel were needed for the administration of 'law and order', to collect customs, excise, rent, and, eventually, in 1921/2, a native adult male head tax to fund the minute public service. In the light of the limited labour source, it was also necessary to see that workers were treated reasonably by their employers, in order to maintain the supply. Development also meant Solomon Islanders stopped killing each other as well as white men, gave up their more repugnant (to Europeans) customs and became willing workers for the planters. If these measures also halted the indigenous population decline – a worry to planters because of the labour shortage and also a concern of the government for humanitarian reasons – all the better.[31]

Maintenance of even 'a precarious peace' was going to be difficult enough for the government.[32] Beyond this and basic medical treatment for plantation labourers, development at this time did not mean 'native welfare'. If it came it would be a trickle-down effect, a happy by-product of European enterprise. And if Christian missions used funds and personnel provided by overseas supporters to convert the heathen and make them more peaceful, healthy neighbours and biddable workers, this was all to the good, providing missionaries kept out of trading, commerce and 'politics'.[33]

The Christian missions displayed a range of attitudes to the involvement of Solomon Islanders in this Eurocentric development process, but within the context of their primary evangelising purpose. The Melanesian Mission (Church of England), based in New Zealand, and the Catholics from France were indifferent to commercial enterprise. The Melanesian Mission had a tradition of criticism of the abuses attendant upon the overseas labour trade, but did not oppose the trade as such. Similar views prevailed in the Queensland Kanaka Mission of the Queensland canefields, later the South Sea Evangelical Mission (SSEM), founded by Florence Young and the Deck family. All these bodies were against the killings they witnessed in Solomon Islands and believed the Protectorate would foster peace. To them, the Australian Methodists who came to the western Solomons in 1902, and the Seventh Day Adventists, peace was a consequence of the European presence which also made evangelisation a less hazardous enterprise. So, most missionaries shared this concept of development with its attendant physical, social and political changes. Those who did not share it saw it as inevitable.

It was this inevitability that caused missionaries to ask, where were the Solomon Islanders to fit into this development scenario? Was their part that of a dying race, vacating land while decreasing numbers periodically worked on plantations? To the Catholics, the answer was a regretful but fatalistic 'yes', because they perceived Solomon Islanders as facing extinction as a consequence of 'civilisation' and the disruption of family life. The mission seemed only able to assist 'at the death agony of this race formerly so strong, so deeply attached to the land'.[34] The Catholics believed in teaching the basic religious tenets. All else seemed superfluous, especially in the first two decades of the century.[35] However, the Marist priests, strapped for funds, encouraged plantation, as well as garden work at their stations in return for schooling, although young 'scholar-labourers'[36] often ran away to plantations in preference to the meagre schooling and unpaid work they experienced. [37]

The Melanesian Mission encouraged a Melanesian Christianity and initially deplored the impact of plantations on village life.[38] However, some saw industrial mission as an answer:

> It is obvious that in the changing conditions of the islands, when their resources must be developed for European trade, a system which shows the natives the advantage of doing regular work in the neighbourhood of their own homes, and at the same time attracts them to Christian influences, must be beneficial, and should be a valuable contribution towards the solution of the native question of the future.[39]

So in 1910, the Mission set up a plantation at Maravovo, Guadalcanal.

> The object of it is not to trade in competition with the regular traders, but to provide the people with work, and in doing so bring them under Christian influence, and give them an opportunity of attending prayers and school.[40]

The outcome was not quite what the mission intended. Despite the payment of the legal wage, the local people merely worked casually, with the majority of workers being drawn from elsewhere.[41]

In 1905 the SSEM cleared fifty acres for coconuts at Onepusu on the west coast of Malaita to provide work for Queensland returners who were afraid to return to their homes and to earn money to support their school.

A Great Coconut Estate

Four years later, in 1909, the SSEM encouraged Christian investors, including Florence Young's brothers, to open plantations where Solomon Islanders could attend school and church under missionary guidance.[42] The first plantations on Malaita, owned by the Malayta Company originated in this way. When the school gardens were exhausted from cropping by 1913, the pupils assisted on the plantations in return for imported food. Although other planters and traders were not impressed with the Malayta Company's connection with the SSEM, Solomon Islanders found the arrangement congenial. So did the government in the early years when it needed help to persuade Solomon Islanders to accept pacification and disarmament.[43] The SSEM also provided some basic industrial instruction at their training centre at Onepusu, but the emphasis remained on evangelisation.[44]

The most systematic expression of industrial mission came from the Methodists. As early as 1905, the Australian church enunciated its support of plantations: the progress of 'Civilisation' was such that the mission had to help their followers keep pace with it.[45] When the pioneering work of evangelisation was well underway in the Roviana area by 1910, Chairman John Goldie's aim was to extend 'the development of our mission lands' and 'industrial mission work'.[46] In the face of what planters and traders were doing, as well as the expansion of the mission and concomitant need for contributions from Solomon Islanders, this broadened to 'teaching our people that the development of their land is a Christian duty and that labour is honourable'. The Methodists believed, 'Land Development means character development'.[47] Through education from its several schools and the Training Institute at Kokenggolo, in 1913, the mission encouraged Solomon Islanders to cultivate their lands and manage their business affairs along with practicing Christianity.[48] The 'economic independence' of the westerners was a consistent theme of Goldie's chairmanship to 1951. Most other Methodist missionaries shared this objective, although it was not always feasible in some areas, such as Choiseul where,

> [the] people are land hungry in the midst of untold acres of virgin bush.... Probably one reason for this is that the hills in most places come right down to the sea, and there is little good coconut or any other kind of land [49]

The last-comers to missionary work were the Seventh Day Adventists from Australia, who started at Marovo lagoon in 1914. The Adventists, though placing emphasis on practical education, particularly in regard to hygiene and the teaching of English along with religious tenets focused on the imminent return of Jesus Christ,[50] were uninterested in any particular economic development.[51] The missionaries sought to site their stations away from the steamer ports where the people had contact with commerce. They saw their work as evangelisation, although when invited to send teachers to villages, they taught habits of living, of industry, and time organisation that were complementary to the Western capitalist work ethic, as well as encouraging tithing to support the mission's work. Adventist dietary restrictions put certain customary food resources, such as shellfish, crabs, sea fauna without scales or fins, turtles and pigs, beyond utilisation. This may have encouraged greater dependency on acceptable foods bartered from other areas and/or the use of imported food, both fostering the strong trading talents of the Marovo Lagoon people.[52] The Adventist missionaries did not actively oppose commerce, they were indifferent to it. Their motto was, 'Traders' business is copra, mission's business is men',[53] a distinction that the Methodists did not make, nor indeed did Goldie who had private interests in a plantation on Vella Lavella. And he held on to them despite protests from planters and traders.[54]

Not only was missionary involvement in commerce an issue, so too was missionary education.[55] In part, this arose because of the labour shortage, unrelieved by the steadfast refusal of the Colonial Office to allow the importation of Asiatic workers. So labour was at a premium, although planters did not want to pay premium rates.[56] Educating Solomon Islanders was not desirable because they would be less likely to recruit for plantations, and just might overcome their traditional suspicions of tribal neighbours as well as learning effective ways to articulate their demands.[57] Europeans believed that the cessation of tribal fighting had deprived the natives of their major interest in life. In this thesis, the new peace made life both so easy and dull that the natives became apathetic and were dying out.[58] The cure for them, in the planters' view,[59] was 'a certain amount of regular work'.[60] So, if mission schools encouraged 'the cultivation of habits of industry'[61] that was acceptable because it could stop the population decline as well as train the natives to be efficient workers.[62] This was too much for some planters who believed there could be 'no better education than the discipline and productive labour' of two years' indenture on a plantation.[63] For the more liberal, industrial education was tolerable, providing it did not foster intellect and, in the words of W. H. Lucas, islands manager of Burns Philp, 'the employer is assisted to utilise [Islanders] in the material development of the latent wealth of these tropical islands'.[64]

Opening the land: from virgin to wife

Like Lucas, Europeans perceived the Solomons as being full of latent wealth and potential fecundity. They could see undeveloped, unlimited resources which the natives had failed to notice or were too supine to utilise fully.[65] The green uncultivated lowland forest land appeared to provide almost unlimited opportunities to the newcomers. It was land waiting to be 'opened up',[66] land which to all appearances was largely untouched, if not neglected, by the often lazy and diminishing native inhabitants and as such was 'waste', of 'no use',[67] and 'of no practical value'.[68] It was thus rightfully the possession of those who could actualise its fertility, those who could open it, tame it, use and develop it.

Because the land was 'heavily timbered with virgin forest'[69] it had to be cleared. The government did not even consider possible environmental effects, so there were no legal restrictions on clearing. The planter and his labour gang constructed basic shelters for themselves, their rations and equipment on cleared, well-drained land close to the beach landing and a water supply. The next clearing was for a garden. Even well capitalised companies were not aware of the dangers from erosion in clearing land to the water's edge until they saw it for themselves and abandoned the practice.[70] This was one of the first adaptations by Europeans to the environment; the land was teaching them how much it would tolerate. However, land-learning would take time, as longer-term effects, such as reduced soil fertility, were not always immediately evident, especially for coconuts on newly-cleared land.

Leaving much of the strand vegetation, the labourers began opening up the forest, by marking out the boundaries of a big block.[71] They slashed their way into the scrub to get at the bigger trees. At Boroni, on north Makira, the planter found, 'The timber is mostly hardwood … it is very heavy clearing big timber and creepers'.[72] There, sizeable trees numbered about 500 to the acre, many around 80–90 feet in bole height with diameters of 4–5 feet.[73] These were felled using axes and a springboard to avoid buttresses, a technique very different from the downward strokes of adzes. More experienced planters had their labour use only three and a half pound axes with a thirty-inch handle more suited to the Islanders' physiques.[74] Planters encouraged 'liberal use of the felling saw',[75] favouring lopping off and packing the top of each tree immediately after felling as the green timber was softer to cut than dried. Using 'bush knives' or machetes the gang 'brushed' the remaining scrub. The slash, the lop-and-top, as well as the big tree trunks were left to await the first burn. Rain was a problem as F. Rigby on Boroni, noted:

> The rains have delayed us a little … We cannot of course burn off in this weather so we have a big area stacked ready for the fire stick. This first burn is very desirable as it kills off the undergrowth and small branches and leaves but a real good burn off is rare down here …[76]

In the 1900s many planters had only one main burning off and then planted amid the fallen logs. Although this left much of the biomass, they abandoned the technique because of insect pests proliferating in the rotting wood.[77] Inexperienced planters sawed or chopped up the fallen trunks, a very labour-intensive practice.[78] Most, after the initial burn, packed dry debris at the butt of the cut log and lit it, when there was wind to drive the fire into the butt. The logs were then repeatedly re-stacked and burned. Hardwoods often were saved for firewood and fencing needs.

Like Solomon Islanders the planters saw fire as their most effective tool.[79] Most planters had come from Australia and soon realised that the tropical trees were 'not so easily burned as the Australian eucalyptus trees' and had to discover the characteristics of the forest timbers, if only to burn them.[80] As one planter, F. J. Hickie, advised:

> Do not fell banyans; burn them down. Do not fell andila [*Cordia subcordata*?] on the foreshore; burn, they burn better green. There are trees in the scrub that burn well green; find these out and save labour.[81]

> Make your fires do the work in burning-off. Don't let your labour use the axe too much – given their own way they would chop every thing into short lengths. Keep the fires going and, weather permitting, the block will soon be clear enough to line.[82]

Solomon Islanders displayed a remarkable affinity for felling the forest. The planter would have two gangs working at opposite sides of a forested block and start up competition between them, as many, especially the Malaitans, took pride in their abilities with the wonderful axes.[83]

> Short and stockily built, the native of Malaita is as strong as a horse and is full of 'ginger'. He will, when in the mood, cut down trees and clear 'bush' at an astonishing pace and with tremendous vigour …[84]

Adventist Pastor H. B. Wicks, at Viru, recorded a similar verve,

> The ground is full of great hardwood trees, from three to five feet in diameter, which they [Solomon Islanders] tackle with jollity. Every now and again one of these great trees falls, then there is the loudest and most weird screaming you ever heard, renewed every few minutes for the sake of encouraging one another as the cutting goes on. Then another monster falls, intertwining each other on the ground …[85]

By 1908, under the steel axe and fire, 'the virgin forest is making way for coconut plantations'.[86] In 1900 there had been about 1,000 acres under cultivation, by 1913, 25,000, and about 50,000 in 1922. The planters had a vision of what they wanted and achieved much of it:

> [T]his group is nothing more or less than a great coconut estate of perhaps 50,000 acres. The white population is engaged in the supervision of the estate, and employs as many of the native population as care to work.[87]

A Great Coconut Estate

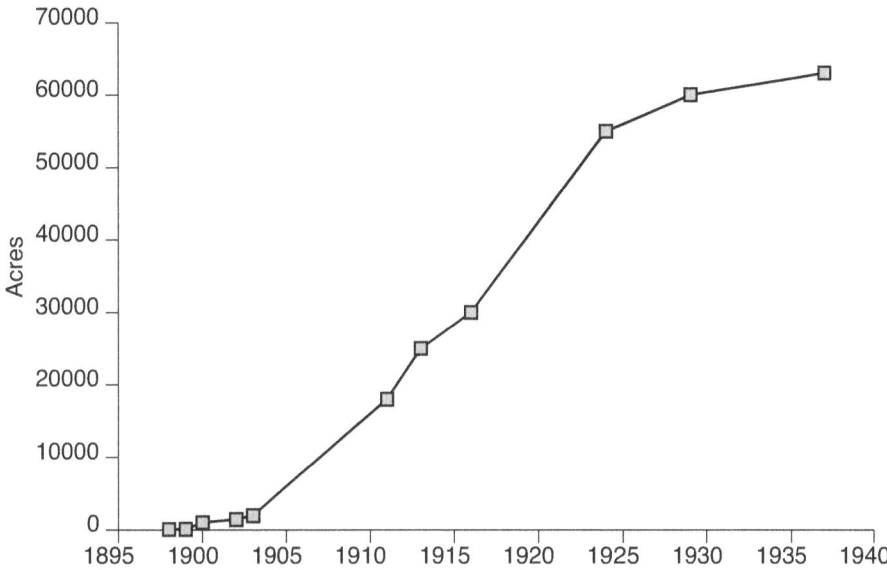

Figure 8. Acres cultivated for plantations, 1898–1929.

Sources: AR-BSIP 1898–1913; WPHC 1779/16; WPHC 827/30; *Handbook of BSIP* (Tulagi, 1911), 34; C. Y. Shephard, *Report on Agricultural Policy for Fiji and the Western Pacific High Commission Territories* (Suva, 1944), 20.

This acreage continued to increase, levelling off during the Great Depression, at 63,000 acres or almost 95 square miles of plantations.

In time, the planters built 'wide-verandahed bungalows' on 'the nearest available eminence' and other structures.[88] They furthered modified the ecology by importing cattle to graze under the coconuts to prevent regrowth of the forest, to reduce labour costs and to feed their workers. By 1932 there were over 5,000 head on Guadalcanal and 3,000 on the Russell Islands. Some, as early as 1913, carried the cattle tick. When beetle (*Brontispa froggatti*) and moth (*Tirathaba rufivena*) invaded the coconut monoculture, Lever's brought in magpies (*Gymnorhina tibicen*) and the myna bird (*Acridotheres tristis*) in the hope they would feed on them.[89] Traders and planters also introduced superior varieties of pig and the dog, much appreciated by Solomon Islanders.[90] Some Europeans brought in goats and even a few sheep and probably new grasses to feed them, along with weeds. Lever's introduced cover crops, including the velvet bean, cowpea (*Vigna* spp.), and the passion flower, to try to prevent the re-establishment of the *Themeda australis* and *Imperata cylindrica* grasslands on north Guadalcanal.[91] Woodford inadvertently was responsible for the Japanese clover now common in the islands, when he discarded packing that protected horticultural specimens from Japan. Soon the clover took the place of the coarse tropical grasses in many cleared areas, assisted by the Adventist mission which favoured the neat appearance it gave to villages.[92] Horses provided transport for planters along the Guadalcanal plains.[93] Here, horse-and bullock-power was soon supplemented by tractors and steam ploughs. By the late twenties, a few cars and trucks were being used to traverse longer plantation roads and a steam locomotive operated on Vanikolo.[94]

Missionaries too cleared land for structures and sometimes for plantations. The Methodists embarked on a building programme after arriving at Roviana on 23 May 1902. Led by Reverends George Brown and John Goldie, the party of 18, predominantly Fijians and Samoans, were able to shelter with traders, Norman Wheatley and Frank Wickham. The next day Brown resolved, 'I have decided to erect a small house which I brought down at Nusuzonga, the island which we bought at Sydney'.[95] The party cut timber at Mbanga for temporary houses and unloaded timber which had come with them on the steamer from Australia.[96] Nusuzonga proved too small for the mission, so Goldie bought land on the Munda promontory known as Kokenggolo after the trees which were found there that bore small sweet-scented flowers. The flowers may have smelled sweet, but the area was a taboo place and the abode of spirits, which may well explain the alacrity of the chiefly vendors, Mia Bule, Gumi and Hingava in selling it.[97] Certainly like most of Roviana, it was not particularly fertile as the soil was shallow over the upraised coral limestone.[98]

Goldie opened his 'hole in the bush', clearing a space for a house on Kokenggolo hill and the surrounding area with the help of labour found locally and from other islands.[99] The carpenter, Martin built the first mission house on the mainland for Goldie and his 'girl-bride', Helena.[100] By 1907, a new arrival, R. C. Nicholson, rhapsodised the transformation:

> ... on reaching the Mission house I was struck with its magnificent position. Situated on a solitary hill, about 400 yards from the beach, it presents a most picturesque aspect. Stretching round the front of the dwelling and along the sea shore, in a semi-circle, there is a fine cocoa-nut [sic] plantation which has been planted by Mr Goldie's own hands, and will in days to come be a source of great financial help to our district ...[101]

A Great Coconut Estate

Figure 9. *Australian timber for Methodist mission buildings being unloaded from Titus at Roviana lagoon, May 1902, for storage on Hombu Pecka, Wickham's trading station.*

Figure 10. *'Letting in the sunshine': land clearing for the Methodist mission base at Munda, 1900s.*

Pacific Forest

Mitchell Library, State Library of New South Wales

Figure 11. Methodist mission land cleared for coconut planting, Munda, c. 1905, from Kokenggolo hill.

Figure 13. Reverend George Brown, the Methodist Mission General Secretary, who regularly visited the mission's stations in Fiji, the Solomon Islands, Papua and New Guinea revealed his

A Great Coconut Estate

Figure 12. 'The hole in the bush': the residence of Rev. J. F. and Helena Goldie on Kokenggolo hill, Munda, c. 1905.

abiding interest in clearance for plantation establishment in this photograph taken by him in about 1906. The patch of forest on the left shows the density of the original strand vegetation.

Expansion continued. In 1911 a clearing was cut across Mbanga island, and by 1917 half the 200 acres at Kokenggolo planted. Of the 1,000 acres on Mbanga, 400 had been planted and another 100 felled and cleared. The Training Institute opened in 1913 and in the early twenties an electric light plant and a saw-mill were operating.[102] Development extended with the opening of the Mbilua and Choiseul circuits. By 1925 mission assets were valued at almost £65,000 (see Appendix 1).

This development was much the same with the other missions. The Catholic Marists first utilised local materials, but soon constructed dressed timber buildings with imported galvanised iron roofs at their regional stations. These were central places of religious teaching. By the 1920s, where the bishop resided at Visale, Guadalcanal, there was a cathedral, schools and pupils' houses, a printery and a workshop, houses for the nuns, priests and brothers respectively, boat sheds, and store houses. Smaller stations existed at Rua Sura, (later at Rua Vatu) and Avu Avu, Tangarere, and Marau on Guadalcanal, at Mbuma and Rohinari on Malaita, at Wainoni Bay, Makira and at Chirovanga on Choiseul, and Nila on Shortland Island. Concrete bridges were built over streams, swamps drained, and paths cut around the stations, with coconut plantations and gardens nearby.[103]

The Adventist mission in the Marovo Lagoon did much clearing for gardens to feed its pupils and staff. The following describes a typical establishment in 1918:

> ... we cleared the remainder of Telina Island which belongs to the mission, and started planting it all with sweet potatoes and bananas. As the island has not the best soil, and is insufficient to supply food for the mission school, we obtained a lease of sixty acres on the mainland a few hundred yards across the water. Having explained to the young people that it was improper for them to depend on the mission rice ... that very hour we all went over and began chopping down the thick bush and timber. Five acres or more have been cleared, but the wet weather has hindered us from burning off ... This is now being planted with sweet potato and taro ... In addition to this the new leaf roof was put on the new church, a launch house built over the water, a landing wharf rebuilt, walks made, a native kitchen going up, etc.[104]

Paths, relocation and administration

The Adventists and the Methodists encouraged the making of paths between villages and to the local mission station. In 1919, the Adventists cut a track across Nggatokae island although some had been 'afraid because of the spirits and devils that are supposed to dwell in the mountains'.[105] On Choiseul in 1927, the Rev Metcalfe at Methodist Tambatamba reported,

A Great Coconut Estate

By permission of the Ellen G. White, Seventh Day Adventist Research Centre, Coorabong, New South Wales

Figure 14. Clearance and path construction, at Seventh Day Adventist village of Penjuku, Nggatokae, 1924.

> In May we had a big working bee and reopened the track down the coast. We are anxious to complete it and connect up with the other villages but I expect it will take us to the end of our present term to complete the seven miles of track....[106]

The government did the same, suggesting villagers cut paths to lessen suspicion between neighbours.[107] In the pre-metal days, most paths had followed natural breaks in the forest, such as rivers and ridge crests. Maintaining even narrow paths through the forest's dark fastnesses had been hard work, but axes and machetes greatly eased the burden.[108] On Shortland Island, by 1908, the people had cut ten miles of 'road' from the Korovou government station west towards Soipa.[109] The district officer on Makira, F. M. Campbell, instructed headmen thus in 1918: 'Small roads should be made from village to village where there are a good number of people and villages not too far apart'.[110] Later, people in north Malaita made paths linking up with the government station at Auki.[111] Once the head tax was introduced in 1921–22, some government officers wanted paths between villages to facilitate collecting.[112] However, the construction of the path from Auki to Su'u slowed with the imposition of the tax as it seemed to the people that 'road'-building was an added exaction of a greedy government. By 1922, there were paths around Uki, Makira, Simbo, Savo, and San Jorge (Santa Isabel) as well as sections of Santa Isabel itself.[113] On Ulawa, despite passive resistance to Campbell's directives to demolish 'filthy hovels', rebuild and clean up the villages, the people did clear two miles of path from Hara'ina to the European plantation.[114] By the mid-1920s, a rough path had been cut round much of the Guadalcanal coast.[115]

As well as path building, government and missions, except for the Catholics, encouraged the bush people to re-locate on the coast and consolidate their villages. Opportunities for trade were greater there. It would make the work of evangelisation and administration less onerous. On Makira, Campbell urged this along with his ideas on hygiene to arrest population decline, building on the efforts of missionaries, like the Rev. Drew of the Melanesian Mission. The missionaries, like the planters and government officials, introduced their ideas of aesthetics and hygiene. In the Arosi district, Drew had

> ... worked on their [villagers'] pride by offering prizes for the best-built house and the cleanest house. He helped them lay out streets, and the dirty village with its houses stuck anywhere they could find ground to put them, gradually took on a healthful, systematic look, and natives from other villages came in and built neat houses.[116]

Peace made this feasible, as did the *kumara*, sweet potato which thrived in coastal soils.[117] This process was furthered by the spread of a taro blight in the Shortlands where it broke out in 1930. Possibly a form of *Phytophthora*, it destroyed the taro, the staple of the bush people in New Georgia in the 1930s. Taro diseases spread as far as Malaita by 1940. Besides substituting sweet potato, Solomon Islanders, mainly in the west, adopted another crop, tolerant of poor soils, cassava (*Manihot* spp.), which had been introduced from Fiji by a Solomon Islander in the 1930s. Relocating themselves on the coast was usually linked to acceptance of Christianity, but it did not mean the people forgot the ancient religious sites, though this knowledge became less common as old 'pagans' died.[118]

The government also altered the landscape. After Woodford had purchased Tulagi for his capital, he used prisoners and police to clear the island which was 'densely clothed with heavy virgin forest'[119] of trees up to 80 feet in height.[120] By 1903, about 120 acres of the 600 acres was planted mainly with coconuts. Woodford experimented in his botanical garden with various plants[121] (see Appendix 2). His clearing and drainage works were not solely for agriculture; but 'to get the fever out of the ground'.[122] Although unfamiliar with the aetiology of malaria, he certainly limited the breeding ground of malaria's vector, the anopheles mosquito, which threatened the health of Europeans as well as Solomon Islanders.

Construction progressed: the residency, an assistant's house, gaol, police quarters, boat-house, as well as a causeway, stone steps, paths, a water-supply and the draining of the swampy foreshore after the mangroves were cut out. The timber used was Australian blue gum (*Eucalyptus globulus*), red gum and Oregon pine (*Pseudotsuga taxifolia*). Each year saw new

A Great Coconut Estate

1. Cemetery.
2. Police Lines.
3. Treasurer's residence.
4. Resident Commissioner's residence.
5. Wireless Station.
6. Tulagi Club & Tennis Courts.
7. Golf Course.
8. Cricket Ground.
9. Footbridge over "The Cut".
10. Gov't employees bachelor quarters
11. Hospital.
12. Doctor's residence.
13. Superintendent of Police residence.
14. Gaol.
15. Elkington's Hotel.
16. Hotel jetty.
17. Judicial Commissioner's residence.
18. Gov't Offices.
19. Treasury & Customs Office.
20. Post Office.
21. Gov't Wharf.
22. Gov't Boatshed.
23. Lands Office.
24. Court House.
25. Butcher Johnstone's residence.
26. Butcher Johnstone's Ice & Meat Works.
27. Chinatown.
28. Stirling's Hotel.
29. Sam Doo's Hotel.
30. W.R. Carpenter's Store.
31. Carpenter's wharf & store shed.
32. Sasape. Jack Ellis' home & sailmaking shed.
33. Burns Philp's steamer wharf.
34. B.P.'s copra shed.
35. B.P.'s office.
36. B.P.'s bachelor quarters.
37. B.P.'s Manager's residence.

Figure 15. *Tulagi in early 1930s.*

From Graeme Golden, *The Early European Settlers of the Solomon Islands* (Melbourne, 1993). By permission of Greme Golden.

building at Tulagi as well as engineering works like the extension of the causeway, the building of the wharf, and the making of 'the cut' through a hillside to connect the club, bachelors' quarters, and the hospital on the west side of the island with the main government buildings on the east side. It had become a township by the late 1920s, a centre for administration and commerce, as Burns Philp's headquarters were based on adjacent Makambo and Lever's on Gavutu. The Chinese carpenters and tailors had built trade stores on the reclaimed land near that of merchant-shippers, W. R. Carpenter. Tulagi boasted a golf course, a cricket pitch, tennis courts, a club, two hotels, an ice-making plant, a light engineering shop, and a wireless station.[123] For the

Metcalfe collection

Figure 16. 'The cut', Tulagi, 1920.

Palmer family, Gizo

Figure 17. Tulagi hospital with nurses, early 1930s.

A Great Coconut Estate

Europeans, the re-creation of a familiar built environment was both an achievement and comfort in an alien world.

Gizo, Korovou, Aola and Auki were smaller, but significant government centres. By the 1920s, the Europeans – government, merchant, planter and missionary – prided themselves in what they had done in and

WHPC

Figure 18. Government buildings, c.1920. Note copra plantation, Woodford's pride, in the background.

SDA Research Centre, Cooranbong, New South Wales

Figure 19. Burns Philp wharf and warehouse, Makambo, 1920, cleared hill in the background.

to the islands. Woodford saw himself as father of the copra industry.[124] S. Knibbs, government surveyor, spoke from wide experience in 1929 when he compared the growth of the group with that of Tulagi:

> The development of the Protectorate as a whole has proceeded in like manner. Each year has seen large additional areas of forests brought under cultivation; business has expanded in proportion; and now the exports demand larger and more frequent steamers than before were necessary.[125]

Planters had firm ideas of their role:

> ... this land, previous to the advent of the white man, had no practical value ... we have literally hewed our plantations out of the living forest,... have borne the 'heat and burden of the day' and have brought this group to its present state of development.[126]

For the missionaries, their greatest achievement had been planting the seeds of Christianity in the hearts of the Solomon Islanders.[127] Some, like Goldie, reflected their Protestant uneasiness with idle land:

> I may claim that no other mission has done so much to instruct the natives and encourage them to plant up their undeveloped lands.[128]

All these Europeans believed they had tamed the islands and their people. Whether it was opening the virgin forest to 'let in the light and air' for government stations and plantations[129] or 'letting in the sunshine' and paths of Christianity across land and sea to the places of dark savagery, the process was one of great change, at least on the littoral.[130] The land had been made to bear more fruit, just as the transformed lives of the Christian converts did among their fellows.

Among Solomon Islanders their 'great vague hopes' of their own development and 'prestige that would make their foes look small' had been only partially realised.[131] This was most marked in the eastern Solomons, particularly labour-rich, but resource-poor Malaita. Development in the archipelago had occurred by the 1920s, though hindered by the labour shortage. It was based on the alienation of vast tracts of land for the production of a single crop. With the exception of the missions, this capitalist development was primarily for the white planters, though many Solomon Islanders did not perceive it thus, seeing plantations, as well as missions, as signs of their participation in a wider world and an expansion of their cultural horizons.[132] The government, the missions, and the planters had facilitated paths across land and sea which often linked people and places in a novel configuration, becoming the way new ideas about religion, government, social relations, food plants, technology and commerce entered the lives of Solomon Islanders. In the view of Solomon Islanders, the newcomers had certainly set their seal on the land. All who lost power and worth took it badly: 'This was a white man's country now', as the 'pacified', former

A Great Coconut Estate

head-hunter chief, Hingava of Roviana ruefully stated to Bishop Wilson in 1900.[133] Most accepted, having no alternative in the face of European superiority, and some in time admired, albeit often with ambivalence; but the loss of power and conquest by a few Europeans aided by 'friendly natives' left resentment, shame and often a sense of inferiority.

The European settlers had domesticated huge acreages. In their eyes, these areas were no longer part of nature, but of the incoming power culture, no longer 'virgin' bush, but crop-bearing land. The Europeans had mobilised major human, plant, new animal, and technological resources to 'open' up vaster areas of contiguous forest in 25 years than Solomon Islanders, led by their big men and chiefs, had done in centuries.

—4—
From the Shortlands to Santa Cruz: Beginning the Timber Industry

> These islands rise out of the ocean with their long, massive backbone heavily wooded. From top to coast is one great forest of big timber, with dense undergrowth beneath the trees.
>
> A. I. Hopkins, *In the Isles of King Solomon* (London, 1928), 19.

Early visitors to the Solomons made sporadic use of timber, mainly for firewood, but, except for a few areas, this was insignificant. With the advent of the Protectorate, the establishment of a timber industry was not the prime consideration of government, planters, merchants, or missionaries. But each group was aware of the potential of the forests and each tested that potential. An export trade in logs or cut timber would have to focus mainly on Australia, if only because of proximity and limited shipping links. Within the Solomons, demand, though steady, was for building timbers as there was no major mining, railway or industrial processing. Because the fate of the Protectorate was so dependent on coconut plantations and the demand on the forests remained relatively small, most pre-World War Two administrators did not consider that legislation for the protection of the forests was urgent. The forests covering the archipelago still seemed unlimited.

Island prospects and trades

Long before European planters contemplated clearing the forests of the Solomon Islands, the explorer, Alvaro Mendaña, arrived at Santa Isabel in 1568. Prospecting for gold for Spain and souls for God, the Spanish decided to build a brigantine, so as not to risk the loss of their three ships. At Estrella Bay, 'We found such a quantity of good timber in this island that many ships might be built there'.[1] The brigantine served them well until near Santa Ana high seas forced them to cut it adrift from the leading ship, to be

From the Shortlands to Santa Cruz

wrecked, it is said, at Kahua Point on north Makira.² Mendana's expedition was counted successful, but two centuries later, that of the Frenchman, Laperouse, was less fortunate.

In 1788, during a great storm, both the *Astrolabe* and *Bousolle* were wrecked on the encircling reefs of Vanikolo, in the eastern Solomons. Peter Dillon, searching in 1826, learned that survivors had struggled ashore near Paeu on the south-west coast. Although some were killed by the Vanikolans who saw them as 'ship spirits' beckoning to the sun and stars, the rest cleared a spot on the bank of the Lawrence (Paeu) river and built a vessel from local timbers. All but one of these Frenchmen were said to have sailed away.³ Their fate remains a mystery, although some evidence suggests the boat, made of kauri, was wrecked on Tinakula.⁴

Almost a decade later in 1835, the captain of the *Anastasia* called collecting more relics of Laperouse's expedition. Like the *Anastasia*, many ships that visited Solomons from the turn of the century were whalers, seeking wood for fuel to process whale blubber. Most was gathered, usually with the assistance of the coastal people, in the whaler 'ports', such as Makira Harbour, Simbo, or Mono. Venturing into the bush beyond the coast was risky,⁵ so whalers collected much of the wood from the beach forest. A handful of bêche-de-mer traders also gathered wood for processing at temporary shore stations, from the 1840s, on an intermittent basis. In areas of relatively little land and forest such as the atolls of Ontong Java and Sikaiana, the periodic fishing out of the reefs resulted in heavy demand on wood for the drying fires.⁶ On Sikaiana, in 1845, Americans had collected and processed 15 tons of bêche-de-mer and, in 1847, Captain Andrew Cheyne, 16.75 tons.⁷ As it takes about nine tons of firewood to process a ton of bêche-de-mer, then in 1845, 135 tons of wood were burned and ten years later another 150 tons went up in smoke.⁸ Moreover, at least 35 coconut trucks or about a half an acre of palms were usually used to construct a drying house and other structures.⁹ This posed less of a difficulty as several uninhabited atolls near Sikaiana were covered in coconut palms. By 1858, although Sikaiana's 'entire area of habitable land' was only 80 acres (32 hectares), towards the pumice-littered interior, 'there begins an exceedingly luxuriant growth of lofty forest trees with huge trunks …'. Demand on wood was intense for a time, yet the isolation of the atolls coupled with the limited supplies of bêche-de-mer and the relatively small human population of around 180 meant that the atoll was not deforested to any serious extent.¹⁰

Solomon Islanders wanted the metal goods proffered by the coconut oil and copra traders. A certain amount of deforestation occurred as Islanders cleared land for coconut groves, especially from the late 1870s, although population decline probably meant former coconuts for food became coconuts for trade. The few resident traders utilised wood for fuel or occasionally to build a boat.[11] Steamships were plying the waters, most fuelled by coal, but the occasional one, like the Melanesian Mission vessel, *Southern Cross*, in the 1890s, by 'she-oak' (*Casuarina*), a strand hard wood.[12]

Traders avidly sought one timber in the western Pacific. The perfumed sandalwood (*Santalum* spp.) flourished in Fiji, the New Hebrides, and New Caledonia and was in demand in China. The sandalwooders investigated the Solomons. Dillon himself had been a pioneer of the trade in 1825 on Erromango. Cheyne, familiar with sandalwood, traded in the western Solomons and Sikaiana in the 1840s.[13] Sydney-based sandalwooders working in the New Hebrides came to the Solomons to get tortoiseshell to trade to the Tannese for pigs to exchange at Santo for sandalwood.[14] The notorious sandalwooder, Captain Ross, was in Solomons waters in 1858 and abandoned a sick seaman on Sikaiana.[15] An apocryphal tale tells of contemporaries of 'Bully' Hayes collecting both ebony and sandalwood on Guadalcanal[16] and 'so-called sandalwood' was exported from the Protectorate in 1901–1909: All wishful thinking, as there is no botanical record of the tree in Solomons.[17]

Nonetheless C. M. Woodford, the first administrator, extolled the qualities of timbers that were there, offering samples to English furniture makers in 1898. The Solomons may have been a British Protectorate, but Sydney dominated trade and it was there a market had to be found, if only because of the transportation costs in exporting to Britain. Breaking into the market would have to come from private enterprise, though Woodford deplored the prejudice in Australia against 'Island timber', warmly praising the special qualities of '*Afzelia bijuga*' (*u'ula* or *kivili*)[18] as impervious to borers and resistant to *Teredo navalis* (the teredo worm). And he looked toward the day when saw-mills in the Solomons would supply local needs.[19]

Trial shipments: logs for Australia

When representatives of the Pacific Islands Company went 'land-looking' with Woodford on the *Rob Roy* in 1900, they remarked on the abundance of timbers, especially on Kolombangara. The Company's concern was more with the costs of clearing, than with logging trees, although, 'It seems a pity to burn them, but this will have to be done unless some market can be found for the timber'.[20] Lever's, the company's buyer, faced the same dilemma. In 1908, as it cleared the land on the Russells for planting, it shipped some of the logs to Sydney.[21] At their Balmain soap factory, Lever's used these softwoods to manufacture soap boxes, but they caused the soap to discolour so the export ceased.[22]

From the Shortlands to Santa Cruz

Burns Philp's islands manager, Walter Lucas was alive to the potential of most resources and followed Lever's lead, albeit with caution. While inspecting company properties he had remarked on the potential of timber on Tetepare.[23] In July 1908 when surveying the island, an area of 55 square miles,[24] he found along the north side:

> Very fine timber was abundant on the low lands averaging three or four fine trees to the acre with trunks from 40 to 80 feet to the lower forks and girthing 10 to 18 ft. a few feet above the ground ... thousands of fine trees are growing within a few hundred yards of the beach and could easily be cut and shipped at an absolutely nominal expense ...[25]

Burns Philp wasted no time. They sent a 'timber man' to examine Tetepare. His presence in the group crystallised interest among planters clearing land, several approaching him to survey their timber. Work at Tetepare did not get underway until mid-1910.[26] The company meanwhile tested the market with a shipment of logs from near their new Makambo headquarters. For each big tree the company paid the Nggela people two shillings, which was 80 per cent of plantation labourers' minimum weekly wage of two shillings and sixpence. Along with Lever's shipments, the total number of logs exported was about 500, valued at £1,627, during the period 1907–1909.[27] However, the project met opposition from 'the combination existing between large timber firms in Sydney, to discourage every thing but Australian timber'.[28] Burns Philp toyed with the idea of opening a sawmill in Sydney and even in the Solomons.[29] Although Lucas predicted, 'in time, when they have been introduced to larger markets abroad, some of these beautiful hardwoods should prove a valuable asset', the company decided against it.[30]

Coffins and cheap furniture

Where Goliath failed, David succeeded. Eric Monckton, an Australian planter, exported a shipment of logs from his 2,500-acre holding at Kokonai, Shortland Island in late 1910. Like all planters, when he first arrived in October 1909 he tried to find something to earn money while he cleared, planted, and waited the eight years for the coconuts to bear.[31] He found it, as he told his uncle Parry,

> I ... have a shipment of 30 logs 12 ft by 1[ft] 6 [ins[to 2 [ft] 6 [ins] diameter which I hope to get £1 a log from £30, cost of getting them to the steamer, cutting and everything £5/10 ... This is an experiment. I have thousands on my plantation and if they sell well I have a few thousand pounds lying in the ground at the present moment.[32]

As he had to clear his land anyway at about £4 10s an acre,[33] the timber was a kind of 'axefall' profit, if he could sell it. Monckton found a buyer, the Island Timber Co. Ltd., in Sydney. He estimated the logs on his own plantation which extended about one and a half miles along the north side of Magusaiai channel would be cut out by late 1914. So in 1912, he applied for the first official timber concession in the Protectorate over the lands of the Alu people on the 'native reserve' on southern Shortland Island, from Maleai to Nuhu on both sides of the channel, reaching inland a quarter of a mile. He was willing to pay one shilling, the same price as a coconut tree, for each 'dhelo' (*dalo* or *bunibuni masa*), the *Calophyllum inophyllum*.[34] Monckton told Woodford that 'the natives have no use for this timber'.[35] In the light of their declining population and a government edict imposed that year, making it a closed district for labour recruiters, Woodford could hardly object and supported the application to his superior, the High Commissioner in Fiji.[36] In Monckton's case as clearing generally was not being done to plant coconuts, there could be problems if the land were left bereft of trees. The High Commissioner favoured replanting of the same species, but Woodford maintained that natural regeneration would be adequate. Woodford increased the fee to two shillings a tree, limited the agreement to five years, with the area to be logged all of Magusaiai by agreement with the Alu. After the Colonial Office approved, the lease, when finalised in December 1912, required Monckton to pay the land holders a total of £10 a year rent as well as two shillings a tree royalty. An amendment followed a year later which affected a subsequent lease with Monckton, requiring him to pay to the government an additional fee of ten per cent of the royalty's value. This was the first tax or duty on export logs in the Protectorate. The original lease was for *Calophyllum* and these had to be no less than 12 inches diameter at six feet above the ground. In terms of 'timber control', Monckton had to keep a register of trees for inspection every six months, by the government on behalf of the lessors, and for a check against exports.[37] In mid-1912, another planter, Hugh Scott of Orlofe tried to steal a march on Monckton by offering the highest-ranking chief of Shortland, Ware and his son, Alisai Kogiau, a boat worth £37 in return for logging rights, but Woodford upheld Monckton's application.[38]

The negotiations with the Alu appear to have been straightforward as the society recognises powerful chiefs and it was they who gave permission for logging. The chiefs spoke for their people, but the rents and royalties collected by the district officer were allocated by the chiefs to those people who had clear rights over the land being logged, since the ownership of a large number of non-food and non-canoe wood trees was a novel element. The son of Ware, Glassa of Maleai who was successor to his childless

From the Shortlands to Santa Cruz

brother, Alisai Kogiau, was the representative lessor to Monckton's lease,[39] and he always made 'a fair division of the rents and royalties'.[40] Problems arose in 1937 when Glassa died and his successor, a minor chief, Kimale of Maleai,[41] kept the bulk of the money, until the district officer, responding to complaints by other claimants, dealt with the matter in court, and paid 'the money direct to the various natives with rights over the land in question.'[42]

Monckton continued logging throughout the twenties.[43] He usually exported about 40 log sections, equivalent to about 15–20 trees, on each six-weekly steamer, mainly Island Cedar (*Callophyllum*) and later Swamp Oak (*Terminalia brassi*), *dafo* or *homba*. He set up a saw-mill at Faisi, but his logs were in 12–14 foot lengths to fit in the ship's hold. Monckton had an advantage in being close to Faisi where the steamer called, resulting in a rapid through-put of logs. This, coupled with spraying the logs with Waxoil immediately after felling, meant they were less susceptible to infestation by the pin-hole (lyctus) borer and, when tied on rafts in Kulitana Bay, by the teredo worm.

By permission of Graeme Golden

Figure 20. Eric Monckton's logging team with the children of Georgina and Carden Seton, Faisi, 1935.

Pacific Forest

By permission of Graeme Golden

Figure 21. Loading log sections into the hold of Burns Philp's steamer at Faisi, 1935.

All planters were hard hit by the Great Depression of 1929. As the copra market collapsed Monckton concentrated on logging.[44] However, Australia imposed a tariff on imported logs. Logs were the only Protectorate product so taxed, and a jump from 10 per cent to 30 per cent *ad valorem* in June 1930, forced Monckton to rely solely on his plantation to eke out subsistence.[45] Strapped for cash, he sold his saw-mill to the Catholic mission at Nila in 1932, and fell back on logging.[46]

Monckton's business got some relief in February 1933 when Australia recognised British Protectorates as eligible for 'empire preference' under the Ottawa agreement and reduced the import duty to 10 per cent, plus 10 per cent primage, paid to the vessels' owners.[47] He negotiated a lease for four acres at Lahiai, Poporang and, after logging them, returned to working Magusaiai in 1937.[48] His logging team of Malaitans and Shortland Islanders, led by the Malaitan *bosboi*, Tondonga, felled the trees on the relatively flat terrain and hauled them by manpower or bullocks to the sea out of the bush and swamp.[49] This simple technology of extraction did no great damage to the surface soil and drainage. From 1932 on, he exported up to 370,000 superfeet a year,[50] most for the manufacture of 'coffins and cheap furniture'.[51] The value of this was about £1,200 – less costs of labour, shipping, Australian duty, rental and royalties. His plantation at Konokai was being choked by 'Koster's Curse' (*Clidemia hirta*), a noxious weed he had introduced unwittingly from Fiji, growing among potted flower plants.[52] Copra by then was virtually worthless anyway and Burns Philp's held the mortgage and managed this plantation and those leased by Monckton's wife Minna, at Taukuna and Saeghangmono Islands. Monckton, like all the small planters, was in financial trouble. Surviving on his logging, he was a 'bad debtor' of Burns Philp, owing their Faisi store £4,229 by 1939. To side-step their pressure to reduce this book debt, he tried unsuccessfully to get another shipper for his logs.[53] By the mid 1930s, the terms of Monckton's timber licence had altered. He was paying 5s a tree, probably for *Terminalia brassi*.

> The work [of logging] is carried out on swampy ground in which this type of tree thrives....The trees that provide this timber are beautiful to look at, very tall and straight and the first branches are met with at a height of about seventy feet.[54]

Instead of a government royalty per tree, proposed new leases provided for a royalty of 1s per 100 superficial feet (superfeet).[55] In 1940, this meant an additional £62 10s on Monckton's export of 125,000 superfeet which would have ruined him, but World War Two intervened.[56]

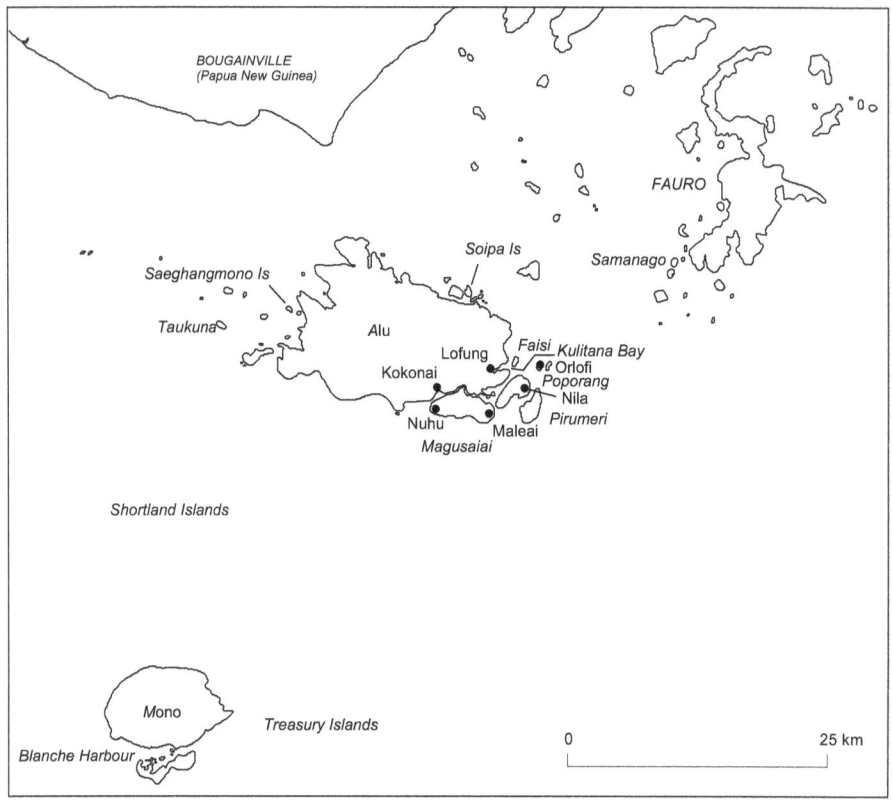

Map 3. Shortland Islands.

Although Woodford had not required reforestation – the artificial establishment of a new forest on logged ground – in the initial licence, by the 1930s, some in the government thought it necessary. The district officer of Shortlands inspected Magusaiai in 1937 and noted that saplings were present around the stumps of trees logged earlier.[57] These saplings appear to have been *dafo*, *Terminalia brassi*, which regenerates easily and rapidly, especially in near-pure stands.[58] However, the *dalo*, *Calophyllum*, did not regenerate so well, at least in the conditions found on Shortland Island.[59]

Milling on Malaita and Guadalcanal

Planters like Monckton had become aware of the forest, even if to know how best to clear it. On Malaita, the plantation manager of the Malayta Company, Neil MacCrimmon in 1923 decided to leave his job and set up as a trader and timber merchant.[60] He took out Crown leases on two acres at Su'u for his base and a 99-year lease on 100 acres at Olusu'u, near Sa'a as well as Native leases on land south of Su'u at Afutala, Weanura, and Hauhui on a total of 930 acres in January 1925.[61] Trading as the Mala Timber Development Company, MacCrimmon and R. W. Jackson also began negotiating for a timber lease similar to Monckton's at the rate of 3s a tree.[62] The negotiations stalled, but MacCrimmon had his own leased lands to log anyway. He floated the logs north to Su'u where the steamer called. Initially the logs, including *Pometia pinnata* (*dawa* or *mede*), found a market in Australia, encouraging MacCrimmon to bid for a lease of distant Tevai, near Vanikolo where the valuable kauri grew.[63] However, the Australian 'timber rings' moved against MacCrimmon to such an extent that the export business 'became a losing proposition' late in 1927.[64] The company turned to trading as well as starting

> ... to cut and sell dressed timber at Su'u for local building purposes; I have seen this timber and it is excellent. They have already received orders for their timber from local firms and plantations. They are employing a Chinese boatbuilder to make dinghies and whaleboats for the Europeans and the natives.[65]

The business survived until early 1930 when it became clear that the contracts for hardwoods MacCrimmon anticipated in Australia had fallen victims of the Depression-driven increase in import duty.[66] MacCrimmon shut down his saw-mill and in July 1930 sold it to Boldery and Cheetham and went to manage the Vanikoro Kauri Timber Company where his wife, Dr Lily Holt-MacCrimmon, became the medical officer.[67] His Native leases on Malaita lapsed, but MacCrimmon had already transferred his lease of Olusu'u, Sa'a to the Mala Development Company, whose director, G. H. Robinson obtained the saw-mill when Boldery and Cheetham went broke. Before this, Boldery had reached a deal with the Vanikoro Timber Company to mill a shipment of 13,000 superfeet of kauri logs, brought up on a W. R. Carpenter's ship. Robinson found several were defective and not suitable for milling. No money was made on this experiment because Carpenters, potential buyers of the milled timber, were dissatisfied with its seasoning.[68]

Neither MacCrimmon nor Monckton had been the first to mill timber for local sale. A saw-mill was milling mainly *Callophylum* in 1911 at Mamara on north Guadalcanal, one of Oscar Svensen's plantations. It produced timber for Mamara and Tasivarongo and for sale locally at 20s per hundred superfeet. It ceased in 1913, once the land had been cleared, though another small mill was operating elsewhere.[69] In the mid-twenties a 'few logs' were exported from north Guadalcanal, probably from Ndoma where the manager, Leif Schroder, took out a timber lease in 1926, but these efforts, like MacCrimmon's on Malaita, did not have a guaranteed Australian market and petered out by 1927.[70]

Exotic eco-tourism and forests of ebony

All these men – Monckton, MacCrimmon, Svensen and Schroder – worked within the colonial legal framework. But others had dubious schemes for the forests. In 1913, Peter Neils Sorenson, an American national, appeared in the Solomons to claim the whole of Mono (Treasury) Island, allegedly purchased from chief Mule Kopa in 1885. Sorenson had a history of criminal activities in the Solomons, dating back to 1877 when a British naval vessel received complaints of his ill-treatment of people on Cockatoo island, off Santa Isabel.[71] Following his involvement in kidnapping on Santa Isabel and later on Makira, he was imprisoned in Queensland in 1886.[72] In 1898 he led a party of Americans in a fruitless search for minerals on Rennell and Makira and was run out of the Protectorate, a process Woodford repeated in 1913.[73] After 36 years, it seemed the Solomons was rid of him.

Not quite. Sorenson had plans for Mono, premised on the timber industry. First, he intended to start a saw-mill there, getting the timber from Bougainville, 'as the king from whom he bought the Treasury Group owned half of Bougainville and agreed to supply him with all the logs of red and white cedar that he might need'. Once this was done, as well as an exporting business and general trading, Sorenson planned to build docks and a hotel in Blanche Harbour. To embellish paradise he 'intended to place various kinds of game animals on part of the island' so that tourists could go hunting. Because Sorenson believed Woodford had deprived him of all this, he claimed in 1925, that the British government owed him at least 'Ten Million ($10,000,000) Dollars'. Officials in Britain were unimpressed.[74]

Sorenson persisted. At age 80 he managed to convince three Americans to join with him in a company, United States Treasury Island Incorporated to pursue the claim once more in 1929.[75] The British did not recognise the flawed claim. Sorenson's 50-year scheme proved as much an illusion as his analysis of the societies of the Bougainville Straits.

Seemingly more reputable opportunists received much the same treatment by the government. Audley Coote, one-time honorary Hawai'ian ambassador and a member of the Tasmanian legislative assembly, represented himself in 1900 to Woodford as owner of St. George (San Jorge) island off Santa Isabel. Coote had plans, but whether to sell the island off at £200 to the Pacific Islands Company or to 'develop' it himself remains uncertain. Coote claimed he had purchased it from a trader who had bought from a chief 23 years earlier, the claim being subsequently recognised by the Germans when Santa Isabel was German territory. Woodford found no title in the German records. Coote was interested in 'deposits' (sic) of tortoiseshell and minerals as well as 'forests of ebony'. There was obvious

From the Shortlands to Santa Cruz

confusion between the trees *Pemphis acidula (Ngirasa)*, a small tree or shrub of very hard, brown wood found on the sandy littoral and *Xanthostemon* spp. (*Tubi*), the true wood of which is black and favoured by carvers. *Tubi*, though present, was not in dense stands on San Jorge, as Coote thought. Woodford did not recognise the claim as the area was occupied and the people knew nothing of the supposed transaction.[76]

Mission milling: building God's house

Missionaries built their first structures from either imported timber or from local timber, largely adopting the small undressed log, sago-palm leaf and split palm construction of Solomon Island dwellings. Timber from Australia was very expensive because of shipping; whereas local material, while usually inexpensive at the outset, did not last more than five to seven years.

SDA Research Centre, Cooranbong, New South Wales

Figure 22. The first pre-fabricated house brought to the Solomons from Australia by Seventh Day Adventist missionaries, G. F. Jones and his wife, at Viru in 1914. It was later moved to Ughele where Mr and Mrs Archer lived in it.

Florence Young, *Pearls from the Pacific*, c. 1925

Figure 23. On the verandah of mission house at Wainoni Bay, Makira. Nurse Clarke and Miss Spedding of South Sea Evangelical Mission, 1923.

Woodford Papers

Figure 24. Roman Catholic cathedral at Visale, Guadalcanal, opened in 1930 to replace the stone building destroyed by earthquake in January 1926.

From the Shortlands to Santa Cruz

The missions in the western Solomons were the first to use local timber on a large scale. The Methodists at Roviana found themselves in an area where leaf for roofing and walls was scarce. Goldie argued that the cost of buying and bringing it to Kokenggolo, as well as its limited life, was almost as much as a house of dressed timber with a galvanised iron roof.[77] Goldie got his iron from Australia and by 1914, timber from the Kokenggolo area:

> Some of the boys under Mr Oldridge [the builder] have felled the giants of the forest, and by hard work in the saw-pit, obtained the materials for three fine new houses for the labour boys and the teachers.[78]

By 1917, there were two saw-pits working.[79] Such was the demand for timber throughout the three circuits that by 1920, W. Oldridge had a saw-mill in operation. With this, an extension of the church's industrial education was also possible.[80] Another lay-worker, E. F. Chivers, arrived in 1922 and kept the saw-mill running with logs from the Mbanga mission land. Assisted by Chinese carpenters, the construction programme continued and included boat-building.[81]

The Depression hit the missions as it did the planters. The Methodists economised on wages and on items like fuel. Finally, the saw-mill had to be shut down in 1934, leaving great hardwood logs to rot where they fell on Mbanga and at the saw-mill.[82]

Methodist Mission Archives, Auckland, by permission

Figure 25. The saw pit team near Kokenggolo in the 1920s. The woman is Lina Jones, a talented teacher at the mission's school.

A few years after the Methodists had installed their saw-mill, the Seventh Day Adventists did the same at their headquarters, Mbatuna. It and the mission's electricity supply were powered by a diesel-fuelled English Gardner engine, purchased in 1924 by Pastor G. Peacock. Although the Adventists encouraged the Solomon Islanders to take pride in the workmanship that went into the large churches, this, plus the practice of villagers consolidating settlements around them, meant a heavy demand on leaf supplies, the same problem the Methodists had encountered. So, often the mission stations and the local church people opted for more permanent buildings, bringing timber at first from overseas for Mbatuna until the mill opened, using trees from mission land.[83]

The process of getting logs to the mill was laborious. Firstly, the Marovo people brushed the scrub, felled the trees, cleaned the trunks, and manoeuvred them by wooden handspike as a lever. They then carried them one by one down to the water's edge, floated them out to the mission ship, and tied them to make a raft of logs. The ship towed them to the shore at Mbatuna and the people hauled them ashore. Logging here and elsewhere (except Vanikolo) was along this pattern at least until World War Two. Thus the bulk of logging was done as close as possible to the water's edge or river banks.

So great was the demand on the saw-mill that it was extended and refurbished in 1929 and in 1931 a planing machine was added, allowing for timber dressing. The mission set priorities in costing in 1931: 37s per hundred superfeet to 'outsiders'; 29s to native churches (presumably including other denominations); 17s 6d to white workers; 7s to departments and districts.[84] Some of these 'outsiders' were as far away as Rabaul, New Guinea and had had at least one consignment of milled timber shipped on Burns Philp's steamer, *Malaita* in 1935.[85]

Solomon Islanders showed aptitude for 'adding value' to the forest product. Even before they had a saw-mill the missionaries were using local timber, not only for building, but also to make oars for dinghies, a skill they imparted to their students.[86] This continued at their training school at Batuna where the students produced cabinet work. The mission carpenter, H. R. Martin, who built the first permanent buildings at Mbatuna, appears to have been the first supervisor of the saw-mill.[87] In the 1930s, Pastors L. Borgas and W. Broad, saw-millers, took charge, but there were times when the mill was underused due to lack of someone to supervise.[88]

From the Shortlands to Santa Cruz

SDA Research Centre, Cooranbong, New South Wales

Figure 26. The following sequence of photographs show Seventh Day Adventists gathering logs for the first Mbatuna saw-mill, Vangunu, 1925. This method of carriage certainly did not compact soils.

SDA Research Centre, Cooranbong, New South Wales

Figure 27.

Pacific Forest

SDA Research Centre, Cooranbong, New South Wales

Figure 28.

SDA Research Centre, Cooranbong, New South Wales

Figure 29.

From the Shortlands to Santa Cruz

SDA Research Centre, Cooranbong, New South Wales

Figure 30.

SDA Research Centre, Cooranbong, New South Wales

Figure 31.

Pacific Forest

SDA Research Centre, Cooranbong, New South Wales

Figure 32.

SDA Research Centre, Cooranbong, New South Wales

Figure 33.

From the Shortlands to Santa Cruz

Figure 34. The refurbished saw-mill, first logs sawn, 1929, Mbatuna.

Figure 35. Bringing up sawn timber from the mill for construction, c. 1920s, Mbatuna.

Pacific Forest

SDA Research Centre, Cooranbong, New South Wales

Figure 36. Making oars from local timber, at Seventh Day Adventist mission, Telina, Marovo Lagoon, 1922.

SDA Research Centre, Cooranbong, New South Wales

Figure 37. Pivu cutting a new stern for a mission launch, Mbatuna, Seventh Day Adventist mission, c. 1925.

From the Shortlands to Santa Cruz

SDA Research Centre, Cooranbong, New South Wales

Figure 38. Furniture made by Marovo pupils, Mbatuna, 1925.

By 1931, the accessible timber at Mbatuna was becoming scarce. As the mission had need of a new vessel the Adventist people nearby generously donated logs for its construction.[89] Three years later when the Mission Committee meeting at Mbatuna calculated it needed 83,000 superfeet for its building programme from Choiseul to Malaita, the 'Marovo landowners' gave their logs free of charge for the saw-mill.[90] Much of this timber was *Calophyllum kajewskii* (*baula* or *buni soloso*).[91] Some was used for boat-building and repair, which had become a feature of the mission under the guidance of Mangyimum Snanaga, a Japanese boat-builder.[92]

Because of the consistent tithing of Adventists in Australia and in the Solomons, the mission did not have to retrench as much as others during the Depression.[93] A new vertical saw was added to the mill as well as a new roof in 1938. However, by 1939, the saw-mill was not operating within budget. In addition to requesting a special grant from the home church, the mission committee increased the timber prices from 7s for dressed to 12s 6d per 100 superfeet or 10s rough, for mission projects, including the large Amyes Memorial hospital on land donated by the late Norman Wheatley at Ariel (Meresu) Cove, Kolombangara.[94]

By 1938, the Catholic mission at Mbuma, Malaita was cutting and milling timber from its land, producing more than by the pit-saw. Brother

Anthony, (John Burke) and Father Dan Stuyvenberg ran the mill. As well as using it on mission buildings, the Catholics sold timber to Carpenter's and Burns Philp where some was hand-dressed for further sale. Most of it was *Calophyllum kajewskii* which substituted for imported Oregon pine. The Melanesian Mission also sought to sell timber to help maintain their work in the Depression. They offered trees at Siota to the Melbourne-based Kauri Timber Company, but the limited quantity of about 50,000 superfeet did not make an attractive proposition.[95]

Mission milling helped to subsidise educational and medical work. Particularly among the Adventists and, to a lesser degree, the Methodists, it also taught skills which affirmed the worth of their converts and which added material comfort to their lives. In its small-scale extraction of forest trees, it provided an alternative model of development, of the person and of community through low to medium Western technology. Though some may argue that introducing Western-style oars, boats, houses and furniture was cultural imposition, the missions at least offered the Solomon Islanders the courtesy of choice; most other Europeans, unless there was money in it for them, thought natives' desire for such things to be pretentious.[96]

Santa Cruz: of scams and schemes

Soon after Monckton's Shortland lease had been granted in 1912, Woodford had sought a way to rid the Solomons of French influence in the eastern-most Santa Cruz district. French trading and recruiting ships, often more kidnappers than recruiters, frequented the area, even though it had been annexed to the Protectorate in 1898.[97] Woodford was also aware that the Société Française des Nouvelles Hebrides, through its predecessor the Compagnie Caledonienne des Nouvelles Hebrides, had claims to land on Vanikolo.[98]

French traders also had an interest in the timber there. Dillon in his reports to the French government in 1825 had remarked, 'The mangrove tree lines the shores in great abundance, and a kind of pine, of which we procured some spars for boats' masts'.[99] Dillon was being coy. He was by no means ignorant of timber and, as an ex-sandalwooder and a former dealer in New Zealand kauri (*Agathis australis*), he was probably aware that the 'pine' was a species of kauri (*Agathis macrophylla*), though koila, a hardwood was more suitable for ship's timbers.[100] Still, there was enough information in Dillon's book and elsewhere to warrant further investigation.[101] Moreover, by an odd coincidence, in May 1898, the same month as the British warships arrived to extend the Protectorate over the Santa Cruz group, an advertisement appeared in the Noumea newspaper, *La Caledonie* offering for sale or lease the island of Tevai near Vanikolo, with 'une immense Forêt de

Kaoris'.[102] The French certainly knew of the timber resources of Vanikolo.

Added to his concerns about the French, Woodford was worried about Japanese trochus-shelling vessels avoiding customs with few to tell the tale. There was only a single trader-planter, Jack Matthews, managing Levers' plantation at Graciosa Bay, Santa Cruz. The *Southern Cross* visited the few areas where there were Christians and recruiting vessels from further west in the Protectorate called occasionally at Ndene and the Reef Islands.[103] In 1913, Woodford verified a report that there was kauri on Vanikolo, following an application by a French trader, Hagen, who wanted to cut 'certain hardwood timber (mangrove)' there as well as gaining exclusive fishing rights for trochus and green snail shell.[104] As he was returning from Vanikolo, Woodford called at Makira, outlining 'a timber proposition' to Rigby, a planter and a partner in the Melbourne-based company of Fairley, Rigby Co. Ltd.[105] Woodford foresaw no problems with the native inhabitants of Vanikolo who were 'perfectly quiet and rapidly dying out'.[106]

The company sent an expedition to inspect the island and applied for a licence 'to cut, fell, and remove kauri timber' in October 1913.[107] Stretched for money, they recapitalised as San Cristoval Estates Ltd in 1916 and renewed their application, but the British Colonial Office postponed finalising it until a claim by the French New Hebrides Company had been settled. Since France was an ally during World War One, any contentious issues between Britain and France were put on hold.[108]

The conclusion of the war saw the Australian government looking to New Guinea and the Solomons for a source of timber for a massive house-building project for returned soldiers. A certain J. T. Caldwell, representing a private syndicate learned of the kauri on Vanikolo. He, posing as lessee of the Vanikolo land, won the confidence of the Australian War Service Homes Commission, selling them the lease for £50,000. In 1919, Caldwell had created a positive impression in the Solomons and applied for a lease which Workman had recommended to his superior in Fiji, but the lease had never been granted, so the High Commissioner informed the Australian authorities. The San Cristoval Estates had priority, so the lease was awarded to them on directions from Downing Street, after an Australian parliamentary inquiry in 1921 exposed Caldwell and the War Service Commission's gullibility.[109]

Caldwell returned to the Solomons, touting a paper pulp manufacturing project and claiming he was backed by 'very strong Canadian interests'[110] and 'an extensive timber business in Melbourne'.[111] Caldwell went to survey Utupua in September 1922. Instead of the expected '200 million superficial feet of Kauri pine', Caldwell discovered only *one* kauri tree.[112] Moreover, there was no other timber in commercial quantities.[113] On

Santa Isabel, a planter at Kia, W. H. Bennett, claimed he could obtain 'big leases' over land there to supply the timber for the wood pulp for 'British investment'.[114] But although Caldwell expressed interest, he and his plans for development of the Solomons timber industry vanished.

Meanwhile, Caldwell's visit to Vanikolo in 1919 had stirred French interests. In December 1920, they notified the British authorities of the original claim of the Compagnie Caledonienne des Nouvelles Hebrides, dating back to 1884 and 1887, to over half of Vanikolo and all Tevai. This came as a shock to the High Commissioner, but was an even greater one for the Vanikolo people.[115] In October–November 1922, Judge Phillips headed an inquiry, but the French refused to tender the original documents so their claim failed.[116]

At last the way seemed clear for the San Cristoval Estates to take up their lease. Earlier, in January 1922, the government had ascertained that the declining local population of 83 on Vanikolo and Tevai used land for gardening only on the coast of the main island.[117] The government in November 1922 signed a preliminary agreement with Tua, Fazano and Tomu for timber rights to Vanikolo and Tevai. Tua and the other chiefs, now aware of the magnitude of rights claimed in the French documents, were adamant that 'there should be no alienation of the soil'.[118] The Resident Commissioner was at liberty to allocate the concession to anyone, except of French extraction.[119] San Cristoval Estates finally obtained the licence in January 1924.[120]

The conditions of the licence stipulated a 20-year term with the right of renewal for a further ten years. Royalty to be paid to the government was six shillings for every tree cut, felled or removed of a diameter not less than six inches at three feet from the ground. The annual rent of £10 each to the three chiefs was the sole remuneration for the people. There was no provision for reforestation.[121] When the Secretary of State considered the licence his concerns had been the French and Australian claims and the level of revenue. Conservation of the Protectorate's forest was not an issue for the Colonial Office, although Resident Commissioner R. Kane had raised the question of reforestation.[122]

Timber policies: underlying assumptions

Much of Woodford's policy-making was based on the assumption that the Solomon Islanders would become extinct: hence his adoption of the Waste Land principle and his encouragement of plantations. It was in the Shortlands and Vanikolo where depopulation was obvious that Woodford supported logging. Still, he did take cognisance of the needs of the existing population when contemplating leases and licences. He was wedded to the concept of development that would bring some benefit for the local people, even if it was only the cessation of killing. Given these premises, Woodford saw men like Sorenson and Coote at best as careless speculators and at worse as potential exploiters. Such Europeans were not to be encouraged, whereas those in for the long haul, like Monckton, were.

The lease to remove timber between Monckton and the land-holders, represented by the government, under King's Regulation (KR) No. 9 of 1913 had stipulated: first, a rent plus a tree royalty to the owners, secondly, a royalty to the government based on that paid to the owners, thirdly, a trunk size and species specification, fourthly, the requirement of a register of all trees felled.

However, Woodford had retired in 1915 and the new agreement between the government and the timber company was negotiated by his successors, who doubted that KR 9 of 1913 was 'sufficient for the purposes' of the Vanikolo application.[123] Under KR 12 of 1922, the Resident Commissioner had the power to acquire the timber rights over land from the owner or holder of these and then grant these to a third party under set conditions. It seems, too, that this regulation enabled the government to acquire timber rights by compulsion, but there was no provision for compensation. This, in theory, gave power to the government to control the disposition of the timber resource on native lands.

The essential difference between the two regulations was that the timber rights in the 1922 regulation went first to the government; whereas in the 1913 one, they had gone to the logger, at least for specific species and size. The 1922 regulation gave the government the control as to who obtained the rights, whereas the 1913 one left this in the hands of the resource holders. Though not specifically conferred in 1913, the government by implication probably had power of veto. Disregarding the specifics, the general form of the 1922 regulation sprang first of all from a conviction that the Vanikolo population was 'a degenerate one' which was 'rapidly dying out'. So, in the Resident Commissioner's words,

> To hand over the proceeds of the proposed licence to these people would in my opinion be absurd and I suggest the Government acquiring the timber right and granting the natives an annuity.

It was an opinion echoed by the Colonial Office.[124] The balance was to go towards funding a district office on Vanikolo.[125] This then was the context of the 'negotiations'.

The preliminary agreement between the Vanikolo leaders and J.C. Barley, district officer, was signed on 4 November 1922. Barley was on the island to assist Phillips regarding the French claims being examined between 31 October and 5 November. This had two notable consequences: first, the inquiries as to who had rights over what lands was thorough;

secondly, regarding the future disposition of timber rights, as Phillips later made clear, he was no more influenced by that than he had by the possible diplomatic embarrassment, alluded to by the French company, in finding against their claims. He stated,

> The Kauri timber concession referred to by the Société Française was not within the scope of my Commission of Inquiry. The Lands Commission has had no part in the negotiation for that concession.[126]

Phillips maintained his impartiality, but the 'Deputy for the Natives', Barley less so. The preliminary agreement, negotiated by Barley specifically mentioned the French as being excluded from gaining the concession. This attests as much to the Protectorate's continuing fears of French influence, as to those of the remnant Vanikolo people's distrust of the French traders and recruiters' proprietary attitude towards their womenfolk[127] and towards their land and 'the Woods, Mines, Waters and Forests ... and also the villages'.[128]

Given the depressed state of the Vanikolo people and their awareness of their rapidly decreasing numbers following a terrible dysentery epidemic in 1914,[129] it may have seemed to them that there was a likelihood the French would get control of the islands, given the apparent arbitrariness with which European powers divided and redivided Melanesia before 1922. Moreover, the opportunity of having a district officer, a medical centre on Vanikolo, a trade store, and calls by the steamer, as well as rent money, may have appealed.[130]

By the 1920s the government was less high-handed about alienating land – a consequence of increased objections by Solomon Islanders, resulting in the Lands Commission of 1919–1925 which returned half the land taken under the Waste Lands regulations.[131] This situation had arisen because the government perceived unoccupied land as a sort of free good for its purposes. That kind of thinking was present to some extent in its view of the forests. They were there for the development of the entire Protectorate and the extension of colonial control, rather than belonging solely to Solomon Islanders in the logging area. However, the government clearly recognised some rights accrued to them as it negotiated a lease and it did consider whether exploitation of the resource would be detrimental to their subsistence.

Nonetheless, the government gave no consideration as to who the particular owners of the trees might be, if indeed there were 'owners' of non-traditionally valuable trees. It did not attempt to discover if rights of a non-specific kind in regard to land use, necessarily paralleled the rights to own and dispose of trees, despite the fact that in most societies a differentiation was made between rights to the land and rights to things on it. If a chief representing a group had clear rights to a territory vis-à-vis others and their leaders, then the government assumed the land belonged communally to the entire group. The same thinking applied to the trees. The chiefs could work out who in the group had rights to specific trees and thus royalties and rents, although the district officer might be called in if claimants felt aggrieved with their distribution, as happened in the Shortlands in 1937 with Kimale. On Vanikolo, a well-known woman, Navanora of Tenema[132] came to the district officer in 1934 wanting 'a cut out of the money for the timber concession'[133] and he sent her off to Ben Ramoli, the chief and son of Tua who was a signatory to the lease.[134] The outcome is not recorded, but distribution of the rents was public knowledge in these small communities.[135]

From the government's British legal perspective, this vesting of the rights to trees in the land holder was common practice. And, as regards missionaries and planters, if they had a freehold or leasehold title or a Certificate of Occupation to the land, they had the right to log it, because they had the right to clear it. There were no legal constraints except where 'tambu' places had been itemised in title documents as not to be cleared.[136]

In regard to leasehold land, the government, by the 1930s, realised it had created a situation in which a 'planter' could lease land,

> where the minimum is cultivated, the eyes of the timber country cut out on the excuse that it is necessary to fell before planting, and the lease finally abandoned.[137]

This happened with MacCrimmon's land leases on Malaita and with Monckton's on Poporang, although it would have been expecting a lot of these men to establish plantations during the Depression.[138]

The government had considerable power over native-held forests under the regulation of 1922, as no Solomon Islander could negotiate directly with a non-native for the disposition of timber rights. Thus it was a limitation on property rights, but then so too was the regulation of 1912 to prevent direct dealing in land and, no matter its original intent, few today can deny that this prevented the people from losing their land. Even under the earlier timber regulation, no logging could be carried on by a non-native on native land (excluding leased land) unless the government issued a licence.

Woodford's assumption that the future of the Solomons was as a plantation economy had left its imprint on policy, geared to the clearing of forests for plantations, not for logging and certainly not for sustainable forestry, even though he, Resident Commissioner Kane and a few other officers were aware of the principles of sustainable forestry. There would be replanting, but with coconut palms, not timber trees. Forest conservancy was a non-issue. If anything, Woodford and succeeding administrations before the mid-1930s saw the forests as a hindrance to the extension of plantations. The timber legislation left the government with no control over the forest resource on non-native freehold, leasehold or Certificate of Occupation land, an outcome it did not fully appreciate until the early thirties. By chance rather than management, it never became a major issue, because of the Depression and the impact of the Australian log tariffs.

Neither Timber regulation, to say nothing of development provisions on leases and Certificates of Occupation, was concerned with sustainability of the resource, including reforestation, or with effective control of exploitation: both were aimed at extending the government's control over property rights of the indigenous people. Both were 'regulations of expedience', which were to leave their mark, not only on the forests, but also on the attitudes of both the people and the administration towards timber resources.[139]

There was another gap in the government's control of the forests. Both timber regulations dealt solely with the acquisition of timber rights. The government did not have the legal power to stop Solomon Islanders indiscriminately cutting down trees, 'a course which may well result in soil erosion, with possible disastrous effects on natural water supplies'. When Resident Commissioner Marchant realised this in 1941, he began drafting legislation to insure both utilitarian conservation and control through 'Protected Forests' which could be declared over major forests on native as well as Public land, thereby vesting cutting rights in the Resident Commissioner.[140] This preliminary draft legislation had to be set aside during the Pacific War.

— 5 —

The Vanikoro Kauri Timber Company

> ... any industry which involves machinery and skilled white labour is too risky in a malarial climate, and we would at times find the whole venture stranded owing to sickness and resignations, this is the reason so many saw-milling ventures, etc. have failed in various parts of the Islands.
>
> BPA: Walter Henry Lucas, memo to Chairman of Directors, Burns, Philp and Co., 2 Nov. 1917.

When the district officer and the logging company's overseas employees arrived at Vanikolo in late 1923 they were probably the first group of Europeans to live on the island for any length of time since the few survivors of Laperouse's wrecked ships. The Santa Cruz district was not an easy group for Europeans to live in, but the most difficult place was Vanikolo. As Laperouse's men had found, it was an island that refused to accommodate strangers for long. For the first major logging company, climate, disease and location combined to reduce efficiency against a background of fluctuating demand and access to the market, Australia. For the British administration, the company's operations revealed serious flaws in its forest policy.

From Australia and New Zealand

After Peter Dillon had recovered relics of Laperouse's expedition, a French naval vessel commanded by Dumont D'Urville visited the island in 1828. They found a population of about 1,200–1,500, who seemed fairly healthy, but with warring divisions. D'Urville's visit was not without its ironies for he brought with him lengths of kauri (*Agathis australis*) from New Zealand in order to construct a memorial to Laperouse and crew.[1] Had D'Urville managed to break through the dense bush and climb inland he would have found enough kauri (*Agathis macrophylla*) to build a million monuments. In an indirect way, New Zealand and its kauri were again to be linked with the re-appearance of Europeans in Vanikolo almost 100 years later.

Accompanied by a government officer, representatives of the Melbourne-based San Cristoval Estates had first visited Vanikolo in 1922 to select the site for the logging settlement.[2] At its extremes, the island, including Tevai, is about 13 miles from west to east and seven miles north to south, with an area of about 72 square miles. Reef-encircled, with little flat land, it rises rapidly from the mangrove fringed coast to its highest point (Mt Popokia), 3,031 feet above sea level. The site the party selected was at the head of Saboe Bay (Sapolombe Bay).[3] This proved a most insalubrious place for the first group of loggers, hired in New Zealand. The New Zealanders encountered a trying, humid environment. The daily temperature is about 32 degrees Centigrade, falling to 22 degrees at night. Vanikolo is the wettest place in the Solomons, and in fact the island Pacific, receiving 250 inches (6,350 mm) a year on the coast, and probably far more in the inland peaks and ranges, with rainfall daily for two-thirds of the year, so logging was not easy in the rough terrain.[4] Of the 15 experienced bushmen from New Zealand who arrived on board the *Houto* to commence operations in late December 1923, only four remained in June 1924, the rest having been invalided out.[5] Along with the Melanesian workers, they suffered from the recurring scourges of this isolated settlement, malaria and malnutrition.

Within a year, the district officer, and then the company, moved their respective bases to Paeu, 'a good site with streams of good water' on land purchased by the government from the local people.[6] Here the company had also built a saw-mill to mill timber for export and for its own requirements.

Capital expenditure for the establishment of the Vanikolo operation amounted to more than £40,000. Calls were made on shareholders, the last in 1925 being unsuccessful. A new company was formed, the Vanikoro Kauri Timber Company. San Cristoval Estates had a half share and the Melbourne-based Kauri Timber Company had the other. The consideration was the transfer by San Cristoval Estates of their assets on Vanikolo, including the logging rights, while the Kauri Timber Co. put up £40,000 working capital and paid San Cristoval Estates a further £10,000 for agreeing to the deal. Most of this was used to pay off their Vanikolo debts.[7]

As its name implies, the Vanikoro Kauri Timber Company wanted kauri (*Agathis macrophylla*) from Vanikolo for the Australian market. From the 1850s, Melbourne, flush with wealth from Victoria's gold, supplemented by the export of wool and wheat, had purchased a huge volume of kauri from New Zealand to build its mansions; so much so, that a group of Melbourne capitalists bought into the kauri (*Agathis australis*) logging in New Zealand in 1888.[8] By the 1910s, the New Zealand supply was all but cut out. Belatedly, the New Zealand government restricted exports to

The Vanikoro Kauri Timber Company

By permission of Business and Labour Archives, University of Melbourne

Figure 39. Paeu from the sea, site of the headquarters of the Vanikoro Kauri Timber Company after the abandonment of Saboe Bay, in 1925.

By permission of Business and Labour Archives, University of Melbourne

Figure 40. Clearing at Paeu for company headquarters and cattle grazing, looking inland (north).

a maximum of 500,000 superfeet a year; hence the Melbourne-based Kauri Timber Company's interest in supplies of a related species in the Solomons.[9]

As well as kauri, there were other valuable trees noted by company timber men, including the *ba'ula* (*koila*), *u'ula*, and *liki* (*Calophyllum* spp., *Intsia bijuga*, and *Pterocarpus indicus*, respectively). These were in mixed forest as was most of the kauri. Unlike its New Zealand congener, the Solomons kauri was not in great stands. On Vanikolo, it was scattered: in the east, where it was relatively sparse, mainly on or near the top of rocky razor-back ridges and spurs, in the south and north along rugged slopes and crests, and in the south-west on the heights of more undulating country.[10]

Of these timbers, the company was mainly interested in the kauri and koila, estimating in 1925 that there were 10,000 acres of kauri and 40,000 of koila which would give 150,000,000 and 60,000,000 superfeet respectively.[11] The bulk of the koila, a hardwood, was for specific purposes such as ships' keels and timbers, while the kauri, a softwood, was used for flooring and building[12] as well as for 'brewers' vats, draining boards, boat building, wash troughs and other liquid containers'.[13] In the late 1920s, the Kauri Timber Company in Melbourne was investigating the economic potential of plywood, first as a supplier of the log material and by the late 1930s as manufacturer. Kauri was found to make excellent plywood. Trials were also made on the koila. Another timber found on Vanikolo, *karamati* or *kete kete* (*Campnosperma brevipetiolata*), from 1929 on became part of the log exports from the island to Melbourne as did 'rosewood' (*Xylocarpus granatum*), for veneers and furniture and odd lots of so-called 'oak', 'whitewood' (possibly *Endospermum medullosum*) and 'Brian Boru'.[14] Kauri made up more than 85 per cent of the exports from Vanikolo with koila at around 5 per cent. The advent of log peeling for plywood saw '[h]undreds of thousands' of Victorian butter-boxes in the 1930s made from Vanikolo kauri as it did not taint the butter.[15] With the existing technology, the karamati was not suitable for plywood and ended up being used for general purposes, particularly furniture-making.[16]

Assisted by government

The logging operation planned for Vanikolo seemed like a joint enterprise between government and the company with each trying to get as much out of the other as possible. The government saw the company as a means of diversifying the economy, reducing its almost total dependence on copra, at a time when the markets for it were beginning to weaken. At the start, the company sought and received concessions, such as a waiver of duty on a saw-mill they wanted to import to Vanikolo.[17] To set up a district office

The Vanikoro Kauri Timber Company

to monitor the number of trees cut and royalty assessment, customs, and the contracts of indentured labour, would cost the Protectorate initially £55 and an annual expenditure of £1,930, so the High Commission stipulated conditions: that the company transport and pay for the erection of the government buildings; as the company had to provide a medical officer, the government workers and any prisoners were to be treated free of charge; government stores were to be supplied at landed cost; the district officer was to be transported from Tulagi and accommodated free of cost until his quarters were ready; the island was to be a port of entry; the government would grant a licence for a wireless installation on condition that it be used to communicate with Tulagi and that all communications go through the station; and the district officer was to be transported free of charge on the company's recruiting trips.[18]

Paeu became the administrative centre for the Santa Cruz district in 1925, but its raison d'être was the presence of the company. The steamers coming for the logs, the Melanesian Mission vessel, the *Southern Cross*, doing the New Zealand–Melanesia bi-annual circuit, along with the occasional tourist yacht or scientific party, all called at Paeu as it was the port of entry. Because the district officer had to be on hand to attend to customs formalities and to check the number of trees felled by the company, he was often tied to Vanikolo for months or even years at a time, so the administration of the rest of the scattered district was sporadic indeed. He relied on recruiters and traders, such as Fred Jones, Charlie Cowan and Norman Sarich and the odd missionary to bring information from the distant islands, since he rarely had a ship at his disposal.[19]

Labour matters were a major concern of both the government and the company. The district officer had the duty of supervising the indentured labour on Vanikolo: that men were recruited according to the Labour Regulations; that they understood the nature of the two-year or, more commonly on Vanikolo, the one-year contract, were of age and were correctly paid a minimum of £12 a year; and that the employer fulfilled his obligations to feed, clothe, house and provide medical care and repatriation for the recruits. The indenture system provided penal clauses by which labour could be fined or even imprisoned for such things as failure to obey a lawful command or desertion. Employers could also be fined if they failed to follow the regulations.[20]

Before World War Two the bulk of labour for Vanikolo came from within the Santa Cruz district and from Malaita.[21] The company had problems initially with the recruits from the Santa Cruz district because they were young and inexperienced and most knew little Pijin, the Solomon Islands' lingua franca, and neither did the New Zealanders. The new recruits did not know how to use shovels or picks, let alone timber-jacks; such tools were totally new to them. However, in time they learned the logging skills. The company preferred Malaitans, but these were far more expensive to recruit and repatriate because of the 400-mile trip each way to Malaita, so before the War they numbered no more than a quarter of the work force.[22] They did not always get along with the Santa Cruz district labourers, which reduced the risk to management of concerted resistance by labour, but made for tense times when the Malaitans were accused of sorcery by the Santa Cruz district men.[23]

One alliance between different groups of labourers was unique at Vanikolo. It was the only workplace in the Protectorate where Europeans regularly did heavy manual work alongside Melanesians. On average, there were about 100 Melanesians employed beside 15 Europeans.[24] The European bush workers were practical men, inured to the hard, isolated conditions which they shared with the Melanesians both in the bush camps and back at Paeu, where their accommodation (corrugated iron buildings, *haos kapa*) was similar. They joined each other for football games and leisure time activities.[25] Some of the overseas workers had socialist sympathies and

By permission of Business and Labour Archives, University of Melbourne

Figure 41. The football team at Paeu, c. 1927.

Tentative identifications are that the tall European at the rear is Andrew Black, a New Zealander; the European to his right is the manager, C. A. Curtis, whose right hand rests on the shoulder of the leading 'boss-boy' and hauler, thought to be 'Black Mack' (Mack Saviot), a Bougainvillean. Curtis took 'Black Mack' to Australia for some primary education.

The Vanikoro Kauri Timber Company

a few were Maori and these had an input into the Melanesians' relations with management.[26] In fact, the first and probably only strike by European labour in 1929 occurred at Paeu when a manager tried to reduce wages to an hourly rate instead of a monthly one in an effort to reduce absenteeism. Although the ring-leaders were sent out of the Protectorate, the European workers were not the 'capitalists'' men, and sympathised with the Melanesians in 1934 when the minimum wage for indentured labour was halved by the government to £6 a year because of the Depression's crippling effects on copra planters.[27]

European workers rapidly fell victims to malaria and infections such as pneumonia that came in its train. Of the 21 Europeans attached to the company at the end 1926 only three had been on the island longer than 12 months. Over the years, most lasted only eight to nine months instead of the two years they had signed up for.[28] And these were 'mostly huge men'[29] 'of fine physique'[30] and 'in the prime of life'.[31] Many were invalided out as had been the near-paranoid company doctor, Kelly and his predecessor, M. O'Shea who was addicted to drugs and alcohol. O'Shea had taken R. B. Martin's place as medical officer. Martin too was an alcoholic, who neglected his work and assaulted the manager, being shipped out soon after. It was not just the company Europeans who broke down. District Officer H. MacQuarrie, while not insane, had lost his balance and had been dismissed in mid-1925; his successor, N. Heffernan was taken out in 1926, a raving lunatic and alcoholic; the energetic A. Studd died of malaria in 1928; and F. Filose was mentally disturbed when relieved of his duties in 1931.[32]

Although illness was common, the Europeans enjoyed a better diet than the Melanesians. And at different times the Europeans had access to their own garden vegetables, and a few cattle and chickens as well as a great variety of fish to vary the tinned food diet.[33] In 1928, the company hired two Hong Kong Chinese, Chow Chui and Ah Chu, to cook for the Europeans at Paeu, which improved morale. They could further supplement the rations the company provided by purchases from the company store which, as was almost universal practice, sold at cheaper prices to them than the much lower paid Melanesians. But malaria could strike the well-fed as easily as the malnourished.[34]

Melanesians, like the Europeans, feared illness, but all the more so because they believed it to be the result of spiritual forces. In a particularly unhealthy year when they were looked after by Mervyn Deland, a conscientious young doctor, but with a 'peculiar manner', seven out of 70 died; which convinced some potential Malaitan recruits not to sign up for Vanikolo until Deland had gone.[35] Although the Melanesians periodically lobbied for higher wages, their strikes in 1934 and 1936 resulted from deaths which to them were occasioned by spiritual intervention. Vanikolo's unhealthy climate, exacerbated by inadequate food supplies, saw men refusing to work in the bush where their companions had died. Although the district officer fined the men in 1934 they still would not return to the bush camp.[36] A severe cyclone in early 1936 resulted in a shortage of fresh fruit and vegetables. When 15 men, who were mainly Malaitans, out of 82 labourers died that year from beri-beri due to bad rice and lack of fresh food, friends of the dead simply refused to work and were allowed by the government to cancel their contracts and go home.

In general, the role of the district officer was as go-between in relations between management and labour. Except in 1934, the penal clauses of the indenture system were rarely invoked, the district officer arbitrating most difficulties. In the main however, this was within the context of labour shortage, especially of skilled labour for logging. By 1941 the company was heavily dependent on the Santa Cruz district for its labour, half the men signing up for work having been previous employees of the company. Each party had to accommodate the other to sustain the production of logs and have them ready for the ship when it came.[37]

Getting the logs out: from stump to tidewater

Logging in the terrain and climate of Vanikolo was difficult. Usually, the bush gangs went up a river to find the stands and set up camp nearby. The fellers chopped down the trees which were often sawn into smaller logs Where the log had to be hauled any distance to the loading place the leading end was sniped or cut obliquely around the circumference, to prevent the rough square edge catching on rocks and trees. The men used timber-jacks as well as primitive handspikes to lift, push and generally manoeuvre the logs. Various methods, depending on terrain, were employed to get the logs out of the bush. Where the soil was free of stone, the logs were hauled along the slippery wet surface. On firm ground, they were dragged along a narrow 'road' or 'broad', by a steel wire rope attached to a steam (or later, a diesel) hauler, or, where there was some continuous slope, pushed down a mud 'shoot' or, where the soil was unsuitable, down a chute made with rickers or narrow, smoothed tree trunks or laid lengthwise. Sometimes the bush gang constructed a rolling road, particularly in swampy areas. Here, rickers laid lengthways to the direction of the road, made a surface for the logs to be pushed over one by one and held while the 'road' was taken up and re-laid in front of the log bank. In very wet areas these were raised by the addition of a foundation of several smaller trees laid across the path. Depending on location, logs were sometimes dragged as much as three miles to the nearest river and floated to the coast, as was the case with the Lawrence (Paeu) River in 1927–1928. By such means the bushmen got the

The Vanikoro Kauri Timber Company

Year	Melanesia district total	Malaita Cruz Australia	Santa Zealand	New	Others Santa Cruz	From for all BSIP
1923/4	40	40		15	1 Bougainville	184†
1925	50	40	10	7		91
1926	21	20		21	4 Japanese	223
1927	72			27	1 Chinese	171
1928	160	28 +	(132-)*	26	3 Chinese	120
1929	164	40	(124)	24 +		116
1930	100	70	(30)	20		?
1931	77	50	27	3		23
1932	45	15	16	5		41
1933	64	27	37	9	1 Chinese	26
1934	111	23	(88)			62
1935	127					23
1936	82	32	(50)			38
1937	80			12	1 Chinese	57
1938	120					?
1939	160	22	(138)			23
1940	115					85
1941	120					?
1942	23	23		12		?
1943						
1944	20	nil	20			
1945						
1946						
1947						
1948						
1949						
1950	100	nil		22		
1951			85	20		
1952	100					
1953					6 Fijians	
1954	130					
1955						
1956	150			15		
1957						
1958	150					
1959	150					
1960						
1961	110			11		
1962						
1963	134			8		

Table 2. Number and origins of Vanikolo employees.[38]

† These are the recorded figures for all indentured labour coming from Santa Cruz to work anywhere in the Protectorate before the war (averaged for years 1923 and 1924).

* Estimated number from totals in columns to left.

Figure 42. The reason they were there: Agathis macrophylla.

The Vanikoro Kauri Timber Company

By permission of Business and Labour Archives, University of Melbourne

By permission of Business and Labour Archives, University of Melbourne

Figure 43. Axemen felling the kauri, 1920s.

Figure 44. Hauling the log through the mud with a steel cable winch.

timber sometimes directly to the seashore, but more often to a loading bank. Ideally, the logs were sprayed with a creosote-based compound as soon as possible after felling to prevent attack by the lyctus borer. At the loading bank, the men, using timber-jacks or winches, loaded them onto carriages on a tramway, pulled by a Shay locomotive. The tram lines went about a mile into the foothills, depending on local conditions. Where the fall was too great they used winches to lower the loaded trucks to, and raise the empty trucks from, steel tram lines. Sometimes winches were also used to pull the trucks, as the company had only the one locomotive in use in the late 1920s. The logs would be carried around on the tramway to the loading area at Paeu or down to the coast then towed behind a boat in floating rafts inside the reef around to the port.[39] The men attached the logs together by means of dog spikes and rafting wire with the 'floaters', kauri supporting the 'sinkers', koila.[40]

By permission of Business and Labour Archives, University of Melbourne

Figure 45. Europeans and Solomon Islanders hard at work moving log section with jacks, 1920s.

Because of the susceptibility of the cut trees to the borer and forms of rot, logs could not be stockpiled for months or years in gullies to await a 'drive' when a dam built across a river was filled then released, and the mass of logs carried to the sea or adjacent flats, as was practice in rugged areas of temperate countries.[41]

Before the ship was due, the men moved the log rafts to Paeu during the north-west season or at Saboe Bay when the winds came from the south east. However, as logging moved north-east, rafts were held near river mouths until the ship was due at Paeu. When the ship came they were floated out and winched on board. The company had difficulties with shipping, contracting at various times with Burns Philp, W. R. Carpenters, and from 1929 on, the Austral-China Navigation Company and chartered Japanese vessels, with irregular shipments going on the South African Millar's line in the mid-1930s. It favoured a five-weekly service such as Burns Philp offered in the 1920s, but when it switched carriers to get cheaper freight, the company was lucky to get a log carrier every two to three months, because Vanikolo was not on a regular route. The time between ships lengthened during the Depression of the 1930s when most shippers were economising.[42]

The Vanikoro Kauri Timber Company

Figure 46. Shay locomotive and logs at landing.

Figure 47. Log 'rafts' ready to be hoisted aboard the James Cook *off Paeu, 1937. Rafting was consistent throughout the company's existence.*

Setting the pattern of extraction

On Vanikolo, the mixed kauri stands were at least half a mile from the coast. The pattern of extraction around the island, broadly, was clockwise from Saboe Bay. However, there was some doubling back over time, not so much due to a lack of overall directional plan, as to failure to fully exploit more inland areas. People from Melbourne, experts in logging in temperate areas, such as Joseph Butler,[43] visited the island periodically and indicated the direction and often the means of getting the logs out but, once these advisers were away, local managers logged the most immediately accessible areas.[44] For example, logging started in the bush east of Saboe Bay under the axes of Andrew Black, Bill Wightman, the Foley brothers and their unnamed Malaitan 'bush Boys' in January 1924.[45] The re-location of the settlement at Paeu focused logging on 'the handiest bits of bush' near the flat, three-quarters of a mile up the Lawrence river from Paeu.[46] Working continued into the Laurence river area, but in 1929 the company decided that Saboe Bay was a better place for a loading point for the ships than at Paeu. Construction of the track for the locomotive pushed east from the logging site in the Laurence river area. The bridge 'was almost completed when heavy floodwater ... swept it 150 yards down the river', and the plan went with it.[47] The company revisited the area in mid-1933 and established a bush camp inland on the western side of the bay, the logs being brought to the bay by hauler and tramline.[48] Within a year this had been cut out and the operation moved on to the Kombi river area, returning to the westerly locus of working. When another timber surveyor, Kidson, examined the island in 1937 he found there was still about 1,800,000 superfeet or a year's work, left in the Saboe bay area, behind the original settlement site.[49] Although the company returned to the south-east of the island in 1963 to log the Emua (Emwa) area, the inland western ranges of Saboe area and the upper Lawrence river apparently were never logged. This example of inefficient working was repeated in other locations.[50] It was clear to the district officer in 1939 that

> successive managers,... possibly realising that their stay would be a short one, concentrated on getting out the easiest timber to make a good tally, consequently the eyes have been pulled out leaving large quantities of timber behind the old workings which may not now be profitable to obtain.[51]

The first company inspection after the war in 1947 confirmed this,

> Practically all the readily accessible timber along the entire coastal fringe has been logged for a distance of one to three miles inland. Future access routes for exploitation must traverse this now barren area in order to reach productive forest.[52]

The Vanikoro Kauri Timber Company

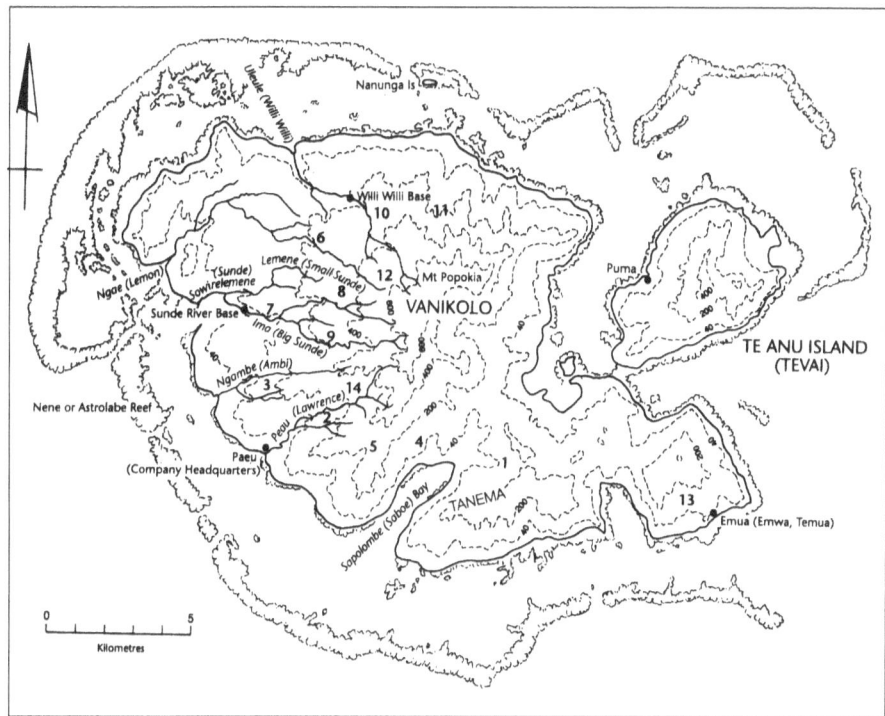

Map 4. The costly pattern of log extraction by the Vanikoro Kauri Timber Company at Vanikolo, 1923–1964. Logging commenced at Saboe (1), then shifted in 1924 to up-river base at Paeu. Several inland stands were bypassed in the process. Variation in place names depends on sources.

Management's failures and missed opportunities

The company made a shaky start with the poor quality of local labour as well as the sicknesses that plagued all employees. It created a financial time-bomb for itself with its erratic logging patterns. The company continued to make avoidable mistakes. For example, in 1926/7, it shipped up from Australia two locomotives and 200 tons of tram-rails. As no wharf had been built, the rails were punted from the steamer outside the reef and thrown into the shallows. It then took nine Melanesians and two Europeans using the steam hauler weeks to drag ashore and stack the rusting metal. The engineer who came to assemble the locomotives considered them to be an unsuitable type and 'were far too heavy for the soft ground on the flats of Paeu Bay and their coal consumption would make them a failure financially'. He awaited the next steamer to Sydney, the same steamer bringing the man to erect the tramline, which needed ballast in its foundations and as there was no rubble available a stone crusher would need to be imported and so on. No one in Melbourne had done his homework and this cost the company thousands of pounds in this one episode alone.

The same could be said of the saw-mill venture. The mill plant comprised 'a series of afterthoughts',[53] due to the financial problems of the San Cristoval Estates and was unsuited for large-scale milling.[54] The company milled some timber for its own use and exported a little to Australia until 1929, with a conversion loss of around 50 per cent. In the light of the cost of upgrading the mill and the difficulties of servicing it, the company thereafter concentrated on the export of logs to be processed in Australia and the mill equipment was left to rust among the accumulating scrap metal at Paeu.[55]

The company's attempt to supply millers closer to likely Solomons markets also failed. Neil MacCrimmon, manager from July 1931, had sent kauri logs to the Mala Development Company to mill at Su'u. Crooked logs, inadequate milling equipment, and MacCrimmon's return to Australia in 1931 meant the experiment was not repeated.[56] Here and even more so at Vanikolo, unsuitable equipment, breakdowns and dependence on often erratic steamers caused delays while replacements or experts were sought. This pattern continued until the Pacific War, and beyond.

The cut from logging: the company's and the government's

In 1926, the company had commenced exporting logs as well as milled timber. This was the first time milled timber had left the Solomons. From 438,857 superfeet in 1926 production increased to 1,553,314 superfeet in 1929 in log only. For much of the pre-war period this was about the limit, with a drop in the years 1932 to 1934, to about 900,000 superfeet, because of the depressed market in Australia (see Fig. 48).[57] Another factor which indirectly influenced production was the increased Australian combined tariff in mid-1930 from 10 per cent to 30 per cent on logs from the Solomon Islands, as it decreased any profit margin the company might achieve and caused it to seek economies. The company reduced staff at Vanikolo as quickly as possible to three Europeans and, when their two-year contracts expired, 45 Melanesian workers. Production fell. After the Australian government reduced the tariff to 20 per cent in February 1933[58] the company in Melbourne 'secured large orders for logs', mainly kauri, koila and rosewood for peeling, so more labour was recruited and output gradually increased.[59]

The Vanikoro Kauri Timber Company

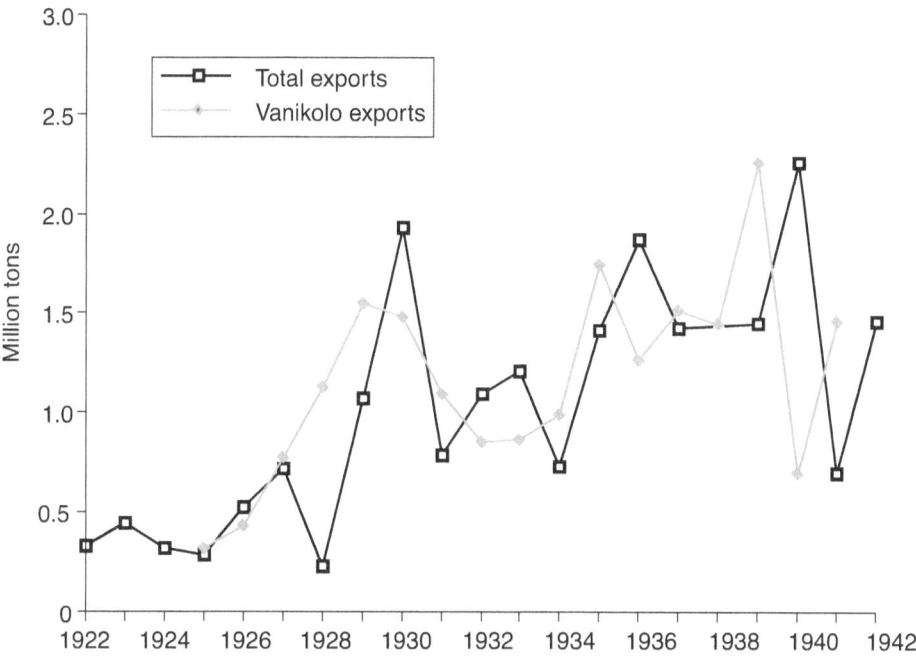

Figure 48. Exports of logs, 1922–1942.

Sources: AR BSIP 1922–1944, AR Santa Cruz district, 1923–1940. Inconsistencies result from the lag between records kept at Paeu and those in the BSIP annual reports.

In July 1930, just before Australia announced the increased tariff, the district officer at Paeu, Filose, advised the company it was in arrears with royalty payments set at six shillings a tree. By the end of the year, Filose stated the company owed £2,323, or more than ten times the royalties paid in any previous year. Filose claimed that 7,744 trees had been cut down for all purposes between November 1929 and 31 December 1930, and that all attracted royalty.[60] The calculation of this hinged on the definition of the word 'tree' in clause 4 in the company's original licence:

> The Licensee shall pay to the Commissioner by way of royalty the sum of six shillings for every tree of not less than six inches in diameter measured at three feet from the ground cut felled or removed by the Licensee as aforesaid.[61]

Both the government and the company sought legal advice. The company was aware that it was liable for timber used in construction, '[i]f the Government terms for Royalty are strictly adhered to', but was not going to pay it unless the government enforced the agreement.[62] The company had on several occasions sought official clarification. To avoid just such a conflict it had wanted royalty be calculated on superfeet exported rather than on the felling of trees of a particular size, a change supported by at least two district officers at Vanikolo.[63]

Since 1923, the company had cut down or removed thousands of trees besides those exported. It cleared paths for the tramway; it cut down mangroves along the water frontage and cleared trees all around the settlement to try to discourage mosquitoes and sandflies; it destroyed understorey trees when big trees were felled; and it felled trees for its own building purposes, such as for the pier, sleepers for the railway, for rickers in log 'roads' and, particularly in the 1920s and early 1930s, for firewood to power the sawmill, the refrigerator, the steam haulers and the locomotives as well as for domestic use. The firewood cutting particularly annoyed Filose as *tovoleko*, a hardwood timber that sold in Fiji at £4 to £6 per 100 superfeet to the Singer Sewing Machine Company of America for cabinets, was going up in smoke. And all this without any recompense to the government or, except for the set rent, to the landowners. So acute was the firewood shortage, especially near Paeu that by the 1930s diesel and petrol were introduced to fuel the larger engines, adding to the Company's costs.[64]

The government's lawyer advised that the company was liable for any trees felled that could be used in construction.[65] However, the legal waters were muddied because the company could show evidence of the government's having excluded small timber used in tramline construction and its failure to contradict the company's assumption that the royalty did not apply to trees used on the site. The government may have been able to counter this, but the fact that its representative at Paeu had waited until 1931 to try to enforce the more inclusive interpretation weakened its case. It was Filose who had precipitated this issue and it was he who further compromised the government's position. Evidence cited by the company, from its employees, including the universally-respected Dr Lily Holt-Mac-Crimmon, as well as the Bishop of Melanesia, indicated Filose was mentally disturbed with a paranoiac hatred of the company's management. He had tried to foment trouble between management and employees, had probably misrepresented the number of trees cut, and his wife had been involved in a questionable relationship with the company's manager, Cook. In any legal proceedings, Filose would have been the Crown's primary witness. His testimony would have been regarded as biased and unreliable.[66]

The issue went as far as the Colonial Office. When the initial agreement was to be signed with the San Cristoval Estates, it had expressed some disquiet in 1925 at 'the inadequate return of 6s per tree',[67] and the need for

The Vanikoro Kauri Timber Company

periodic review, but had not realised the latitude that the clause implied, in spite of warnings from Resident Commissioner Kane to the High Commissioner in Fiji.[68] By the 1930s, the Colonial Office was regularly seeking advice from a leading colonial forestry expert, R. S. Troup who had been a forester in India and Burma, taken up the chair of Forestry at Oxford University in 1920 and, four years later, the directorship of the Imperial Forestry Institute.[69]

Troup suggested sending R. A. Sykes, then on a fixed term secondment from Nigeria to Fiji, across to Vanikolo to investigate. Troup recommended that the basis of any royalty be a flat charge on the volume of timber exported. Sykes' report supported this. Troup and other forest advisers went on to recommend forestry officers for both Solomons and Fiji, but the High Commissioner and Governor of Fiji, Murchison Fletcher, argued against this in 1934 amid the Depression on financial grounds – the outcome, at least in the Solomons, being in effect a moratorium on any new commercial enterprise because it could not be supervised properly. Logging of kauri species at Vanikolo as well as Nadarivatu in Fiji had highlighted policy and staffing inadequacies. The Colonial Office was not to forget the lesson.[70]

The issue of the royalties due on trees felled at Vanikolo resurfaced as a prelude to a revision of the agreement. The company wanted clarification because with the requirement of minimum diameter being six inches at three foot from the base of the tree, practically every tree it felled or removed would attract royalty and this would prove too costly.[71] Ashley, the resident commissioner, proposed a royalty of 2s per 100 superfeet and an export duty, but the high commissioner in Fiji would not support the additional duty because it was not in the original agreement. The company provided figures to show that it could not sustain the royalty of two shillings; in light of its losses, this was probably accurate.[72] The compromise in October 1937 was that the government waived prior claims and would allow the company to cut all small timber needed for its operations free; the company being liable for royalty at 1s per 100 superfeet, calculated by Hoppus measure.[73] The district officer could do checks for timber control, but relied to some extent on the company for figures. Fortunately for the Protectorate, Australian Customs were willing to send their figures on log imports from Vanikolo and these could be compared with the Protectorate's export figures.[74]

The new agreement came into operation in late 1940. It encouraged a minimum output of one million superfeet, since royalty accrued automatically to this amount, except at the discretion of the resident commissioner.[75] The old six-inch clause disappeared. The resident commissioner now had the power to alter the minimum felling girth of any species: in the case of kauri this was to be no more than seven feet and no less than five feet; with other species the minimum felling girth was not to be more than six feet or less than five feet – the rationale being to preserve seed-bearing trees and enforce conservation.[76] Unable to enforce reforestation or undertake it without a forestry officer, the government inserted a provision enabling it to 'close in the interests of silviculture, areas deemed to be logged over', providing these areas were no greater than 20 per cent of the total area under licence. And there was a scale of fines for breaches of the agreement.[77]

Reforestation

These clauses were the closest the government got to enforcing passive reforestation at least, through natural regeneration. The original agreement of 1924 between the company and the government lacked a reforestation clause, much to the disgust of subsequent resident commissioners.[78] When S. J. Kajewski, a botanist from the Queensland herbarium, had surveyed the flora on Vanikolo in 1928, he urged the Protectorate government to 'replant each block as soon as the natural timber is finished'. He also condemned the logging practices of the company because of its 'removal of the large heavy cone bearing trees' that would have provided seed for new growth. He believed that the kauri would grow at least an inch in diameter a year and reckoned that in 50 years 'a good substantial log' would have grown for the time 'when the Solomon Islands will want all the durable timbers for their own development works, so provision should be made for the future'.[79]

R. H. Garvey, the district officer, suggested the company reforest the cut-over lands and hold the timber rights in perpetuity. High Commissioner Fletcher was supportive of the general principle as he was concerned about concessions in Fiji which also had no reforestation clauses, but the company refused to entertain the idea and had no legal obligation to do so. The company knew nothing of the silviculture of the kauri, as they believed the trees were 1,000 years old – a selling point for the timber in Australian advertising, but clearly, not an attractive proposition for a return from investment in reforestation![80]

The Colonial Office could take some comfort in Sykes' advice in 1932 that he had seen promising signs as regards kauri in the regeneration of logged areas, but he made the point that this was only noticeable where there were parent trees nearby.[81] The district officer was less sanguine because he noted the number of 'self-planted seedlings crushed by each fall' of the mature trees.[82] The regenerative capacity of the kauri was evident to F. S. Walker in 1945 when he conducted his survey of Solomons forests.[83] Given the haphazard nature of the extraction processes over the years and the company's failure to cut out valuable trees in difficult places, this natural

The Vanikoro Kauri Timber Company

regeneration resulted from the company's inefficiency and the terrain, not any conscious positive strategy. This was reforestation by happy chance.[84]

Land owners and leases

When the first negotiations for the timber lease on Vanikolo were made in early November 1922, the government officer Barley had obtained a tentative agreement from Tua, the chief of Tevai, for the logging of that island. However, the island was temporarily forgotten and subsequently the lease was never ratified, despite the company's attempt in 1925 to persuade the government to use its influence with the Tevai people in return for all the milled timber it wanted at cost price. The government was not impressed.[85] By 1928, it described Tevai as a 'government reserve', but it seems this was set aside for the government's future use, rather than the people's.[86] Although the kauri was less dense on Tevai it was relatively accessible to the sea and periodically, as the company's operations moved further inland on Vanikolo proper, it revived attempts to obtain the lease to Tevai.

In 1923, the remnant population of Vanikolo had had no idea at all of what a logging operation meant. When the ships came to Saboe to bring the first team of bushmen, Tua came around from Tevai to see what was going on. The attempted French land grab, the investigations of the Lands Commissioner in 1922 and tales from men who had worked elsewhere in the Protectorate (and probably in the New Hebrides and New Caledonia) had given Tua an inkling of what land alienation might mean. He made it clear that he and his fellow heads of clans wanted to retain control of the land.[87] He and others were astonished to see the men, machines, and supplies disgorged from the ships at Saboe and later at Paeu.[88] Both Tua and his son and heir, Ben Ramoli soon had ample evidence of what logging could do and how nominal the control was which they now held over the land and its resources. In 1929, the district officer noted that he had tried,

> to persuade Ramoli and his people to part with timber rights, but Ramoli [remained] adamant, stating that the 'white men' had gained possession of Vanikoro and that he wished to keep the island of Tevai entirely for his people.[89]

The district officer did not have many to persuade. The local population of 83 in 1922 had fallen to 56 in 1930, 39 on Tevai and 17 on Vanikolo.[90] Ben Ramoli had rights on Vanikolo too and Tua had been one of the three signatories to the lease for the island. The company persisted with its attempts to get the Tevai concession, but in 1935 the resident commissioner was 'unable to entertain' the request because of Colonial Office policy as well as local feeling.[91] In the flurry of wartime interest in timber in the islands, the Tevai lease question again arose in June 1945 regarding another company, but the government still could not help because all future timber concessions hinged on the outcome of a forest resource survey then being conducted.[92]

The Vanikolans could be adamant on other issues. The district officer paid out the rents of £10 yearly on the lease to the signatories or their heirs, representing the three clans. The people on the western side of Vanikolo were the least numerous; so much so that by 1934, the only representatives of one of the clans were two males. The district officer suggested a re-distribution of rents on a numerical basis rather than by clan. This astonishing suggestion was 'not welcomed' by those most concerned.[93]

The original lease was up for review in 1942, in the midst of wartime uncertainty, and it was extended by mutual agreement for ten years.[94] In 1947, as the company began to re-assess its operation on Vanikolo after five years of non-production, Ben Ramoli wanted to renegotiate the terms. His understanding of English may have been limited, but he knew what he wanted when he wrote to the district officer,

> Dear Sir master one thing I want to ask you about the list for timber because Company's Promised for 25 years Finish now. and we like to make the Newspaper more again about the list for timber. We like for 20 pound every month by year. That is way we want ask master for help us. Only that Price we like master. (sic)[95]

Ramoli as well as the other clan representatives, Teilo and Opola got their wish in 1952, with an increase to £240 or £80 each. The resident commissioner believed,

> In view of the fact that the Protectorate's revenue from this source is estimated ... at 2000 pounds per annum to-day,... I do not consider that these demands are either excessive or unjustified; indeed I venture to suggest that morally they should be entitled to a much greater proportion of the revenue accruing from Royalties than the 10 per cent which this represents, although I do not propose to suggest this to them at this juncture.[96]

Not only did the government get the rights over all timber on the entire island of Vanikolo, it also obtained the right to allow the company to cultivate 'unused land' for the production of crops for the company's employees.[97] This loosely-worded clause, reminiscent of the concepts of the Waste Land regulations of the early 1900s, would have enabled the company to use the land as plantation land had it so wished and, in fact, this was seriously considered in the early 1960s.[98]

Although the company had its eye on Tevai, it had looked no further afield before the war, unlike its major labour recruiter, Fred Jones, who also had a trade store on Vanikolo. Jones knew the islands well, having traded in the district since 1925 and married a woman from Taumako in the Duff Islands. In 1940 he approached the government for a timber lease on Ndene

The Vanikoro Kauri Timber Company

(Santa Cruz) where there was some kauri.[99] He planned to investigate the possibilities of logging there by bringing in a timber 'expert from Australia'. He asked the government for a first option on the island, but the war put the plan on hold, as it did to most things in the Protectorate.[100]

Company performance

By the 1960s, the plantation option was one of the possible remedies considered for the Company's financial problems. These were present from the outset, but were probably at their most acute in the pre-war period when production remained low relative to the supposed volume of timber available. The scattered, mixed stands, along with poor management, difficulties of communication and transport, added to a trying and unhealthy environment, all hampered production, resulting in losses, as the parent companies poured more and more money into the operation. By 1934, the San Cristoval Estates could not provide any more funds and the Kauri Timber Company bought out its interests. Losses mounted, and in 1939 the Kauri Timber Company wrote off £43,000, with £50,000 still owed by the Vanikoro Kauri Timber Company to the parent company. Finally, in 1941, the Vanikoro Timber Company was put into liquidation and wound up. Its plant was taken over as a branch of the Kauri Timber Company, with the money still owing to the Melbourne company capitalised as establishment expenses. The amount of £76,000 was then put into a 'Forestry Properties' account to be written off over the estimated remaining 100,000,000 superfeet of timber (a considerable over-estimate) on Vanikolo.[101]

Like all commercial operations in the Solomons, the Company was badly affected by the war. The Japanese invaded the Solomons early in 1942, capturing Tulagi in May, to be forced out of most of the islands in late 1943 by the Allies. The Japanese did not occupy the eastern Solomons, so Vanikolo escaped, although both Japanese and American air and sea craft reconnoitred the district.[102] Anticipating the worst, the Company evacuated most of its European staff via the New Hebrides in March 1942. There was no shipping going west into the battle zone to return the Malaitan employees, but Skov Boye, the manager, sent the Santa Cruz men home with the trader-recruiter, Jones. Coastwatcher Ruby Boye who was running the wireless for the Allied network, elected to remain with her husband. He kept the Malaitans employed in maintenance work at Paeu and on the log stock, their irregular food supplies being supplemented by a big garden, until late 1943 when they were shipped home by the government.[103]

Shipping shortage meant the logs produced in 1941 remained at Vanikolo. Boye was relieved when the American military showed interest in purchasing the stock. An officer from the US Army Board of Economic Warfare inspected them and, after verbal consultation with the Americans based in New Caledonia, the resident commissioner requisitioned the 840 logs for military purposes in October 1943. A year later when rot had got into the sapwood, the Americans returned, but were not willing to pay the original price. An attempt to sell them to the Australian government failed for lack of shipping. When the logs were 'de-requisitioned' in late 1944, they had deteriorated, representing a loss of over £4,000 to the Company. Logging resumed briefly in 1946, but there were no exports until 1949, when contractors took over the logging in an era when the government and the people were becoming more conscious of the possibilities of the logging industry in the Protectorate.[104]

—6—
World War Two: Focus on Forests

... the history of timber is largely that of the sawmill platoon.

Third Division Historical Committee. *Pacific Pioneers: The Story of the Engineers of the New Zealand Expeditionary Force in the Pacific* (Dunedin, 1945), 43

The grievous mistakes of the Vanikoro concession ought not to be repeated.

CO 225/338/ 8655/1A: Minute, L. Robertson, 3 Aug. 1944.

World War Two had an enormous impact the Pacific islands. This was so for the Protectorate. The war changed Solomon Islanders' concept of themselves, their islands, and the rest of the world. In many areas Islanders were reluctant to accept the re-establishment of the colonial order and aspired to the material prosperity that they had sampled in the generosity of the Allied troops, particularly the Americans.

They were not alone in revising their perceptions of their islands. What, to Europeans in the Depression decade, had been the Sorrowful Islands, seemed for a brief couple of years to be once again, as in the 1900s, islands of opportunity with unlimited timber resources just waiting to be 'developed'. However, now the Colonial Office required assessment of those resources before development could begin.

The island invasion

After Rabaul fell to the Japanese in January 1942 they moved south-east from New Guinea. In May they captured an abandoned Tulagi. Soon after, they landed on north Guadalcanal and commenced building an airfield on the plains. This would have greatly extended their airstrike range into the Pacific. To counter this, the Americans landed on Guadalcanal on August 7. By February 1943, Guadalcanal and the adjacent islands were in Allied hands and the arena of battle moved to the western Solomons where the Japanese were well established. The Allies eventually secured control of the New Georgia group by October 1943, with the last of the Japanese being driven from the Shortlands and west Choiseul by early 1945.[1]

War is ravenous for all earth's products, including human beings, and it wastes them in every way. The forest was often the first casualty. Trees were felled by armies to create clearings for airstrips, camps and roads. Trees were blown up by bombs, weakened by shrapnel, bullets and mortars, and chopped down or had their crowns lopped to clear the field of fire for artillery. Near Tulagi, for artillery practice, the Americans blasted a small islet and its mangroves out of existence.[2] As well as being swallowed by the maw of destruction, trees were felled for construction purposes. As soon as the armies set foot on the Islands they needed timber. The Allies brought some with them, but as troop numbers swelled in the Solomon Islands, accommodation for men and equipment, along with bridges and air-raid shelters, had to be built and, with shipping shortages, demand outstripped supply.[3]

Demand for timber was proving a strain on the Allies' domestic economies. Unlike the United States of America, Australia and New Zealand had been at war since 1939 and substantial amounts of timber had been sent to the European theatre. Once the war entered the Pacific, the timber trade between Australia and New Zealand was severely reduced. Their imports from North America for civilian purposes largely ceased because of shipping and dollar exchange considerations. New Zealand's main supply of hardwoods from Australia was needed for the war effort elsewhere, so New Zealand's public works like bridge building were deferred. Timber control authorities in these countries curtailed civilian building to send timber to the Pacific front and to provide for 'the phenomenal increase in the use of box-making timber for packaging foodstuffs to the forward area'.[4] In New Zealand alone during the war years the people sacrificed the equivalent of 20,000 houses, or half of those that normally would have been built, for defence purposes at home and abroad. Families of mainly women and children often lived in squalid conditions as a result. Exclusive of their own camps, New Zealand had sent enough timber to build 2,200 houses into Pacific Islands camps.[5] Consequently, as the offensive against Japan in the western Solomons began in 1943, the Allied armies brought in forestry or saw-mill units consisting of sawyers and 'bushmen' from New Zealand (or 'lumberjacks' from the United States) to utilise local timbers.

World War Two: Focus on Forests

National Archives of New Zealand, Head Office, Wellington*

Figure 49. New Zealand engineers at work cutting logs for bridge and culvert construction on Vella Lavella, 1943. Platoons of the 20th Forward Company built semi-permanent bridges over the Uzama, Manavari and Joroveto rivers, using 'jungle logs' and 'coconut logs'. These replaced temporary fords and damaged bridges, built earlier in the war by the American engineers.

*WWII 7/13 C1 No. 2, Photographs–2 NZEF Personnel in Pacific Theatre – Vella Lavella, Dec. 1943– Mar. 1944. Photograph no. L 1464/5/43.

Some units were created in the field. For example, the Sawmill Platoon of the 37th Field Park Company (New Zealand) was formed in November 1943, from men in the Company itself as well as others from the New Zealand Third Division (Army). This platoon in turn, co-operated with personnel of the Royal New Zealand Air Force (RNZAF) mill already located at Kukum, Guadalcanal, making roads and getting logs out of the bush by tractor and truck.[6] In some areas, engineering units, the 'Seabees', the Construction Battalions (CBs) of the US forces and the New Zealand Engineers did their own felling and forming on the spot, particularly for bridge construction and for 'corduroy' roads through the mud.[7]

National Archives of New Zealand, Head Office, Wellington*

Figure 50. New Zealand engineers at work cutting logs for bridge and culvert construction on Vella Lavella, 1943.

*WWII 7/13 C1 No. 2, Photographs–2 NZEF Personnel in Pacific Theatre – Vella Lavella, Dec. 1943– Mar. 1944. Photograph no. L 1468/5/43.

On north Guadalcanal, although lacking sheltered anchorages, the plains became a key area for command personnel, storage of supplies and a transit station for reinforcements and the wounded. During 1943, Point Cruz to Lungga was a maze of maintenance workshops, command centres, depots and hospitals with a 135 mile network of roads reaching in the east towards the three major airstrips near Lungga and the logging areas beyond. By September 1943, the Allies had six saw-mills operating: the United States' Navy had three and the U.S. Airforce, the Service Command, and the RNZAF each had one. At this time, all except the RNZAF No. 1 mill, about a half a mile from the coast up the Mataniko river near Kukum, were located to the east of Henderson's Field (Henderson airport), where shrapnel in the trees

World War Two: Focus on Forests

was less common than in the western area where fighting had been intense. In August 1944, the New Zealanders set up another saw-mill, No. 2 mill, to the east, about five miles up the Tenaru (or Ilu) river, near the Americans who were logging on the river flats to the south of the New Zealand zone where, by the second half of 1945, they had cut out approximately 300,000 cubic feet of 'Pacific Mango' or *ailai* (*Mangifera solomonensi*s), mainly of trees two feet in trunk diameter.

The timbers were new to most loggers so the names were approximations to familiar timbers. Much of the timber cut was softwood, though 'mahogany' (probably *Dysoxylum* spp.) was used. A bridge built across the Lungga river was called the Million Pounds Bridge because of the mahogany stringers. The most common timbers in this area were *tawa* or *tauna* (*Pometia pinnata*) and *koilo* (*Calophyllum*), both suited to construction work, while the Americans used most of the *ailai* as dunnage.[8]

USA National Records and Archives Administration

Figure 51. American soldiers at work, sawing plantation coconut palms into blocks for foundations of Quonset huts, north Guadalcanal, January 1943.

Pacific Forest

Because of the limitations of equipment and personnel, various kinds of supply arrangements evolved. For example, in June 1944 the RNZAF No. 1 mill supplied timber for the Allies on Bougainville as follows: 187,411 superfeet to the RNZAF, 2,000 to the Fijian Military Forces, 2,350 to the NZ Army, 1,000 to the NZ Navy and 42,489 to the American forces.[9] When the RNZAF mills were closing down on Guadalcanal some New Zealand bushmen were destined for Los Negros, Manus where the Americans planned to give them 'a mill in working order with a block of millable timber for the use of our saw-mill personnel. Of the finished timber the RNZAF will get its share'. Other New Zealanders went to Bougainville to cut logs for an American CB saw-mill and received in return half the out-turn of the mill.[10] Later, another group, mainly from No. 2 mill, helped the Australians at Jacquinot Bay, New Britain.[11] The assistance was mutual as the American forces supplied most of the saw-mill plant. The New Zealanders, for example, had produced their first timber in 1943 using a captured Japanese 'tank' to

National Archives of New Zealand, Head Office, Wellington*

Figure 52. Captured Japanese tractor, 6-ton artillery, full-track, type 98 (Izusu). Fitted with 5-ton winch, usually used as a prime mover for howitzers and field guns, being refitted to power Royal New Zealand Air Force No.1 saw-mill at Kukum, north Guadalcanal, 1943.

* Air 118/64, No. 13, Photograph Album – Guadalcanal, Bougainville, 1943–1945, Photograph no. 1287.

World War Two: Focus on Forests

power the mill at Kukum.[12] As the RNZAF saw-mill (No. 1 Islands Works Squadron), Guadalcanal was packed for shipment to Manus in June 1945, the Senior Works Engineer noted,

> Approximately 70% of the equipment was supplied from New Zealand and say 30% ex-American sources in the field, e.g. a D7 tractor, Pacific Saw Bench and Hyster Logging Arch held in Custody receipt for approximately 8 months. (These have been returned now to US Forces.) In the final analysis practically all of the mill equipment could be deemed to be of American origin – even that procured in New Zealand for use here. The SWE [Senior Works Engineer] states without any hesitation or reservation that without the active co-operation of the US personnel and repair facilities the amount produced by RNZAF Mills would have been practically nil.[13]

What the New Zealanders had lacked in equipment they had made up for in production.[14] From mid-1943, the weekly output at the Guadalcanal mills was about 200,000–250,000 superfeet with the New Zealand mill

National Archives of New Zealand, Head Office, Wellington*

Figure 53. Loading logs cut from north coast of Guadalcanal, by saw-milling unit, RNZAF, 1944.

* WAII, 1 DAZ 130/1/18, war Diary HQ 3 Div. Enmgs Mar., 1944, Appendix 1 – 37th Field Park Co. NZ, official Force photographer of Unit Activities and Groups of Personnel 5–6 Feb., 1944, Photograph no. B 1910/5/44.

National Archives of New Zealand, Head Office, Wellington*

Figure 54. Logs arriving at one of RNZAF mills, north Guadalcanal, 1943. The truck is GMC 2.5–5 ton, American make.

*Air 118/64, no. 13, Photograph Album – Guadalcanal, Bougainville, 1943–1945, Photograph no. 1288.

producing about a third to a half of this.[15] Much of the timber for the New Zealand mill (No. 1) near Kukum came from the area between Mt Austen and the coast. Of a conservative estimate of nine million superfeet (27,017 cubic metres) cut from northern Guadalcanal from mid-1943 to early 1945, the New Zealanders cut over 4,111,641 superfeet. Most of this came out of the No. 1 saw-mill, producing about 2,635,749 superfeet, with No. 2 at Tenaru, running from August 1944 to March 1945, producing 1,475,892 superfeet.[16] Between June 1943 and January 1944, the two saw-mills of the US 61st Construction Battalion, located at Ilu, about three and a half

World War Two: Focus on Forests

National Archives of New Zealand, Head Office, Wellington*

Figure 55. Sawing logs at RNZAF mill, Guadalcanal, 1943.

*Air 118/64, no. 13, Photograph Album – Guadalcanal, Bougainville, 1943–1945, Photograph no. 2057.

miles inland from the Tenaru river mouth, produced more than 1,000,000 superfeet. The output of the saw-mills of the 46th and 26th Battalions on Guadalcanal is unknown. Similarly, in 1943 the Seabees, 27th Battalion, cut hardwoods on Nggela for wharf and harbour facilities at Sesepe, Tulagi for a PT-boat base, at the seaplane base at Halavo, and at the landing craft repair bases at 'Carter City' and 'Turner City' on mainland Nggela (Map 5).[17] At the American base on the Russell Islands, in late 1943, the Marine engineers of the 4th Base Depot, under the forester, Major Donald Burdick, logged and milled local timbers. Many of the engineers were black troops from the southern United States so 'were suited to logging in the Tropics'. The saw-mills turned out 5,000–7,000 superfeet daily of timbers including *Calophyllum* spp., *Vitex cofassus* (*vata*), *Mastixiodendron smithii* (*mumu*), *Albizzia falcata* (*fai*), and *Chryosophyllum roxburghii* (*ghaimelo*).[18]

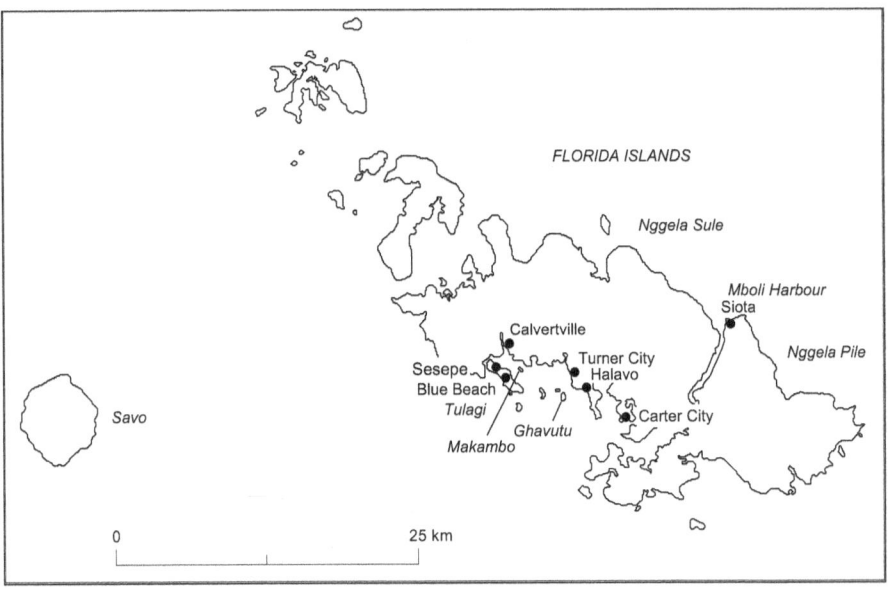

Map 5. Savo and Florida, with some wartime bases.

Mobile saw-mills were the choice of the Allies. The alternative had been to centralise on Guadalcanal and ship timber to where it was required. This was attempted in 1943/4, but the shortage of shipping meant delays, so saw-mills were located nearer construction sites. For example, as construction of the American camps expanded in the south New Georgia area, the New Zealanders set up a saw-mill at Arundel (Kohinggo), on the west side of Noro passage in November 1943. By 22 May 1944, it had produced 800,844 superfeet. No longer needed, it was shipped back to Guadalcanal where it became the Tenaru (No. 2) saw-mill. Two American mills were moved to the New Georgia area in mid-1943.[19] Mills on Guadalcanal had ceased military production by March 1945, as others had been set up closer to the forward area in New Guinea.

The Americans had several saw-mills in the New Georgia area in late 1944. There was one on Roviana Island at Gega, and another at Koloboa, north-east of Munda. On Vella Lavella, there were two: one at Dolebei, near Barakoma, and another at Lambulambu. A small one at Seghe cut timber for the camp buildings. Earlier, during September and October 1943, the US

World War Two: Focus on Forests

Navy also cut about 20,000 superfeet in logs at Seghe, Mborukua Island, and Titirona and towed them by sea to the abandoned Seventh Day Adventist saw-mill at Mbatuna.

After Japan had invaded the Solomons, a skeleton British administration went into hiding, as coast-watchers in the BSIP Defence Force. It did what it could to govern and to organise the Solomon Islands Labour Corps (SILC) in which Solomon Islanders assisted with stevedoring, carrying behind the lines, and basic construction work. The Defence Force, headed by the Resident Commissioner, worked closely with the Allied Command on Guadalcanal. So, when the Americans left the Mbatuna area the government, represented by Martin Clemens, requisitioned the saw-mill. He had difficulty getting timber, as pre-war logging had cut out the big trees close to the water within a distance of five miles of Mbatuna. Clemens relied on the man-power of paid local workers to cut, carry, and canoe-haul logs to the mill. While the government ran the Mbatuna mill from January to May 1944, under the supervision of the Chinese shipwrights, Chan Wah and Chan Yiu, it turned out 18,886 superfeet. About a third of this timber went into temporary headquarters at Hombu Hombu, Roviana lagoon and the rest was used for the Gizo government buildings.[20]

The government also ran the mill at the Catholic mission at Mbuma, Malaita. A former Vanikolo employee, Lieut R. A. Dethridge of the SILC was in charge of rebuilding the mill early in 1945. Demand for sawn timber was certain as the war drew to a close. Following the re-establishment and extension of local government councils, the council courts would hear minor criminal cases and fine the guilty. So, district administration would lose the labour of prisoners who had built local-material buildings at the stations. Leaf buildings were to give way to buildings of milled timber. Dethridge employed some of Monckton's ex-employees and used prison labour.[21] He persuaded the government to pay more than the going wage (one pound a month plus rations which were valued at two pounds) which the SILC received since the 'SILC do very little real work … natives are inclined to shy at saw- mills which will be real hard work'. His Malaitan 'boss boy' Isofiala was paid five pounds a month, a high wage for those times.[22] The mill started producing in July 1945 and by November 1946 it had milled in excess of 263,994 superfeet of good quality timber, mainly *koila* (*Calophyllum kajewskii*), from the stands in the freshwater swamp. In September 1946, the administration signed an agreement with the mission, allowing timber cutting to continue in return for one tenth of the mill's out-turn and a rent of equipment at £100 a year, paid to the mission. This arrangement continued with Chan Wing as mill superintendent until early 1948 when the mission purchased the government's saw-mill for £1,000.[23]

Pacific Forest

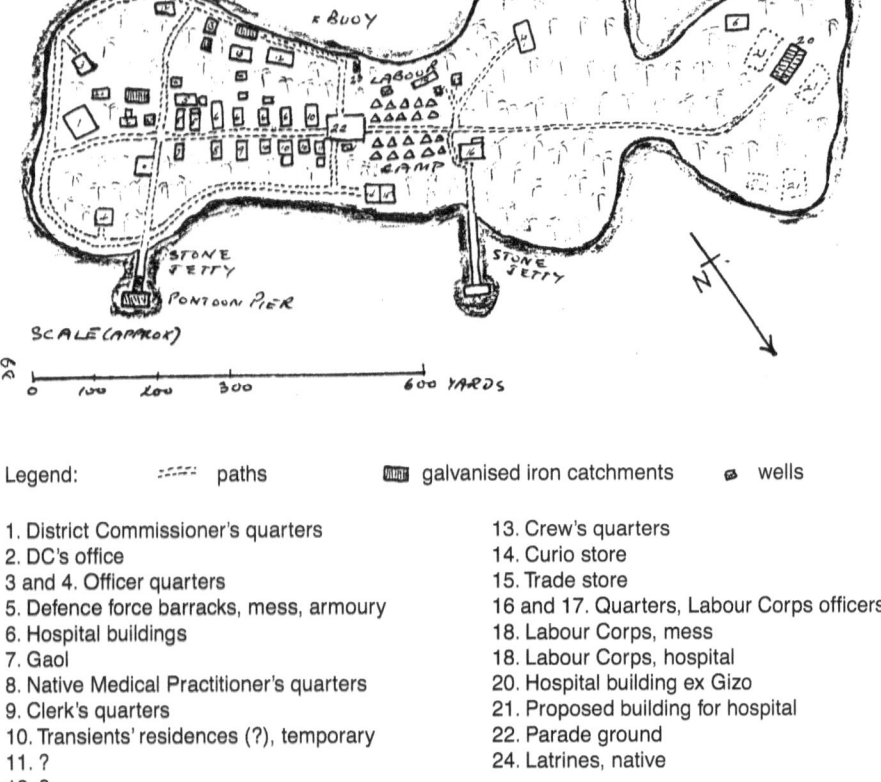

Legend: ┈┈ paths   ▦ galvanised iron catchments   ⊚ wells

1. District Commissioner's quarters
2. DC's office
3 and 4. Officer quarters
5. Defence force barracks, mess, armoury
6. Hospital buildings
7. Gaol
8. Native Medical Practitioner's quarters
9. Clerk's quarters
10. Transients' residences (?), temporary
11. ?
12. ?
13. Crew's quarters
14. Curio store
15. Trade store
16 and 17. Quarters, Labour Corps officers
18. Labour Corps, mess
18. Labour Corps, hospital
20. Hospital building ex Gizo
21. Proposed building for hospital
22. Parade ground
24. Latrines, native

Figure 56. Sketch map of Hombu Hombu, c. 1944 (Enclosure, BSIP 1/III/14/19).

Most military mills had closed down by mid-1945 when shipping problems had become less acute and the timber of north Guadalcanal more accessible. A visiting colonial forester noticed that, from the Ngalimbiu area,

> a considerable number of logs are being transported to the Kukum landing beach, ostensibly for shipping abroad. Some three loads a day are rumoured to be going out in this way. Species include Pometia pinnata (Island ceder), Terminalia sp., Palquium sp., Canarium grandistipulum (nut trees) and others ... This removal of logs would hardly appear to be a necessary military requirement nor classifiable as war damage.[24]

Preceded by samples, this 'furniture' timber was going on order to buyers in Honolulu.[25]

World War Two: Focus on Forests

The American commander ordered this to be stopped, but as these last logs were being cut as the war ended, it highlighted the costs to the people and the forest. Although many loggers were trained in their craft, the exigencies of wartime evaporated any considerations of conservation. Operations had been carried out in

> anything but a workman-like manner and consequently have resulted in a good deal of waste ... over the areas first worked only certain species have been removed, leaving quantities of good timber remaining, but which would not be a commercial proposition ... the mango species now being worked has been very badly handled ... trees have been felled in all directions without regard to the destroying of trees to be left behind, the result being considerable waste.[26]

On north Guadalcanal, most of the plantations destroyed by the fighting belonged to Europeans. Some land was native or customary land which the people left as fighting started. War damage in the Protectorate amounted to millions of pounds. Neither companies nor local people received any direct compensation. A broken Japan could not pay, although a token of overseas assets went into the post-war government Trade Scheme, preventing total collapse of commerce.[27] At the time Solomon Islanders were silent on this, although claims have been and continue to be made for compensation for lives lost.[28] There are a number of possible explanations for their silence. In former times, conquest by occupancy simply had to be accepted. Clearing the land was a sign of possession and power, thus only the foolish would quarrel with great armies. Fear was not the only motivation: islanders viewed the Allies, especially the Americans, as benefactors whose generosity put the British to shame. For Melanesians, permitting logging was reciprocity in action, and the honourable way to treat such friends. Moreover, as the Americans pulled out they left a considerable amount of material that professional salvagers did not bother about. Miles of piping, acres of Marston matting on airfields, corrugated iron, and drums sat beside the crippled aircraft and abandoned wharves. This filtered back to villages and some of it can still be seen there. During the war little timber was taken to make money out of it; most was being used for shelter in the Solomons or nearby. Its use fitted the Melanesian subsistence ethic. As the government sought timber for re-building in 1945 and early 1946 it sought permission of the landholders, but most did not ask for payment.[29]

The war was like an extended cyclone. It came like a fury, causing upheaval. Then the armies departed and only the rusting, abandoned wrecks and Marston matting reminded people of the tumult of battle and the busy routines of the airfields and camps. The forests were stripped in a few places, but natural regeneration soon closed the gaps and covered most of the eroded paths of tractors and trucks. It was only where the forest was not able or not permitted to re-assert itself that cleared areas remain. On Guadalcanal, between the Mataniko river and Kukum and south to Mt Austen, post-war urbanisation has meant that the forest is gone. Where extensive earth-moving occurred in geologically susceptible places, such as at Munda, New Georgia where the Japanese then the Americans bulldozed away the thin layer of soil and graded and filled the sub-stratum of up-raised coral to make airfields, the land and vegetation still have not fully recovered. About 11 square kilometres in the archipelago fall into this category.[30]

Maasina Rulu

Wartime conditions in part stimulated Maasina Rulu or Marching Rule, a movement in which the Malaitan peoples sought self-determination and acknowledgment of cultural values. They demanded higher plantation wages, refused to pay taxes and resisted the government. By December 1946, this movement affected parts of Guadalcanal, Makira, Ulawa and for a time eastern Santa Isabel, as well as most of Malaita. Maasina Rulu was both cause and result of a more critical attitude to the colonial government, reinforced by contact with the generous Americans and their political ideas, but stemming too from an old dislike of elements of the colonial regime, and from pre-war movements which wanted more recognition of Melanesian culture and economic worth.[31] Some realised that the 'development' demonstrated in clearing the forest in which Europeans had taken so much pride, had been underwritten by the labour of Solomon Islanders: 'We make the paddock for them, then it bears fruit, they make copra, they take the money; they don't give us good money or good wages'.[32]

Maasina Rulu advocated island-based self-government. It levied a tax, symbolising both rejection of the British and an intention to finance its programme. Malaita had never been a major producer of island products – copra, trochus and pearl-shell. Its cash came from its men's wage labour on plantations, just as it did when Malaitans worked in the SILC. Taxes needed money and the 'dollar invasion' of free-spending Americans was over. The young leaders of Maasina Rulu in 1946 called people to leave their scattered inland hamlets, migrate to the coast, 'to clear away the bush', build 'towns' and large gardens so that they could grow cash crops of the sweet potato, and to be where they could 'live well'.[33] Initially, in mid-1945, the administration was supportive:

> Throughout Malaita there is evidence of a revival of interest on the part of the people ... in the management of their own affairs. A growing sense of alertness and awareness has prompted this hitherto parochially-minded and reclusive people to contemplate the building of communal villages. In some areas this coagulation of communities and rebuilding of villages has already begun. Likewise as keen interest is being directed towards exploring the possibilities of communal farming ...
>
> The possible results of this on native timber requirements are as yet hard to gauge, but in Ariari [sic] at least plans seem to envisage a year of garden-building and clearing of sites for new villages. This would be followed by another year of village-building and extension of communal gardens. An abnormally large native demand on timber resources therefore may well be experienced during the next five years.[34]

In some places paths were built and wharves constructed as well. Malaitans had the model of the huge military 'towns'. And there was the precedent of the administration urging them to consolidate settlements on the coast, as had most Christian missions. The Americans had met their need for fresh vegetables by establishing the largest military farm in the South Pacific on 1,800 acres at Ilu in 1943–1944, employing 240 Melanesians as well as a high level of mechanisation. Requiring generous dressings of fertilisers, it produced vast quantities of vegetables at large capital cost. The economics would have escaped the Melanesians, but the size of the gardens further convinced them that America was powerful and wealthy.

The war and the largesse of the Americans reawakened the 'great vague hopes' of the Malaitans. Within Maasina Rulu those hopes burgeoned, and it seemed as if the Melanesians involved were about to take responsibility for the kind of development they had craved. This was demonstrated in their efforts to turn significant areas of lowland forest into gardens for cash and towns of up to 500–600 people.[35]

Yet there was ambivalence in this movement. The Malaitans wanted to be culturally and politically independent, to live according to their tradition (including more recent tradition, Christianity) or 'Custom' *(Kastom)*. But they also wanted support from their most recent benefactors, the Americans, as they tried to be free of the British. There was no talk now of reciprocity; simply a desire to tap the source of the greatest power they had ever experienced to obtain material improvement. In light of the lack of basic infrastructure and adminstration, the British viewed the wish for a dependant alliance of the Malaitans with the USA as particularly inappropriate at this time. However, it also demonstrated a desire for change which could be turned to British purposes. The Malaitans too believed they could press the advantage of their new-found unity.

USA National Archives and Record Administration

Figure 57. War's ways of clearing land. Tunnel entrance under Kokenggolo hill in the centre of Munda's wartime airfield. A dead Japanese lies at one tunnel entrance.

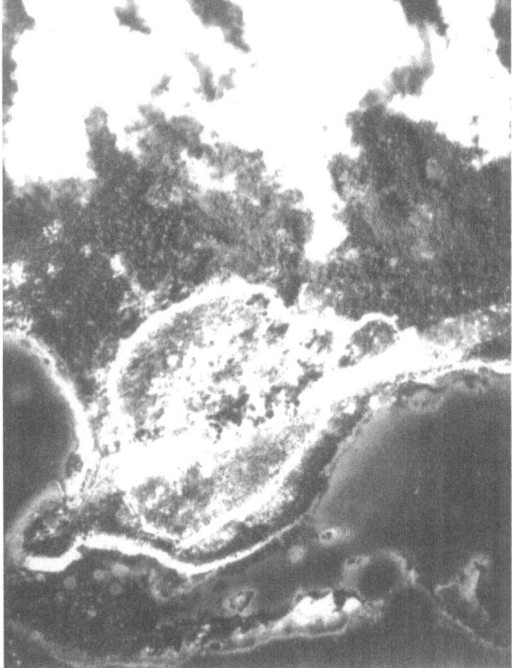

Figure 58. American bombs exploding on Japanese-held Munda airfield, 1943. Kokenggolo hill is inland from Munda point, at lower left of the upper half of the irregular semi-circular road. Nusuzonga is at bottom of photograph.

USA National Archives and Record Administration

World War Two: Focus on Forests

USA Natioanl Archives and Record Administration

Figure 59. Munda airfield, looking eastwards. Vegetation is beginning to encroach on the airfield by 1946, when this photograph was taken. Kokenggolo is to the centre left.

USA National Archives and Record Administration

Figure 60. American Seabees repairing Munda airfield after bombing, August 1943.

USA National Archives and Record Administration

Figure 61. Lambete landing east of Munda field, September 1944. Note bomb craters and exposed uplifted limestone. Considerable waterfront alteration is noticeable. Compare these wartime photographs with Methodist mission clearing of forest, Figures 9–13.

Five Australian and British warships visited Malaitan waters in 1947, the last demonstration of colonial sabre-rattling in the Protectorate. On Malaita, Makira and parts of Guadalcanal, the government arrested and imprisoned the leaders of the Maasina Rulu movement in 1949, when it became more militant and sometimes coercive. By mid-1950, the High Commissioner had made several concessions and reached an understanding with them, though this was more an outcome of changes in colonial policy than of Maasina Rulu resistance, despite the coincidence in timing. Meanwhile, without their leaders, most of the big Maasina Rulu villages collapsed in 1949/50 and their inhabitants dispersed to inland areas. The vegetation soon reclaimed the ruins and the gardens. From 1950, the government began to seek the means by which Malaitans and other Solomon Islanders could be aided in developing themselves and their resources.[36]

World War Two: Focus on Forests

The quest for timber concessions

Military timber men recognised the worth of many Solomons timbers. Even before the war ended the New Zealand government investigated the Protectorate's forests. It was seeking timber because of decades of careless exploitation and wartime logging of its own forests and the anticipated demand when service-people returned to civilian life.[37] The New Zealand government, with the permission of the Commander, South Pacific Area and the British Resident Commissioner, initiated its survey under forester C. T. Sando in late 1943.[38] From December 1943 to April 1944, J. Billings and K. B. McCrae surveyed selected areas in Guadalcanal, Malaita, Santa Isabel and New Georgia until their health failed. Sando briefly returned to the Protectorate in September 1945 and collected specimens, as had Billings and McCrea, for classification and testing, particularly for use as veneers.[39]

Even in wartime, Britain had no need of 'colonial timber'; for reasons of distance, cost and lack of a regular supply.[40] Adhering to its 1930s policy, the Colonial Office, aware of the growing interest in Solomons' timbers, sounded a note of caution as early as October 1944,

> Considerations of any proposals for the commercial exploitation of the Forest Resources of these areas should be deferred until the report of the reconnaissance survey has been examined and a forest officer can be appointed to ensure that any exploitation is carried out on a systematic basis … the future forest policy will need to be based on the broadest grounds of the long-term interests of the Protectorate.[41]

The war and the Atlantic Charter had forced the British government to review its colonial policy. Moreover, when a Labour government came to power in Britain after the war, colonies became less fashionable. A countdown to self-government and eventually independence started in the bigger colonies. Although the Western Pacific was not to be buffeted by the 'winds of change' that were beginning to gather velocity elsewhere, it could not avoid the occasional light air, even if only from the self-righteous huffs and puffs of Uncle Sam. Britain saw no likelihood of independence for the tiny Solomons in the immediate future, but amid scarcity of personnel and plant, put in place policies that aimed at creating an economic base for future developments.[42] Regarding forestry, the Colonial Office was not going to allow more mistakes like Vanikolo. It funded F. S. Walker, from the Malayan forestry department, who arrived in June 1945, and it also provided finance under the Colonial Development and Welfare Act (1940) for a forest survey, at last realising some of the recommendations Troup had made in 1932.

Informal inquiries into logging possibilities seem to have been made during the war. American news-magazines told of stands of mahogany, rosewood and teak and military-related publications on timber species aroused interest.[43] However, the first official request came not from newcomers to the Protectorate, but from those who knew something of the timbers, but had no market before the war. In April 1945, Fred Jones, trader at Vanikolo, Lieut R. A. Dethridge, who had worked at Vanikolo and was at Mbuma, Lieut Fred Archer, planter on Buka and Bougainville for 22 years, and Lieut J. Claiborne of US Ordnance applied for logging rights for the Dethridge Timber Syndicate on Santa Cruz. They proposed a scheme whereby the people would be allocated shares in lieu of payment for timber rights. The government was exceedingly sceptical of a share issue unless fully backed, as a company could declare a no-dividend and increase the salaries of the expatriates. Share-holding and transmission were not covered by Protectorate legislation.

There were others interested in logging. In May 1945, a New Zealand ex-serviceman applied for a concession in Beaufort Bay, west Guadalcanal; in June there was an inquiry regarding Tevai; in August, an ex-Marine wanted to cut hardwood on Vella Lavella and offered to replant; in March 1946, former New Zealand airmen expressed interest in logging New Georgia; the same month saw a request for a concession on Savo.[44]

One entrepreneur who seemed to have the ear of the government was Captain W. E. Goodsir, the manager of the Fiji Kauri Timber Company, a man of some influence in the timber industry there and apparently with the Melbourne timber trade.[45] He was the 'commercial representative' and an expert on the technicalities of logging for Walker. Goodsir's time with Walker had given him an opportunity to assess timber potential. He wanted to log *dafo* (*Terminalia brassii*) on Malaita in a swampy, uninhabited area at Osi Lagoon. Though the Colonial office valued his advice, it restricted his activities. Moreover, the Resident Commissioner and Walker were concerned that Goodsir's operation would extend to *dafo* south of Auki among village settlements. They were suspicious of his attempts to persuade them to ease restrictions on Walker's proposed extraction techniques.[46] Reflecting the Colonial Office, Walker was not going to be persuaded:

> Virgin forest will not depreciate greatly in volume, and may appreciate in value, if it is necessary to wait for increased demand to make restrictions acceptable.[47]

Postponement of the logging was likely to be long-term because by February 1946, the Malaitans around the Osi Lagoon area were involved in Maasina Rulu and did not want to negotiate with any European.[48] When later asked by the High Commissioner (and Governor of Fiji) in April if he would be prepared to let Goodsir log the *dafo*, Walker agreed, providing the stand was delineated carefully and that the forest department was a reality.⁴⁹ Goodsir could not wait that long.

Similarly with the Dethridge Syndicate, Walker postponed the granting of a concession to Santa Cruz. Fred Jones, one of the Syndicate had a strong case as he had applied for the concession in 1938, and had asked the Resident Commissioner in 1940 to give him six months' grace while he brought in an Australian expert to assess prospects, but the war had intervened. Although Jones had no binding agreement he was first in line. Walker's survey of the area indicated about 555,000 cubic feet of kauri, but he wanted both Santa Cruz and Tevai to be reserved from logging until plans for regeneration were formulated, and he convinced the government of this.⁵⁰ So the Dethridge Syndicate failed.

Walker advised against logging until the survey had been completed and warned that there was a danger in selecting one or two species and leaving the rest unlogged. Moreover, he rightly predicted that the immediate post-war demand for timber in Australia and New Zealand would be short-lived. In the light of a skilled labour shortage, Walker thought logging might prove more expensive than was thought, as indeed the Vanikoro Kauri Timber Company's record had proved.⁵¹ He believed the existing timber legislation was too limited, giving the government little control, and recommended a consolidated regulation. A forest department must precede logging.⁵² All would-be loggers received the same answer: 'The question of timber concessions must await the formulation of a forest policy as a result of the present survey of forest resources'.⁵³ High Commissioner Grantham in Fiji made it clear that the proposed forestry department was to control any loggers.⁵⁴

Rather than fritter away a potential sustainable resource, the three tiers of British rule were prepared to wait until the Protectorate had a plan for forestry development and control, despite the fact that the economy was struggling. This decision had its roots in the Vanikolo lesson of 1932 and not in Maasinu Rulu unrest. The largest timber supplies definitely were not on Malaita or Makira; in fact Walker found that Malaita had only a few 'patches' of productive forest, including the Osi Lagoon; Makira had none worth exploiting. Guadalcanal did have about ten per cent of the forests, but the western part of the island which had some extensive stands had never been a Maasina Rulu stronghold. The finest forests in the Protectorate, amounting to an estimated 75 per cent of the total, were in the western Solomons, a region unimpressed by the Malaitan movement.⁵⁵

Disciples of India

F. S. Walker had come to the Protectorate with plans based on the setting aside of large areas of forest land for sustainable extraction and replanting with valuable species under government control. He had worked in Malaya where almost 20 per cent of the land area was reserved forest.[56] This system was the offspring of British experience in its older colonies, for Britain had failed to establish sustainable forestry at home and, from the seventeenth century, sought colonies, in part, to met its timber needs.[57]

There were also wider imperatives of political economy. Deforestation in the wake of European plantation agriculture in the Canary islands and Madeira after 1300 and in the West Indies from 1560 had exposed them to erosion. Theophratus of Erasia had warned of the effects of deforestation on erosion, soil loss and climate in the fifth century BC in Greece, about the time the Tikopians, as well as the Greeks, were transforming their respective environments, but it was not until 1483 during the Renaissance that Theophratus's writings became more widely known. In eighteenth-century Europe, the growth of the sciences, especially botany and geology, raised awareness of the process of species extinction, which Darwin's *Origin of Species* endorsed in 1859. Scientists were interested in species depletion, soil degradation and decline in rainfall as subjects of intellectual curiosity, but these had social consequences, as their medical colleagues were quick to point out. Administrators were concerned about practical ramifications: their own economic loss and the unrest among subject peoples suffering because of the destruction of the means of existence. On Mauritius, French colonial scientists experimented in 'systematic forest conservation, pollution control and fisheries protection'. These conservationist ideas were transported to the British East India Company's colony of St Helena and on to India where, under the British Raj, professional forestry emerged by the late 1840s.[58]

Since medieval times in Europe, the landed classes had employed their own foresters. By the early nineteenth century, their knowledge had begun to be systematised and taught at universities in Prussia and France, countries that did not have vast colonial sources of timber and feared wood shortage. The earliest American and European foresters were educated at these institutions. The Germans also influenced both British and Dutch colonial thinking, providing the men to set up scientific forestry and forestry departments.[59]

In Burma and India, the British employed German foresters, such as Dietrich Brandis and his successor, William Schlich, to establish the forestry service in the 1850s and 1860s. Their students and colleagues in the Indian

World War Two: Focus on Forests

civil service fed ideas back into Britain, as did Schlich when he left India to establish forestry studies at the Royal Indian Engineering College at Coopers Hill in 1885. After the College closed in 1905 both Schlich and the forestry course went to Oxford University.[60] Though its practicality was denigrated by academics of the older disciplines, Britain's colonial governments, along with most of Europe's, were finding professional foresty interesting, in the light of its financial success in India.[61] By 1922, products of Oxford University and the Indian Forestry Department were leading contributors to *Empire Forestry* which, as the journal of the Empire Forestry Association, had a seminal influence in Britain's colonies.[62] 'In forest matters, the Colonies may well be regarded as disciples of India'.[63]

Walker's intellectual forefathers came from this background. When Schlich died, R. S. Troup, a colleague from Indian days, took his place at Oxford in 1920 until his own death in 1939. Troup was one of the leading thinkers on forestry in the British empire and his influence had reached as far as Vanikolo and Fiji in the early 1930s.[64] He emphasised the long-term economic significance of forests, advocated the establishment of reserved forests through legislation, and developed a method of and rationale for forestry and its administration. Part of the rationale for scientific forestry was that it had to make a financial return to government, albeit for the 'public good'. Troup's constituency by the late 1930s was considerable:

> Britain's Colonial Empire comprises more than fifty administrative units, with a population of nearly 50,000,000. The principle of trusteeship for its peoples applies to the Colonies and Protectorates as well as to the Mandated Territories. Forest conservation is an important part of this trusteeship, for it affects the welfare not only of existing populations but, even more, of generations unborn. The value of forests ... lies not only on the produce which they furnish but also in their protective influence, which may be a matter of vital importance to the welfare of a country. The total area of the Colonial Empire is about 2,000,000 square miles, while the forest area is estimated at over 600,000 square miles, but of this under 9 per cent has as yet been permanently reserved as forest. Whereas in some Dependencies forest reservation is well advanced, in others it is very backward, and there is much leeway to make up before a condition of safety is reached. The economic development of the forest resources of the Colonial Empire is another direction in which progress may be expected in the future.[65]

When Walker's *The Forests of the Solomon Islands* appeared in 1948 it was a practical application of the principles enunciated above and throughout Troup's *Colonial Forest Administration* of 1940. Its ancestry was not only in the protection of forests for commercial purposes, but also, if more distantly, in conservation to obviate social suffering and disorder resulting from the hydrological consequences of tropical forest degradation.

The forest survey: suppositions and recommendations

As requests for timber concessions poured into the administration, Walker had to come up with recommendations. He hastily produced a preliminary survey in February 1946, before most of the data had been assembled, to provide some basis for addressing these. The preliminary report revealed several assumptions which were to conflict with the findings of the district commissioners and much of the thinking of Walker's superiors. The final report of 1948 was based on more comprehensive investigations. Besides studying existing partial forest surveys Walker, under Colonial Office direction, had asked the district commissioners for information on: population distribution, trends and size of families; native names for timber; number of canoes and rate of replacement; reputed life of various timbers; number of buildings and amount of timber normally used in house construction; size of area under cultivation per family and number of years before new bush is cleared; possibilities of persuading natives to clear only secondary growth for gardens; views regarding the participation as share-holders by natives in timber firms and compensation to natives in regard to ownership; and native land tenure and how land-use rights were obtained. The survey was not only of forest with commercial logging potential; it was also an attempt to assess the subsistence value of the forest and the likely needs of the people in the future – a major Colonial Office consideration.[66] The published report concentrated on indigenous use of the forests relative to gardening and timber, and only peripherally on the political question of who had rights to it.

Walker had initially believed that the population was decreasing and supported the 'Appropriation of natural forest resources to the Crown as trustees for the public, much in the manner of mineral resources'. The Mining Regulation No. 15 of 1940 had deemed all minerals in or under the ground in the Protectorate to be the Crown's, though no Solomon Islanders knew of this.[67] In 1939, those people whose land was being mined had complained that it did them no good, so such a regulation would almost certainly have been opposed.[68] Walker admitted that Crown reservation 'might meet with difficulties', but thought education as to benefits would persuade most. Failing that, the logging firm which had the concession could make an ex-gratia lump sum payment to land claimants.[69] Yet, as one district commissioner warned, the issue of dormant rights was sure to emerge:

> over much of the ground which will come within the scope of forestry development there is in existence substantial claims of ownership by the natives. That in some areas that ownership has not been made manifest in a more definite ... manner is simply because there has not been any real need for it to be so.[70]

World War Two: Focus on Forests

When the survey results from Malaita came in, the situation was put forcefully:

> The Community recognises an actual ownership vested in the 'line' of all ground and trees found on that ground. In cases where this [Government] Station requires timber, trees cannot be cut on undeveloped land without first obtaining permission of the 'line' owning the ground. Many instances can be quoted of individuals paying up to 5 pounds for a single special tree for canoe building.
>
> The natives do not recognise any Crown ownership of land and would resent such a suggestion. If a European firm desires to exploit timber on Malaita it can only be suggested that the timber desired be the subject of a direct deal between the firm and the 'line' concerned.[71]

Walker's superiors contested his notion of an ex-gratia payment. All payment had to be of right, as native ownership of trees had always been recognised; 'to deny ownership now by assuming the right of appropriation would be both inequitable and unfortunate politically'.[72] High Commissioner Grantham added, 'I go further and say it would be immoral'. Grantham disputed population decrease. For that reason too there had to be a forestry department, because the forest would come under more pressure as population grew.[73] Because reservation was necessary to protect replanting Walker modified his recommendations so that if education failed and the reservation of land under the Crown proved politically unacceptable payment would be made to purchase land, compulsorily if need be.[74]

Although Walker conceded a slight increase in the population, it was so small that he favoured immigration of 'a more energetic race' to boost labour, which would 'result in an improved local market for forest products'.[75] The Colonial Office continued to reject immigrant labour, seeing what political consequences this was having in Fiji.[76]

There was another way the future forest department could proceed. Fiji had recently set up a Native Land Trust Board to control commercial operations on native lands. Many administrators, including Resident Commissioner Noel saw such a board as a solution to the difficulties with native land under an emerging 'development' scenario.[77] Walker perceived native land rights as being in communal ownership, so adopted Noel's suggestion of a Trust Board as an effective substitute for the Crown.[78] Communal ownership, in Walker's view, had to be fostered in order to make the pill of the Trust Board more palatable. He explained,

> clans could be permitted to derive direct benefit only from those lands which they are actually using and controlling. In this sense should be excluded those areas not needed for the actual garden plots of the community. All the remaining 'unoccupied' land might be administered by a Trust Board for the common benefit of all natives residing in a particular island or island group in which the land were situated.[79]

This Trust Board, as in Fiji, would be the body to give permission for the future conservator of forests to declare any area a forest reserve, prior to any logging. Following this, the procedure was to be (i) exploration, enumeration, demarcation and mapping; (ii) consultation with natives to explain benefits to expect; (iii) preparation of draft working plan and statement of licence or agreement conditions; (iv) distribution to firms interested, requiring tenders for premium by a certain date.

Walker also broadened the scope by allowing for logging on 'native' land with the royalty going direct to the rights-holders rather than to the government in the first instance, as was the case with royalties derived from the Forest Reserves. Presumably the loggers could negotiate directly with the people concerned.[80]

The report recognised shifting cultivation as an aspect of forest use. Walker acknowledged, 'Agriculturally the practice is effective, and must be considered the most suitable for local conditions, until alternatives are evolved.'[81] However, like many other colonial foresters, he believed that agricultural officers ought to work to find 'some other rotation for the present "bush fallow"'.[82] Whatever, 'a sound land policy' was the basis of any forestry enterprise, as indeed it was for agriculture. The Trust Board concept led inevitably to the need to define who had what interests in what land and whether the principles underlying them were universal. Once this was known the land could be registered and recorded in the names of the 'owners' or rights-holders within their respective 'tribal' units after the Fijian model. This project was beyond Walker's brief.

Even before Walker's final report was completed, Noel in January 1946 advocated a lands commission to inquire into native tenure.[83] Thus, almost all that Walker had recommended was laid aside until the commission delivered its report.[84] In September 1947 the first steps were taken, but momentum flagged in the face of resistance from Maasina Rulu. Sending an officer to probe land matters in Malaita, Makira or parts of Guadalcanal would have been unwise, so the government postponed the commission until the unrest subsided. In 1951 it applied for British funding, which was granted, but the Special Lands Commissioner, Colin Allan, did not take up the post until 1953.[85] Of Walker's recommendations, the only ones realised before this were the establishment of the Forestry Department with a staff of three in 1952 and the beginning of silvicultural research.[86] The rest of his recommendations awaited the completion of the lands commission and legislation that would be compatible with the forestry legislation.

Walker's report was not solely concerned with policy and legislation. Its value was that it attempted to place botanical and utilisation data within a socio-economic context, reflecting a colonial government supportive of the subsistence value of the forest, and of utilitarian conservation. His botanical collection included over 300 specimens, primarily of species of

World War Two: Focus on Forests

'particular scientific or economic' importance.[87] He made meticulous records of his methods of survey, collection and preparation of specimens, as well as careful descriptions of the species. Walker recorded a large number of indigenous names for the trees, no mean feat in an archipelago of more than eighty languages. Although his report was a milestone for the Protectorate and remained a standard reference well into the 1960s, it provided only a notional estimate for commercial operations. Loggers or later foresters would have to do their own site surveys.

The war had spotlighted the forests. From 1943 to the end of 1946 when reconstruction began, they produced a conservative 10,000,000 to 11,000,000 superficial feet – about 8,200 cubic metres a year, more than had ever been exported in any preceding year. The exigencies of war had countenanced no moratorium on logging. Walker's report outlined what had fallen under the beam of the war's spotlight, describing in part the potential of the forests. However, his policy recommendations, though modified on advice from government personnel, reflected a lack of knowledge of the Protectorate's history. Walker missed the lesson of the 'Waste lands', which were notionally Crown lands as the government could assign them to others for 999 years. Solomon Islanders had won back half of these in the 1920s and remained dissatisfied about the rest. Maasinu Rulu too, though not focused on land matters, signalled that many Melanesians resented cultural imperialism. Any future government demand on land was bound to cause controversy.

Perhaps Walker was misled by the ease with which the Mining Regulation had slipped into the Protectorate's law books. The fact that virtually no Solomon Islander knew of it explained the lack of reaction. However, Walker did shift on the appropriation of land for forest reserves, by recommending the establishment of a Native Land Trust and payment, if unavoidable, for land. Depending on its statutory basis, it could represent indigenous interests in the face of government expediency. Yet the trust concept could lead back to what Walker wanted to avoid, costly surveys of tiny patches of land to register the several rights of the claimants involved. The next step was the lands commission.

Walker's report provided the new forestry department with its structure, schedules of royalties, regimes for loggers and silviculture objectives. His work on tree species, their qualities, description and occurrence was a foundation for future developments. Yet, the legislation needed to put all this practice pivoted on the central point of human subsistence in the Solomon Islands, the rights to land.

—7—
Domestic Needs and Overseas Markets, c. 1950–1963

> Either the export market must accept a proportion of lower grade material than prime, or it must find a local market. The local market for sawn timber can never be large ... Government, plantation and Mission demands might account for a considerable proportion, particularly if ... Government encourage utilisation of local resources, in preference to imported, by precept and example.
>
> F.S. Walker, *The Forests of the British Solomon Islands Protectorate* (London, 1948), 25.

The demand for timber as the war ended was for export, but as the administration, a few commercial interests, and the missions re-established, they all needed buildings. Imported milled timber was expensive and scarce. Although nowhere as great, there was also a demand among some Solomon Islanders for milled timber. Australian companies – Nielsen Pty Ltd. and W. R. Carpenters Ltd – had been interested in the timber in 1947.[1] In 1949, New Zealand's Fletcher Holdings wanted to log in the western Solomons.[2] A few years later, a pre-war planter, Carden Seton, had the ingenious idea of logging from place to place using a war-surplus LST boat to transport a portable mill.[3] But these, like the earlier interested parties, had to wait for the Lands Commission, a forestry officer and appropriate legislation. And the unrest occasioned by Maasina Rulu meant the administration proceeded with caution on land and forestry matters.[4] Inquiries for concessions declined as world trade began to normalise. For all the early Australasian demand for Solomons timber, the two log-exporting companies that dominated until the early 1960s had trouble making a profit on the Australian market.

Domestic Needs and Overseas Markets

The local market

Timber was needed as the government rebuilt. In purchasing over 5000 acres of land from Lever's at Lungga, in 1947, the government committed itself to siting the capital at Honiara where there was an airfield and room to expand.[5] Wartime structures such as bridges, being of untreated timber, were decaying.[6] The saw-mills at Mbuma and Mbatuna had met some of this demand, as had American war-surplus Oregon pine and imports.[7] However, both mills were reclaimed by their respective missions. At Mbatuna, the Adventists had difficulty maintaining production because of sickness among their personnel. Moreover, *Calophyllum* logs had to be brought from further afield – the Lever Harbour region on north New Georgia – though the closer Nggatokae people were clearing it from within 100–200 yards of the beach for the mission saw-mill.[8] The Catholics at Mbuma continued to sell some of their output to the government, but, at 130s per 100 superfeet, the price was 'a bit high'.[9] By 1951, the High Commissioner, Gregory-Smith, summed up the government's predicament:

> Without timber … we will quite definitely have to considerably curtail our development programme not only in Honiara, but in the Protectorate generally.[10]

A solution seemed close at hand in Tenaru Timbers. When the Dethridge Timber Syndicate failed on Santa Cruz, Dethridge looked elsewhere. He and Ken Dalrymple-Hay, Norman Wallis, and Harold Davies had formed a new syndicate, Tenaru Timbers Pty Ltd in 1950. They negotiated with Lever's who held the land at Tenaru, north Guadalcanal for a lease for five years and an option for another five years at £3,000 annual rent.[11] The government was in a quandary. It wanted timber and the syndicate was willing to sell milled timber locally, as well as exporting logs to Australia. Yet, as the Secretary of State of Colonies reminded the Protectorate, the principles of sustainable logging were to be the foundation of the industry, with replanting essential.[12] However, under the existing legislation, the government had no power to demand reforestation on privately-owned land and neither Lever's nor Tenaru Timbers was willing to undertake expensive replanting.[13] In discussions with the logging company, the Resident Commissioner noted,

> I asked about re-afforestation and whether he considers natural re-growth would replace the trees being cut. He said, no, it never does but from a forestry point of view a company to protect itself would not cut indiscriminately, and a sawyer would not cut the young trees as immature timbers are more expensive to handle.[14]

This was not quite the picture presented to the High Commissioner. The company estimated that the 11,000 acres it wanted held 50,000,000 superfeet. Under selective logging of mature trees only, so the company claimed, on 240 acres a year for ten years, the exploitation of the area could go on indefinitely with a second cut after three years, and subsequent ones every five to six years. The company felt natural growth of the immature trees would replace the felled trees. If this did not eventuate, then they would replant – so they said. The Resident Commissioner believed self-interest would dictate sustainable extraction, while in Fiji, the High Commissioner believed that 'some cutting of mature and over-mature timber can be safely carried out'.[15]

They took no cognisance of the silvicultural needs of the various species.[16] *Pometia* was a hardy grower which thrived on abandoned garden land and so would probably regenerate, but little was known of *Mangifera salomonensis* as it was rare, found mainly on north Guadalcanal.[17] Moreover, there was no mention of extraction techniques, working plans and erosion control. And self-interest was unlikely to sustain the company (or the forest) if, after several years of logging, it went broke and withdrew; or if it 'mined' to clear the forest within five years to escape paying the £3,000 yearly thereafter. Other than completely refusing the transfer of the timber rights, the government could do nothing but accept the company's unformalised guarantees. The only ace up its sleeve was the probability that the company would want concessions elsewhere and an unblemished record at Tenaru could influence future government support. The Secretary of State conceded,

> If the exploitation of this area is proposed as [a] pilot study which would lead to large scale exploitation of [a] much larger area of forest, [the] proposal would appear to be reasonable.[18]

With Lever's approval, Tenaru Timbers had sent a trial shipment of 50,000 superfeet of *Magnifera salomonensis* and *Pometia pinnata* logs to Australia in November 1950. The success of this shipment meant Tenaru Timbers signed the agreement with Lever's in August 1951 to lease 11,000 acres between the Tenaru and Naliumbiu rivers.[19]

Like governments everywhere, the Protectorate sought to impose some tax on the operation. Like private enterprise everywhere, Tenaru Timbers sought to avoid it. The company proposed that it would sell at between 105s to 110s per 100 superfeet milled timber, providing the government imposed no levy or tax on their output for five years. In July 1951, the High Commissioner in Fiji signalled he was willing to consider no export tax for five years and no royalties to government for three.[20] The Secretary of State demurred, insisting on some immediate tax on exports because the company could log rapidly, 'creaming' the forest, before the tax moratorium was over, thus destroying any chance of successful natural regeneration.[21]

Imposing a tax was necessary for the Protectorate purse and to keep Tenaru Timbers honest. But the Protectorate's financial secretary faced a dilemma. The only payment in respect of timber extraction at the time was the royalty of the Kauri Timber Company at 1s per 100 superfeet. Placing a royalty on Lever's sub-lease was of dubious legality, so the only solution was an export duty. About 3s per 100 superfeet was on a par with duties paid in Fiji. However, this would conflict with the understanding the government had with the Kauri Timber Company, which paid no export duty. The government sought a compromise. The Kauri Timber Company was concerned as the new levy would cost it between £3,000 and £4,000 extra a year, at a time when it was experiencing difficulties with its contractors at Vanikolo. Tenaru Timbers argued it was paying the equivalent of royalties in rent to Lever's. The government planned not to impose the duty on the company until 1 July 1953 and, in equity, had to exempt Tenaru Timbers, burdened with establishment expenses, until at least 1 January 1953.[22]

Both the Kauri Timber Company and Tenaru Timbers continued to make representations to the government, resulting in the postponement of the duty imposition. They had a good case. Their export market, Australia, had in early 1953 introduced a quota system on imported timber and re-imposed import duties of 7.5 to 17 per cent on logs and sawn timber, except for its territory, New Guinea. Already exempt from Australian income tax, New Guinea-based timber exporters thus had a further advantage in selling timbers similar to Solomons' species on the Australian market. Moreover, the timber exported from the Solomons was of a lower grade than Fiji's, so that the same rate of duty of 3s per 100 superfeet, would render the Protectorate's timber uncompetitive in Australia.[23] Consequently, in August 1954, the government decided to shelve the proposal because,

> the imposition ... might now lead to withdrawal by both Companies from their present operation ... it would certainly remove all hope of getting anyone else interested in timber exploitation in the Protectorate. As we do get a certain amount of indirect revenue from the two existing Companies I should prefer to make certain that we retain all the revenue ...[24]

In its representations to the government, Tenaru Timbers had outlined the impact the duty would have had on intending overseas investors. There was 'no available surplus cash' in the Protectorate for investment in any successor to Tenaru Timbers.[25] In 1951, when the company had been formed, shares had been offered to the public. Six civil servants were interested and under Colonial Regulations 53 (2) they applied to the Resident Commissioner for permission to buy. The first to apply was R. L. Barnfather, government pharmacist and medical storekeeper, who wished to buy 1000 shares at a pound each. R. Firth was the officer in charge of income tax returns, Miss K. S. Poole was a senior clerk in the secretariat, J. D. Davies was the chief wireless officer, and Miss D. Thompson was on a two-year contract. In the view of Resident Commissioner, the duties of Poole and Firth could bring them into contact with confidential documents which might possibly have a bearing on the operations of the company. In the British colonial as well as legal system this was known as a conflict of interests – that is, the civil servant's duty to the government could conflict with his/her own personal interest, particularly in relation to pecuniary gain. It would take a great stretch of the imagination to see the government pharmacist ever having a conflict of interest in matters pertaining to logging. However, 'to be consistent and fair to all concerned' the requests were turned down by the High Commissioner.[26] Thus, not only did the government keep its employees beyond taint of partiality, it also confined the limited supply of investment capital to non-government personnel and overseas interests. In spite of this, the company's capital came from within the Protectorate, although the investors were expatriates.[27]

Tenaru Timbers Pty Ltd in operation

The Tenaru Timbers negotiations highlighted several issues for the government: first, the urgent need for a forest officer, someone who would understand 'technical aspects' relating to silviculture and forest science; second, the inconsistencies and gaps in forestry legislation. As Tenaru Timbers began production, other issues emerged. So many logs were being exported that the first company saw-mill at Tenavatu, owned by Davies, ran short. He tried to persuade the government to allow him to log the *Terminalia brassii* on a 400-acre section to the east of the Threlfall property near Dodo, on land once part of the Ilu land occupied by the Americans. Davies proposed to sell to the government at a lower price than the timber milled for Tenaru Timbers and was willing to pay rental or royalties to the land holders. Although the government was disappointed at Tenaru Timbers' preference for exporting to the neglect of the local timber market, Davies' scheme failed to get its support because of land disputes. In 1949 the government had tried to buy 1800 acres in the Ilu area for agricultural purposes from the Tadhimboko people, along with adjacent land of Burns Philip and Lever's that was held under Certificates of Occupation (Wastelands Regulations, 1900–1904). During the negotiations, the Tadhimboko people had become aware of the extent of the partly developed holdings of these companies, subsequently sold to the government. They disputed the sale of land held

Domestic Needs and Overseas Markets

under the Certificates, so the government was not inclined to inflame the situation by allowing logging there. Their new awareness exemplified the increasing 'land consciousness' of Solomon Islanders and aroused fears, well-founded on their pre-war experience, that the government still wanted their land.[28] Had they read Walker's report, their worst suspicions would have been confirmed.

In August 1952, Tenaru Timbers was experiencing periodic log shortage. The company asked the government where it stood with the purchase of logs cut by the people of Hauhui, west Malaita.[29] Hauhui had been one of the areas selected by the government, mainly in the wake of Maasina Rulu, for a community development scheme involving rice-growing and the extension of coconut plantations.[30] All the government's propaganda on how to make money and 'develop' was taking effect.[31] The Hauhui people had seen Neil MacCrimmon cut and float logs to his place at Su'u in the 1920s; they and their neighbours had supplied labour to him and the Kauri Timber Company. So they did not have to look far for a resource. They suggested selling the logs at £7 each, in shipments of 300 log lots until their new crops matured.[32] The District Officer was supportive, although the best logs closest to the water had been cut before the war, especially near Su'u. They wanted some of the payment in milled timber to build 'permanent houses'. The District Officer noted that his 'initial investigations into the ownership of the trees tends [sic] to show that we need anticipate little trouble on this point'.[33] This suited the company as it did not have to negotiate cutting rights and avoided 'responsibility for any ensuing observance of the law, extant or later introduced'.[34] The government approved the Hauhui initiative, but since the arrival of the first forestry officer was imminent it awaited his advice. He supported it, but in 1953 Tenaru Timbers could not spare anyone to go to Malaita to obtain the logs.[35]

Throughout the 1950s, Tenaru Timbers continued to experience log shortages, mainly because of poor extraction patterns. Initial prospects had been good, as the company achieved a level of mechanisation unheard of in the pre-war Solomons. The Solomon Islander employees felled the trees with axes, then cut them into 15 to 25 feet lengths with Danarm petrol-powered chain saws. From the felling site, the logs were loaded by crane onto trailers (ex-wartime torpedo carriers, caterpillar tracked) and hauled along rough tracks by trucks. In time, tractors replaced some of the trucks, although the trailers often got bogged during the wet season. In 1954, at a cost of £10,000, the company introduced a railway system with two petrol-driven locomotives to bring the *Magnifera salomonesis* logs to the beach and thence by truck to Tenavatu, where they were floated out to ships. The company also upgraded its saw-mill, installing a planing machine to dress the *akwa*, *Pometia pinnata* and *Calophyllum* spp. for local sale. However, improved technology had its limits. As there was no natural harbour on northern Guadalcanal, the company often lost logs in trying to raft them to the ship in rough seas.[36]

By 1954, Lever's mill on the Russells was also producing timber, mainly for its own use though some seems to have been sold locally. The problem with all the timber was that it was unseasoned and some of it badly cut. Poor extraction did not help log quality. By 1956, Tenaru Timbers had abandoned the rail system and was hauling logs by tractor for up to three miles, taking a toll of time, plant, labour and log. The company went into voluntary liquidation in late 1957, when virtually all the *Magnifera salomonesis* had been logged. Harold Davies continued running Tenavatu saw-mill as Tenaru Sawmills Company, using mainly *Pometia pinnata* logs from Levers' land. Standards were not high, although in 1957 Davies constructed a primitive drier which produced better timber. Improved roading to good stands saw production recover and, at the close of the decade, the company was

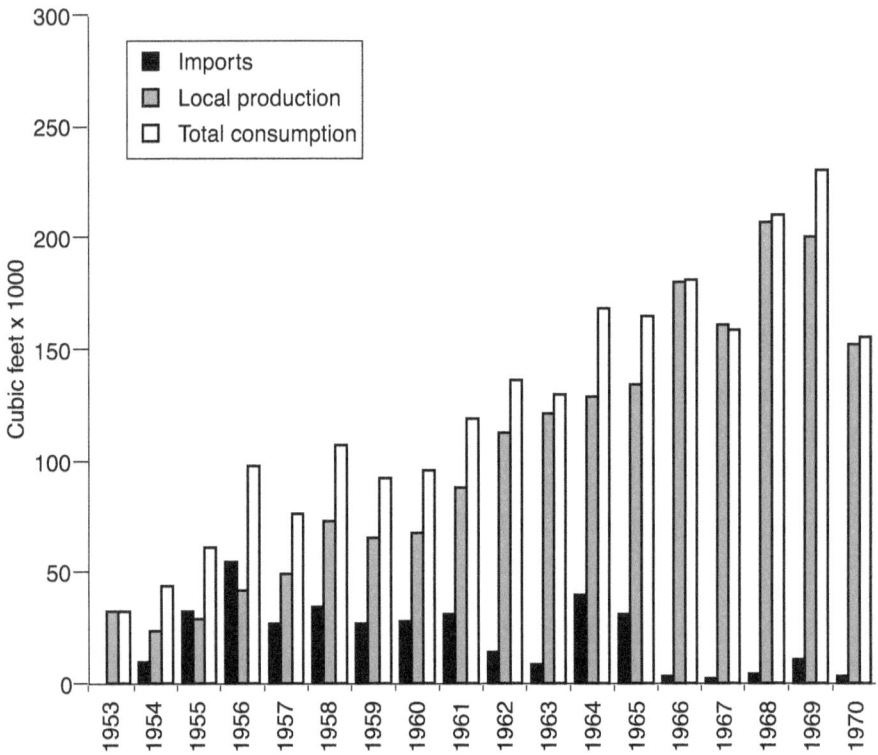

Figure 62. Total consumption of sawn timber, 1953–1970.

Source: ARFD 1953–1971

Domestic Needs and Overseas Markets

seeking a new concession. Before it found one, difficulties with the band saw equipment halted production. Much of its plant had been second-hand, which pushed up servicing costs.[37] In 1962 the British Solomons Timber Company purchased the interest and built a new mill. Timber production for the local market had increased from 12,938 cubic feet (true measure) in 1956, when figures were first kept for Tenaru Timbers, to over 53,000 in 1961. By then, local saw-mills were producing almost three-quarters of the Protectorate's consumption, with Tenaru Timbers supplying over a half of this (see Figures 62 and 63).

Tenaru's production record for sawn timber was challenged by a saw-milling venture at Mongga, Kolombangara in 1961. This operation, too, was on Levers' land, with the cutting rights over 10,000 acres leased to Tichler Constructions Pty Ltd. whose saw-miller, J. B. O' Keefe Pty Ltd, was soon turning out about a third of the Protectorate's total. Tichler used much of this in his construction business around the Protectorate, but some

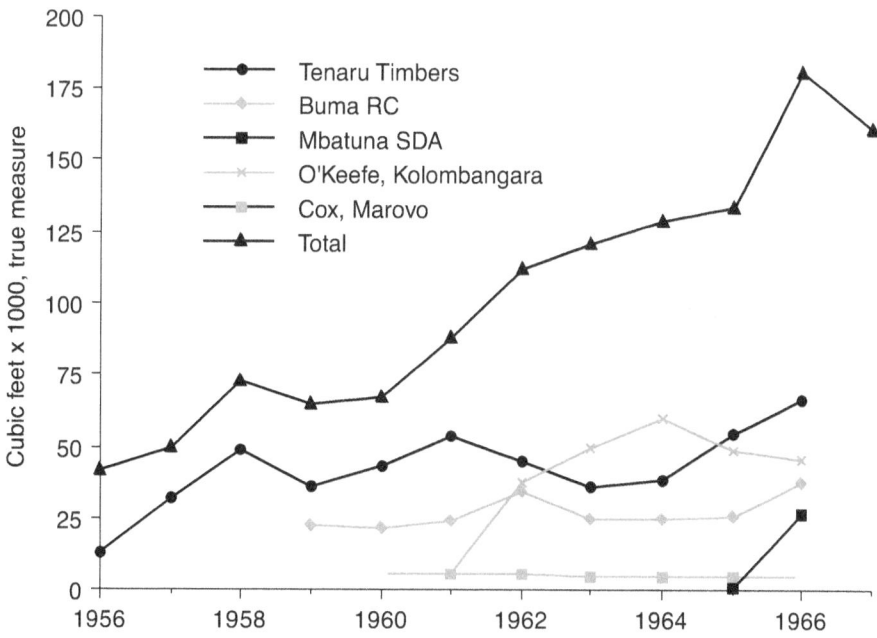

Figure 63. Sawn timber, local production, 1956–1967.

Note: Tenaru Timbers became British Solomons Timber Company in 1962. Production by saw-mill is not itemised after 1966. Source: ARFD, 1956–1969

sawn timber was exported in 1961–1963 to the Gilbert and Ellice Islands, the first, with the exception of wartime requirements, since the Vanikoro Kauri Timber Company's experiment in the late 1920s.[38]

The Forestry Department established

As part of a ten-year development programme in the Protectorate, in 1949 the British government set aside from the Colonial Development and Welfare fund £14,700 sterling (A£18,370) to establish a forestry department. Although the Protectorate in the interim had sent two Solomon Islanders to Fiji for training as forest guards and rangers, it did not engage its first forester until October 1952, when John Logie arrived on secondment from Kenya.[39] These employees were men, a colonial assumption which meshed with Melanesian practice of tree clearance as a male domain, but ignored the use of vast range of forest products that were the preserve of women and essential to subsistence.

Based in Honiara, Logie was soon scrutinising Tenaru Timbers.[40] Thousands of cubic feet of timber were being logged and the department had no control over it. Although Logie believed there was some legal basis for demanding that the company follow a working plan, this would be of little significance if the company pulled out when its initial five-year lease expired. Unlike his immediate superiors, Logie realised that, '… it would pay the Tenaru Timber Company to cut out the forest as fast as possible. The faster they can cut, the less the payment to Lever's per unit of volume'.[41] If the government spent money at Tenaru on regeneration there was no guarantee that Lever's would keep the area permanently as a forest. Logie wryly observed that the annual rent Lever's was getting was half the estimated recurrent expenditure for his fledging Forestry department, currently being funded by Commonwealth Development and Welfare grant. All these were good arguments for the establishment of forest reserves under government control.

Logie was committed to the establishment of reserves on Crown or government land. As well, he wanted legislation framed to create reserves on native land, reserves on some private land with a rebate of royalties or tax when it was logged to provide for silvicultural operations, the prohibition of cutting on protected areas, and the collection of royalties on all logs cut.[42] He rehearsed Walker's ideas:

> It is normal for governments to take over unoccupied forested land even where it is at the time economically inaccessible so that the soil and water may be protected and so there will be a timber reserve when the country develops later.

Domestic Needs and Overseas Markets

Logie soon accepted that most of the forest was claimed by Solomon Islanders, so he had learned a fundamental reality of Melanesian society in a shorter time than had Walker. Development of forest reserves was going to be expensive and some land would have to be purchased. He suggested that the native forest reserves be set aside by the Native Councils, 'leaving them to deal with the original owners' and dividing any profits from logging between them and the government, although the High Commissioner would retain the power to ratify this as well as to declare native forest reserves.[43] Some of Logie's superiors were unwilling to concede native 'ownership' of all unregistered land. However, the government favoured the involvement of the Councils in the management of the native forest reserve, although experience in Kenya was not promising. The intention was mainly political since such involvement 'would help to reduce suspicion' of the government's intentions towards Solomon Islanders' land.[44] Logie believed the 'native habit' of shifting cultivation would have to undergo 'adjustment' to protect the forest. He had seen, for example, in the seven years since Walker's survey there, the transformation of 'a timber getter's paradise' on Catholic mission and native customary land at Beaufort Bay, Guadalcanal, as people cut both secondary and primary forest for gardens.[45] The Resident Commissioner gave Logie the task of preparing draft legislation and to begin mapping out possible forest reserves.[46]

Logie prepared the draft legislation for the Resident Commissioner and the Advisory Council in 1954, enshrining the ideas of forest survey and reservation before logging.[47] It gave the government considerable powers over Crown land, private leaseholds and native reserve land. Timber could only be worked on a commercial scale under a licence system which required the logger to have a working plan and to contribute to reforestation.[48] The draft had taken Logie nearly two years, as he found,

> [t]he unusual system of land tenure and land consciousness of the Melanesians and the small areas of Crown Land are among the factors that make the drafting and application of a forest legislation unusually difficult here.[49]

The government accepted much of the draft, but it could not be finalised until the Lands Commission had made its findings. The concept of native reserve land was shelved in case its enactment prejudiced the Commission's work.[50] In Logie's draft, the native councils could request of the High Commissioner that 'any ownerless or unclaimed or unoccupied native land be declared a Native Forest Reserve', though allowance was made for those claiming rights to 'forest produce' thereupon.[51] Thus, the idea of vacant land and rights to trees apart from land had to be considered – complex issues best left to the Commission.

Since the Commission was still investigating, Logie turned his attention to more practical matters, despite shortages of government land, accommodation, and equipment. He conducted preliminary surveys on Shortland Island, the Kira Kira hinterland on Makira, Kolombangara, Vangunu, and the Marovo lagoon area. Research began with the establishment of a small nursery at Kukum, not a particularly suitable site, given the atypical micro-climate of the hilltop grassland of north Guadalcanal. About half a dozen endemic and exotic species were planted on a hill above Rove, not with great success, but it was a start on data collecting on species suitability for north Guadalcanal.

Logie had a technical staff in 1953 of two Solomon Islanders, A. L. Naqu, a forest ranger and Silas Ronoe, a forest guard. A forester from England, Chris Hadley, joined the Department in early 1954 and was posted with Naqu to Vanikolo, to work on kauri and to monitor the loggers.[52] The Vanikolo operation remained the only other exporter of logs beside Tenaru Timbers until the early 1960s (see Figure 66).[53]

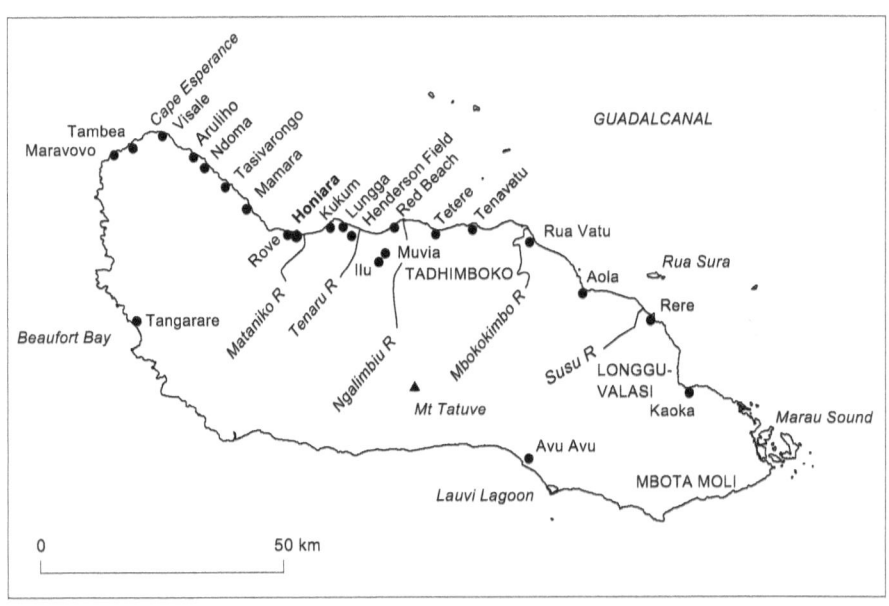

Map 6. Guadalcanal.

Domestic Needs and Overseas Markets

Domestic needs of Solomon Islanders

Prior to World War Two almost all milled timber, whether imported or indigenous, had gone into structures for European use or built on European instruction for local usage. This had stimulated some demand among Solomon Islanders for similar housing. In the 1930s, local Adventist pastors had become so attached to their houses of dressed timber that they purchased them from the mission, the cash to be refunded if they were transferred to another district.[54] The rapid 'building boom' of the Allies, demonstrating the potential of local timbers as well as more comfortable and often more durable structures than the leaf and pole native houses, increased demand for milled timber. The desire for this had seen ex-Solomon Island Labour Corps men salvaging wartime milled timber for building material, although the British allowed very little into the boats taking the men back to their home villages.[55]

In the late 1940s and early 1950s, the Marovo lagoon people towed in logs to sell to the mission at Mbatuna, many preferring to take their payment in the form of half the timber milled from their logs.[56] On Santa Isabel too there was interest.[57] Late in 1952 when Dethridge was planning to leave Tenaru Timbers, Hubert Sele of Liliga approached him with a proposition to log 1000 acres of 'useable land and workable forest'. The people wanted 'to have a boat-building concern on Ysabel and an available supply of sawn timber for house-building'. Sele and 'his fellow owners of the forest land' wanted Dethridge to work the forest of mainly *Campnosperma* and *Burkello hollrungii* in return for boat and house timber. Dethridge had boat-builders at hand, but he was unable to get a bank loan and asked the government to supply two-thirds of the A£30,000 capital. The government was wary since Dethridge's scheme, as it had over-estimated domestic timber demand. Although supportive of a partnership in the boat-building operation, the government did not proceed, as Dethridge failed to produce details of costs and marketing.[58]

As well as supplying mills to get their building timber, some Solomon Islanders were interested in running their own. A group approached the western District Commissioner, but when the costing of the equipment came to about £15,000, the plan folded. Two other native-owned mills, according to Logie, had

> lost money which caused ill-feeling by those concerned against the government. The reason money was lost at a time of high timber prices was that the running of a saw mill is a skilled job.

Logie was supportive of local involvement in the timber industry. He, like Walker, saw increased domestic consumption of milled timber as an adjunct to the export trade, absorbing the lesser grades of timber. Local involvement in milling was a legitimate aspiration, but Logie remained cautious because of the range of skills required:

> the best way to introduce the natives into the timber industry would be by partnership with a skilled sawmiller. In this way some of the local people would be able to learn the management and trading side of the business besides the maintenance of a mill.[59]

Local participation in saw-milling could foster a favourable attitude among the councils to the 'native reserves' in the proposed legislation since the royalties would add to council revenues.[60] As a result of Maasina Rulu, the government wanted to involve Malaitans in 'useful commercial enterprises', 'for the exploitation of forest produce, possibly on a co-operative basis'.[61] Before any progress was achieved, Logie's secondment ended in April 1955, and forestry matters remained in abeyance under the Department of Agriculture for over a year.[62]

Besides timber for house construction, Solomon Islanders needed firewood. Scarcity of firewood for villagers was unheard of before World War Two, although European settlements and their industrial needs had begun to cause localised pressure on resources, including mangrove forest.[63] When in 1920, coal supplies ran short in Tulagi following industrial action in Australia, the Resident Commissioner sent prisoners to cut mangroves around Tulagi to fuel the government steamer, *Belama*.[64] On Vanikolo, in 1928, after less than five years of operation, the manager requested a diesel engine for running the freezer and engineering works, because '[f]irewood supply is an extremely difficult matter, and becoming worse'.[65] Wood was abundant, but at increasing distances from the settlement so carrying it was costly.[66] By the early 1930s, the Nggela people made money gathering firewood to sell to residents and captains of inter-island vessels at Tulagi.[67] During the Depression, European planters were experiencing 'a shortage of wood fuel' for their copra driers to supplement the coconut husks and shells burnt for heat. To reduce costs on labour for firewood collecting, several considered converting to oil, but imported fuel was no cheaper.[68]

Following the expansion of post-war district centres, wood for their needs became increasingly difficult to obtain within easy reach. In mid-1951, Honiara's 'firewood gang', who supplied the fifty European households, the government barracks and the Police training school, had denuded the hills north-west of the Police Depot.[69] Logie thought that the forest around Mt Austin with its old wartime road was an obvious source of firewood for Honiara, but his superiors had it slated for reservation as a park. Although there was ample wood in the foothills behind Honiara, their broken nature and the lack of roads were problems. Logie wanted to plant quick-growing Australian eucalypts suitable for coppicing for firewood, but the lack of government land within easy access of Honiara precluded this. Eventually, Tenaru Timbers contracted to supply the wood.[70] In 1958 Auki too was experiencing fuelwood shortage and inquiries were made of the forestry department with regard to quick-growing species. As settlements and population expanded so too did the demand for firewood.[71]

The Kauri Timber Company

In 1949, after the wartime hiatus, Queensland contractors Haling Brothers took over logging for the Kauri Timber Company on Vanikolo. The parent company in Melbourne estimated that an annual output of 4,000,000 superfeet from Vanikolo was needed to meet Australian demand and to return a profit. Volume was value.[72] After six years, the contractors had failed to achieve this: their maximum never went beyond 2,600,000 superfeet, although they claimed they could produce 3,600,000, under ideal conditions.

Halings attributed their lack of success to labour problems.[73] There was a chronic shortage of labour in the Santa Cruz district. Malaitans preferred employment in Honiara and other centres to Vanikolo, with its isolation and bad reputation.[74] As the indenture system gave way to civil contracts in 1947, the Santa Cruz labour force was in an ideal bargaining position which they exploited to drive up real wages above those elsewhere and obtain a 40-hour working week. These developments were the outcome of disputes, strikes, and inducements offered by Halings.[75]

The brothers' interactions with the 15–20 Europeans employed were not always conducive to a solid working relationship. Halings provided no amenities for these bush-workers, who lived most of the time in huts or caravans at the timber head. Logging vessels with mail and supplies called only every three months. The rare emergency visit by a flying boat underlined their remoteness, just as cyclones and earthquakes demonstrated their vulnerability. They resented high prices charged for 'European' foodstuffs and the time the brothers spent on their trade store and not 'on the job' in the bush.[76] Good workers became frustrated. One overseas engineer,

> … went because he could not stand the … method of working. He was expected to keep trucks running continuously, by patching them up from day to day, until it [sic] fell to bits. This was anathema to him. To his way of thinking each truck should come right off the road and be serviced properly at regular intervals, instead of going on until it was nothing but a battered chassis six months after delivery.[77]

From the time Halings started at Vanikolo in 1949 through to early 1953, 54 employees from Australia cancelled their contacts before having worked the agreed two years: this certainly made for lack of continuity and supervision.[78]

Poor supervision meant that basic servicing was left to Solomon Islanders who had little understanding of the nature of wear and tear on mechanical parts, which then jammed, resulting in lost production.[79] The labourers lacked expertise for some of the logging operations and, without consistent supervision, produced many defective logs. Inexperienced fellers cut trees so they fell badly, resulting in 'draws and often bad shakes'.[80] Costs increased when hauler drivers failed to anticipate 'trouble looming up and will wait until a log is hopelessly jammed before signalling stop'. Log quality was diminished by borer or toredo as a result of poor spraying and prolonged storage on land or in the sea.[81] The kauri plywood cut from this was full of holes and was therefore either rejected or sold as low-grade ply, as 'backs and centres ... much in excess of face veneers'.[82]

Although post-war mechanisation had the potential to make logging less labour-intensive, the Kauri Timber Company often selected unsuitable equipment. For example, tractors had been scarce after the war, as most of the supply came from the United States and there were various restrictions on currency outflows to America. In late 1948, the company could not get tractors for Halings in less than 12 to 18 months, so settled for 'rebuilts', war surplus D8 Caterpillar tractors, weighing 22 tons. Three were quickly obtained, but they proved too heavy for the water-soaked terrain. Known as 'Gutless Wonders' when going uphill, these frequently broke down and delays occurred while parts came from Australia.[83] It took the Vanikolo company over four years to adopt the labour-saving chain saws which had been used by Tenaru Timbers since 1952.[84]

Road construction was poor. Gradients were 'extremely steep being as much as 1:5 in one place and 1:7 and 1:8 are common'.[85] Under tropical conditions, acceptable gradients range from 1:14 and 1:20. Safety of personnel was an issue, as was the permanence of the roads, particularly as the Forestry Department wanted access for regeneration experiments. Tracks soon became badly eroded, yet, as Logie found, there was no legislation to address these concerns and the company admitted no responsibility for maintaining disused roads.[86]

Halings logged kauri (*Agathis macrophylla*), *karamati* or *ketekete* (*Campnosperma brevipetiolata*) and *koila* (*Calophyllum* spp.) The koila was a hard, tough wood, but was used for much the same purposes as Australian hardwood. Even when jarrah (*Eucalyptus marginata*), the hardest Australian wood, was brought to Melbourne from Western Australia, it was still 25 per cent cheaper to produce than Vanikolo koila, so the demand for the latter remained confined to ships' timbers and lifting sticks. Improved technology, involving the steaming of the wood, enabled it to be peeled for plywood, but during the 1950s, its colour lacked market appeal. Karamati was particularly susceptible to teredo worm, and in 1954 the Melbourne firm was getting only 40 per cent recovery from the Vanikolo logs. The company was permitted to leave the karamati and koila unlogged when it showed the Forestry Department the costings, promising to continue research into the commercial viability of koila ply.[87]

When Halings left in late 1956, the Kauri Timber Company had to write off £150,000. Reaping the rewards of the post-war Australian building boom and the increasing scarcity of indigenous pine species, the parent company in Melbourne had made profits from 1949 to the end of 1955 from plywood and sawn timber cut from the Vanikolo logs, but that was at the expense of the branch company.[88]

After the Kauri Timber Company resumed control of logging there was little improvement. Labour issues persisted, though with no strikes.[89] Because it held the Australian import licence quota for Vanikolo and feared heavy capital loss if it abandoned the branch, the company continued operations. Initial production figures of over two million superfeet were encouraging. However, the company's agreement was due for renegotiation in 1957. The government was willing to extend it for a further ten years, but planned to increase the royalty from 1s to 6s 6d per 100 superfeet to fund regeneration research.[90] The company thought this rate excessive and managed to get a continuance of the old royalty to late 1957 when an interim rate of 5s for kauri came in. Both Keith W. Trenaman, chief forestry officer from September 1956, and the British forestry adviser C. Swabey recognised a form of transfer pricing whereby the Melbourne company purchased logs from the Vanikoro branch at a rate which suited its purposes for taxation and sold the sawn timber at a healthy profit in Australia. An independent assessor was consulted to recommend a royalty based on financial analysis of the company. Much to the company's dismay, the assessor presented an analysis easily justifying a government royalty of 9s per 100 superfeet. This was phased in, beginning at 6s 6d in 1959 to allow for reorganisation of the company's operations; finally in 1961 the rate of 9s per 100 superfeet kauri prevailed.[91] Along with Tenaru Timbers, the company had escaped the proposed export duty on its logs in 1953. Hopeful of a predicted expansion of exports, Trenaman supported a duty which came into operation in late 1961. However, in order that the company would not be taxed twice for the same log, the export duty, though levied, was subtracted from the royalty. This gave the company parity with others working government-owned land. So, say the company had to pay 3s per 100 superfeet in export duties, then this 3s would be deducted from the royalty of, say 9s per 100 superfeet. When duty on kauri increased in 1963 to 4s 2d per 100 superfeet, then the company still paid only 9s, but technically 4s 2d of it was export duty and the remaining 4s 10d was royalty. Thus the royalty rate, 9s, was the ceiling of 'taxation' to government.[92]

Meanwhile, the Kauri Timber Company had allowed its plant to deteriorate: by 1959 it faced costly replacements or declining production. Most of its costs were fixed, so it had to produce 2,750,000 superfeet yearly to get the unit-volume cost to 'break even point', but under the new agreement there was a limit of 3,000,000, any excess incurring a higher rate of royalty of 19s 6d. Regeneration trials at Vanikolo indicated that kauri was a faster grower than first thought. With a coupe of no more than 3,000,000 yearly it was possible that the existing mature kauri would last about 20 years from 1957, when a second cut could start in the first-logged areas. Trenaman wanted to slow extraction so that Forestry's regeneration work could catch up. He conceived of Vanikolo, Tevai and Santa Cruz as a management unit for the sustainable logging of the kauri forest.[93] The company hoped they could persuade the government otherwise and by 1961 was aiming to produce over 4,000,000 a year, which would mean 'cutting out ' Vanikolo in three or four years, if its estimates of remaining kauri were accurate. But the government had to be convinced that the logging was being conducted efficiently – that the company had logged the several stands bypassed earlier to get the most accessible logs. This had to be done before a new area was opened. To achieve production of 2,750,000 and more, the company needed new equipment. To take the example of tractors again, the company abandoned to the bush the D8 Caterpillars and replaced them with four D6 ones, only to find in 1963 that a single D6 was not powerful enough to haul some of the big logs. The company needed at least one D7, useful also in roading.[94]

There was the possibility of logging Tevai and Santa Cruz to increase production, but the company now considered the Tevai forests inadequate and the Forestry Department would not let them log Santa Cruz until they had finished Vanikolo. And Ben Ramoli's son and government headman, Ben Tua, continued to refuse timber rights to Tevai, no matter the company's or Trenaman's plans.[95] Moreover, the company was cautious about estimates of 39,000,000 superfeet said to be on the island, particularly as the terrain was broken and there were no accurate topographical maps of the island.[96] There was still kauri on Vanikolo; the only trouble was that decades of lack of working plans, countenanced even by the Forestry Department since 1954, had left the bulk of this inland or in isolated pockets, so increasing extraction costs.[97] Methods of extraction were still inefficient: roading did not keep up with logging, so logs were often hauled long distances. Embedded rock in

Domestic Needs and Overseas Markets

these logs frayed both saws and tempers at the Australian plywood mills. So exasperated did the millers become that they sent up photographs and a critical analysis of 55 unacceptable logs in one shipment alone to Brisbane in 1960. Although about 60 per cent of the kauri ended as sawn timber, its value too was considerably reduced if the logs were faulty.[98]

By permission of Business and Labour Archives, University of Melbourne

Figure 64. What upset the Brisbane plywood millers in 1960: logs that were not only unsuitable as ply logs, but also of limited value even as saw logs. Log no. 333, (a) Grading: Bad saw-mill log; (b) One-third of log being decayed material, reduced timber recovery as well as a four-foot drag.

The company chaffed under the new royalty scale. It was also being squeezed by the Australian government. In 1961 Australian duty went up on 'saw' logs, but not on 'peeler' logs. The company could only shelter for a short time under Customs' misapprehension that their logs imported were all peelers. Quarantine was another issue. Now the Australian authorities wanted a hygiene certificate from the Solomons end, but there was no appropriate officer on Vanikolo and the manager's certification was all it could supply. When any sign of insect infestation was found in Australia the company had to pay for fumigation, adding 2s per 100 superfeet to the costs.[99]

By permission of Business and Labour Archives, University of Melbourne

Figure 65. Log no. 379, (a) Grading: Good saw-mill log; (b) Knots will affect output quality.

What managers at Paeu did not see was the larger picture of the parent company. Since about 1944, the Kauri Timber Company in Australia had taken over many milling and plywood companies. It then diversified into other enterprises. Over-extended, it was caught when the economic recession of 1961 resulted in a down-turn in the building industry, which forced it to sell many of its interests. In the process of reconstruction, it wrote down its capital by 62 per cent in 1963.[100] Its plywood sales had already received a blow when in 1960, the Australian government granted Japan and other foreign countries export licences allowing cheap plywood into Australia.[101] For Vanikolo, the writing had been on the wall for years; now it was writ large.

Moreover, the company was no longer dependent on Vanikolo for its pine supplies. In 1954, the Vanikolo kauri logs landed 'in the yard' at the subsidiary, Brisbane Newmarket Ply Company cost 116s per 100 super-feet, while imports from South-east Asia, were 99s. Hoop pine (*Araucaria cunninghamii*) from local sources was about 76s.[102] This pattern persisted for the next ten years, with not only red cedar (*Shorea macroptera* and other

Domestic Needs and Overseas Markets

spp.) and white cedar (*Parashorea malaanonan*) from Borneo, but also ramin (*Gonystylus macrophyllum*) and bindang (*Agathis alba*), all of which remained cheaper than the Vanikolo kauri and were in regular supply. By 1960 cheap klinki pine (*Araucaria hunsteinii*) from New Guinea was available. The more costly pine plywood from Vanikolo was meeting buyer resistance.[103] It was no wonder that by 1963, the company officials in Melbourne could say of the kauri on Vanikolo and Santa Cruz 'we do not now necessarily want the timber'. A cost-benefit analysis in late 1963 revealed production on Vanikolo had not achieved even the allowable cut of 3,000,000 superfeet a year and to continue at the current rates would mean an annual loss of £25,000.[104] The company decided to withdraw in early 1964. It had never been a great success, but had survived mainly because it subsidised the saw-mill business in Melbourne. The Protectorate may have benefited from the duties paid from the company's exports of logs, but other than that, asked a later governor, Colin Allan, 'What was to show for it? Very little indeed – a scarred landscape and much personal suffering'.[105]

Marking time

The 15 years after the war were years of reconstruction for the Solomons. The botany and the stock of some of the major forests had been surveyed, to be followed by a geological survey and the much-awaited survey of land tenure. Major agricultural projects commenced as well, although with mixed results, while the local government councils were consolidated and strengthened. This was a prelude to increased economic development, in which the forests were to play their part.[106] Delays were forced on plans for forest reserves and research into plantation forestry because the status of customary land remained uncertain.

Consequently, the Forestry Department established in 1952 was limited in what it could do. The Tenaru Timbers operation again highlighted the government's lack of control of the forests on lease land. Even if it had had the power, it is debatable whether or not the government would have prevented the virtual cutting out of most of the *ailai* (*Magnifera salomonensis*). Logie's draft regulations provided for government control of forests on leasehold and native land as well as land owned by the Crown. Councils were to exercise considerable control over native reserves and of any income derived from logging there. In the late 1950s, there was a model at hand for disbursing such income when the government allocated some of the royalties from the Vanikolo operation to the local council for spending on community projects, such as a council house, a school, and a teacher in return for reserving the kauri forest for ten years for regeneration experiments.[107] Although this level of local participation had political motives, the colonial government of the immediate post-war period saw a role for Solomon Islanders in the control of their forests, just as it did in joint saw-milling projects. The government intended to prevent direct dealing between the owners of the trees and potential loggers though a system of licensing, with several regulations to control erosion and roading. However, the draft legislation was premised on land that no one owned, which would become native forest reserve land. But if no such land existed, no council would request that such reserves be declared, thus leaving declaration solely to the fiat of the High Commissioner. Thus what appeared an enlightened policy of local council participation, even if sugar coating on the pill of ultimate control of forests by government, had the potential to be a political and financial mine-field.

These issues had to be resolved before any major developments in forestry and logging, particularly on unregistered land. The loggers at Vanikolo had several reasons for withdrawing: their technology and standard of management were never sufficient to achieve the volume of production that might have made the operation viable in the face of more efficient producers elsewhere. However, there were others waiting in the wings. Not long before the Kauri Timber Company left Paeu, its accountant, I. A. Rogers wrote to Trenaman,

> Maybe there was a lot of truth in what your Colonial forestry man said – that the place was so isolated and difficult that it would never be successfully worked by Europeans.[108]

Burns Philp's W. H. Lucas had told his directors the same thing in 1917 and they had the wisdom to take notice. Lucas's assessment had been accurate as well as prophetic. There was also an element of prognostication about the final logging shipment from Paeu. It was destined for Osaka, Japan, the first and last log export by the Kauri Timber Company to that country.[109] It was among the pioneering shipments of 1963/4 which marked the opening of a new market for the Protectorate's timber and the advent of Asian loggers in the islands.

—8—

The Government is but a Stranger ...

> Forest reservation is hardly ever popular and is likely to meet with opposition....
> Reservation is least unacceptable during the stage in the development of a country where land is plentiful and where vacant land is the absolute property of the State. Much valuable forest has been lost through the failure to realise how fleeting this stage can be.
>
> W. A. Gordon, *The Law of Forestry* (London, 1955), 328.

The Protectorate government in the late fifties and sixties introduced laws to create a Forest Estate, implying perpetual or long-term dedication of land for commercial forest use. Most of this was to be Crown or public holdings. In the 1960s, despite the United Nations' views to the contrary, the administration believed that no true political independence could come into being without economic self-sufficiency. The Forest Estate was to be one cornerstone of that self-sufficiency for future generations.[1] Setting that cornerstone in place was to prove more difficult than anticipated. The government attempted by several and often legally-complicated means to obtain land for the Estate, but failed to comprehend the tenacity of Solomon Islanders' attachment to their land and forest. It encountered mounting resistance to the alienation of timber rights and intensifying demands by the people for direct dealing with logging companies.

Demand for trees, demands on land

During the 1950s, tropical hardwood demand expanded with the economic growth of the industrialised countries, especially Japan, Western Europe and the United States, and as supplies of temperate hardwoods lessened. The Far East was the largest producer, led by the Philippines in the early fifties, along with Malaysia (Sabah and Sarawak) in the middle of the decade. At this time, Japan began importing lauan (or meranti) logs from

South-east Asia for processing into both sawn wood and plywood to be sold for foreign exchange.[2]

The economists of United Nations' Food and Agriculture Organisation predicted world timber demand would double by 1975 and continue to rise.[3] Utilisation of tropical forests on an industrialised scale was advocated by bodies such as the World Bank. Few questioned their reasoning and the social, ecological and indeed economic costs.[4] The development gurus foresaw that timber industries would pay a key role in the attack on 'underdevelopment'.[5] In the Protectorate, timber was a renewable resource and surplus to local requirements. Timber industries could provide import substitution and distributed loci of development, attract investment in processing, create employment, increase export earnings and have a multiplier effect throughout the economy. It seemed all that was needed was forest control as well as a data base on the resource.

By 1958/9, the Japanese were seeking further supplies of timber. The Protectorate government was receptive, but had committed itself to a land policy that affected forests. The Special Lands Commissioner, Colin Allan, had delivered his report in 1957. He thought that the Crown should control the occupation of lands acquired under the Waste Lands Ordinances of 1900–1904. It had transferred these 'waste lands' to planting companies for 999 years under Certificates of Occupation. In 1919–1925, the government's Land Commission under Judge Phillips, returned 508 of the 1012 square kilometres alienated to the native claimants, but would not consider claims arising after 1919.[6] These retained lands, recommended Allan, should come under the control of a new Lands Trust Board, with Solomon Islanders members. Allan also recommended that the Trust Board administer 'vacant' land. This was non-alienated land, deemed 'vacant of interests' because it had not been occupied or cultivated or leased for 25 years prior to 1958.[7] Although the administration ignored several of Allan's recommendations, it accepted the concept of land 'vacant of interests'. This went into the Lands and Titles Ordinance of 1959, as did a diluted version of Allan's Lands Trust Board.[8]

'Vacant' land seemed ideal for Forestry Department purposes. However, negotiation for timber rights for logging companies could not proceed because under the prevailing King's Regulation No. 12 of 1922, the government had to compensate the 'owners' of native customary land for the grant of their timber rights. In recognising 'owners', any rationale for subsequently declaring the land vacant was obliterated. So, the government permitted only surveys by loggers, missing investment opportunities but preserving the resource for long-term return to the state.[9]

The Forest Estate: by way of vacant land

Keith Trenaman arrived on secondment from Uganda to become the Chief Forestry Officer in 1956. A pilot during the war, he went on to study forestry at Oxford University. In the Troup tradition, his main objective was to create the Forest Estate as a sustainable asset, to supply both the export market and local needs.[10] Trenaman calculated about ten per cent of the Protectorate's land area was needed for this, a figure twice revised: in 1966 to about 13 per cent or 1,500 square miles and in 1968 to about 4.3 per cent or 500 square miles.[11]

The Forest Estate had two elements: Forest Reserves and Forest Areas. Forest Reserves would be tracts of land dedicated to the production of timber on a sustained yield basis. Reserves could be set up for watershed or even soil protection, but this was not then a major consideration.[12] Ideally, Forest Reserves were to be on Crown or public land because this enabled control, important for trees, a long-term 'crop', with a then-estimated 60-year regrowth period.[13] This land was to be specified after the Lands and Titles Ordinance of 1959 came into effect, providing the mechanism for the declaration of 'vacant' land. Since the population of the Solomons in 1959 was 124,000 or 11 people per square mile (four per square kilometre) on average, it seemed likely that substantial tracts would be found to be vacant of interests.[14]

The second element of the Estate was the Forest Areas,

> to provide for the immediate protection of valuable forest tracts by prohibition of cultivation, settlement and tree-felling for other than domestic purposes, but with no commitment as to the final allocation of the land to forestry or other form of use. The declaration of a forest area is ... applicable to any class of land and with no inquiry or right settlement procedure involved ... forest areas would be declared mainly over 'vacant' lands.
>
> The second function of 'forest areas' is to regulate working of forests ...

The Forest Areas designation gave control to the Forestry Department to say how the trees should be exploited and replaced.[15] This would sew up the loophole in earlier legislation that had allowed anyone with a freehold deed or lease rights to log with impunity, as Tenaru Timbers had done in the 1950s on Lever's land.

Trenaman's Forest Estate elements became law in late 1960 with the passage of the Forests Ordinance. It took a sharp eye to read this legislation. It did not mention 'vacant of interests' land, but defined native customary land as per the Lands and Titles Ordinance, 1959. The other land category defined was registered land and this subsumed land held by the Land Trust Board. The bulk of the land registered under the Board was expected to be land deemed 'vacant' under the Lands and Titles Ordinance, 1959.[16] The designation of 'vacant' land awaited the appointment of a Registrar of Titles to the Lands Department, as did the Lands Trust Board because the Registrar was to be a member. This appointment eventuated in 1961 and the Land Trust Board was established early in 1962. Until then, the full effect of the Lands and Titles Ordinance, 1959 and consequently of the Forests Ordinance, 1960 was largely suspended.[17]

Most Solomon Islanders resisted the concept of vacant land.[18] Member of the Advisory Council and war-hero, Jacob Vouza of northeast Guadalcanal, articulated majority feeling when he said, '...there is no waste land here. Every bit of land belongs to someone'.[19] That sentiment was echoed in 1959 when the Pacific Timber Company and the Forestry Department investigated timber stocks in the Rere-Kaoka district of northeast Guadalcanal, though Trenaman believed ex-gratia payments to the local Council might mollify their resistance to the area being declared a Forest Area.[20] Throughout the possible Forest Estate areas, the government's past policies loomed large. Its application of the Waste Lands Regulations over vast areas had been unjust and the Phillips' Commission had only healed half the wounds. Moreover, there was suspicion of a government that could allow possession by Lever's of huge tracts of land on Kolombangara which remained in 1959 as they had been in 1905, visibly uncultivated, unused, and unoccupied – 'waste' or 'vacant' land, in fact.[21] As Lever's and other expatriates moved to tenure conversion on the new register after 1959, many Solomon Islanders were surprised to find alienated land that had been abandoned during the Depression or that, like much of Lever's 'Certificate of Occupation' land, had no sign of occupation.[22] Thus the fundamental question of whether or not there was land 'vacant of interests' was lost amid past grievances.[23]

'Vacant' land, as Trenaman thought, proved a chimera for the government, which dropped the idea in fact in 1961 and in law in 1964, with the dissolution of the Land Trust Board.[24] Abandoning his professional dispassion, Trenaman declared,

> In the Solomons the timber industry has advanced but little from pre-war days ... land policy has been either ill-defined, or defined but not implemented ... Forestry ought certainly to be one of the main industries ... of the Solomons. Had the basic land problems here been resolved sooner, it is conceivable that the whole of development planning for the Protectorate would have borne a different aspect ... If establishment of the estate on native or native customary land ... were to be preferred, or regarded as inevitable, the decision would have been appropriately made many years ago. In the light of such a decision timber development could consistently have been encouraged and begun under the

provisions of the old Timber Regulation [of 1922]. The frustrations and delays accepted for the sake of a public land estate would become a fruitless sacrifice if the aim were now abandoned.

The vacant land plan had been a blind alley which lacked 'political feasibility'. Trenaman's view was that the Forestry Department's proper concern was not land policy, but getting the Forest Estate established by whatever means acceptable.[25]

The Forest Estate: by way of purchase

By 1960, frustrated with the search for 'vacant' land, Trenaman sought alternatives. He recommended negotiations with lessees of freehold to which the government held the reversionary title, under the land regulation of 1912 that had established the government as sole purchaser of customary land. Another possibility was buying the rights to freehold and land held under Certificates of Occupation.[26] This could provide an opportunity

> [t]o demonstrate the possibilities of expanded timber development on Government or 'alienated' land before or while seeking this on 'public' land[27]

Throughout 1961/2, the government obtained alienated areas for public land on Rob Roy, Vaghena, Tetepare, Baga (Mbava), and Alu (Shortland) islands.[28] As well, following a suggestion of the Lands Commissioner, in 1963/4, tracts with little or no population were purchased from Solomon Islanders: 35,000 acres on Vangunu, 75,000 acres at Allardyce Harbour, Santa Isabel, and 73,000 acres between Viru Harbour and Kalena Bay, New Georgia, a total of about 300 square miles (777 square kilometres) or about a quarter of the area Trenaman hoped to acquire. The terms of the transactions were

(1) an initial payment at the completion of the land transfer transaction, at ten cents an acre;

(2) payment of ten per cent of timber royalties during each year in which timber working takes place – this payment to apply to the timber standing at the time of the agreement for a maximum of 25 years;

(3) the grant to the former owners, indefinitely, of certain rights of usage within the tract; and

(4) reversion to the former owners of stated acreages after the felling of timber thereon, but in any case within a stipulated period.[29]

Once the loggers had been through, and the reversion of the 'stated acreages' completed, the remainder could be designated Forest Reserve.

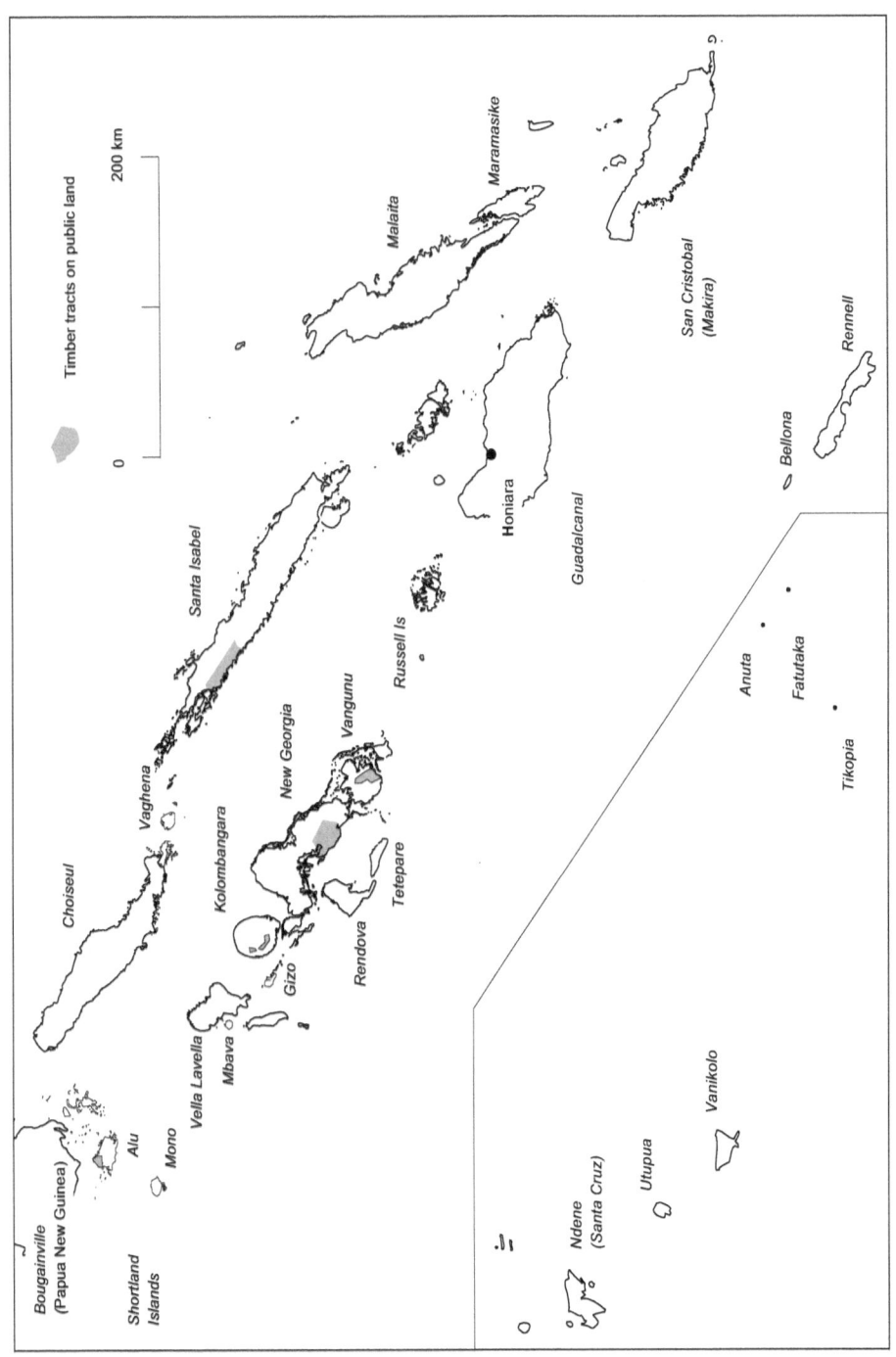

Map 7. Timber tracts on public land of Solomon Islands, 1968, exclusive of declared forest areas.

The Government is but a Stranger ...

Solomon Islanders soon began to question the government's purchasing and its motives. On Vangunu, a leader of one of the corporate groups or *butubutu*, David Livingstone Kavusu, opposed the sale of land in the south-west, once he realised the extent. Although his objection was legally out of time, Kavusu pursued the issue through to the Commissioner who agreed to rescind it. Thus it was only the south-east area that was sold, an area of few settlements along the swampy coast, where the people were mainly Seventh Day Adventists who did not hunt in the bush.[30]

There were objections about the validity of the purchases. Nathan Kera claimed that a parcel of land, Hovilana (Chuvilana) had been sold by false claimants, as part of the Viru, New Georgia transaction.[31] Besides cross-claims to land, there was considerable antipathy to the government, despite a Forestry Department campaign to explain the purposes of the Forest Reserve.[32] As Job Kimi put it, in August 1965,

> It is we who were born in the land and are the right owners of the land here concerned. The government is but a stranger to this native land of ours ... why should it seek to extend its power and wealth and luxury at the expense of the poor in this country?[33]

The predicted demonstration effect of 'expanded developments' on alienated land aroused Solomon Islanders' interest, not regarding the augmentation of the public purse, but of their own. By 1963, on Baga, the British Solomons Forestry Company (owned by Nanpo Ringo Kaisha Ltd) had started logging, as well as having purchased the Tenaru Sawmill Company on Guadalcanal, and Lever's Pacific Timbers Ltd. were operating on their Gizo land, with plans to extend to their Kolombangara land. In 1964, the Australia-based Allardyce Lumber Company started to log their concession on public land at Santa Isabel.[34] Solomon Islanders wanted to sell logs to the loggers. R. T. Kera, a Forest Ranger from New Georgia, Simeon Nano of Vura on behalf of his chief, Sarere, and Paul Alisae of Samanagho, Shortlands all sought Trenaman's consent to sell logs or timber rights to logging companies nearby.[35] At Munda, Milton Siosi told politicians in 1965 that many wanted to sell their timber rights directly to the companies and get the royalty, instead of only ten per cent.[36] The Forestry Department pointed to Section 53 of the Lands and Titles Ordinance (1959) which disallowed Solomon Islanders from negotiating directly with non-Solomon Islanders to dispose of timber rights on non-registered, customary land.[37]

This legislation had two loopholes. First, Section 53 applied only to non-registered land, but Solomon Islanders now could register their land in individual title under the Land and Titles Ordinance (1959). In 1961, the Forestry Department had earmarked as possible 'vacant land' two blocks on east Guadalcanal, one of 20 square miles between the Susu (Rere) and the Bulo rivers and another of eight square miles between the Tina and Toni rivers. With the vacant land concept still on the statute books, Jacob Vouza of Aola, plus headmen from the Tadhimboko area and a European, J. Bryan at Rere called upon the Commissioner of Lands in November 1963, 'with a request that their land be registered, so that they could sell the timber on their land'. The Commissioner assured the party that 'vacant land' was a dead letter, but the government could not object to them applying for a registered estate.[38]

Secondly, the law did not proscribe logs cut by Solomon Islanders for selling to loggers or millers.[39] There was expanding awareness of the forest potential throughout the Protectorate. In the mid-sixties, the Catholic mission at Tangarare, Guadalcanal and Wainoni Bay, Makira introduced co-operative logging and milling schemes, funded by Australian and New Zealander co-religionists. Logs became milled timber and furniture.[40] By late 1965, at Keli Bay, in the Western District, R. Cox was purchasing logs from local people and producing 1000 cubic feet a month of timber with a portable mill.[41] Local people on Kolombangara were selling logs for milling to O'Keefe at Mongga and to Kwan How Yuan (KHY) Pty Ltd, of Gizo.[42] There was a fine line between the sale of individual logs and the sale of timber rights and it was one which both eager logging companies and enterprising Solomon Islanders would cross, given the opportunity.

Prior to purchasing land, the government had presented arguments in favour of the Forest Reserve. It pointed out the need for management and regeneration of the forests; the demarcation of land for gardens and commercial agricultural development to be outside the estate; and the benefits that would flow from the timber industry:

> ... local people stand to gain very considerably. The government will make a substantial initial payment, and thereafter payment of a proportion of the timber revenues collected on the original trees in the forest, for a maximum period of 25 years. Timber and forestry development would mean plenty of employment in the areas concerned, with opportunities of learning skills such as the handling of mechanical equipment. The timber companies would make roads into the forest which would help the people in later expanding their garden and crop areas. The Companies would also bring schools and medical facilities to the area.[43]

These were the development promises of 1963. Trenaman who had worked unremittingly to revive post-war interest of overseas companies was optimistic.[44] 'The colossus appears to be astir and a pattern of timber development appears to be evolving', exemplified in the operations of the British Solomons Forestry Company and the plans of Lever's, Allardyce Lumber Company, and Mears and Ireton.[45] However, the market weakened, especially in Japan, and remained weak into 1966, with further fluctuations in 1967. A more significant disincentive to the expansion of the industry was the availability of gigantic, well-stocked timber concessions in Indonesia in the late 1960s.[46]

The Government is but a Stranger ...

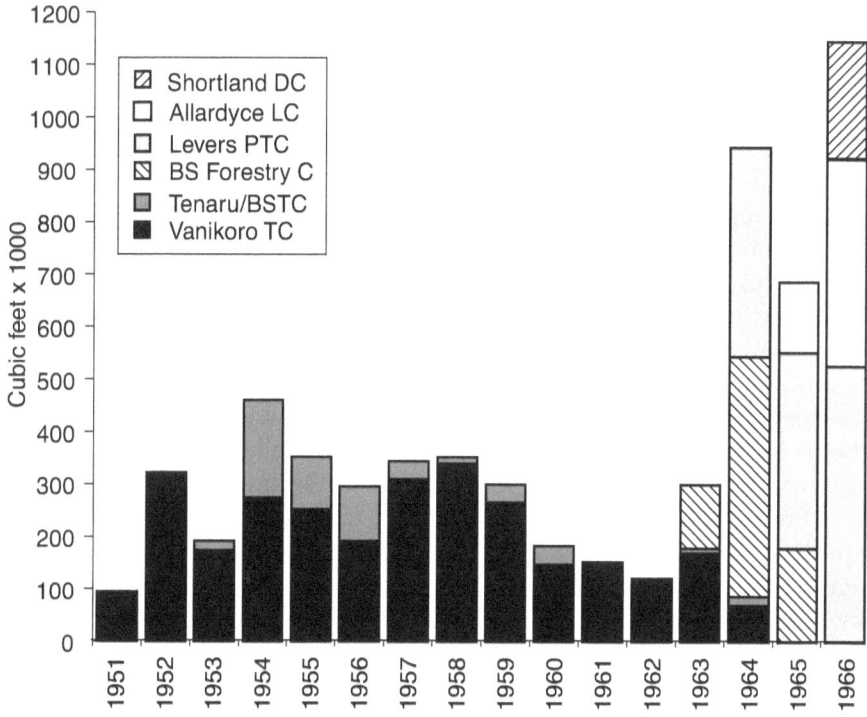

Figure 66. Exports by logging companies, 1951–1966.

Units are cubic feet true measure. (One superficial feet of log is one twelfth of a Hoppus cubic foot and 0.106 of a cubic foot true measure).
Sources: FDAR 1953–1965

In 1964, not only had the Kauri Timber Company left Vanikolo, but the company on the Viru-Kalena concession, Kalena Timber Company also failed to start and produced no logs the following year. The Shortland Development Company, with 50–50 Japanese-Australian backing, completed its base in 1964, yet had shipped no logs in 1965.[47] Moreover, British Solomons Forestry Company's production from Baga included much *Dillenia salomonensis* or *mudi*, a hardwood, which also made up about 60 per cent of the stock on Vangunu. The market for the little-known *Dillenia* was limited, so the company left Baga in 1965 and declined a concession on Vangunu.[48]

In the Protectorate, loggers faced many problems. All had to contend with almost total unavailability of local labour skilled in the use of chain-saws, tractors and yarders, the consequent high wage costs of expatriates, and difficulties with machinery and roading in the wet climate and sticky, clay soils. These establishment problems and the unforeseeable decline in the world market did little to bring substance to the government's earlier bright predictions and caused some local groups to doubt its words.[49]

For the Forestry Department, the probability that more Solomon Islanders would register their lands loomed larger because the Lands Department was actively encouraging 'land settlement'. The object of land settlement was to 'to give individuals the power to dispose of their [registered] lands as they see fit', in other words, to assist some to become entrepreneurs and big land-owners.[50] The rationale for such a radical tenure change was the assumption, still held to-day by orthodox economists, that security of title was essential to economic development. Starting with a successful pilot scheme in New Georgia in 1965/6, more and more land could be outside the scope of Section 53 of the Lands and Titles Ordinance.

Locating and mapping land suited to the Forestry Department's requirements had taken time. By 1967/8, the Department had reasonable data of timber areas based on ground enumerations and aerial photography. Yet, the government did not want to rush willy-nilly into purchasing. Reforestation on logged government land had only just begun. Time was needed to continue experiments in silviculture techniques. The market was also a factor – was it better to invest in enrichment planting of fast-growing, good quality hardwood species or in close, concentrated planting of very quick-growing, 'cellulose' producing timbers for pulp with a short rotation of six to ten years? Such matters were highly technical, requiring intensive research and extensive trials.[51] The worry for Trenaman was that while all this was going on, through the registration of native title and the sale of logs, much of the potential economic forest could be cut down, without any possibility of systematic reforestation.

Buying time: declaration of Forest Areas

To deal with these concerns, the government invoked Section 23 and 24 of the Forests Ordinance (1960), so that suitable customary land could be declared Forest Areas, thereby setting the land aside 'until timber working arrangements were needed therein' or the people needed garden and agriculture land.[52] As more difficulties had been put in the way of the creation of the Forest Reserve, Trenaman had become increasingly pre-occupied with the production rather than the protection function of forestry. Beyond the Forest Areas, though he foresaw that only 'patches' of customary land would

The Government is but a Stranger ...

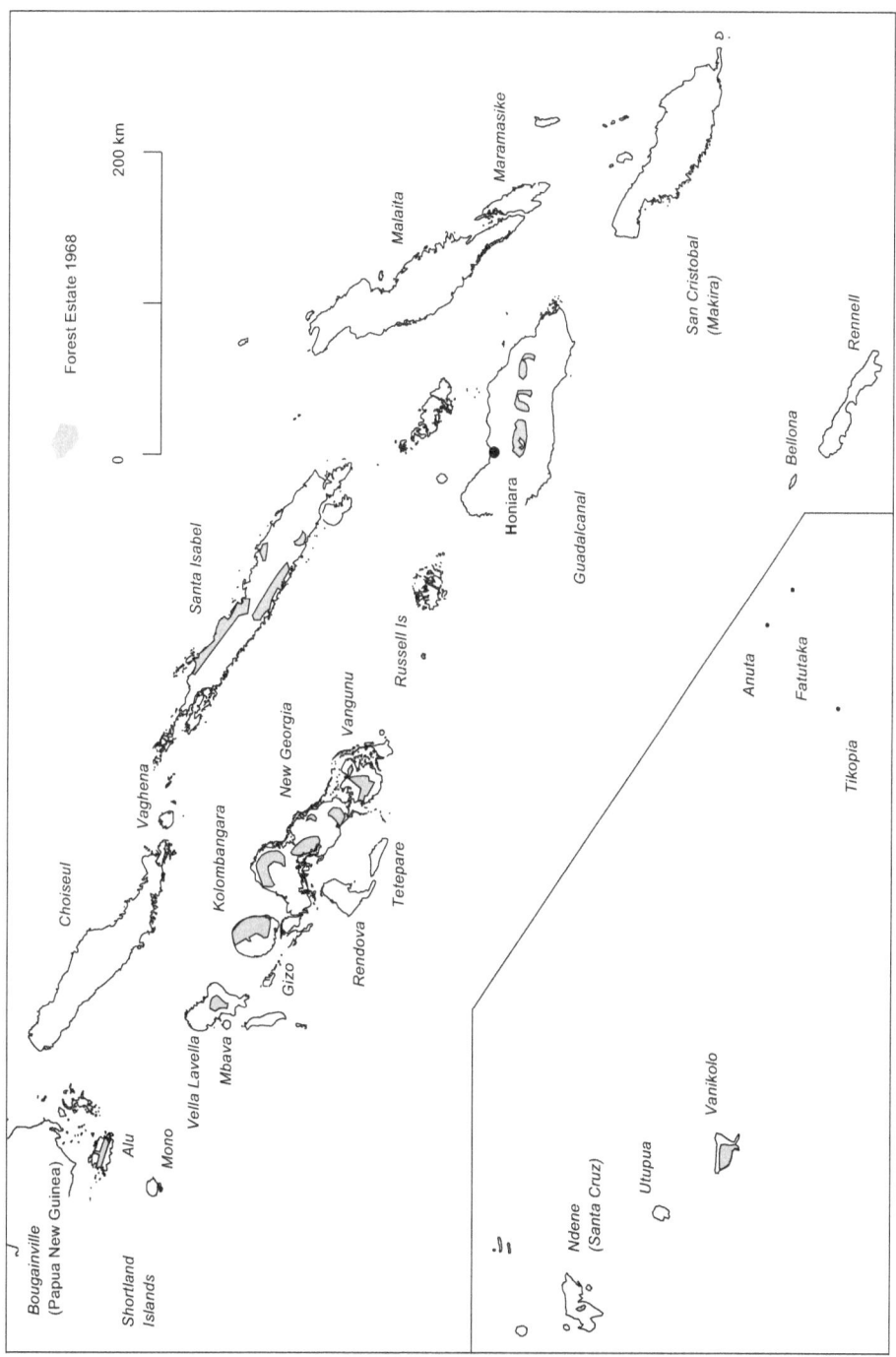

Map 8. Forest areas declared in 1968, exclusive of land purchased by the government.

be involved, he was even prepared to allow the unfettered exploitation of trees without reforestation.[53]

The government realised that invoking Forest Area controls could have political ramifications, but Trenaman's rationale had a compelling logic, provided Solomon Islanders accepted the premise that the government had the authority to impose forest laws.[54] The government's initial Forest Area declarations in 1966 had received no publicity, since the first was on part of Lever's' Kolombangara and Gizo land which it assigned to the government, and the second, in the Shortlands, involved an exchange of customary for alienated plantation land. The people were closely consulted.[55] The same cannot be said for the 17 areas declared in May 1968. Mainly on the New Georgia islands and Santa Isabel, with pockets on Alu and inland north-east Guadalcanal, the 17 Forest Areas made up 760 square miles (1,968 square kilometres). In its notices (Appendix 3) and in radio broadcasts, the government stated,

> that the Forest Area declarations do not affect or alter the ownership of the land and have nothing to do with land acquisition. Any land purchase proposal by the government will be made quite separately.[56]

Most Solomon Islanders believed the government was taking the land, not simply reserving trees. So the People's Protection Party, the Protectorate's second political party (see Appendix 4), demanded:

> Do we own the land? If so, don't we own the timber? Why can the government steal out timber and our land in this manner? Government has told us that they are not confiscating our land, but Section 25 [of the Forests ordinance] contains a provision that the Chief Forestry Officer may order the people who work the timber on our land to replant. This could go on forever. Isn't this the same as stealing our land? Why didn't the Government consult us first before making use of Section 23? What moneys have the Government ever spent in the past on the areas now being expropriated? Can we trust the Government to spend the royalty for our benefit? Why can't we make our own contracts for the sale of timber in what is unquestionably our forest country? Is it because Government thinks we are not fit and proper people to make such contracts?[57]

They had their own ideas about development,

> We want to sell our timber for our own benefit so we can use the money we receive to pay for our own development ... It seems to us that outside of Honiara, Auki, and Gizo, nothing exists in the eyes of the Government. With the money that we should get from our timber ... we can clear large sections of our land in a proper and modern manner, buy tractors, pay for fencing, cattle and other livestock; and look forward with confidence to a better future.[58]

Balesi of Boreko, New Georgia told the government it could not take the land:

Reason – Because the ground is the life for me and for my whole people. The piece of land is not increase but my peoples are. Timber or trees would be allowed to any one who wants to buy.[sic][59]

Direct dealing was developing into a major issue as Solomon Islanders became aware of returns from log sales. Lever's, for example, paid on average 23 cents per cubic foot while the government paid in royalties to former land rights holders an average of 0.125 cents per cubic foot, ten per cent of the royalties it collected from the loggers. Memories of government's cash payments for the land and trees faded in the light of this differential.

Reflecting these concerns, the Legislative Council threw out the notion of Forest Areas. So intense was the feeling that the government was forced to abandon aspects of its policy enunciated in a White Paper on Forestry and take on board the recommendations of the Select Committee on Forests of August 1968.[60] Although Trenaman saw the rejection of the Forest Areas as a retrograde step, in order to save the rest of the White Paper, he sought a compromise.[61]

The Forest Estate: by way of 'profits'

Again, the government was forced to change tack. Under the new Forests and Timber Ordinance, 1969, 'State Forests' replaced Forest Reserves, the lands owned or leased by the government. 'Controlled Forests' could be declared, after consultation, on any land, but only to protect vegetation 'in a rainfall catchment area for the purpose of conserving water resources' (see Appendix 5). The Forestry Department adopted a system of control of forest exploitation by licence on government land and tried unsuccessfully to augment the Estate through purchase or long-term lease. Companies in the Protectorate wanted further tracts and long-term security before they would consider the possibility of processing timber. Trenaman suggested the purchase of the timber 'profits' – relatively short-term leasing of the timber-cutting rights which, he still hoped, would precede subsequent government purchase or long-term lease.[62] He based his hope on the fact that the rights acquisition programme involved 'a form of registration of ownership'.[63]

By 1970, the Department was busy reforesting the logged areas of the Forest Estate and could not manage any more, so the new scheme had the potential to create a pool of forested land with defined boundaries and owners.[64] This became more urgent with the government's Sixth Development Plan (1971–1973), with its stress on economic development and diversification. Forestry was to make a major contribution, an increase of log exports from nine million cubic feet in 1971 to 15 million in 1973, and

20 million in 1975.⁶⁵ The enabling Amendment for 'profits' came into law in 1972.⁶⁶ It was not the hermetic seal that the Forest Areas would have been, but with ownership of the trees and hence the land clarified, the Forestry Department could negotiate with the recorded owners without any complications.⁶⁷ This was the theory, but like many colonial assumptions, did not necessarily reflect Melanesian reality.

Negotiations for timber rights took time, as did the submission of 'bids' by interested companies after they made their own appraisals. The first success of the government under this system was for 'profits' over a large tract on Santa Isabel, but the devastation of cyclone Ida dramatically cut off the proceedings in 1972.⁶⁸

When the Government switched its attention to New Georgia, negotiations slowed because of shortage of funds and experienced Lands Officers. However, there were other developments which were to impinge on transactions for timber as well as land rights. Late in 1972, the Lands and Titles Amendment Ordinance came into being, altering the law in two important features. First, for purposes of 'land settlement', title no longer had to be individual: where groups of five or more people had joint ownership, title could be recorded under the group's customary name with the names and descriptions of the persons appointed to exercise rights of ownership on its behalf. Secondly, instead of disputes over customary land being dealt with by the local native courts, with the right of appeal to the high court as had been introduced in 1959, there would be customary land appeal district courts to hear appeals from the native court. The government hoped this would speed up land settlement while conciliating Solomon Islander pressure for a simpler system of land registration. In fact, the procedure was much the same as before, with rigid demarcation of boundaries and recognition of only certain interests in land, but with a clan's name tacked on instead of an individual's. Moreover, although much disagreement was settled informally, dissatisfied plaintiffs from the customary land appeal court still had the right of appeal to the high court. The Amendment appeared to remove the partiality of the small localised native courts by creating a district level one, but, in keeping the tier of courts, it adhered to an adversarial British system with no reconciliatory mechanisms necessary for what were often extended-family and clan disputes over land use.⁶⁹

The introduction of clan representatives, trustees in effect, seemed to the government to betoken an end to protracted negotiations by officers with anything from dozens to hundreds of individuals. But clan heads' and chiefs' authority had been weakened by introduced institutions over the decades. Many with the knowledge of rights did not have the personality or formal education to assert themselves beyond the family or village level. Opportunities had arisen through church and government positions and the cash economy, creating new contenders for local power. Thus even within groups who had a common descent, there was manipulation of genealogies and clan history to create a new orthodoxy, benefiting sectional interests.[70] By 1974, the government's negotiations for forest land on north New Georgia ran foul of these internal power struggles and protracted court hearings. Many did not trust the 'trustees' and feared that these would claim ownership of clan land through registration.[71] The problem was sent back to Legislative Assembly which wanted the government to consider

> enabling timber extraction rights to be secured in New Georgia in such a manner as to ensure that customary land involved will remain as customary land [72]

Opening the forest to direct dealing: saw-millers and the Forestry Department

For three years at least, the Forest and Timber Ordinance, 1969 had restricted commercial logging to government land. However, it allowed the felling of trees for milling under licences, requiring the applicant to specify where trees were to be cut, and the quantities sought, and these trees could be on customary land. By regulating saw-milling, several goals should be achieved. First, no Solomon Islander could sell individual logs to the saw-miller unless they were from within the area designated by the licence. So, the government's hand was stronger than under the 1960 legislation and it could stop trees being cut from 'important timber areas'.[73] Second, the Conservator of Forests had the power to require a mill be erected and maintained or some other 'timber manufacturing plant' built.[74] This would eliminate the need for imported milled timber and could lead to 'more local industry per unit volume of the resource'.[75] Thirdly, saw-milling licences would be issued 'to smaller operators to cover working of relatively unimportant timber areas', also spreading the milling benefits.[76]

Experience had convinced the government that only expatriate companies or operators could run saw-mills efficiently, though local interest grew along with the success of such mills. Throughout the 1960s and 1970s, these mills had produced more than four-fifths of the domestic demand. In the late 1960s, the Forestry Department's initiatives and the government's refusal to purchase untreated timber saw the introduction at most mills of the boron-diffusion process, which produced timber resistant to fungus and termite attack, although the standard of milling remained variable. The British Solomons Timber Company, successors in 1962 of Tenaru Timbers Ltd, was the major producer, along with O'Keefe's mill on Kolombangara (until 1970 when it closed) and the mission saw-mills at Mbatuna and Mbuma. These produced mainly for the local market and missions, but for a few years from 1966 some sawn timber went to New Caledonia and the New Hebrides. In 1973, Foxwoods, a Queensland-based company, bought out British Solomons Timber Company. By 1976, as well as producing timber for export, Foxwoods was manufacturing veneers for Australia, despite high freight which added 25 per cent to costs.[77]

An indigenous saw-milling business had started on Malaita, the West Kwara'ae Sawmill Company in 1956. Unlike Logie, Trenaman had not supported the training of Solomon Islanders through joint venture saw-milling, but was at least willing to give advice. Seeing the demand for timber in Auki, the company raised money for equipment. The power saw was too slow and could not keep up the supply of flitches to the circular saw, as Trenaman had warned. The logs or hand-hewn flitches had to be manhandled increasing distances to the mill. Accounting was sporadic and, after the clerk drowned when the *Melanesian* sank in July 1958, little sense could be made of the books. Soon after, the shed housing the equipment mysteriously burnt down.[78] The district commissioner estimated the losses at £900 and considered the business, 'a local (and dead) white elephant'. Malaitans were discouraged from setting up similar enterprises lest they get their 'fingers burnt'.[79] Trenaman had seen the lack of mechanical skills among local timber workers in the mid-1960s. Over 100 expatriates were working in the logging industry in 1967 because of 'the dearth of technically skilled personnel in the Protectorate'.[80] Lever's manager described the situation more bluntly:

> There was no industry here except copra, where you merely picked up a coconut, cracked it open and scraped it out. Men have no background in machinery. [81]

Under the aegis of its South Pacific Technical Assistance Programme, Australia sent its first resource person in 1967 to advise companies in the planning of '"on the job" training of heavy plant operators'.[82] Lever's brought in Chinese, Fijians, and West Africans to make up the deficit and teach the local workers, who, by 1966/7, were operating and maintaining machinery with minimal foreign supervision.[83] It was not until the mid-1970s, however, that the replacement of overseas operators and technicians by local people became common.[84] So Trenaman's hesitancy in supporting local enterprise was not racism: at best, it was prudence, at worst, it was paternalism. Paternalism may have prevented pain, but it did not further the learning process or find popular support.

By 1974, the Governing Council, with half its members elected, had been replaced by an almost totally-elected Assembly with a ministerial system and a Chief Minister. Elected members, 'in a very clear majority and obviously with the feel of power', demanded more local participation

The Government is but a Stranger ...

Ted Marriott, FDAR 1969, by permission of Forestry Division

Figure 67. Yarder for high-lead log extraction, 1969. Note Asian driver. Many foreign staff were employed by logging companies in the 1960s because they had the technical skills needed.

in logging and wider geographical distribution of its economic benefits.[85] Several Solomon Islanders continued to express interest in running portable saw-mills, seeking action from their politicians.[86] This was difficult for Trenaman. His view of economic benefit was largely that of the bureaucrat looking at the Protectorate within the world economy and, being a forester, within a long time frame. A forest industry that was of benefit to the state provided all its citizen with benefits 'though it is not easy for local people to appreciate all or any of them unless closely connected with the industry' in places where logging was operating.[87] By 1970, the Forestry Department's main objective was 'resource and land management' concerned with the 'farming of tree (wood) crops' on a large scale.[88] Production forestry was a business, thus the Forestry Department was not interested in service or extension work, such as showing villagers how to replant a couple of acres or setting up saw-mill schools.[89]

Trenaman had long believed attempting large-scale reforestation on customary land would be fraught with complications.[90] Between 1965 and 1973, 12,000 acres (4,000 hectares) of government land had been replanted, financed by British aid. Future reforestation was almost certain to be aid-derived or in co-operation with commercial interests and this meant negotiations at a central government level. Small was not beautiful to Trenaman; it was a waste of scarce departmental and national resources.[91] Consequently, he did not consider alternatives to plantation reforestation on government land.[92]

Trenaman, by 1973, had to bow to increasing political pressure for local development at some level, but deflected some of this on to the Agricultural and Industrial Loans Board. Although it financed chain-saw and winch purchase to assist log-cutters to fell and move logs out for sale, it too had reservations:[93]

> Solomon Island participation ... has been very limited. Opportunities are presently confined to areas providing market outlet for logging operations. Requests for portable sawmills have been generally discouraged because of doubts as to expertise available to operate mills and to market outlets for timber produced.[94]

The Malaita Development Council, with former forest guard David Kausimae Nanu and businessman Mariano Kelesi as members, urged more involvement.[95] Trenaman observed,

> ... there has for many years been interest in small business or local authority enterprises operating small sawmill units. The technical and managerial skills required are high in relation to the scale of operation and returns, and experience with such enterprises ... has seldom been happy.[96]

To fill the technical and managerial void, Trenaman favoured a partnership between local 'satellite' saw-mills with a logging-milling company, and had both Allardyce Lumber Company Ltd and British Solomon Island Timbers expressing interest.[97] He held that any 'geographical extension of timber working' was predicated on

> the continuation of the timber rights acquisition programme to the additional tracts concerned. Even with local equity and/or other participation, it will remain as important to ensure that cutting rights are available to operators on a completely secure basis.[98]

Trenaman now was prepared to provide surveys of stands of native-held timber to foster interest from companies, and he supported European managed local mills such as one based at Onepusu, Malaita run by Atasi (Alliance Training Association Solomon Islands, the business arm of the South Sea Evangelical Church, set up by New Zealander, David Harry). In 1974, the Legislative Assembly gained the right of assent over the issuing of licences. Solomon Islander insistence was such that the Forestry

Department supplied a supervisor for portable saw-mills, a demonstration mill at Dala, and short-listed operators for possible loans. On Malaita, the Rafea and Kwaleunga Sawmill Co-operative was chosen. Sponsored jointly by the Department and the Development Bank of the Solomon Islands, the co-operative started milling at Bina with a Forestmil in 1975, upgrading to a Forester band saw in 1979/80 with another loan. In its early years, it employed an experienced New Zealand manager and did well. The Forest Policy Review Committee echoed the demand for small mills. All this interest came at a time when FAO consultant A. Leslie was advising the Department that log exports, as opposed to processed timber, were likely to be of greater economic benefit for several years to come.[99]

The Forest Policy Review, 1974

In 1968, the uproar concerning the Forest Areas declarations had overshadowed the White Paper on Forests. This document indicated where policy decisions were required, regarding the extraction rate, the size and distribution of the Forest Estate, and regeneration programmes.[100] Implementing a proposal of the Paper, the Legislative Assembly set up a Select Committee on Forest Policy in 1974, to investigate a range of concerns, including 'how timber rights on customary land can best be made available for large-scale timber working requiring high-level investment'.[101] It received submissions advocating direct dealing between timber rights' holders and the logging companies, but came out in support of government purchase of timber rights, recommending that the Area Committee of local government have the power to settle disputes over rights to customary land, with the right of appeal to the Customary Land Appeal Court (CLAC). Timber rights could be assigned, without registration of land ownership. Log exports were to be no more than ten million cubic feet per year. All new licences to loggers incorporated a provision for local processing. Trenaman's hope that the Department would become a statutory body, freeing itself from administrative practices unsuited to estate management and from possible political interference, was not supported by the committee. The committee recommended replanting of customary land with local consent. It opted for local financing of replanting, but the Assembly rejected this, preferring to depend on aid.[102]

Land matters concerned this Committee. Knowledgeable Solomon Islanders, like Fred Osifelo, objected to the government retaining the power to declare Controlled Forests, but others were to go further.[103] Almost before its findings could be digested, another, the Committee of Inquiry into Lands and Mining, was hearing submissions. Its radical recommendations in 1976 urged the return of all alienated lands, including government land, to the original 'owners' and the abandonment of 'land settlement', except where land-holding groups were concerned. Individualisation of title was rejected. It recommended direct dealing between timber-rights holders and loggers.

Forthcoming independence constrained both the departing colonial administration and the emerging Solomon Islands government. The British would not supply aid indefinitely and a future government needed some assets, including state forest. An amendment to the Lands and Titles Act, supported in the Legislative Assembly by Chief Minister Peter Kenilorea in November 1977, introduced a compromise. Privately held alienated land would become 75-year government leases, but the government would keep its lands, including those for forestry.

Direct dealing between the forest holders and the logging companies was a consequence of this and the Forests and Timber Amendment, 1977. A company had to seek consent from the Conservator of Forests to commence negotiations with the owners. Area Committees would adjudicate on customary claims to trees and would be the bodies that certified who owned the rights. The Conservator could veto agreements on purely procedural matters before logging commenced. He could then recommend to the Minister of Natural Resources that the agreement be approved. The Minister had the power to make regulations, but this was mainly confined to procedural matters, too. The customary owners could seek the Conservator's advice. His powers to control loggers in relation to the agreements and logging methods, once extraction had commenced, were practically non-existent on customary land. For some, the new law seemed a solution to the bottleneck in the government's negotiations for access to customary land on Kolombangara and north New Georgia.[104] How an untried procedure would function amid the many demands on the new state had yet to be seen.[105]

The legacy

The attempts to establish a Forest Estate revealed two different perceptions of the forest. Through the Ordinance of 1960, the British colonial government could set aside land of no other apparent use to create a sustainable asset for the future Solomons' state and people. Underlying this legislation was the assumption that the Solomon Islanders needed to be protected from cheats among potential loggers and the forests from unsustainable logging regimes, but a deeper sub-text was the Western settler view that idle land was anathema.[106] The British perceived much of the Protectorate as idle land and even in the 1950s and 1960s they believed, just as Woodford had thought in the 1890s, that the Solomon Islanders were not making the most of their resources and that, through no fault of their own, they lacked the skills to do so. Self-government and independence were not far off and Britain could not underwrite the Protectorate indefinitely, so development had to be pushed. There was too much danger in putting this into the hands of 'aliens', such as the Chinese, and significant European settlement was a lost cause. If the people could not do it, the government – eventually an independent one of educated leaders – would have to, and thus it should have control of significant assets for the public good.

Within the broader colonial context of Africa and South Asia, Britain was simply in the habit of seeing forests as a financial asset for the colonial state, and had introduced forest policies by coercion in an age when this was politically acceptable – to the colonisers at least. The old habit persisted, reinforced by a cadre of professional foresters who saw scientific forestry as a justification for land acquisition, but who often understood more about the culture of the forest than the culture of its occupants. In the post-war context, the Forest Estate project seemed not only scientifically and economically rational, but also commendable because it was to benefit a soon-to-be independent Solomons, not colonial capitalist ventures as was the case, say, in nineteenth century India.[107] Unlike much of India, it seemed even simpler in the Protectorate as there was no indigenous tradition of commercial timber trading and marketing;[108] the forest, from the purview of most of the British administration, was almost a free good.

However conceptualised, the original estate of 'vacant' land represented a form of enclosure, of designating as government-controlled lands which formerly had been to Solomon Islanders clan forest commons and potential garden sites.[109] Economically, it had been a form of capital that the forest dwellers and farmers could draw upon when needed. For the many small societies of the Solomon Islands, there could be future localised need and government control would lock up the land. To assign dormant resources therein to a colonial government was almost unthinkable. Moreover, to some Melanesians, land and its life-giving forest cover were not only economics, a view that the unimaginative administration never really understood. Even where groups no longer continuously occupied an area, land still for them embodied and defined them. It was the source of history and of what made each society unique. To renounce claims to land and forests demeaned the living and the dead who had taken their identity and sustenance from it.

The failure to obtain the basis of the Forest Reserve – the elusive and, to Solomon Islanders, the illusory 'vacant' land – retarded the progress of the Forest Estate. The government then became preoccupied with the purchase of alienated and customary land. Customary interests in or rights to land thus became the equivalent of ownership of land, and the chance of the government exercising state or public rights for the common good over all forested areas, was increasingly contested. By 1966, Trenaman's vision of the role of his Department had narrowed. He was prepared to allow access to customary lands by loggers through direct dealing with the rights holders because he thought that exploitation on customary land would be confined to isolated areas. Anyway, this was not likely to be a great threat, as he still had the trump card of the Forest Areas section of the 1960 Ordinance by which the government could declare any forest protected. A significant outcome of this assessment was lack of attention to development of administrative structures and procedures suited to reforestation of customary lands.

There had been no need to declare Forest Areas in 1960 as there was no new logging going on, but towards the close of the decade there were four companies involved and as many millers. The time taken in building a forestry database as well as in government negotiations for the purchase or lease of land gave many Solomon Islanders opportunity to realise from the logging going on that they, as individuals or groups, would obtain more immediate return from the direct sale of timber rights to loggers while still retaining their lands. Desperate for time and desperate to preserve the commercial forests, Trenaman activated the Forest Areas legislation in 1968, signalling another form of government enclosure, and incurring the wrath of Solomon Islanders. The government was indeed 'a stranger' and did not appreciate fully that Solomon Islanders claimed enduring rights to their forests.

Bowing to vociferous objections, Trenaman hoped to protect the forests through the Ordinance of 1969, setting aside Controlled Forests to control, via the saw-mill licensing requirements, felling of trees on customary land and to safeguard water catchments. Any customary land the government wanted to protect could be excised from a licence. Since the Department expected large-scale loggers to provide saw-mills, they had neither anticipated nor had they encouraged Solomon Islanders getting into the saw-milling business on any significant scale, because of lack of expertise. They also assumed on past performance that companies would co-operate with them. But Solomon Islanders saw the forest as their resource and wished to use it for their development, not through government mediation. Their interest in local saw-mills and cutting their own trees, as well as the frustration with not getting the bulk of royalties from their timber rights, reflect their desire to be in charge of their own development. On the eve of Independence, this was an obvious popular cause for an ambitious politician.

When Trenaman left the Solomons in 1975, the State Forest was about 250–300 square miles, about half of what he finally considered the minimum for a sustainable asset.[110] The creation of what had been the Reserve largely ceased in 1969. From then too, the colonial government, despite the intention of the saw-mill licensing clause, largely ceased to control the forests on customary land. In order to assist the logging companies and their increasing demand for logs in areas beyond government land, Trenaman applied the timber 'profits' concept, which still interposed the government as broker between local people and loggers. As an outcome of these policy assumptions, the government had never considered setting up structures and legislation whereby the logger and resource owner might deal directly in a mutually acceptable and predictable manner.

Legislation had been forced full circle, partly mirroring the rise and retreat of the control of the colonial government: from virtual direct dealing in King's Regulation No. 9 of 1913 for the minor Shortland logging operation, to government as lessor of timber rights and licence-giver in King's Regulation of No. 12 of 1922 when a major operator came to Vanikolo, to the Forest Ordinance of 1960. This Ordinance gave the government enormous potential power over the forests at a time when the log market was expanding, but, as local pressure mounted, the government retreated to being lessor and licence-giver under the timber profits leases Amendment of 1972 and then, in 1977, substantially back to direct dealing between loggers and landholders on customary land.

The nascent independent government set the seal on the weakening of state control of forest use for the public good when it introduced this legislation in 1977. Unlike the departing colonial government, Peter Kenilorea and his supporters must have believed that the people of this new nation of villagers were indeed 'a fit and proper people to make contracts'.

—9—
The Forestry Department and Control of Species, c. 1960–1980

> Forestry policy in the long term remains aimed at stabilising the growth of log extraction, cautiously increasing local processing and replacing natural forest timber with plantation grown material of similar quality.
>
> OF, F 2/3, 306, Vol. 111: Statement: Forestry, c. May 1978.

The government's thrust for the Forest Estate was premised on the economic sustainablilty of the forests. Though some timber was for local use, the dominant concern was with creating a viable export industry. For the Department this meant that timbers grown had to be suited to the market as well as plantation conditions. At this early stage of planning for an inherently long-term 'crop' and a distant future, policy focused on production forestry, rather than protective forestry. This dictated the direction of the Forestry Department's research programme on matters ranging from reforestation techniques to botany, a direction reinforced by the need to attract aid grants from Britain and other donors.

Functions of the Forestry Department: production, protection?

The first five years of Keith Trenaman's service had been dominated by waiting for legislation and for 'vacant' land. Trenaman looked to putting likely land for the Forest Estate 'to early productive use'.[1] Policy broadened at the turn of the decade to stress,[2]

> ... the significance of the 'protective' function of trees and forests under the topographic and climatic conditions prevailing in the Solomons, and ... the prime importance of securing a Forest Estate adequate for both the protective and productive needs of the territory. Attention has also been drawn to the

The Forestry Department and Control of Species

importance of conservational land use generally and the need for co-ordination between the various land using agencies.³

This received more prosaic expression in a 1960 statement to Solomon Islanders to explain the new Forest Ordinance. This gave the government powers to control the exploitation of the forests. Its rationale incorporated the dual protection aspects of forestry: first, as an adjunct to conservation of soil and water; second, as a way of protecting the forests from exploitation that would interfere with their sustainability as a resource, including the provision for reforestation by natural or artificial means. Trenaman, from 1960, did not consider the former aspect an issue because of lack of population pressure on the land, and if it were so in odd places, it was more appropriately the domain of the Department of Agriculture.⁴ Though a margin of 800–1,500 metre coastal fringe was excluded from Forestry Department calculations of timber stocks in the western areas from 1968, this was to allow for village cash cropping, rather than species, shoreline or reef protection.⁵

By 1969, soil degradation was reported in West Kwara'ae, Malaita from too short a bush fallow, induced by increasing population. This had been noted for parts of Malaita as early as 1947. The Kwara'ae study had involved C. Nash, a senior member of the Department, but only rated a brief mention in the Forestry Department Annual Report of 1969.⁶ In 1975, Acting Chief Forestry Officer K. Marten noted, regarding the role of protection forestry in maintaining clean water and soil fertility: 'A critical situation may well have been reached on parts of Malaita.'.⁷ The following year, with the Agriculture Department, Marten set up trials to test for suitable species to grow on the exhausted hill soils of Malaita, on old garden sites.⁸

More apposite to the Forestry Department, a 1970 study of logged land at Allardyce on Santa Isabel had revealed that ten per cent of the area (apart from roads) had serious topsoil damage (which was still apparent and a hindrance to planned reforestation 20 years later!). Although the Department required logging plans from companies and monitored extraction, it accepted that soil damage was inevitable.⁹ A study by the Agriculture Department on Kolombangara in 1973 revealed that the use of tractors in logging had re-distributed and compacted the top-soil, reducing the growth rate of replanted trees. Excluding the gravelled roads, 15 per cent of the logged area was so damaged.¹⁰ Yet Lever's were only taking trees of six feet girth and over.¹¹ If this damage worried Trenaman, it was in terms of the dangers to the economic, not the ecological, sustainability of the timber resource and soils.¹² The Department judged it 'of little consequence' because line-planting (see below) was the preferred method of reforestation and, with a space of 10 or 12 metres between rows of seedlings, the poor spots could be avoided.¹³

Ted Marriott, FDAR 1969, by permission of Forestry Division

Figure 68. Log extraction by crawler tractor in the late 1960s. The several passages of the tractor over the one path result in compacted soil and often erosion trails.

Although the term 'protection forestry', like 'sustained yield', features in the Forestry Department's reports of the decade, Trenaman's primary concerns were not ecological, except insofar as economic species of trees were concerned.[14] In 1963, when Levers' Kolombangara survey by W. (Bill) R. Evans revealed more forest type variability than expected, Trenaman commented, 'grist for the Botanist/Ecologists' mill' which indeed was confirmed soon after by the British forest botanist, T. C. Whitmore, who recorded many unique features of Solomons' forests and urged the creation of nature reserves on Kolombangara and elsewhere.[15]

Earlier, in 1954, the Protectorate's first nature reserve, Queen Elizabeth II national park, in the Mt Austin–Lungga River area of north Guadalcanal, had been declared, becoming the responsibility of the Department of District Administration. Its geology meant it was a fertile area, originally covered with closed forest dominated by *Pometia pinnata* and *Vitex cofassus*. John Logie had thought this area would have been ideal for supplying Honiara with firewood. It did, at least for some. The park, year by year, suffered degradation as squatters, mainly Malaitans,

The Forestry Department and Control of Species

gardened, cut wood, lit fires, built houses, kept pigs, directly or indirectly killed birds, and generally denuded the hills, now dominated by thermida grass and the 'weed' tree, *Brousonnettia papirifera*. Although Geoff Dennis of the Forestry Department, fought for years to preserve the park, within sight of the Department's office in Honiara, Trenaman did not consider the increasingly-populated area suitable for the preservation of the forest and was concerned only with the fate of the Mt Austen experimental forestry plots, while the local Guadalcanal people increasingly were concerned about the squatters.[16]

Courtesy of Marion Trenaman

Figure 69. Keith Trenaman and assistant, with Trenaman's sons, on way to Mt Austen forestry plots, early 1960s.

As directed by the Colonial Office in 1961, he was, however, more supportive of Whitmore's proposed reserve on Kolombangara, from the central crater edge with its upper montane forest to the lowland tropical forest near the seashore, because studies conducted there would almost certainly provide data relevant to silviculture. Whitmore was aware of the limitations of economic sustainability and reminded administrators of what was to become known as the principle of intergenerational equity:

> Predictions of future demand are leading to the establishment of new tree crops of only a few species; the ecosystem is being simplified. There is a very strong scientific case for the retention of adequate representative samples of primitive rain forest in which the full diversity of the natural ecosystem is left undisturbed. Such areas act as a control against which managed forests and other uses of the land can be measured and as a reservoir of plant and animal species, at the lowest level to be available to the potential use of future generations....[17]

Although Whitmore's research on Kolombangara between 1964 and 1971 provided valuable information on disturbance (mainly by cyclones) and elevation as ecological factors affecting forest composition, there was no progress on this or other reserves.[18] By 1975, in the light of heightened Solomon Islander sensitivity to land reservation, the moment and momentum for 'nature' reserves had vanished.[19]

Trenaman initially may have seen protection for economic sustainability as a long-term goal for the entire forest, but once the possibility of 'vacant' lands disappeared in 1961, his focus constricted to the public lands for the Forest Estate.[20] Trenaman stated in 1962,

> Any appreciable timber development in the Solomons must, for the foreseeable future, be largely for export, which entails large-scale working to meet shipping requirements and secure favourable freight rates. Against this background it will not often, if ever, be possible to arrange management of individual forests tracts on a sustained yield basis. It will be necessary to group areas into units of sustained yield working ... Such arrangements will be feasible with the establishment of the forest estate on public land.[21]

Trenaman believed that future logging on customary land was not likely to be extensive, and he was considering opening it up to timber exploitation because companies might want to log adjacent areas after the first cut of the Forest Reserve. This situation first arose in 1962/3 when the government, unbeknown to the people concerned, had offered Lever's the first option on rights to native customary land on Kolombangara and north New Georgia. Lever's wanted these timber rights because of planned investment – including a timber processing, plywood and pulpwood plant – and needed a guaranteed supply, beyond their lands on Kolombangara, to recoup that investment. The government believed investment would stimulate the timber industry, creating employment and services. So it concluded a deal with Lever's: Lever's would recognise local squatters' rights on sections of their retained Kolombangara land of 25,000 acres, and pass to the government 95,000 contiguous acres of it for the Forest Estate, after logging. In return, the government promised: first, a waiver of regeneration requirements on the 95,000 acres as they passed from their possession; second, the conferral

The Forestry Department and Control of Species

of freehold title to replace the Certificate of Occupation title on Lever's 25,000 acres; and third, the option of the timber rights on the customary land in Kolombangara and north New Georgia.[22]

The forests on all customary land 'would be regarded as wasting assets' because it was impossible to require Solomon Islanders to reforest their own lands.[23] The visiting British forest expert, C. Swabey, advised the Protectorate in 1964 that ten per cent of the land area for the Estate 'would almost certainly not represent the potential of the timber industry in the Solomons'.[24] As late as 1970, on revised figures of 500 square miles for the Estate, Trenaman thought that, as this represented the best of the timber area and stock, the rest was not a concern; the 'rest' constituted more than 90 per cent of the Solomons' forested land.[25] The emerging Forest Estate seemed a secure asset, but the price had been risking the rest of the resource and the acceptance of the view that this forest was expendable.

Reforestation: techniques, targets and tree species

Before 1964, the Department had not attempted replanting as it was still assembling information on replanting regimes and species. Logging firms were rarely in the expensive business of reforestation. Those pioneering the logging of government land had not been required to replant as it had been difficult enough to attract investment in the first place.[26]

Consequently, the first reforestation was assisted natural regeneration of kauri on Vanikolo. It is ironic that this was on customary land – the *bête noir* of the Forest Department/Division – but the Department was planning to acquire the area.[27] Commenced in 1954 under Chris Hadley and his team headed by A. L. Naqu, William Piaito, and Nelson Kausimae Nanu, and continued after 1958 under Piaito, combinations of weeding and poisoning were tested to free self-sown kauri seedlings in areas logged by the Kauri Timber Company.[28] By 1966, of the 13,000 acres logged, 5,600 acres had been treated to eliminate competition in the early years of growth and to manipulate the canopy, providing considerable data on different treatment regimes. The Company's closure in 1964 had removed logistical support, but it was the Department's focus on reforestation of hardwoods in the mid-sixties which resulted in the kauri forests being considered as a 'special problem', and receiving only sporadic attention until 1977, when the Department began to reforest with kauri seedlings on Santa Cruz (Ndeno).[29]

Elsewhere, from the late fifties, species trials had been confined to odd plots of land belonging to the government, sympathetic Europeans and Missions. The Department had conducted research on over 70 species by 1963.[30] Taking what it had learned in the plot trials it applied the data

Figure 70. Forestry workers and others at Vanikolo, 1954. Back row: Wete, A. L. Naqu, Nelson Kausimae Nanu (David Kausimae); Centre: Clement, Chris Hadley, Front: Claire Hadley, William Piaito.

The Forestry Department and Control of Species

to field-scale silviculture on its logged-over land, starting in late 1966 on Gizo, then in 1967 at Allardyce, Santa Isabel, in 1970 at Viru, south New Georgia, in 1975 on Kolombangara and in 1977 on Santa Cruz.

The first plantings at Gizo were the most varied. Softwoods tried were *Agathis macrophylla* and the exotics, *Agathis robusta, Araucaria cunninghamii,* and *A. hunsteinii.* The hardwoods planted included native *Terminalia calamansanai* and the exotics *Maesopsis eminii, Terminalia superba, T. ivorensis,* and *Cedrela mexicana,* followed in 1968 by high quality timbers, *Cedrela odorata* and *Sweitenia macrophylla.* At Allardyce, the Department planted natives – predominantly *Campnosperma brevipetiolata*, which was dominant in the wild, as well as *Gmelina molluccana, Calophyllum kajewskii,* and *Terminalia* spp.[31] On Kolombangara, the selected species were

Year	Area (hectares)	Cumulative total
1966	9.7	9.7
1967	133	142.7
1968	173.7	316.4
1969	279.2	595.6
1970	232.7	828.3
1971	940.7	1769
1972	937.3	2706.3
1973	130.2	2836.5
1974	1433.5	4270
1975	881.3	5151.3
1976	1395.7	6547
1977	1529.3	8076.3
1978	3433.9	11510.2
1979	2757.8	14268
1980	1645.1	15913.1
1981	198.7	16111.8
1982	514.9	16626.7
1983	1011.1	17637.8
1984	918.2	18556
1985	1227.7	19783.7
1986	957.8	20741.5
1987	1120	21861.5
1988	1037.8	22899.3
1989	1316.7	24216
1990	685	24901

Table 3. Annual replanting by Forestry Department.

Sources: FDAR 1967–1989; Martin Bennett et al., *Forestry Project Identification Mission, May-June 1991,* 22; See Fig. 87, Chapter 12.

Campnosperma brevipetiolata, Terminalia brassii, T. calamansanai, and the introduced *Eucalyptus deglupta*.[32] By late 1974 over 4,270 hectares (10,550 acres) of logged government land had been replanted on several sites (see Table 3).

Species that had fared badly in trial plots were monitored further, but not always put to field tests. *Cedrela australis*, for example, suffered from shoot-borer attack by the caterpillar, *Hypsipyla robusta*. Many of the species trialled in the late 1960s were dropped as plantation species, but some underwent further plot trials on various sites. For example, *Maesopsis emenii* showed its susceptibility to wind damage in the cyclone of November 1967 while the pines, *Agathis* spp. and *Araucaria* spp. displayed slow growth and proneness to bud-rot in the western Solomons.[33]

Seedlings need a place to put their roots. Until 1968, site studies were the 'least satisfactory' area of research, although some soil samples had been sent to England for testing.[34] This was rectified on an impressive scale by a British-aid funded study of soils. Although the final report did not appear until 1976, possible reforestation sites received priority, with data emerging as early as 1969.[35]

Ted Marriott, FDAR 1969, by permission of Forestry Division

Figure 71. Experimental plot of 12-year old pencil cedar (Cedrela mexicana), *a quality hardwood timber. These trees were planted in 1957.*

The Forestry Department and Control of Species

Ted Marriott, FDAR 1969, by permission of Forestry Division

Figure 72. Experimental plot of Eucalyptus deglupta, *a fast-growing hardwood trialed by the Forestry Department in the late 1950s because of its suitability for both timber and pulp. These 11-year old trees were located at Mt Austen.*

In the regeneration of forests for commercial production, there are three main techniques. First, assisted natural regeneration of a selected species, as was the case with kauri on Vanikolo. Secondly, closed-line or close-planting where the distance between the planting lines or rows of seedlings is four to five metres (13–16 feet), with a variable, but lesser distance between the trees in the rows, usually three metres (almost ten feet). A prerequisite to this planting technique is clear-felling of the site, with about six months for the slash to decay. Usually, after planting, the seedling is weeded within a metre radius. Planting and weeding are followed by intensive maintenance until the canopy closes. Generally, selective thinning occurs at this time.[36]

The third technique, commonly known as enrichment planting or the line-planting method is similar in grid form, except that the distances between the rows is much greater, from ten metres (33 feet) to 22 metres (72 feet), depending on species (and expected mean crown diameter), with two to three metres between the trees within the rows. In the Solomons since 1974, the most commonly used grid has been ten metres by three metres. Any overhead canopy remaining after logging is poisoned. After about three months, two-metre wide lines are cut at 12 metre intervals through the logged area. Planting is carried out soon after at spacing along the lines of three metres. Ideally, three maintenance operations are carried out in the first year of growth, two in the second and three in the third. During the third or fourth year, thinning is carried out to reduce the number to the final stocking per acre.[37] This enrichment planting technique is advised when five conditions are met:

1) production is aimed at large timber and veneer logs, not thinnings (i.e. saplings, reject logs, and even branches off saw logs) for pulpwood;

2) the species planted must be fast-growing, naturally straight and self-pruning; that is, a colonising or gap-filling, light-demanding type;

3) there must be no existing upper canopy, though low secondary forest is acceptable;

4) the forest must be non-flammable;

5) there must be no browsing animals.[38]

Based on his Uganda experience, the enrichment planting option appealed to Trenaman.[39] He reasoned that close-planting would involve clear-felling and burning, the latter difficult to achieve on a plantation-scale when there was no clear dry season. Existing logging operations generally were selective, based on size of the trees, so clear-felling would be rare.[40]

The Forestry Department and Control of Species

Close-planting, Trenaman held, would have been feasible if a *taungya* regime had been introduced, but population pressure was not yet great enough to require intensive agroforestry.[41] If the five conditions were met and the maintenance schedule followed, reforestation by the enrichment planting method would result in a crop costing less than a third of that by close-planting. The lower cost was the appeal of enrichment planting to Trenaman, not its potentially higher biomass and thus soil fertility and the greater retention of natural ecosystems.[42] Trenaman believed that the five pre-conditions existed in the Solomons, providing that fast-growing 'gap' species were selected to plant.

In relation to the first condition, there had been no taker for the thinnings from the logs: Trenaman's hopes for a processing plant were dashed in 1964 when Lever's abandoned the idea.[43] Acting on the recommendations of the British forestry adviser and an FAO specialist in 1969, the Department concentrated on fast-growing, good utility timbers as the best market option. When the Japanese Overseas Afforestation Association (JOAA) visited the Protectorate in 1972 they were interested in plantations of timber for pulp and paper, but wanted large tracts of 50,000 to 100,000 acres, which was impossible in the 1970s.[44]

In 1966/7, Trenaman had realised that, on the available information, the entire natural forest of the Solomon Islands would be worked out in about 35 years, if extraction was ten million cubic feet (283,000 cubic metres) yearly. Thus, 'if a sustained-yield working is to remain the objective', systematic reforestation had to commence as soon as possible.[45] The Forestry Policy White paper of 1968 agreed.[46] The replanting programmes fell broadly into three periods. The first was planned to go from 1969 to 1972, an interim programme to meet the recommendations of the White Paper until more information was available regarding the role of forestry in the economy. The goal was to replant 10,200 acres (4,130 hectares) on logged government land, partly as a response to increased extraction rate goal of ten million cubic feet (283,000 cubic metres) by 1972.[47] The planting of 1968–1972 would not have replaced the logged timber volume expected to be extracted, but Trenaman had to compromise between the need for compensatory reforestation and for data on species performance.[48]

The Department altered both goals of replanting and extraction as a result of the country's 1971 Sixth Development Plan, which set the second replanting programme at 5,000 acres (2,020 hectares) a year and extraction at an eventual 20 million cubic feet (567,000 cubic metres).[49] The doubled extraction rate was considered sustainable because of increased estimates of the timber volume available, the planned increased volume of replanting and the revised rotation age for fast-growing hardwoods from 35–40 years to 25 years. By 1974, neither goal had been attained. The extraction rate from 1971 to 1975 averaged only eight million cubic feet (233,000 cubic metres) per year because of severe damage on Santa Isabel by cyclone Ida in 1972, difficult market conditions in 1972 and 1974, and over-estimation of the timber industry's capacity for expansion, in the light of the uncertainties engendered by problems with obtaining timber rights. Less government land was logged than estimated, thus the area available for replanting was smaller. Although, with the exception of Kolombangara, replanting schedules on government land were on target, replanting at its yearly maximum in 1974 reached only 3,542.3 acres (1,433.5 hectares).[50]

In 1975, the Legislative Assembly, acting on the Report of the Forest Policy Review Committee, recommended that annual extraction rates be set at a maximum 14 million cubic feet (about 400,000 cubic metres) to prevent too-rapid exploitation. The Committee also wanted increased local processing as a condition of new logging licences. The annual target for this was ten million cubic feet (about 285,000 cubic metres). A political issue in the early 1970s, the Assembly pushed for more local and localised participation in the benefits of the timber industry.[51]

Local processing had concerned the Department for some time, and had influenced the selection of species and silvicultural techniques. In the late sixties, without the likelihood of local processing or even large-scale export of chips, Trenaman had doubted the wisdom of committing reforestation to very fast growing (6–10 years rotation) cellulose timbers. He had sought more information on future market trends while continuing work on experimental plots of close-spaced trees, particularly softwoods. A. Leslie, a consultant from the United Nations Development Programme and FAO during 1972–74, gave somewhat equivocal advice. Besides recommending planting of high-value quality trees with a rotation of 40 years or less, Leslie urged plantings of general purpose timber suitable for construction, veneers and pulp.[52] Leslie initially had favoured the high quality timber to general purpose timber, but costings supported enrichment planting and the faster growing general purpose timbers. G. Watt, a forest economist in the Department, did several studies of the timber industry's potential during 1972–1974.[53] Leslie and Watt, along with G. Chandler, another FAO consultant, did not support local processing beyond about 25 per cent of the timber cut, a view reiterated in 1979 by yet another group of FAO consultants who recommended no more than 20 per cent of a slightly increased extraction rate of 10.6 million cubic feet (about 300,000 cubic metres). This 20 per cent desideratum should become mandatory only when a consistent forestry policy operated to control rates of extraction, safeguarding the

The Forestry Department and Control of Species

high level of investment necessary for milling and processing, and when logging companies could make arrangements among themselves to meet the processing requirement. Overall, the advice was that there was more economic benefit in the export of logs. Any greater level of local processing was way in the future, given the level of existing local expertise, equipment, and the market demand.[54]

As a consequence of the increased reforestation recommended in the Forest Policy Review Committee's Report in 1975, Trenaman applied for further British assistance. The Reforestation Project of 1975–1979, the third replanting programme, set a target of 19,500 hectares (48,186 acres), funded by a grant of $2.5 million. On the basis of the economic data and market projections, Trenaman continued to plant the three main native species – *Campnosperma brevipetiolata*, *Terminalia brassii*, and *T. calamansanai* – as well as *Eucalyptus deglupta* from 1975 on. Until 1981, much of the planting programme was based on wildings of uncertain genetic quality, so often the proximity to a regular supply was a consideration in species choice. For example, Leslie had not recommended *Campnosperma*, mainly because of the destruction of stock by cyclone Ida on Santa Isabel. Moreover, Whitmore's research on Kolombangara showed that this species was cyclone-susceptible in the natural forest. However, in plantations, young trees proved more wind-firm. It was also an active coloniser of disturbed forest and, in its natural state, regenerated quickly, with good form in dense stands and was practically disease-free – apparently an ideal candidate for reforestation.[55] Like many a forester, Trenaman stayed with the species whose silviculture was most familiar and least demanding, although none of the native species had yet a complete plantation performance record, as opposed to knowledge of growth pattern and form in the natural forest.[56] Trenaman planned that appraisals of high-value timbers such as the exotic, *Swietenia macrophylla*, would continue, that indigenous *Agathis macrophylla* would be replanted on Santa Cruz, and that market trends would be monitored.[57]

The selection of species was determined by resistance to cyclones and pests, site suitability, and above all, the relative costs of different planting techniques. Besides these obvious silvicultural and economic parameters, a major consideration was the predicted market for the timbers. Trenaman gave this much thought in the late sixties and early seventies.[58] He has been criticised for lack of attention to 'balancing growth rates with the potential end use and marketability of the timber', but being able to predict the likely market 20 years ahead was as difficult in the early 1970s as now.[59]

Research: utilisation, silviculture, and botany

'Marketability' was the rationale for utilisation and botanical studies. The sixties were years of intensive data collecting and research, made possible by Commonweath Development and Welfare (CDW) grants and overseas laboratories. Under Trenaman, systematic utilisation work commenced, with sample logs of 35 indigenous species and ten plantation exotics being sent over the years to Melbourne for testing by the Commonwealth Scientific and Industrial Research Organisation (CSIRO).[60] The purposes of these studies were:

1) to assist the timber industry, mainly by investigation of the properties of individual timbers, in marketing as wide a range of species as possible and in promoting local manufacturing;

2) to secure better use of timber in the Protectorate – for example by the study of problems of preservation and seasoning; and

3) to provide data to guide the Forestry Department's regeneration and management policies.[61]

The CSIRO data along with local results on the durability, preservation and seasoning of the individual timbers were complied in the late sixties and early seventies and published as *Major Species, Timber Booklet No. 1* (1977) and *Minor Species, Timber Booklet No. 2* (1979). Lever's Pacific Timbers had done considerable research itself on timber properties from 1963 to 1965, in order to develop the market for the logs from Gizo and Kolombangara.[62] Drawing on this and the Department's research, G. T. Pleydell's *Timbers of the British Solomon Islands* (1969) provided the industry with data on timbers, their working properties and uses.

From the early 1970s, the Department regularly recorded its silvicultural research. Of use to the Department and investors, these 'notes' contained matters on specific species such as natural distribution and habitat, seeding and flowering, nursery techniques, site selection, planting method, growth rates, yields, pests and diseases. Bound as an unpublished report in 1980, the revised version by K. D. Marten and B. R. Thomson appeared the following year as *The Silvics of Species*, dealing with eleven species and the technique of enrichment planting.

Before 1980, the Department had produced a series of volume tables that gave the industry the average cubic contents of a species, thus enabling calculations of yield from a particular natural forest tract. Another useful research area concerned the volume of timber recovered from extraction operations. Studies in the late 1960s pointed the way for better instruction

The Forestry Department and Control of Species

and control of logging teams as there was only an 85 per cent recovery rate of volume of the merchantable trees. Losses were due to felling too high up the trunk, leaving considerable timber in the stump.[63]

F. S. Walker's pioneer study of 1948 had put the major tree species on the world botanical map, but more systematic study of the flora was needed. The British government sent the forest botanist T. C. Whitmore who trained Solomon Islander collectors, mainly from Kwara'ae, Malaita: Jack Arifanata'a, Baea, Iromea Gumu, Nakisi, L. Ramodia, Susui, as well as Jim Herahura from Are'are and two Guadalcanal men, William Masu'u and John Sore. Starting in 1963, Whitmore and his energetic team of tree climbers had collected and identified over 4000 specimens.[64] The team provided the foundation collection in the Honiara herbarium and augmented collections overseas. Moreover, these were given their Kwara'ae name as well as, in some cases, other Solomon Islands language names. His *Guide to the Forests of the British Solomon Islands* appeared in 1966, and became the standard reference for decades. He continued assembling specimens until the final collection reached 20,000 in 1969, representing a large proportion of the archipelago's trees and shrubs.[65]

Although the Department opened a new station in Munda, New Georgia, for the Silvicultural Research and Utilisation Divisions in 1969, the research programme, except for silviculture work, was less intense. Trenaman considered that the botanical studies had achieved their main objectives, despite delays in publishing timber research towards the end of his service, mainly due to senior staff shortages and the demanding reforestation schedule.[66]

Funding the forest

As new loggers started operating in the early 1960s, the question of the government's share arose, particularly as it was buying land for the Forest Estate and would eventually fund its reforestation. The export duty and royalty system that had developed in the late fifties was inconsistent. The best policy solution seemed to be a reforestation levy on log exports, along with royalties.[67] The levy would be an alternative to the regeneration requirement the government was to impose on future loggers on government land, once the Protectorate had proven a viable timber source. It would discourage a company logging private land or native registered land from offering a better price for logs/timber rights than that paid by loggers on government land where the regeneration requirement was in place.[68] A royalty (stumpage) was to be paid to the government if it owned the land, or to the customary land holders if they had, through the government, assigned their rights to a company. These royalty payments were to be uniform. The government believed that the levy,

> ... by its generality, and being the predominant element in direct timber revenues, ... will bear out the point that the main forest resources are to be treated as primarily a national asset; and establish that all sectors of the timber industry contribute to forest revenues and the costs of regeneration.[69]

This meant the entire forest would contribute to reforestation of the public forest only.[70] Should the circumstance of direct dealing arise, the company would have to budget accordingly in its offer to the rights' holders to cover payment of the levy to the government.[71]

Although when first considering the levy in 1964, Trenaman believed that it and other funding would underwrite reforestation, once the costs became clearer he altered his opinion. The massive initial outlay commenced at time when revenue depended on British grants and the export of mainly copra. In 1965, timber exports contributed only 4.1 per cent of total domestic export earnings.[72] Self-funding would be feasible when the first planted timber was cut. Trenaman saw the reforestation levy as establishing the principle which he believed would pave the way for a self-financing forestry industry in 20 to 30 years.

The levy, replacing export duty, came into operation in late 1970 at 2.5 cents per cubic foot of log export and on sawn timber above certain levels for local sale.[73] The levy went into general revenue at this stage, bringing in between 2 per cent and 3 per cent of the government's recurrent budget until 1977 when, in an attempt to boost the marketing of sawn timber, it was abolished temporarily. As the levy went, the government reinstated an export duty at 6.5 per cent per cubic metre f.o.b. which produced a similar contribution to its revenue. Through all this, royalties continued to be collected, for either government or timber rights' holders.[74] During the early 1970s the local government councils were looking further afield for income than the 'rate', a head tax, and licences. The Santa Isabel Council wanted a share of the royalties/levy from logging the Allardyce area, an idea put on hold by cyclone Ida, but indicative of local aspirations.[75]

The White Paper on Forestry in 1968 had reiterated the policy of seeking external funds for reforestation. From the outset, Britain had supplied this. By 1973, with independence inevitable, Trenaman tried to attract major 'timber-working or wood-consuming interests' and other development agencies.[76] Trenaman wanted an investment mix, to prevent a monopoly which could attempt to influence policy (as Lever's Pacific Timbers had demonstrated). This was not forthcoming. Although it gave grants to fund research plots mainly on Kolombangara from 1973 on, JOAA wanted extensive field-study results on *Eucalyptus deglupta*, *Gmelina arborea*

The Forestry Department and Control of Species

and *Terminalia brassii* before making substantial investment. In 1976, the Japan International Co-operation Agency (JICA), showed interest in pine planting on Guadalcanal. The Department started trials there, but these were burnt out regularly by grassfires. Negotiations with New Zealand Forest Products Ltd came to nothing. The Asian Development Bank and the International Bank for Reconstruction and Development (later the World Bank) would not provide funds for local costs. By the time Trenaman left the Protectorate, he had won continuing support with CDW funds, as well as aid from New Zealand in the form of the salaries of two officers, equipment to supplement the Reforestation Project of 1975–1979, and the establishment of Poitete base on Kolombangara.[77]

In part, the Reforestation Project was designed to meet the expanded reforestation targets recommended by the Forest Policy Review Committee in 1975 and enunciated in the National Development Plan. The Committee believed greater volume of log export and local timber processing necessitated an expansion of reforestation. As well, the Committee recommended that customary land, after it was logged under a timber rights lease, be replanted like government land. The Department investigated prospects and continued small-scale trials, but neither investor nor aid agency could be persuaded to fund reforestation of customary land, because there was no guarantee it would be left to mature.[78]

Another recommendation of the Committee was that some of the reforestation finance needed should come from revenue from the timber industry, along with overseas investment and aid.[79] This was the only recommendation modified by the Minister David Thuguvoda, on the grounds that 'with the shortage here of revenue generally there would seem strong arguments for continuing to obtain all funds for replanting work from acceptable overseas sources …'.[80] Dependency on aid was to continue.

The major funding source continued to be CDW and its successor, British Overseas Development Administration (ODA), with aid totalling more than SI$4,300,000 from 1966 to 1980, resulting in the replanting of 17,787 hectares on government land.[81] Although the Department had failed to meet its replanting targets by almost 3,000 hectares during the 1970s, this was a considerable achievement in the view of ODA's auditors. The Australian Development Assistance Bureau (ADAB) made a foray into a forestry project with a 1976 pilot of a scheme called Cattle under the Trees (CUT), which was to combine pastoral and forestry activities. Despite a poor database, the project went ahead on government land on Kolombangara in 1979, the first such project for Australia and an independent Solomons.[82] It was on Kolombangara, too that the Commonwealth Development Corporation maintained a 'watching brief' on reforestation with an eye to a possible future venture with local landholders.[83]

Problems and changed directions

Trees are host to myriad insect species and fungi, which are part of the forest ecology. These rarely reach plague proportions, unless there is substantial damage to the forest or some other disturbance. In some respects, monocultures represent both damage and disturbance and thus are prone to attack by unwanted insects and fungi, which then become designated by foresters as 'pests'. The plantations in the Solomon Islands began showing signs of attack by insect pests early, but most noticeably by 1974. Exotics were particularly susceptible. *Hysipyla robusta* (a shoot borer) attacked some of the Meliaceae, particularly the seedlings of the exotic *Swietenia macrophylla* (mahogany) and *Hyblaea puera* (caterpillar) damaged *Tectona grandis* (teak).[84]

However, the major problem in the mid-1970s was undiagnosed decay on *Eucalyptus deglupta*. Under the Reforestation Project of 1975–1979, the Department made substantial plantings of the exotic hardwood on Kolombangara and at Viru. The major plantation native species, *Campnosperma brevipetiolata* was affected also.[85] The culprits appeared to be *Amblypeta cocopharga* causing die-back, and the wood-borer, *Oxymagis horni*.[86] The Department appealed to the British ODA for a forest entomologist. During 1978–1981, M. Bigger's investigations formed the basis of the first entomological specimen collection, now housed at Dodo Creek Agricultural Centre, Guadalcanal. Besides the confirmation of *Amblypeta cocopharga* and *Oxymagis horni* as the problem in the two plantation species, many other pests were detailed and recommendations made on their control. While all this was going on, the Department ceased to plant *Eucalyptus deglupta*, except on the CUT project. Bigger's recommendations indicated ecological control – varying silviculture techniques to replicate natural forest conditions more closely, rather than any insecticide regime.[87]

The greatest threat to the Department's replanting programme was not insects, but plants. When old trees collapse or after a cyclone has swept though, the climbers, including *Merremia peltata* and spp. and *Operculina riedeliana* usually proliferate in the gaps, along with fast-growing, soft-stemmed, short-lived species. Usually longer-living species of the high forest establish under these fast-growing trees that have escaped the climbers. In some cases, providing climbers are not too dense, a few high forest species, including *Campnosperma brevipetiolata* and *Terminalia calamansanai*, colonise the gap directly. Although, to the unfamiliar eye, the devastation caused by a cyclone on natural forest may seem total, many upright living trees do survive and, within a couple of months, have sprouted new growth and create a certain amount of shadow, which climbers do not like. Moreover, except for land slips from heavy rain, soils are little damaged by cyclones, and retain the fertility necessary for the growth of the pioneers of secondary

The Forestry Department and Control of Species

succession. The Solomons forests, unlike those of many other areas in the tropics, are cyclone-adapted. In such circumstances, the climbers rarely dominate for long.[88]

The first logging in the Solomons, especially on Gizo and parts of Kolombangara and north New Georgia, to a large extent mimicked natural disturbance. Logging was selective, not because of an ethic of sustainability, but because Lever's had not found a market for their 'small' and 'super-small' trees. Lever's left standing trees unwanted by the market or below a girth of six feet, so there remained a disturbed canopy.[89] Because of this, the level of compacting and stripping of the soil by tractors and other heavy equipment was relatively limited. There were areas of shade that the soft-woody pioneers could colonise. Climbers were not a problem, except on the skidder tracks and roads. These patches could be cut back or poisoned. Seedlings were established by the enrichment planting method, with the low scrub-controlling weeds and climbers in the areas outside the immediate radius of the seedling. The usual schedule of three weedings in the first year saw most seedlings survive. Initial results were promising, especially at Gizo. When the CDW Reforestation Project commenced in 1975 its aim had been to increase replanting from around 700 hectares a year during 1971–1974, to reach an average of 5,000 hectares by 1979.[90] The Project used enrichment planting, a spacing of ten metres by three metres, in order to meet larger volume targets of almost 140 cubic metres per hectare on a 20-year rotation.[91]

Almost as soon as the Project started on Kolombangara, Lever's changed their logging regime, as other loggers had done or were doing. Lever's had found a market for more species and, mainly in Japan and Korea, for their super-smalls – good for business, but not good for the reforestation scheme. Logging became clear-felling, except for the huge banyan trees and a few other rejects.[92] Not only was the forest stripped away, so also was the soil compacted and stripped of humus. In some areas, such as Barora on north New Georgia disturbance affected as much as 70 per cent of the area. The soil's first defences are the climbers, particularly *Merremia* spp. These spread rapidly, taking advantage of the massive canopy hiatus. Assisted by a rapid climb up exposed banyans, from where their seeds can disperse over large areas, they act to prevent soil erosion. Eventually, seedlings grow up though them, but often become bent and twisted in the process. And the climbers did their work with a vengeance, as the Department found when it followed up with replanting.

By the late 1970s, *Merremia* was such a problem that plantations required up to nine maintenance operations a year, increasing costs. Other problems beset the Project. Major species were proving to be unsuited for plantation growing conditions. This was the case with *Campnosperma brevipetilata*, which showed poor form, and *Terminalia calamansanai*, which was particularly susceptible to climber damage. Some seedling stock was inferior because, in order to try to reach planting targets, wildings of uncertain quality were used; nursery practice was often sloppy. Considerable research still needed to be done into nursery techniques, seed storage and stock control, but the Project had no research component. The Department's own research unit could not cope with its own programme plus the demands of the Project. Moreover, experienced senior staff, like Trenaman, had left the Protectorate in 1975/6. Lower to middle management staffing was inexperienced or simply lacking throughout the Project.

The Department was struggling with a variety of duties. The Department's large staff were scattered over several sites, with almost 95 per cent planting or maintaining trees.[93] Estimated wage costs ballooned to 17.5 per cent over estimates, and well above the inflation rate, and output suffered. As the ODA evaluation team, led by Wood and Watt, observed

> Virtually all silvicultural work was on a fixed task basis, taking no account of site variation or weed intensity; many teams of workers apparently complete these tasks by 11 a.m. which leads not only to high unit costs, but also leads to unrest on forest stations.[94]

All the more so at Poitete Camp, where the 800 men were difficult for one Range Officer to control and monitor.

The Replanting Project might have been more successful if Trenaman's advice regarding the establishment of a statutory body or corporation to replace the Department had been followed. Trenaman saw the function of the forest service mainly as estate management. The re-organisation of administration that went on prior to independence meant the service was no longer a discrete identity: for example, the head of the service did not control payment of staff. The absorption of the Forestry Department by the Department of Trade, Industry and Labour in 1974 and then its transfer a year later to the Department (and later Ministry) of Natural Resources where it became the forestry division, cost it considerable autonomy. Trenaman believed that the tasks of estate management were not difficult, and he was struck by the efficient performance of Lever's in its operation in comparison to his own Department.[95] A. Leslie affirmed Trenaman's approach in 1975:

> But forestry, once it goes beyond a purely custodial function, is much more a matter of public business than public administration, and the traditional form of organisation – a state department of forestry – is very poorly adapted for exercising a production role. The situation is even worse when forestry is only a part of another department or ministry.[96]

The Forestry Department and Control of Species

The Wood and Watt report highlighted the problems of a government department attempting to function as a commercial venture. A statutory body could initiate varied and flexible management and employment regimes; for example, employ a local contractor who could organise his work team on the spot. Had a statutory corporation been established to concentrate on government plantations, there might also have been a role for a government division to service the need for forestry extension, for advice and support of community forestry involving customary landholders. And as Wood and Watt suggested, individuals or groups could contract the corporation to plant up customary land.[97]

By 1978, the Forestry Division had fallen behind in its plot inventories: it did not know the stocking levels on its plantations. When updated in 1980/81, some sobering figures emerged. On the then-current figure of 17,787 hectares planted, instead of the projected 111 trees per hectare – the objective at the end of the 20-year rotation – there were, on average, only 82 trees likely to be suitable for saw logs at maturity; and the rotation was nowhere near the end for the bulk of plantings. This would produce about 60–90 cubic metres per hectare after 20 years instead of the predicted 100–140 cubic metres. In early 1980 The British forestry adviser R. H. Kemp urged a shift to close line or close-planting, which had seemed successful in early problem areas at Viru. The Forestry Division of an independent Solomon Islands had less certainty of continued grants. Confronted by the high costs of maintaining plantations as it drew up another application for ODA aid, the Division under Terry Kera reviewed the enrichment planting technique and decided to abandon it for close-planting at spacings of five metres by three metres for most species. The objectives were quicker crown closure to assist in controlling *Merremia* and provide greater stocking. As weeding and maintenance would be intensive and costly for the first three years, after which it was supposed to cease, the Division reduced its planting targets from the previous 3,000–4,000 hectares a year to 1,000–1,500 hectares.[98]

Achievements and limitations

As with all research in forestry, this change in sivicultural method was premised on the establishment of a commercial asset for the future Solomons government, the enduring aim of the Department/Division under Keith Trenaman. Trenaman was a utilitarian conservationist, deeply concerned with the scientific development of selected forests as a resource to be rationally and profitably exploited in a pattern of logging and replanting. The wider forest ecology was of little significance unless it impinged on the central concern of commercial production. This assumed a plantation model and a forest estate on government-owned or leased lands. This could be a real asset, provided it was run on business principles and not primarily to manufacture paid employment under public service conditions, but this model was never tried.

The most fundamental limitation on the Department's programme, however, had been looming independence. In the Pacific, by the late 1960s, 'Britain was in a leaving mood'.[99] It wanted to exit decently, leaving commercially viable assets for the Solomons. Yet commercial forestry, by its nature, is a very long-term investment. Work only began a mere 25 years before independence and reforestation had been underway for only 11 years. Vagaries of soil, climate, timber species and pests, to say nothing of personnel and markets, affected reforestation schedules. On top of this, there was political turmoil aroused by the creation of the forest estate. Though much was achieved after the Department's establishment, there were no returns expected from the plantation forests until about 1990 – in the interim someone had to fund reforestation. In 1968 and 1975, the Solomons politicians backed off from the reality of self-funding solely by levies and duties on export logs and assigned future reforestation to dependence on aid. The real cost was never faced – at least not then. The colonial government, honouring the principles of self-government, had to bow to this. The best Trenaman could do was to try to diversify funding sources, because Britain's aid could not go on indefinitely.

Trenaman and his successor, Kera, had not seriously advocated re-planting of customary land, partly because little had been logged systematically to 1980 and partly because they had problems enough trying to raise funding for re-planting on secured government land. Despite calls for a programme of reforesting customary land and for more extensive and intensive extension work, neither the colonial government nor the newly-independent Solomons government attempted to involve villagers in social (community-based) forestry to any significant degree. Vis à vis the likely returns on plantation forestry, the costs in time and personnel for social forestry were simply too great as independence neared. Thus, although the Department/Division made huge strides in the technical and scientific aspects of commercial forestry and left an impressive research legacy, the involvement, understanding and empowerment of the original holders of the forest lagged behind. What appeared, in the 1960s and 1970s, the most efficient use of aid and public money, would cost both the forest and its dwellers dearly in the 1980s and beyond.

— 10 —

Contest for the Forests, c. 1963–1985

Caliban:
>I'll show thee the best springs; I'll pluck thee berries;
>I'll fish for thee, and get thee wood enough.
>A plague upon the tyrant that I serve!
>I'll bear him no more sticks, but follow thee,
>Thou wondrous man.
>
>   W. Shakespeare, *The Tempest*, Act 11, scene ii,
>   ed. Northrop Frye (Baltimore, Maryland, 1959), 70.

The logging companies in the Protectorate from the mid-sixties were those that had market connections in Australia, Europe or Japan. Their capital, as did most of their principals, came mainly from either Britain or Australia. Thirty years on, in the mid-nineties, the most powerful companies involved in logging in Solomons were owned by ethnic or national South Koreans, Japanese or Malaysians. In the interim years, some Solomon Islanders established forms of partnerships with foreign companies to extract logs, but usually the balance of power was with the latter. As logging moved from government to customary land in the early 1980s, the 1977 Amendment to the Forest and Timbers Act coupled with Melanesian patterns of reciprocity proved vulnerable to the loggers' extractive regimes and business methods.

Colonial companies in transition

By 1962/3, Trenaman had invited four companies to join Lever's in logging. Forest Estate lands were to be leased under specified controls on logging in return for the security of government land. Initial outlay on a logging venture, as Vanikolo had shown, was high. Most companies wanted 20 years' guaranteed production to recoup this and make a profit. Along with possible access to contiguous areas, reforestation and regrowth, in theory, could make a sustainable cycle.[1]

Four areas were offered: parts of the Shortland Islands, Baga island (Mbava), Viru on south New Georgia, and the Allardyce Harbour area of Santa Isabel. The first interested company, the Shortland Development Company, was a Japanese consortium. The second, Nanpo Ringo Kaisha Ltd (The British Solomons Forestry Company), started on Baga in 1963 and had an option on Vangunu, but establishment problems, falling demand in 1965, and market resistance to the unfamiliar *Dillenia* saw it withdraw from the western Solomons in 1966. The third, interested in Viru, was the Kalena Timber Company, with directors G. E. C. Mears and D. A. Ireton involved in logging in South-east Asia.[2] The fourth company, Allardyce Lumber Company's parent, was Cullity Timbers of Western Australia, associated with another whose principals included Lester and Devon Minchin. Lester Minchin was a former major in the Allied forces and had seen the potential of the ramin (*Gonystylus*) on the floating islands in the Rejang delta, Sarawak. By 1963, the Minchins had the largest timber concession in the colony. Britain's departure from Sarawak saw the company planning to pull out and diversify sources of supply.[3]

Allardyce Lumber Company – from west to east and back

Allardyce shipped men and materials from Sarawak, commenced operations on government land on Santa Isabel in 1964 and soon struck problems. The coast was lined with mangrove swamp, no gravel for roads could be located, and the tree counts from coast to ridge indicated only about 600,000 cubic feet true measure over bark per square mile (about 65 cubic metres per hectare), at least 70 per cent stocked with *Campnosperma brevipetiolata*. The log floated, a plus for loading it onto ships. It was soft and pale-coloured and peeled well.

For five years the company exported *Campnosperma* and the tall swamp species, *Treminalia brassii* to Australia and Japan at a loss. Roading and transport costs per cubic metre alone were A$8–$9.[4] The clay-like soil was unable to cope with the rain or the traffic, until intensive research was done on surface compaction in 1969. The business recovered and by 31 May 1972, had made more than a million US dollars. That night cyclone Ida struck, destroying virtually the whole resource.[5]

The Forestry Department found the company responsible and efficient. Anxious to keep Allardyce in the Protectorate, once the salvage logging was completed, it offered government land of 30,000 acres at Santa Cruz (Nende) and at Alu, as a consequence of the Shortland Development Limited's withdrawal.[6]

Contest for the Forests, c. 1963–1985

By permission of Devon Minchin

Figure 73. The impact of cyclone Ida. The 'million dollar' consolidated road that led nowhere. The bare tree trunks of large defoliated Campnosperma. *As many trees were uprooted as were left standing.*

By permission of Devon Minchin

Figure 74. This had been part of the road built at great cost by the Allardyce Lumber Company.

Allardyce started logging kauri on Santa Cruz in 1973, the timber destined for Japan, France, and New Zealand. But stocks were sparse and scattered so by 1980, roading costs and ageing equipment forced the company to pull out, leaving a mixed reputation. It had set up a nursery of kauri wildings and established plantations on government and some church land, providing a model for later companies.[7] Unfortunately, it also left several sizeable felled logs to rot. It promised the Eastern Islands Council infrastructure which never eventuated. There were also allegations of unauthorised logging on customary land.[8]

The company left just before the national parliament passed the Provincial Government Act in 1981, setting up seven provinces, each with an elected assembly, and providing for the devolution of powers and functions to them.[9] In the early 1980s, under the government of Solomon Mamaloni, provincial governments were consulted on questions relating to their respective areas.[10] The company in 1982 tried to return to Santa Cruz because it had found a lucrative niche market in France for kauri, but was unsupported by Temotu province, advised by the Forestry Division. No licence was forthcoming, nor did the province endorse logging proposals by Costigan Brothers and Walter Jones of Eastern Enterprises.[11]

The Shortland Development Company had withdrawn from Alu in May 1972 because of difficulties with marketing and handling the sinker logs among the *Pometia pinnata*, a species which produces look-alike floaters. Allardyce began logging on government land in 1977, testing each log from a crane; if it failed to float, it was rafted with four floaters, mostly *Callophylum*.[12] Devon Minchin remained a director in 1983, but new and dominant interests were represented, including Hong Kong and Malaysian nationals.

In the early 1980s, the company had extended its logging onto customary land, without complying with the procedures of the Act of 1977. Unlike its predecessor, it had formed reciprocal relations with the community supported by the chiefs. The chiefs were crucial because land rights reside more in them rather than in clans of ancient residence, since the Alu are descendants of nineteenth century conquerors from Mono. Royalty payments were considered fair, the same as Lever's, on a sliding scale from 7.5 per cent to 17 per cent, f.o.b. for each Australian dollar per cubic metre; for an increase or decrease of A$5 per cubic metre in selling price overseas the royalty would increase or decrease by two per cent. As well as employment, Allardyce provided people the means of getting their produce to Noro, New Georgia for sale to Taiyo's cannery workers. It had made gravelled roads and provided materials for churches and schools. The Forestry Division turned a blind eye to the company's breach of the law because there were no serious complaints and the Minister of Natural Resources, Peter Salaka, was the local parliamentarian.[13]

Kalena Timber Company

The Kalena Timber Company made erratic progress at Viru from 1965. Logging of *Calophyllum, Dillenia* and *Terminalia* commenced in 1967, but production was under the allowable 500,000 cubic feet (14,160 cubic metres) a year and its mill set up in 1970 operated below capacity. Roading could not keep up with logging and the timbers were not of particularly high value. There were major pollution problems at Viru in 1975, due to poor logging practice, resulting in siltation of the harbour and severe damage to the reefs and fishery. Plans for Vangunu were suspended because of financial problems. Ataka, the Japanese interest in Kalena, withdrew in 1977. Kalena re-equipped in 1980 and production improved.[14]

By 1985, the company had built a relationship with the local *butubutu* (land-holding group), in line with their concept of negotiations with a 'side' or a party willing to take responsibility for its actions.[15] Kalena joined forces with the local Saikile Development Company and extended logging on to customary land. Soon after, Earthmovers Solomons Ltd headed by James Boyers bought out Kalena.[16]

Lever's Pacific Timbers

Lever's Pacific Plantations had the country's biggest holdings, but relatively little under coconuts.[17] In the early 1960s, Lever's, through the United Africa Company, a Unilever subsidiary, investigated the timber of this land that had been idle for over 50 years. Lever's Gizo project had proved the viability of the logs, mainly *Calophyllum*, in Japan. However, investment envisaged by Lever's on Kolombangara incorporated an option on timber rights to adjacent customary lands and on north New Georgia for a long-term economic through-put. Knowing the suspicions of the Kolombangara and Russell Islands people towards Lever's, Trenaman in 1962 assisted its timber surveyor, Bill Evans to gain access to customary lands.[18] But Lever's Pavuvu forests had 'a miserably low stocking', thus could not supplement any processing planned on Kolombangara.[19] Following Lever's decision in 1964 against processing, the government withdrew the New Georgia option, but reinstated it in 1967 when Lever's unveiled plans for over half a million pounds' investment in roading and plant on Kolombangara.[20]

Pacific Forest

Australian Timber Journal, September 1970

Figure 75. Lever's Pacific Timbers, 1970. Cat D7 tractors, fitted with Young Iron Works integrated arches used for snigging.

Figure 76. Lever's Pacific Timbers, 1970. Kenworth three-axle truck, 35 ton capacity, for hauling logs with average length 65 feet. Note the quality of the road, coral compacted, capable of carrying trucks up to 50 tons gross in 140-inch rainfall area where it rains all year.

Australian Timber Journal, September 1970

Contest for the Forests, c. 1963–1985

Australian Timber Journal,
September 1970

*Figure 77.
Lever's Pacific Timbers, 1970. Dump loading is by spar tree using double drum Carco winch, with 400 feet of 1-inch main line. Capacity is 15 tons.*

*Figure 78.
Lever's Pacific Timbers, 1970. As above.*

Australian Timber Journal,
September 1970

Australian Timber Journal,
September 1970

*Figure 79.
Lever's Pacific Timbers, at Ringgi Cove, Kolombangara, 1970.
A gang of fallers assemble their chainsaws each morning and sharpen them each night.*

*Figure 80.
Lever's Pacific Timbers at Ringgi Cove, Kolombangara, 1970. Solomon Islander felling a tree with a chain-saw.*

Australian Timber Journal, September 1970

Contest for the Forests, c. 1963–1985

Australian Timber Journal,
September 1970

*Figure 81.
Lever's Pacific Timbers
at Ringgi Cove,
Kolombangara, 1970.
After felling, the stump
and log are numbered
identically to provide information on
compartment number, log number, butt
girth, species, acreage
yield, species quantity,
individual faller tally
and a check that all
logs arrive at the wharf
area.*

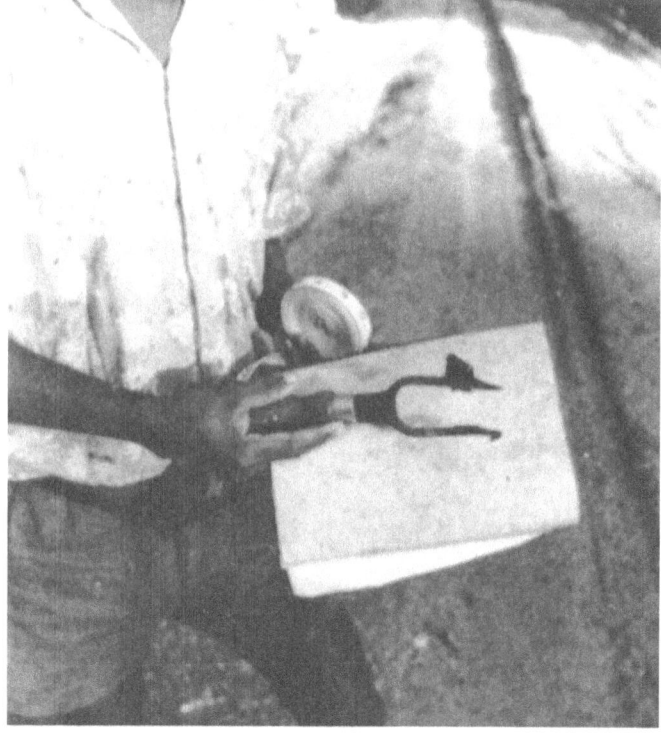

*Figure 82.
Lever's Pacific Timbers
at Ringgi Cove,
Kolombangara, 1970.
A man follows the
faller and, using a
scribing knife, effectively marks all these
details on the log.*

Australian Timber Journal,
September 1970

Eight years on, the government was still trying for the option, a major concern as stocks on Kolombangara would be logged by 1982. Lever's was pivotal to the west's economy, employing 40 per cent of all workers and contributing almost 17 per cent of the country's foreign exchange earnings. It was the most efficient logging operator and annually paid the government $300,000 in corporate taxation, as sole contributor in this category, after an average profit of $800,000.[21] Yet its logging had also upset Kolombangara people by damaging sacred sites and polluting water sources.[22] The New Georgia negotiations had stalled because of disagreements among the people about rights and representation. Many disliked the need for a form of registration, specifying ownership of the land, before the timber 'profits' could be transferred. A possible solution emerged in the Forestry Policy Review Select Committee 1975 report recommending that area committees decide ownership.

Area committees appeared as the 32 small local government councils were replaced in the mid-1970s by eight larger district-based ones. The 'areas' were council wards. Members of the committees were elected, with the elected councillor from the particular ward always a member. In theory, chiefs and big-men had consultative roles, but often members knew little of land matters. Under the Forest and Timber Amendment Act of 1977 the area committees became the bodies which decided who held the rights.[23] Appeals against the outcome could go to the customary land courts, introduced in 1975, but could be set aside, on points of law, by application to the High Court.

By 1978, attempts by two area committees to establish ownership of the rights on north New Georgia fell foul of the Western Council's refusal to recognise their rulings. Appeals to the High Court proved inconclusive, as the court could rule only between the two parties concerned; a further appeal could introduce a claim by a third party. As at least six *butubutu* were involved, to say nothing of divisions within each, court proceedings had the potential to drag on interminably. Already the courts had ruled in favour of men whom many considered not to be leaders. Moreover, there were differences between followers of the Christian Fellowship Church (CFC) of the Holy Mama, Silas Eto and of the Methodist and Seventh Day Adventist churches regarding who should represent the *butubutu*.

The North New Georgia Corporation

Lever's had begun logging government land at Barora, north-east New Georgia in 1978, so the need to obtain rights to adjacent forest was pressing. An alternative to court action was mooted in discussions between the

Contest for the Forests, c. 1963–1985

Western Council, the Attorney-General, and certain representatives of the *butubutu* and their lawyer from November 1978 to March 1979. Some of the leaders of five of the six groups, namely Koroga, Rodana, Lupa, Ndekurana, and Nggerasi, agreed to establish the North New Georgia Corporation to represent the 1,900 claimants. The timber rights were to be transferred to the Corporation, which could licence extraction and dispose of the profits. Provision was made for the appointment of a board of directors, chosen by tribal leaders, *bangara*, on a rotational basis in each logging area. The Corporation required an act of parliament.[24] According to G. Beti, the member for north New Georgia and a supporter of the Christian Fellowship Church:

> ...this bill tries to separate land ownership from the resources standing on the land so that whilst the land is still owned by the owners the resources by right made legal by this bill will be belonging to the Corporation who will then utilise them through the multi[national] company LPT. In fact the government is saying, 'You go ahead and dispute over land ownership, go ahead and fight; kill as you wish; I will not give a hoot as to what you say, I am only interested in your trees'. So regardless of what the local politics is over land the timber operation must go forward.[25]

Despite such opposition, the parliament of 1979, under Prime Minister Kenilorea, passed the North New Georgia Corporation Bill into law. Lever's purchased the timber rights of 45,000 hectares, estimated to contain three million cubic metres of merchantable hardwoods. The royalties to be paid ranged from 7.5 to 17 per cent (averaged at about 12.5 per cent) of f.o.b. price, for a period of 15 years. Although there was support for logging, the fundamental issue of who held the land rights and thus the timber rights was not resolved by the Act.

Problems immediately became evident. The original negotiating board, under the chairmanship of Frank Jamakana, was tardy in resigning in favour of directors from the areas being logged. Quarrels arose about the arbitrary methods of disbursing payments and about which 'tribal leaders' had the right to receive monies and choose directors.[26] When Lever's moved into Koraga, Milton Talasasa with Nginabule (Methodists) who favoured the logging, claimed leadership, as did Reuban Quabule, supported by the Holy Mama and Job Dudley Tausinga, son of the Holy Mama. Many Seventh Day Adventists also supported the logging. Tensions grew as Talasasa, whose son Ian was a director of the Corporation, distributed royalties.[27] Because claims to land rights had never been settled, many felt aggrieved when they did not get their due, seeing this as bias from a man many regarded as an imposter. At Jericho village, followers of the CFC were in this category. They believed that the logging scheme 'took the control of development away from them' and criticised the environmental damage done at Barora.[28] Until the Holy Mama died in 1983, he had refused to attend the meetings leading up to the establishment of the Corporation, so CFC followers considered the agreement with Lever's not binding.[29]

The north New Georgia situation had ramifications elsewhere. Lever's Solomons Ltd (LSL) in the Russell Islands wanted to log their land at Pavuvu, preliminary to planting with coconuts and cocoa. Lever's Pacific Timbers (LPT) applied for an export licence in January 1980, but the government refused in an attempt to force milling for domestic needs. LPT was not interested, so the Forestry Division tried Foxwoods, but it wanted only the valuable *vasa* (*Vitex cofassus*), about 20 per cent of the forest stock. Although LSL had its licence by 1981, squatters were moving in. Because of this and the difficulties emerging in north New Georgia, the parent company, Unilever, did not press its rights.[30]

The Enoghae incident and the Jericho raiders

On north New Georgia during 1981 and 1982, the CFC instigated a series of incidents. In early 1981, Job Dudley Tausinga held a meeting at Paradise, his father's village, and then went to Lever's at Ringi Cove, and ordered the company to leave Enoghae, in the Koraga area, by 2 April.[31] Direct negotiation did not appeal to Lever's, which saw this as the national government's responsibility. In taking this stance, Lever's failed, in the eyes of the north New Georgians, to behave as 'a side to be reckoned with, and with which the local *butubutu* could engage rationally', so lost respect.[32] In June 1981 when Lever's brought their first bulldozer into Enoghae, forty people were arrested for obstruction; in December, people from Jericho interfered with Lever's road-building, but were dispersed by police. Finally, on 27 March 1982, 200 Jericho villagers raided Lever's Enoghae camp, burning and damaging 79 houses and logging equipment, valued around one million dollars. Even as the government's ineffectual commission of inquiry was meeting, the Koraga objectors posted warnings to Lever's workers not to resume logging.[33] The Western provincial assembly, formerly the Western council, called on the government to stop Lever's, a call which Prime Minister Solomon Mamaloni heeded because of Western threats to secede in 1978 and resource conflicts in nearby Bougainville.[34]

Lever's intimidated labour force refused to return to Koraga. The company concentrated extraction on Lupa and Ndekurana, as well as Kolombangara for trees left after the first logging. The company also obtained logs from Rarumana, Vona Vona, under an agreement with the Vuragare Development Association. A local operator appears to have cut trees on Kohinggo for the local people who sold them to Lever's.[35]

Contest for the Forests, c. 1963–1985

As Lever's surveyors moved into the Nggerasi area they faced harassment, culminating in an attack on Puha camp, south of Barora in September 1984, allegedly by CFC followers. To add to Lever's woes, the Heta Heta clan in Lupa objected to the logging. They were not getting their share of the royalties and blamed the Forestry Division which had no power in such matters. And, if the Division was not blamed, then Lever's was.[36]

Attacks on Lever's came from further afield. Paul Scobie, a New South Wales-based conservationist with kinship links in New Georgia, put Tausinga in touch with Australian conservationists. Scobie wrote about the Enoghae incident in *Habitat*, an Australian conservation journal in 1982, which led to the involvement of Friends of the Earth organisation in Britain campaigning against Lever's activities. Unilever, the parent company, did not welcome the publicity.[37]

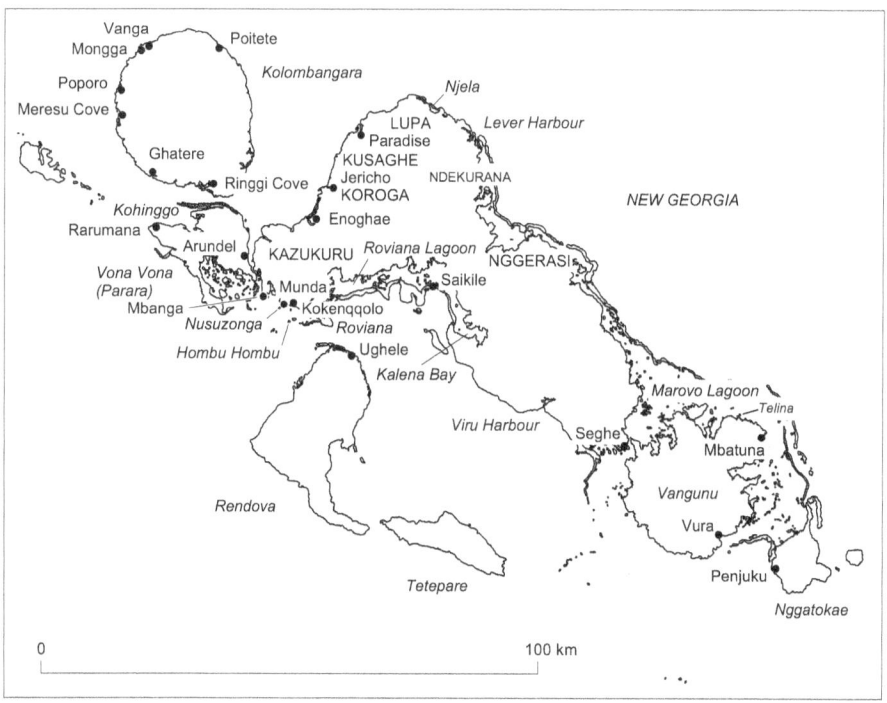

Map 9. Eastern New Georgia Islands.

The departure of Lever's Pacific Timbers

Many politicians supported Lever's. For a few years after independence, Peter Kenilorea and his United Party saw resource use in terms of the public good with the company contributing to that as well as local interests. LPT employed 460 Solomon Islanders and conducted several training schemes. Ringgi Cove was a township. There were schools, a clinic, a community centre, an airstrip, wharves, and well-maintained roads around the island. Lever's Barora base had over 160 houses, a school, clinic and airstrip. Both employees and the public had access to these services. At Enoghae, however, similar developments halted in the face of the Jericho raiders.[38]

Attempts to solve the difficulties inherent in the North New Georgia Corporation Act centred on an amendment in 1984 which, in part, meant that Lever's had to negotiate timber rights afresh with each *butubutu*. By late 1985, little had been achieved. The Forestry Division, sympathetic to Lever's, was hoisted on its own petard, as a moratorium on new licences

Wood World, January, 1970

Figure 83. The logging camp at Ringgi Cove, Kolombangara, 1970. The camp provided accommodation for 300 workers and their families, as well as a community centre, school, clinic, stores and recreation facilities. Electric power is connected to the camp.

Contest for the Forests, c. 1963–1985

Wood World, January, 1970

Figure 84. Children of workers at Ringgi Cove.

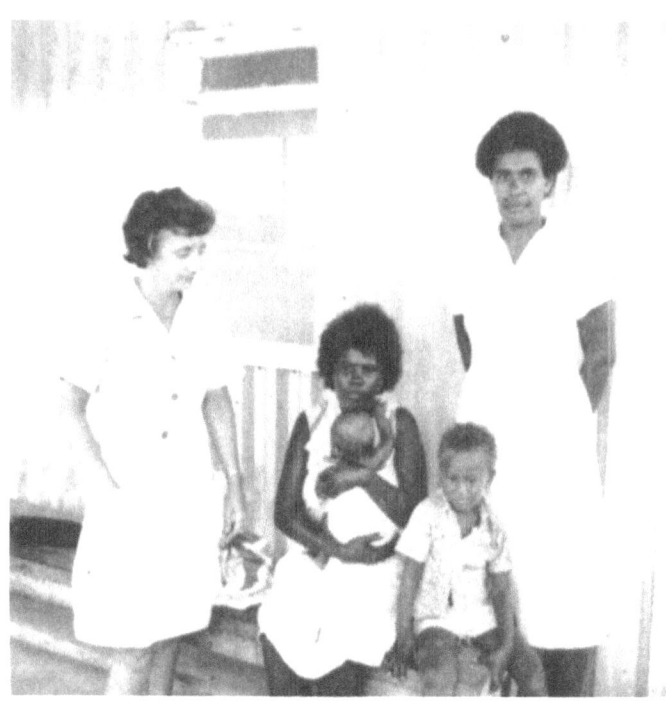

Figure 85. Australian and Fijian nurses with Solomon Islands' patients at Ringgi Cove clinic, 1970.

Wood World, January, 1970

operating since 1983 prevented Lever's from pursuing timber rights elsewhere. In 1986 Tausinga, who advocated conservation and small-scale logging, became the premier of Western province and wanted Lever's out and the Act repealed.[39] Lever's had already invested SI$6 million (with the falling value of the Solomons dollar, this represented SI$12 million in 1986) and were tired of throwing good money after bad, so left the province that year.[40] For Lever's, landholders, and provincial and central governments, the Act had been a failure. Attempts to separate timber rights from land tenure had been questioned in the colonial period as the forest areas fiasco demonstrated, yet history had been ignored. The outcome of the Act repeated the hard lesson to its Solomons creators in the 1979 parliament.

Riding the tigers

Prime Minister Mamaloni saw Lever's and its 'Britishness' as a reminder of the colonial past. Even its settlement at Ringgi with managers on the hilltop, families on the hillside, and the drab labour lines on the flat reflected colonial hierarchies. Mamaloni blamed Lever's for the Enoghae incident, rather than the lawbreakers or the Act itself, and pressured the company to mill 20 per cent of the timber cut.[41] Unlike the government of independence of Kenilorea, Mamaloni felt himself in a position of strength in dealing with Lever's. That strength came from the possibilities offered by Asian loggers.

From the early 1970s, Solomon Islands' biggest customer for 'South Seas logs' had been Japan, which took 80 per cent and more. Throughout the 1950s, Japan's major supplier had been the Philippines, whose exports fell dramatically in the mid-1970s. Indonesia exported logs, especially from Kalimantan, in great quantities from 1970, but its government began to encourage local processing and introduced a progressive ban on log exports in 1978. Exports had fallen drastically by 1982 – except for Indonesia's Irian Jaya province. Malaysia was an important supplier of logs to Japan in 1978, with most coming from Sabah and Sarawak, as Peninsular Malaysia was progressively banning log exports, to cease completely in 1984. Singapore and Hong Kong, efficient small island states, also imported logs from Malaysia for processing and re-export but, after 1982, they began to seek new sources.[42]

During the years of transition from colony to nation, the forestry Department/Division had assumed that the pattern of timber extraction would continue steadily and that 'the existing ratio of logging companies to forest resources' was 'about right'.[43] When customary land was required, established companies would negotiate for timber rights. Until all government land was logged, there was little need for the companies to do 'deals' with communities, such as providing roads, transport, clinics and schools. When it invested in welfare facilities, Lever's had wanted to ensure a stable, resident workforce. Out of necessity, in the 1960s, logging companies provided job training. For them, their role was straightforward. They logged, paid their export duties, levies for reforestation, and company tax. This was their input into revenue for the government to 'develop' the country. Companies had operated in a commercial culture in which, if deals were made with government officials, the benefit – other than for the company – was for the state, not for the officials personally, or for particular groups. The product of empire, it was an impersonal and bureaucratic culture, that sometimes relocated resources from small groups to the entire population for the public good and often dealt coldly with individuals. This was not how small Melanesian societies had developed or operated. A de-personalised, faceless government had no more appeal than faceless logging companies.

With an expanding market and a shrinking South-east Asian supply from 1980, new companies, several with Asian links, began to offer terms to log customary land that seemed attractive and meant that Solomon Islanders could pick and chose among loggers. For the first time in at least 40 years, Melanesians now were in control of allocation of the forest. To compete with the newcomers, older companies had to adopt similar methods. And the Melanesians found that Asian business methods were more like their own, based on reciprocity, personal ties, and patronage.

Guadalcanal: hoodwinked by Hyundai

Guadalcanal was an attractive proposition to loggers. In north-east Guadalcanal, local people had set up a development company in 1981 with plans for agricultural development. Soon after, a paper company gained a controlling interest for Hyundai, a South Korean multinational, worth in total assets about $18 billion. In December 1982, Koo Woong Seo, Kim In Sik, Chung Mong Heon and Wong Jong Young registered as Hyundai Timber Co. Ltd with the aim of logging much of north-east, east and south Guadalcanal. Many north-east Guadalcanal people mistakenly believed this company had taken on the responsibilities of the development programme.

Procedures were irregular. Both the Forest Division and the Foreign Investment Board believed the provincial government should consider Hyundai's logging proposal first. The provincial government met in July 1982, agreeing in principle. Prime Minister Mamaloni wrote in September to the relevant national officials, directing them to expedite the process because the 'landowners' wanted it. The province and Forestry Division read this as central government approval, although the provincial officials were concerned that, once they put the application to the area councils (replacing the area committees), they were signifying support and so there would be 'strong inducements for local landowners and the Area Council to simply rubber-stamp the application in the hope of getting roads and quick money'.[44] The Forestry Division approved the first step, known as Form One, 'Application for Consent to negotiate with landowners', apparently following the urgings of the prime minister.

With the seeming approval of the government, the province sent letters to the area councils to see if they would ensure the landholders understood the offer and were willing to allow the company to have the timber rights. D. Ruthven, the Provincial Secretary, alerted the councils to things to check – a reforestation plan, possible pollution of rivers by the loggers, and the recommended standard of roads and bridges to be built. He advised that the people should get at least SI$8 a cubic metre or 12.5 per cent of overseas market value, the same as Lever's royalties.[45] The three area councils concerned were sent Form Two, 'Certificates of customary ownership'. The area councils of Tasimauri (south Guadalcanal) and east Guadalcanal rejected the proposal, but the north-east council finally posted the form two notices in November. These notices had to be displayed publicly for two months to give people time to consider matters before public meetings to decide. These meetings at the designated places did not occur. The royalty offered was SI$5 per cubic metre and after five years, seven per cent f.o.b. price – a real return of *less* than SI$5. This did not match with a verbal offer, said to have been made, of 'roads, clinics, "development" and an increase in royalties after five years.'[46] A concerned province wanted central government advice. Meanwhile, 'The Company is paying money or gift to important people to help them get through quickly as they wish' (sic). By such dubious means, the Form Two stage was completed.

Minister of Natural Resources Peter Salaka, whose son was a shareholder in the original development company, the front for Hyundai, supported the company's proposed agreement despite the province's concern. Many landholders signed a poorly-worded agreement modified to be even more favourable to the company. The minister ignored Forestry Division's misgivings about the agreement and those of the local member of parliament. The Division's hands were tied because the only person, under the 1977 Forest and Timber Act Amendment, with the power to 'make regulations for the better carrying out of the provisions and purposes of this Part' [Section 5] was the minister.[47] He could set out the *form* of agreement between the landholders and a company, but Salaka and his successors to late 1984 chose not to. Many illiterate landholders signed without full knowledge; self-appointed trustees signed for entire groups. Several of these 'signed' lists were attached subsequent to the agreement. Appeals against the determination

Contest for the Forests, c. 1963–1985

of the area councils failed because of legal technicalities. Minister Salaka directed the Forestry Division in March 1983 to issue the licence while one such appeal was still in process.[48] He was behind a national moratorium on the issuing of new licences. His motive can only be surmised, since the licence for Hyundai was issued on 5 April, the two-year moratorium coming into effect on 19 April 1983. It did serve to give the company a virtual monopoly in the area. It was one of Salaka's last ministerial acts before Mamaloni sacked him, just before the prime minister left for South Korea.[49]

The government and the Division, under the Act, now had virtually no part to play between Hyundai and landholders. As the logging commenced and money flowed, the company persuaded more landholders to sign over rights. Among those who hesitated, a co-operative trustee, fed by favours and Fosters beer, assisted. Meetings, known only to some, regarding determination of rights often turned into drinking parties. Area councils frequently became de facto negotiators with companies, which eroded their impartiality in deciding timber rights.[50] Most agreements after early 1983 were merely verbal or sketchy notes. In train came land disputes, court cases between rights' claimants, but as the licence had already been issued for the area concerned, the appeals were ill-founded in law.

At all levels, officials had compromised themselves. As Ombudsman Isaac Qoloni observed,

> … junior provincial officials and politicians appear to have been bullied or persuaded to sign documents when full drunk. It is hard to blame these people. Their take-home pay may be worth less than a couple of cartons of beer a fortnight and they are easy prey to people who may wish to compromise them by hospitality, food, drink, transport and other favours. Our tradition is one where we work to help each other and gifts and favours must be repaid. These junior officials are in an impossible position, especially when they receive little support or encouragement to stand strong against such things and suffer no penalty for accepting.[51]

Complaints against the company's logging methods, failure to build roads, bridges, schools and clinics, and abuses against property, the environment, and custom poured into the government and the ombudsman. He summed up the government's attitude:

> … the strict legal view has been that logging agreements are private agreements made between the people and the company in which no government office has the right or duty to be involved. The landowners must sort thing out for themselves. The landowners have dug their own grave – 'saed blong olketa' [their own doing].

There were remedies in the law of contract and probably in torts, but these were little-understood British-based legal tools. When the ombudsman advised the aggrieved parties to seek the Public Solicitor's advice, the latter asked for a copy of the agreement, but no one had it and Forestry Division would not release theirs. Although the Public Solicitor had extensive legal powers, he was hamstrung by understaffing. Unable to obtain redress, the aggrieved people obstructed roads. The protests led to Hyundai's action in the High Court which upheld its 'commercial rights' against the protesters. So the weakest were even more victimised. Failing to get action from the Minister of Natural Resources, the provincial authorities in 1983 drew up their own standard agreement for future dealings in Guadalcanal.[52]

There was more. In 1987, after the 1977 Amendment had been further amended and the law supposedly tightened, Hyundai obtained from Forestry Division, under Minister of Natural Resources Kenilorea, an 'extension' of its licence into Longgu-Valasi, Guadalcanal, 'with the identities of the Custom Lands and the Landowners not even considered...'.[53] Complaints of abuses soon emerged.

From Guadalcanal to Malaita: Solmac and Kayuken

Most elements in the above could be repeated in practically every instance of the granting of new logging licences in the early 1980s. For example, a company called Cape Esperance, with Salaka and the McArthur family among its directors, obtained a saw-milling licence and wanted to log for export from west Guadalcanal in 1980. The province refused. The McArthurs formed Solmac Construction and Timber Company in December 1980 and commenced negotiating with landholders, without permission under Form One.[54] Certain west Guadalcanal landholders of Visale, Kolina and Vura had formed their own company, Sagalu Exim Pty Ltd, in September. In early 1981, Cape Esperance and Sagalu invited Solmac to form a joint venture, with Solmac holding about 40 per cent and Cape Esperance 60 per cent, and Sagalu being subsumed under Cape Esperance.[55] The joint venture failed, because Solmac had already been refused a logging licence.

Solmac reappeared in July 1982, seeking a licence to log Burianiasi and Nafinua on north-east Malaita. 'Ships, or landing craft, bulldozers, graders, foreign men from Australia and New Guinea' appeared further north at Ata'a Cove, started logging and destroyed a plantation of Leslie Fugui's clan, in breach of the 1977 Act. Minister Salaka had granted the licence, although Form Two, to ascertain who held the timber rights, was uncompleted. Chief Forestry Officer Sam Gaviro, and his permanent secretary, Leonard Maenu'u, opposed the illegal licence and, along with the Attorney-General's lawyer, Phillip Tegavota, sought a court ruling as part of Fugui's action.[56] These public servants had been under considerable pressure. Nester Bele, secretary to the Investment Board in the Prime Minister's Office and allegedly a director of Solmac, argued beguilingly that,

> ... as there were no Solomon Island shareholders in Solmac, if Solmac was allowed to go ahead in the disputed areas of Wards 18 and 19 and then faced liabilities, Solomon islanders would have nothing to lose.[57]

In October 1982, the High Court fined Solmac \$12,000 for destroying property. Chief Justice Crome also censured Salaka, yet Mamaloni retained him in his cabinet until April the following year. Solmac, with an annual quota of 38,000 cubic metres, appears to have continued logging as the Forest Development Company until 1985 when it was liquidated.[58]

These events have a particular irony since a local company, Atasi Ltd, affiliated with the South Sea Evangelical Church, had been running a sawmill since 1976 at Gwarimundu, Atori in the area Solmac wanted. Atasi sold milled timber to Honiara and trained local men, some in New Zealand. It planned a processing plant at Auki, employed 40 Solomon Islanders and contributed to Atoifi hospital. Atasi's plan was to log 3–4,000 cubic metres yearly over 20 years on leased land while Solmac wanted 72,000 cubic metres yearly. Atasi had completed all the requisite procedures, offered similar royalties as Solmac and, unlike Solmac, had provincial support. It would be hard not to wonder, as Atasi did, what had prompted Salaka to attempt to issue a licence to Solmac?[59]

The Guadalcanal–Malaita nexus re-appeared with Kayuken Pacific. Kayuken's two major directors were Kean Siang Khoo and Hin Siang Khoo, ethnic Malaysians of Australian citizenship. From 1982, the company had tried several provinces without success. In May 1983, they applied for a licence to log West Kwaio, Malaita, an island where a UN Development Advisory Team had found only a relatively small, scattered timber resource a year before. The study recommended the timber be used for local saw-milling only. Even then, without reforestation, it would be worked out in 25 years. The province endorsed this and plans were commenced for a reforestation project.[60] It was not supportive of Kayuken's application, as nearby Bina and Mbuma saw-mills, run by local groups relied on this area for timber.[61] In line with the April moratorium, Forestry Division refused the application.

Back on west Guadalcanal, Cape Esperance had obtained a logging licence in 1981, but lacked capital.[62] It provided the front for Kayuken so the latter could log. Kayuken was hardly a model corporate citizen. In 1984, the division reported that its felling techniques were wasteful of timber and damage to soil occurred because of poor roading and logging excessively high and hilly areas. Some landholders claimed the company had logged their land illegally. The Forestry Division suspended the company's operations, just as the 1984 Amendment to the Forests and Timber Act was gazetted, specifying the duties of loggers regarding logging plans, protection of water sources, and prevention of soil erosion. Yet, when the company appealed to Prime Minister Kenilorea, he obligingly cancelled the suspension order in July 1984 (Appendix 6). In 1985, Forestry Division found the company had disturbed Kesau, Aruligo and Popo water sources. The damage to the Kesau riverbanks caused flooding of Tambea, a locally-operated tourist resort, owned by the Torling family, Solomon Islands citizens. The province forced the company to rectify some of the river damage. Throughout 1984–1985, Kayuken was under Forestry Division censure for failing to process 20 per cent of the logs, but it prevaricated.

Kayuken's former employees tell of the shipment of protected species, such as ngali nut (*Canarium* spp.) trees, and of keeping two sets of accounting books, one for its own purposes with the real prices it obtained for its logs and one that showed prices it declared to the government. Customs officers often signed on at the beginning of loading of the ship, then disappeared to return after loading was complete to sign clearance for the number of logs stated on the manifest. Honest customs officers who checked each log, came across batches with two in each numbered with the same number, yet on the ship's manifest there was only the batch number, say ten, not ten plus one. Multiply this by 100 batches and the company was doing nicely, but the government and the landholder were losing duty and royalty respectively, to say nothing of possible company tax on profits. In mid-1985, Kayuken renewed its attempt to obtain logging rights in East Kwaio, but failed, as the moratorium was still in effect.[63]

Makira: lumberjacks and Japanese

In 1972, a young Solomon Mamaloni, member of the Governing Council had asked,

> Why is there no activity at all in this Dept in the Eastern Solomons especially inner islands, is there no property over there with any good timber?[64]

Trenaman had thought not. Makira had the lowest merchantable volume per hectare in the group, with the exception of Rennell Island. Nonetheless, Integrated Forest Industries (IFI) obtained a licence in 1982. The company had Australian and Fijian nationals as directors, including Paul Freeman. Andrew Nori, a Solomon Island lawyer, briefly joined the directors in 1983.[65] It sub-contracted to Howell Enterprises, which supplied the equipment and a team of American lumberjacks. To welcome the loggers at Kaonasughu village, Premier Ramoni and Minister for Makira and Temotu provinces Tropa were on hand in September 1982. The company promised employment, to 'mak[e] Makira a better place to live', and an American High School.[66]

Contest for the Forests, c. 1963–1985

It was no wonder the first royalty payment in November of $31,594.10 brought smiles all round.

Financial Statement of Integrated Forest Industries (SI) Ltd, 30 June 1983

Figure 86. People at Kaonasughu village, Makira, receiving initial royalty payment of SI$31,594.10 from representatives of Integrated Forest Industries. Tinned goods, cash, matches and cigarettes make up the payment and gifts.

In building its base in 1982, IFI had bulldozed 'tambu place, graveyards' of Kaonasughu and neighbouring people. Soon after, an infection similar to scabies afflicted the loggers. A ship was wrecked nearby and a person killed by a shark. For many, this was the *kik baek*,'kick back' of the ancestors, clear signs of their displeasure with the treatment of clan lands. These were but omens.[67] By early 1983, a royalty of SI$5 a cubic metre for the first six shipments replaced the IFI's promise of a royalty of US$5. At the exchange rate of about 0.83, losses were considerable. The main coastal road deteriorated under logging traffic. Bush 'roads' were mere skid tracks. The company needed gravel for roading, so simply appropriated it. It cut down protected trees without compensation and logged outside its concession. Villagers complained of blocked rivers, wasted timber and compaction of the soil, yet a parliamentarian told them not to complain as this would discourage future investors. Disputes began about rights to royalties. Howell's equipment found its way east to be used by Rural Industries Ltd, a company controlled by the Campbell family, Solomon Islanders at Kirakira, and about to go into receivership. Landholders there voiced complaints of poor logging techniques and unfair distribution of royalties.[68]

Too late it emerged how IFI got its licence. In Makira parliamentarian Ben Kenika's words:[69]

> Notice [Form Two] was given out, put up, bat iu save hao long, not even one month! Then provins hem sei raet kam talem Ministri o everything olraet nao, letem kam, givim kam logging permit ia. Bat agreement signed weitem pipol no stret tru. Go toktok nomoa wan morning finis long morning ia, may be one or two hours then sign go nao. Pipol signem go agreement nao, fast tumas. No brek, jes toktok, pusim go pepa underneath nose blong olketa big man ia, signem long hia, signem go nao. Tekem provins raet go long Ministri issuem go licence, licence go nao. So many people long Makira, even myself hu kam from that area, no save long samting ia, sek nomoa olketa sei o samfela Amerika kam, 65 of them. Bat taem mi harem nius ia, late finis, olketa there finis nao.

His company in financial difficulties, Freeman became unwelcome. Two Japanese, Toshio Hashimoto and Shuda Noda, assisted, buying into the company in 1984. Hashimoto then tried to get logging rights in Star Harbour, Makira through the local Hagaparua Sawmilling Association, but failed, probably because of the moratorium.[70]

Licence seekers on Santa Isabel and Choiseul

Five companies wanted rights in Santa Isabel. One represented by L. Y. Tai, on behalf of Japanese interests, made an agreement with Eti-Eti land holders to set up the Sagawa Development Company in 1979, but government officials insisted on carrying out a survey first, delaying the issuing of the licence.

In late 1982, despite the Solmac scandal, Prime Minister Mamaloni publicly declared that logging companies should be able to go direct to the rights-holders without the procedures stipulated in the 1977 Act. Yet, these procedures were minimal in the extreme; loggers had no restraints put on the agreements they could make and their methods of logging. The division's powers were virtually nil once procedures regarding permission to negotiate and the establishment of landholders' rights were followed. In fact, the Amended Act of 1977 had no protection for landholders or the forest on customary land. Mamaloni's stance drew criticism from Isabel province, as well as one of the national newspapers, and the province disallowed the agreement between the Eti-Eti people and Sagawa.

Contest for the Forests, c. 1963–1985

The Vanuatu-based Fletcher Organisation was interested in logging also, but reconsidered in 1984 when the market faltered. Another company, Taisol, headed by Suny Wun Sai Tong (known as Suny Tong) of Tong Poy and Sons, Taiwan, in association with the Taiwan government, began its quest for a concession there in 1982.[71] James Wong, a ship builder, obtained a licence in 1984 to log his leased lands, formerly held by R. C. Symes. Another company, Island Resources Limited was formed. It was to be managed by Allardyce's manager, John Dixon who was to hold a 30 per cent share with 70 per cent in the hands of Solomons business people, including Margaret Lepping, Nacy Regine, and local landholder, Mary Iro and her husband Sam. The company planned to log for the Maringge landholders, but officials feared it would simply be a front for Dixon, who had been implicated in war relics trading and was not in good standing with the government. Other local landholders, headed by Lily Poznanski, opposed the plan and it faded. On nearby Choiseul in 1983, the South Korean-owned company Eagon applied for a logging licence, but it was not granted because of questions about the legality of the area council and the role of Western province.[72]

Winners and losers

Following the procedures to the letter, the consequences of direct dealing under the Act of 1977 would have seen government agencies under great pressure. The Act, however, did not have the opportunity to operate under optimal conditions. Many, including a couple of provincial governments, did not fully understand it, believing that area councils were to negotiate with the company on the landholders' behalf, instead of just determining who held the timber rights. Several instances of circumvention of the intention and often the wording of the Act occurred at the highest levels. Even Prime Minister Mamaloni in 1982 had advocated the interests of the loggers at the expense of the bureaucracy enforcing the law. Mamaloni liked to project an image of champion of his compatriots' freedoms, a compelling stance to those with recent memories of colonial paternalism and threatened land rights. But the question has to be asked, were such compatriots, mainly illiterate, ignorant of world prices, the subtleties of contract law, and the environmental repercussions of poor logging practices, capable of forming a fair and valid contact with well-capitalised, often multi-national companies which could buy the best of legal advice? Was it a meeting of equals, a fair contest for mutually beneficial returns?

Companies were able to suborn several Solomon Islanders. Low-paid provincial officials could be bought off directly. Isolated junior Forestry

Division officers could be purchased by a few day's hospitality at logging camps. Many of their masters appeared to be unable to resist suasions, too, and some had compromising trips to Asian cities. Influence on senior public servants also came in the form of ministerial command, petitions from provincial politicians demanding concessions from the central bureaucracy, or resource holders who welcomed loggers' promises. And it is to be wondered why no minister ever used his powers under the 1977 Forest and Timber Amendment Act to introduce the remedy of a simple but mandatory standard logging agreement, such as the one recommended in 1978, following the amendment to the act.

Loggers' promises of development were exaggerations, if not lies. Nonetheless, there were examples that gave hope to isolated communities for the means to educate their children, heal their sick, build their churches, open up the land and the bush for cash crops, and to travel more easily to employment, markets and friends. Lever's had provided much on Kolombangara and at Barora as had Allardyce in Shortland. Yet, Lever's carried the burden of its history and business culture which epitomised for many Solomon Islanders colonial disrespect. Lever's – and all other parties – fell victim to the North New Georgia Act, a clumsy construct that tried to legislate away fundamental problems of Melanesian communally-based land tenure confronting capitalist enterprise – where individual rights had a cash, rather than a subsistence value.

Few communities obtained any long-term benefit. On Kolombangara, at Poporo, the Viuru tribe's royalties of about $90,000 dwindled to $25,000 as the money was spread around families or loaned to people to buy outboard motors or to set up a store. Families spent most of the money on consumer goods and few loans were repaid. At Ghatere, the community under Allan Pitakera planned for the future. It set up a cattle project, but also made sure there was forest preserved from logging to met subsistence requirements. Similarly, the Vuragare Association on Vona Vona, advised by Nicol Kuruti, hired Lever's to cut their logs, but withheld forest for subsistence use and preserved the soil by laying down rickers of unwanted timber over which the heavy equipment moved. Logged land was for cash cropping and a cattle project. Money was reinvested in community work and a saw-mill.[73] On south New Georgia, the Saikile Development Company reinvested some of the land-holders' royalties in *butubutu*-based businesses.[74] But these were the exceptions; most were like the Poporo community. Even the North New Georgia Corporation's celebrated $1.7 million housing project at Njela soon fell into disrepair.[75]

Not only did most companies not contribute to economic and human development in more than an ephemeral way; their presence often choked more lasting enterprise. The Tambea resort on Guadalcanal was flooded

Contest for the Forests, c. 1963–1985

because of Kayuken's greed. Beset by financial problems, the Atasi saw-mill at Atori was forced to close the same year as foreign companies moved into the adjacent forest.[76] Local development companies perceived partnerships with loggers as a short-cut to their aspirations. Often, when the loggers departed, they were left without any tangible benefits and without their forest.

Another pattern was beginning to emerge. The surviving pioneer logging companies, as well as new local companies, ended up with Australian, Fiji nationals (often ethnic Asians), and Asians among their directors. By the mid-eighties, Asians controlled most companies wanting licences. To facilitate their negotiation with several claimants, many loggers especially from 1982, opened the forest with what some would call 'gifts and favours' and others would call bribes. Solomon Islanders had seen the wonderful, sleek tigers of promise come out of Asia, and tried to ride them. The majority ended up in their bellies.

— 11 —
Attempts at Control? c. 1980–1990

> What is particularly disheartening is that the only changes that have been made in the Forest and Timber Law have had the effect of making it easier for overseas Companies to come in, without even the pretence of local participation. Nothing has been done to protect people, the environment nor to make the best of our dwindling natural resources.
>
> I. Qoloni, *Report of the Ombudsman, 1991* (Honiara, 1992), 16.

By 1984/5, loggers had tried to gain access to all the provinces. Several had succeeded and those who had not were lobbying and more, to win over landholders and politicians. Weaknesses in the government's forestry policy, legislation, and administration emerged. Some had been anticipated by outsiders familiar with problems in South-east Asia. The rest became apparent as logging intensified in 1981 when the Forestry Division saw what was happening in its own country.[1] As the decade progressed, conflicts between legal powers and political realities in forestry matters mounted.

Timber control

In 1979, FAO consultants warned of practices in the timber industry worrying South-east Asian governments. Transfer-pricing was mentioned. This involves the 'remitting of profits by the manipulation of product or input price to levels that vary substantially from the current market price'.[2] Taking several forms, it most commonly involves the sale of cargoes of logs in the host country by a subsidiary company to an allied company offshore at an artificially low rate that becomes the basis for export duty, royalty and levy calculation, and ultimately corporate taxation in the country of the logs' origin.[3] The logs are later sold at market prices by the allied company and

Attempts at Control? c. 1980–1990

the higher real profit is stashed away in a tax haven, such as the Virgin Islands, making its way back to the parent company.

The Forestry Division had to rely on the companies for price trends information, so was at a disadvantage in calculating tax and royalty returns.[4] In 1980, A. I. Fraser, a timber industry fiscal advisor, urged the government to establish a timber control and utilisation unit to monitor the sale of logs internationally and to prevent timber wastage.[5] The Division checked customs returns and export entries and found discrepancies. As the British Overseas Development Assistance (ODA) was pouring aid into reforestation, their adviser too could see no sense in the Solomons government throwing away revenue by failing to control the industry, so funded an overseas officer to establish the unit.

The unit came into being in late 1980. Its purposes were: to inspect and control logging operations to ensure that wastage of timber during felling was avoided and that destruction was minimal; to monitor export shipments, their species and quality composition; to investigate marketing and pricing of log exports and to promote market development; to investigate methods for improved milling, preservation, and grading of locally sawn timber for domestic and export market; to train more staff; and to advise the government on fiscal matters pertaining to the timber industry. The first, the inspection of logging operations, revealed one of the several deficiencies of the 1977 Forests and Timber Act Amendment. The forestry department, in colonial times, had needed no permission to go on government land. The premise that the bulk of logging was to be on government land had underpinned the 1969 Timber and Forest Ordinance and this carried over into the Amendment of 1977, despite its allowing for direct dealing, mediated by set procedures, which opened the way for the exploitation of customary land. The real powers of the Timber Control Unit were few and circumscribed by inadequate staffing. By 1984 there were 14 logging companies and over forty saw-mills licensed. In its early years, the Unit had an establishment of 18, but often the staff allocated fell to only 12, to police a scattered archipelago and mobile loggers.

It had some small victories. The unit was successful in assisting companies to avoid timber wastage, for example, through splitting; but savings here benefited the company as much as the country. At Alu, in 1981, Allardyce Lumber Company followed the unit's suggestions for more efficient felling. However, the officers found the company was leaving some species, such as *Dillenia* and *Instia bijuga*, (the latter not a protected tree until 1984) in the forest on the excuse that they were not marketable, yet other companies were selling them. More pertinent, these species were sinkers and the company lacked barges to take them to ships. The difficulty was then how to penalise Allardyce, without the requisite regulations.

Occasionally landholders were able to set up their own timber control. The Kalena logging was checked by a former Forestry Division employee who left the Division to advise his people. Terry Kera, former Chief Forestry Officer, resigned to work with Kalena's management while the land of his own people was being logged.[6]

However, the Division's timber control inspections usually were confined to areas adjacent to government lands. Thus Kalena Timber Company improved its felling practices under the eyes of the officer based at Arara. If officers had to stay at logging camps there was the risk of their being beholden to the loggers whom they had come to inspect. Moreover, agreements made between landholders and companies often permitted poor logging methods, and the unit could do little.[7] Throughout the early 1980s, Chief Forestry Officer Sam Gaviro's admonitions persisted:

> The present 1969 and 1977 Forest Acts place the Timber Control Unit at a distinct disadvantage since it has no authority to exercise control over logging and sawmill operations on customary land.
>
> With 90 percent of the total exploitation of the forest being carried out on customary land, control of the resource is negligible, especially in areas of unnecessary waste, lost revenue, poor infrastructure grading and scaling and provision for reforestation.
>
> It has been estimated that almost half a million dollars are lost annually from unnecessary waste of saleable timber.[8]

That did not include losses through transfer-pricing and under-declaration of exports. No tax legislation addressed this or other common forms of tax-avoidance by loggers. Gaviro consistently requested more manpower for inspection, but was ignored at a time when the government's revenues from timber were skyrocketing. By 1985, 95 per cent of logging was on customary land and the Division publicly stated it could effectively police only two of the 12 operating logging companies.[9] From 1982, Gaviro also begged his political masters for a Timber Marketing Board to get the best price overseas. Again this fell on deaf ears, but the Division in 1983 did establish the monthly publication of a bulletin of the previous month's average price per cubic metre of round log species exported.[10]

Moratorium and new policy, 1983

Underlying these difficulties was the rapid increase in the licensed cut. Until 1984/5, forest policy, in theory, followed the 1975 statement, which set the maximum annual extraction at no more than 400,000 cubic metres, with the aim being 280,000 cubic metres of this in sawn timber. In 1981, a review of policy was overdue because of developments in the industry, but when

Attempts at Control? c. 1980–1990

Solomon Mamaloni came to power he dismantled the Select Committee on Forest Policy. By 1983, if all licensees, loggers and saw-millers had taken their annual allowable cut, 948,900 cubic metres could have been extracted and the forest as surveyed by the Division would have been cut out by about 1993, with no chance of reforestation making up the deficit. When Minister of Natural Resources Salaka introduced the moratorium on new licences in April 1983 there was no room for complacency, as those 'in the pipeline', had the potential to take the annual cut to over 1,000,000 cubic metres. In 1984, even with several operators working at under capacity, the cut exceeded 400,000 cubic metres by 57,840 cubic metres. In terms of hectares logged, this was about 10,000, compared to fewer than 1000 reforested that year.[11]

The moratorium gave the Division time. It now could carry out a forest resource survey and assess tolerable extraction rates, as the surveys of 1960s and 1970s did not include species that since had become marketable. Some areas had not been surveyed fully and others had their stocks reduced by subsistence activity and logging. In 1983, the Division sought assistance from the Australian Development Assistance Bureau (ADAB) and a preliminary study began. The moratorium would also enable the Timber Control Unit to develop and the Division to concentrate on reforestation priorities, including upgrading nursery techniques, building a refrigerated seed store at Munda, and setting aside, as state forest, suitable government land.

The Division, other departments, Guadalcanal Province, and the public had been pushing for a revision of forest policy and legislation.[12] In June 1983, Cabinet approved a new policy, based on 'the maximum desirable processing of logs, minimum wastage', 'reinvestment in the forest', and sustainability of yield. The policy statement supported a forest resource study to start in 1984, the moratorium, and revision of the legislation. Timber control was to be strengthened and would consist of a logging inspectorate, a timber inspectorate and a utilisation unit. The export of logs was to be phased out by 1994 when all timber would be milled domestically. The goal was 50 per cent milling within the first three years, to promote added value and local employment. Existing logging licence-holders could follow this or invest in reforestation, but any new ones were bound by the milling requirement.[13]

New legislation, 1984

Amendments to the Act of 1969 and Amendment of 1977 went before parliament in 1984, along with a standard logging agreement (SLA) for future contacts. Minister Tropa proudly stated that the bill had been 'drafted and redrafted over again under my close supervision', but he presented a very confusing document.[14] The 1984 Act listed 24 amendments and two substitutions.[15] To find the law, one had to read the Forests and Timber Ordinance 1969, a minor Amendment in 1972, the Amendment of 1977, and the Forest Resources and Timber Utilisation Act 1984 in conjunction. These were almost impossible to read as one as each evinced different drafting styles, had errors, and together and often singly were disjointed. The biggest hiatus was between the Act of 1969, focused on logging on government land, and the Amendment of 1977, for direct dealing on customary land. The freedom of landholders to negotiate was now limited by a mass of procedural requirements that would confuse them and the logging company. Uncertainty prevailed as to whether landowning rights were identical to timber rights. The effects of these combined acts in terms of owners' rights are unclear.[16] Several parliamentarians claimed the Act of 1984 sewed up several loopholes,[17] but the resulting garment had become a crazy patchwork.

The process for obtaining a licence now involved over 15 steps: its complexity can be gathered from the following simplified version. Potential loggers had to apply to the government's Foreign Investment Board (FIB) which evaluated the project and notified the provincial government concerned and Forestry Division for comments. The FIB also sought the views of other ministries concerned. If FIB thought the project acceptable, various ministries and provincial authorities would be advised. The company could then apply to the Chief Forestry Officer (Commissioner of Forests) for approval for negotiations, using Form One. The Chief Forestry Officer would then send the application, a copy of Form One, to the provincial government and the appropriate area council. The latter's duty was as under the Amendment of 1977. Through several set, but complex, procedures, the area council was to attest to the ownership of the land concerned, whether or not the owners wanted to sell their timber rights, the nature of the rights to be granted to the applicant (logger), and the involvement of the provincial government in the management of the venture on behalf of the owners. This information became the basis for the issuance by the area council of a certificate of land ownership, known as Form Two. This was sent to the Chief Forestry Officer, who would then recommend to the minister that approval be given, unless an appeal to the customary lands appeal court (CLAC) was made against the determination of the area council within a month. If there was no appeal or it failed, the applicant for the licence had to undertake to prepare a five-year plan to give to the landholders and the national government. The negotiation of the SLA then had to be completed as part of Form Four and should be held in public. The commissioner of forests must establish that the procedures have been followed correctly, that the time of appeal has lapsed or the appeal has been determined, and that the SLA has been completed in accordance with the certificate issued

Attempts at Control? c. 1980–1990

by the area council. If all was satisfactory, he issued a certificate approving the negotiation, within 14 days after the completion of the SLA. Despite the confusing order, this certification is Form Three. The company then applies for a felling licence that the Commissioner 'may' reject or accept, providing all the preliminary steps have been carried out.[18]

The SLA was not gazetted until mid-1987, three years after Parliament approved it, and it was a legal document with technical schedules needing 'translation' to village people and probably to the logging companies.[19] It was formulated under the regulation-making powers of the Act, but because of its wide scope, informed legal opinion suggests it is *ultra vires* of the Act and thus invalid. In it, the landholders had to nominate government officers as their agents for monitoring the agreement, a condition that was probably in breach of the country's constitution.[20]

Despite its attempts to define the control of licensing of loggers on customary land, the wording of the Act, for example, with respect to Section 5b, applied only to persons who were engaged in acquiring timber rights with the intention to export or mill logs, and not to any person who already owned rights on customary land. Those who could claim the rights would not have to follow the procedures and could start their own logging operation with – given the history of so-called joint ventures – offshore finance to export the logs. Prosecution might follow under another section of the Act (the principle 1969 Ordinance), regarding the felling of trees for sale, but was less likely to incur a major penalty. This same section in both 1977 and 1984 Amendments applied to log exporters or millers already 'carrying on business' in Solomons, so in theory, no new companies could apply for a licence, and *all* granted since 1977 were illegal, except for the logging and milling companies then in business.

Inconsistencies such as this reduced Forestry Division to hesitancy in enforcing the law, although it did make its own consolidation of the laws for internal use in 1985.[21] The standard logging agreement did not have retrospectivity, despite the wishes of parliamentarians like Ben Kinika and Ezechiel Alebua whose people had been hurt by loggers on Makira and Guadalcanal respectively.[22] When new licences came to be issued and the agreement presumably enforced, it, like the Act, was sure to be tainted with uncertainty.

Policy in practice, legislation in action

In terms of its functions, the Timber Control Unit was the linchpin of the new policy and law. The government's commitment to the unit was abysmal. In 1984 and 1985, Sam Gaviro complained of his 17 staff being unable to carry out effective control on customary land because of the persisting reasons of inadequate agreements, shortage of qualified staff, and lack of housing and transport at a time of logging expansion. Yet, in 1986 his staff fell to six, rose to seven in 1987, and dropped back to six the following year. That year, 1988, the Division was staffed only to 68 per cent of its establishment, so it was not only timber control that was struggling. Logging on customary land, now subject to inspection under section 5j of the new Amendment, could be inspected only sporadically. The unit still had to rely on the industry for log prices, and it was only in 1986, three years after the new forest policy was promulgated, that its few officers were given authority to check ships taking logs overseas.

The forest policy had endorsed the moratorium to facilitate the national timber resource inventory. Further aid was sought in 1984 from the New Zealand government which provided the expertise for an aerial forest survey. Australia was to finance ground surveys. Although the New Zealand mapping adviser did not materialise in 1985, Australia followed up its preliminary 1983 study with another to identify the parameters of the inventory while the Division obtained an extension of the moratorium for 18 months from April 1985.[23] The Solomon Islands government formally requested aid from Australia in 1986 for the forestry sector. That same year, the greatest betrayal of the fine sentiments of the 1983 forest policy with its emphasis on sustainability came, in the Division's eyes, when the first exemption to the moratorium received Prime Minister Kenilorea's support. The exemption was granted to enable Kayuken to apply for a licence to log west Kwaio, Malaita.

Some west Kwaio landholders wanted their forests logged, claiming that development was being retarded, because people did not want to clear land for agriculture and gardening, just to see the trees rot or burn. They had a right to dispose of their trees and government was barring the way – a standard and understandable argument.[24] However, in unauthorised negotiations by Kayuken with landholders and politicians in late 1985, Kayuken told of the benefits it would bring, 'such as a new house for the paramount chief, new roads and money'. The new provincial premier, Maomatekwa, unlike his predecessor, supported logging and was 'treated well' by Kayuken.[25] He asked Minister Sande, in December 1985, for permission to negotiate, which was refused because of the moratorium. Soon after, the premier discussed the matter with Prime Minister Kenilorea, a Malaitan from west Are'are, and both approached Sande in January 1986. Another delegation led by the member for west Kwaio, Jonathan Kuka, drove home the matter to the minister, who accepted their views, consoling himself with the fatalistic rationalisation that as Malaita's expanding population had all

Attempts at Control? c. 1980–1990

but deforested the island for gardening, the remnant forest might as well be logged too! Yet, at this time, the New Zealand government, at the behest of the Solomons government and the Division, was beginning a pilot project to *pay* landholders to reforest their own land, some of it logged, some of it over-gardened, in the north in Kwara'ae, to the tune of SI$ 384,586.[26] Observers could be forgiven for thinking that aid donors and their taxpayers, in such circumstances, are fools.

Although the moratorium was lifted for Malaita in February 1986, this did not mean Kayuken automatically got the licence. Because of its concurrent performance on Guadalcanal, Gaviro refused permission for negotiations with landholders to proceed. The company ignored this and submitted its proposals with the Malaitan premier's support. Gaviro was concerned, particularly as locally controlled saw-mills of Maasina Sawmill Ltd (formerly Onepusu Sawmill Ltd), Mbuma, and Mbina were interested in the timber rights. More Malaitan delegations, including the president of the west Kwaio council, Andrew Foasi, to Forestry Division followed and finally consent to negotiate was given. The ombudsman, I. Qoloni, found that Foasi was an advocate for Kayuken, allegedly in their employ. Qoloni's report is a litany of evasions of procedures at virtually every phase of the company's quest for the licence. For example, one entire area council, Dorio, which was included in the licence as issued in August 1986, was not consulted and their appeal was conveniently mislaid in a drawer of the clerk of the court while the licence was finalised. The short-staffed Forestry Division was not able to send a timber control officer to the area until July 1987, after logging had started, when the officer and overseas visitors found their road blocked by a Kayuken employee.[27]

A key issue in this was the granting of the licence by Gaviro. He not only had to deal with constant pressure from a parade of politicians from Malaita to Forestry Division in July 1986, but also, on taking legal advice from the Attorney-General's office, was told that under the Act, Section 5c he could not refuse to recommend to the minister that the licence be granted *even if he knew the area council had not followed the correct procedure*. In effect, this stripped Gaviro of the few powers he had to protect the forests of Solomon Islands and the landholders who had rights to them.[28] This was the first real test of the Forests and Timber Utilisation Act 1984. That the Act needed a total revision was obvious and Gaviro pressed for expert advice.[29] Support for this came from many quarters including the provinces and ironically, the new Minister of Natural Resources, Kenilorea, whose government had unleashed the juggernaut by lifting the moratorium.[30]

An exception to the moratorium, an exception to the law: Choiseul

When the colonial forest surveys were carried out, only northern Choiseul was among the 'smaller areas' considered. The island's full stocks were unknown, but estimated at 7,000,000 cubic feet (198,000 cubic metres) or less than two per cent of the timber resource in 1974; it had not been a significant factor in calculations when the 1975 figure for sustainable logging was set at 400,000 cubic metres annually.[31] Its forests were a potential counterbalance when weighing the excess allowable cut, double the target, under all the licences issued until the moratorium of April 1983. That was not the logic applied. Because Choiseul had been outside calculations, its forest was to be outside the scope of the moratorium and open for exploitation by the Korean loggers, Eagon Resources Development Company.[32]

Eagon had supporters in high places, including the national parliamentarians representing Choiseul, successive prime ministers, and most ministers of natural resources. It managed to get the approval of the Foreign Investment Board in March 1983, despite the fact that it wanted to extract 200,000 cubic metres a year, half the so-called annual permissible cut for the country. There were also concerns about transfer-pricing, as the company was going to sell the logs to its own veneer factory in Asia. Western province asked Forestry Division to survey Choiseul's forests.[33] They were rich, but the topography was such that intensive logging would cause severe erosion. Forestry Division thought 50,000 cubic metres yearly was the maximum that could be extracted. Meanwhile, the company had been doing its public relations among a people who perceived themselves as isolated, and 'in desperate need of project development, means of communication and other services'.[34] The Asians' 'sweet sugar' dished out to villagers, included sports equipment, soft drinks, towels, cigarettes and food. Provincial officers were offered expensive wrist watches. At its proposed headquarters on Moli mainland, it promised a new school, as well as a clinic with a doctor.

By early 1984, Gaviro had given his consent for the company to negotiate, though gifts had already sweetened some. However, the Choiseul area council was not satisfied with the company's procedures and asked it to repeat some with more attention paid to the landowning groups' identity. Of the 300 landowning groups, 25 wanted the company. They believed the agreement they had with the company was one like the 1984 SLA (still ungazetted), plus extras like clinics and schools. The document signed when a leading spokesman, Clement Kengava was out of the country, omitted not only the extras, but also several of the essentials and incorporated a disadvantageous royalty set at SI$5 a cubic metre. Gaviro pressed the company

Attempts at Control? c. 1980–1990

to reinstate the essential clauses. It complied, although minor clauses were omitted as well as the extras wanted by the landholders.[35]

Forestry Division received, purportedly from landholders, a flood of letters, extolling the company. Minister Keniloria, in Prime Minister Alebua's cabinet, issued a logging licence in September 1987, without a certificate of recommendation (Form Four) from the Division. Although the Division had advised 50,000 cubic metres per annum should be the 'cut', the licence was for 150,000 cubic metres, until the year 2007, and, so Ombudsman Qoloni claimed, 'covered the WHOLE of Choiseul'. This was astonishing as the procedures required under the Act of 1984 had been carried out for only nine per cent of the land and landholders.

Western province objected, but was ignored by Minister Kenilorea. On top of this, the Division found that when trying to enforce the 20 per cent of timber cut saw-milling regulation – the 50 per cent of 1983 policy having been abandoned – that the original conditions of the investment approval had been altered to 'nil conditions' – that is, no reforestation, no milling, and Eagon also had a concession to import its equipment duty free. The company must have been laughing all the way to the bank when Forestry Division sealed the deal with the ultimate in sophistry:

> Areas for which Licences have been granted by the Forestry Division can be regarded as having completed the formalities under the Act – otherwise there would not have been any Licence issued.[36]

Control or conservation?

With the departure of Lever's in 1986, Job Dudley Tausinga and others had the way cleared to attain their professed aims of self-development and the repeal of the North New Georgia Corporation Act. Interest reawakened in plans discussed earlier, in 1980/1. At that time, the corporation had sought advice from the UN Development Advisory Team for post-logging development. Some areas were suitable for plantation agriculture, but the soils and topography made such enterprise marginal. Short-rotation pulpwood planting (*Eucalyptus deglupta*, *Gmelina arborea*, and *Terminalia brassii*), was possible, but the projected returns would not justify the corporation's investment. Hardwoods would return a profit, but the longer rotation meant greater risk for any partner-investor, with the trees on customary land. There was also the *Merremia* problem, which seemed likely as Lever's was almost clear-felling. Maintenance costs of plantations would be significant, because the population was too small to fill labour needs. Advised by finance management specialists, the corporation planned to set aside almost half its royalties for redevelopment, which would be at least SI$250,000 yearly during logging. Another adviser, Paul Boenders, suggested a six-year pilot programme to test the economic feasibility of agriculture and forestry on the government's land, paid for by the corporation and an overseas donor. All this was suspended in 1981/2, as the troubles between the Koraga people and Lever's escalated. Civil disturbance was not an inducement for long-term investment, even for an aid agency.

Despite all this early advice to the corporation, several observers considered that the royalties had been squandered largely on consumables, though the Njela housing project seemed substantial to the villagers. Once the housing projects were completed, the corporation hoped to invest in something more enduring than tinned food and Fourex beer for future income generation, so by 1985 wanted reforestation possibilities on their land to be further investigated, perhaps to feed a processing plant on Kolombangara. Their hopes faded with the demise of Lever's in 1986, as the future of the Kolombangara operation became uncertain.[37]

There was talk of Earthmover's Solomons Ltd, headed by James Boyers, logging on north New Georgia. Negotiations started between Lever's parent company, United African, and Earthmovers, but landholder opposition continued. Both Frank Jamakana, sometime chairman of the corporation and Tausinga, premier of Western province, seemed poised to set up a new company, not governed by the North New Georgia Corporation Act, to log the area. Early in 1987, Tausinga was all for having the Act repealed. He had been vociferous in his criticisms, not only of Lever's, but also the Act, calling it 'an illegal concept, made legal by the Government evading its duty to protect the rights, properties and heritage of the citizens of this nation'.[38] Yet he suddenly did an about-face. If the Act were repealed, any licence for a new company would have to go through the normal procedures of area committee determination of land rights holders and that would revive the ill-feeling between contending leadership claimants and delay the project. Moreover, if the law ran its course, there was no chance of a new licence being issued for New Georgia, as the moratorium was in effect. For Tausinga and many others, keeping in the 'the illegal concept' of the Act meant keeping out Forestry Division.

A number of companies, besides Earthmovers, were interested. The ubiquitous Toshio Hashimoto, of Integrated Forest Products on Makira and more latterly, Dalsol Ltd, offered his South Pacific Timber Corporation as a contender, but his financial viability was questionable. In 1987, the Allardyce Lumber Company, ending its logging in the Shortlands, found that some timber-rights holders in the north-western part of New Georgia wanted to do business. The area included part of the lands held by the Koraga group to the north, and the southern area known as Kazukuru right- and left-hand lands. For the Koraga group to negotiate would require a parliamentary amendment to put them beyond the reach of the North New Georgia Corporation Act. The negotiations stalled for years after a land dispute broke out, mainly between claimants to the right and left Kazukuru land, the former led by H. Paia, who did not want Allardyce logging and went to court to test the legality of its licence application in 1989.[39]

The winner was Golden Springs, incorporated in the United States, whose principle, Kang Wibisono was based in Indonesia, controlling P. T. Sumber Mas Timber Group in east Kalimantan and a shipping company in Hong Kong. With the Indonesian ban on log exports, Kang Wibisono had turned to Solomons. Both Tausinga and Jamakana favoured Golden Springs, which appears to have planned a processing plant as part of its operation. This was odd because the supply of timber from north New Georgia was unlikely to be sufficient, as the more-experienced Lever's had found in the 1960s and the early 1980s. Moreover, there were several requirements under the Act – such as mandatory legal advice for the landholders and the 1984 amendment setting out separate agreements with each major group – that Tausinga seemed to think could be renegotiated between the parties he represented in the corporation and the logging company. This remained a matter for the National Investment Board and the Attorney-General's office, as the Forestry Division could have no input, because north New Georgia, being under a unique Act, was outside its purview.

Since the corporation's major spokesperson was Tausinga, former Christian Fellowship Church (CFC) opponents of Lever's had come to support the Golden Springs, in spite of the objections of Ian Talasasa who still claimed to be chief of Koraga.[40] As Edvard Hviding has observed, the opposition of the north New Georgia *butubutu* to Lever's had been based on dissatisfaction with procedures, rather than 'simple conservationism'[41] and this is highlighted even more dramatically in Tausinga's stance. Despite the input of *butubutu* on logging procedures and Tausinga's previous vehement advocacy for 'locally-controlled, small-scale timber processing' and protection of the forests for the present and future,[42] Golden Springs' logging methods soon proved little different in scale and impact than Lever's.[43] By the time Golden Springs was logging, Tausinga was a member of parliament, achieving considerable notoriety, not only in the CFC and business, but also in politics. The effects of his about-face were written on the land and written about in the *Solomon Star* where a Kusage man, Roqo Denamu, poured scorn on Tausinga's earlier anti-Levers, pro-conservation stance. Tausinga had written an article enunciating his pre-1988 views for a book that was supposed to appear in 1988, for the tenth anniversary of independence. Published in 1992, its contents had dated. Tausinga's article concentrated on condemning and 'describing the side-effects of logging in the areas in the early '80s'. In the light of events after 1988, Denamu saw the article as hypocrisy and asserted scornfully that Tausinga ought to revisit his words by 'Going back to your own vomits'(sic).[44]

Tausinga, though familiar with the rationale of conservation through the New South Wales Rainforest Information Centre in the early 1980s, had been far more concerned with his own people's control of their forest resource development and with value-addition to local timber production, than forest conservation as such. He failed to find a company that offered any more than the promise of such downstream processing. In relation to environmental damage, the tragedy of the resolution of north New Georgia conflict was that it merely resulted in exchanging one foreign logging company for another, albeit one more attuned to local leadership and social relationships.

The lifting of the moratorium

Without an effective party system in the democratic Solomons, political allegiances are exceptionally fluid and brokerage for ministries in exchange for support, commonplace. Politicians are particularly vulnerable to pressure from constituents who want their way and from interests that can pay to have their way or assist a politician to retain support. Politicians become receivers and distributors in a reciprocity network, like big men of old, but with an overlay of patronage emanating primarily from exogenous opportunities. Consequently, a consistent political platform espoused by a stable core of politicians no more exists than the concept of collective responsibility. In such a milieu, the public service's ability to work towards long-term policy goals is not optimal. Of all sections of the administration, forestry, because of the long-term nature of the subject with which it deals, is probably the most vulnerable to political vagaries. Forestry needs forward planning on a time-span of at least 20–30 years, so a minor policy change can have exponential effects on a magnified scale. Political expediency and the interests of forestry are thus usually inimical.

There were landholders and companies in the country who wanted the moratorium lifted to make money. Forceful constituents had ways of showing this. As Peter Kenilorea, sometime Prime Minister and Minister of Natural Resources, explained

> ... I had a visit from a group from a certain part of our country armed with bush knives because they wanted logging in their areas. They came to the office and I had to very kindly ask them to, please, leave their bush knives outside when we discussed the issue that they had come to see me about. But I was saying that, you know, the policy is in a moratorium in this particular area. And they said,

Attempts at Control? c. 1980–1990

moratoriums are the policies of the Government, the laws still says that we can apply for logging licences and the provisions of the law are still current and the moratorium has no effect in that sense.[45] (sic.)

Although further extended from September 1986, the moratorium was lifted by Prime Minister Alebua in February 1987, in the face of the country's increasing economic difficulties, exacerbated by cyclone Namu.[46] The Australian assistance for the resource survey had materialised, its terms of reference widened to include a review of the forestry sector with reference to the national development plan for 1985–1989. The initial report in 1988 with strategic input from the Division promised significant institutional strengthening and provided, in part, the basis for the latter to push for a new forestry policy and legislation. Echoing this were FAO and World Bank reports in 1988/9 examining the directions the government might consider on forestry policy as well as on Kolombangara. They supported a resource inventory, an effective timber control unit, a forestry management plan, reforestation, revision of the forest laws, and sought clarification on the legal standing of the Division to refuse a licence in relation to the country's constitution.[47]

Another policy, another law

Most of the FAO's recommendations on forestry policy were accepted by the Division and the cabinet of Prime Minister Mamaloni. A White Paper, presented to parliament in 1989, outlined six imperatives for policy foundation: protection of forests and soils, flora and fauna and sacred sites; sustainable use of the resource; basic needs provision – food, water, fuel, building, medicine and recreation; development of the potential of the forest to produce wealth; participation of all levels of government and people in management of forests; and distribution of benefits. It also addressed obstacles to achieving these, set objectives and discussed resource constraints. There was stress on ecological issues, a reflection of a growing Pacific region environmental consciousness, particularly since 1982 when the South Pacific Regional Environmental Programme (SPREP) was set up.[48] Most significantly, in this 1989 White Paper, for the first time since that of 1969, actual strategies were detailed. In August 1989, all this was published in a clear format entitled, *Forest Policy Statement*.

The parliament endorsed the policy, acknowledging bribery in the timber industry and the understaffing of the Forestry Division. At least one politician, B. Fa'aitoa, minister of posts and telegraphs, revealed, with no apparent intention of being sarcastic or ironic, that he believed that, prior to this 1989 paper, there was no policy on forests. The former minister,

Kenilorea, admitted that the current law had meant that the Division were 'mere spectators' in logging deals.[49] In many ways, the policy statement of 1983 had been only noble sentiments on paper, which was obvious from the parliamentary debates of 1989 and the admission by the Permanent Secretary for Natural Resources, S. Danitofea, that it had failed because of 'shortcoming in forest legislation, institutional weakness and lack of public awareness'.[50]

In policy-making words were cheap and greater clarity of expression made them no less so, unless actions flowed from them. Some delays were inevitable as the major programme under the auspices of the Division and AIDAB commenced. The incoming Mamaloni government of 1989 almost jettisoned the forest resource inventory, but the Division managed to persuade the government to reinstate it. Negotiations between the government of Solomons and AIDAB on this project and another to strengthen timber control continued throughout 1989 to mid-1990.[51] However, within the strategy of the *Forest Policy Statement*, there were things that could have been actioned immediately, such as 'imposing a moratorium'. Urged by the Division, the cabinet did introduce a limited moratorium in late 1989, but only for those provinces which already had one or more logging companies operating.

One issue that needed attention was the new draft forest legislation, formulated in consultation with Forestry Division by FAO lawyer James Fingleton in early 1989.[52] Fingleton was advised that more provincial input was needed and thus further revisions of the draft. Although not given much time by the Division, the provinces and other interested parties submitted suggestions and criticisms. The timber industry was extremely critical, Allardyce, for one making much of the apparent lack of provincial input and the influence of foreign advisers on the Division.[53]

At the eleventh hour, Mamaloni jettisoned the draft, which he considered failed to recognise his plans for devolution, was too lengthy, and was premised on the establishment of a Forestry Board that had the potential as a statutory body to be more resistant to political pressure – just as the draft would have, if it had become law. With the draft went the concept, because, as Commissioner Gaviro put it,

> Unfortunately, in November 1989, Cabinet decided, in line with current Government policy, that the functions and powers to legislate over forestry would be devolved to Provincial Governments and that national forest legislation was therefore no longer necessary.[54]

In the midst of this, one of the legal time-bombs inherent in the inconsistencies of the existing acts exploded when Chief Justice Ward ruled in favour of the Attorney-General, representing the Division, in the case of *Allardyce Lumber Company Limited and Bisili and nine others* v. *Attor-*

ney-General, Commissioner of Forests, Premier of the Western Province and Paia. Allardyce had claimed that it should not have been refused a licence to log the Kazukuru right-hand land on New Georgia as it had entered into an approved agreement with Bisili and nine others. Paia claimed that as Basili and companions were not customary landholders of the land in question, they could not assert timber rights. It was the recurring question of whether or not rights to trees equated to rights to land and vice-versa. The legislation and associated application forms confused the two so much that the law was uncertain. Both the plaintiffs' claim and the defendants' counterclaim failed. The judge ruled that, as well as the Form Four agreement, there should have been an agreement with the provincial government and the area council as to profit-sharing and management representation. This was music to the ears of the provincial governments as they were constantly trying to find ways to increase their revenue. Thus, all other companies, licensed from 1984 in having only the Form Four agreement, probably had void licences. Furthermore, with this interpretation of the meaning of the 1984 Amendment (particularly Sections 8 and 9, amending sections 5a and 5b of Part 11A of the 1977 Ordinance), logging companies would not only have to pay landholders in royalties, but also with part of their profits.

Mamaloni's government extended the moratorium to new licences and the renewal of existing ones and in May 1990 rushed through an amendment to clarify the meaning of the act, to avoid being sued by companies, landholders and provincial governments, which all could have made much of the mistake.[55] Subsequently, Beti, Bisili and Paia, as representatives of the Voramoli tribe, went to the court of appeal when Allardyce obtained their licence after the amendment came into force in July 1990. The court held that the amendment overlooked the fact that before a Form Three (certificate of approval) can be issued, an approved agreement had to have been completed. None of the two mandatory meetings took place, thus no agreement was possible. The amendment was not sufficient to validate the licence issued, so Allardyce had to start again and submit a fresh Form One application. Unlike some other institutions involved in logging agreements, the court was not amenable to manipulation.[56]

Mamaloni directed the Division to draw up a transitional act in which the power of the commissioner was reduced to veto of a licence application for an area, only if one already had been granted. In effect, the commissioner would be merely a registrar of licences. Mamaloni appears to have abandoned this because in the hiatus between the provinces getting the legislative power and each drafting their legislation for forestry, there would be no effective legislation, national or otherwise, applicable to the timber industry. Given the government's sorry record of legal drafting gaffs and tardiness in forestry legislation, there was no likelihood the under-resourced provinces could do any better. Indications were that chaos would result and an even greater dissipation of the forest resource because, unlike agriculture, the provinces had built up virtually no forestry expertise. As well as validating all the post-1984 licences and re-wording ambiguous sections, the 1990 Forest Resources and Timber Utilisation (Amendment), transferred most of the commissioner's powers to the provincial government, although the commissioner still gave the initial permission to begin negotiations to companies and could refuse a licence on the advice of the area council concerned. However, under Section 5f, the commissioner appears to have the power to withhold recommendation to the provincial government to approve the agreement if the certificate relating to determination of timber rights, the responsibility of the area council, and/or the agreement between the logging company and the landholders were irregular and if the appeals to the customary land appeal court were still pending. This would prevent the issuing of the licence.[57]

The Act may have rescinded the ambiguous sections of clauses 5a–5c, but it also set the seal on all the dubious, if not dirty, deals of the past by its saving clause 5l:

(1) For the avoidance of doubt it is hereby declared that-

(a) any licence granted under Part 11 of the principal Act prior to coming into operation of this amending Act shall be deemed to have been validly, properly and lawfully granted notwithstanding that the provisions of that Part in force at the time of such grant may not have been complied with in every particular or requirement.

The same retrospectivity was applied to timber rights agreements. Little of this exercised the parliament that passed this bill into law.[58] In effect, this was a denial of natural justice, and set in concrete such patent injustices as the Eagon licence and agreement on Choiseul, as the ombudsman pointed out. All his telling reports from 1988 to 1991, critical of crooked practises which tainted most licences, were largely ignored by successive parliaments.[59]

Under this 1990 Amendment and yet another in 1991 to cover two major drafting errors, the *Forest Policy Statement* of 1989 became invalid. It lost its relevance since it was premised on the nation as a whole, not on provincial control of several, if not all, of the functions of the Forestry Division. The policy, while not aborted, was smothered soon after it had been born.[60]

Throughout the decade 1980–1990 some degree of economic rationality was evident in the government's practices, as opposed to their espoused policies, on the timber resource. The government wanted both revenue and foreign exchange for the country's economy. The wider picture reveals, at best, extreme shortsightedness, however. Despite increasing information on loss of national earnings through poor logging practice, undervaluing of

Attempts at Control? c. 1980–1990

species, underdeclaring of volumes, and transfer-pricing, to say nothing of considerable environmental damage and thus cost to both the formal and informal sectors, successive governments did next to nothing to strengthen the hand of the Forestry Division to control logging methods and the export of logs. If anything, they brought the Division to near-immobility as regulator of forest use. Someone other than the general citizenry of Solomon Islands stood to gain from this. What governments did do, albeit reluctantly under Mamaloni, was agree to accept assistance from Australia, but this was not to come into effect until the early 1990s; and many would say this was too late. Yet the bones of a timber control unit had been extant since 1980, but governments had refused to fund it adequately.

The Solomon Islands government had set its seal of acceptance on a more permanent loss – that of the resource itself. The licensed cut for 1989/90 was, as in 1983, over 900,000 cubic metres, more than twice the 'allowable' cut. A timber control unit was not needed to police the cut, just the ability of political leaders to say no. The policy statements of 1983 and 1989 meant little. The ombudsman's several reports detailing disregard for the legal process and the long-term interests of the resource owners; the cries of landholders, post-logging; the lack of commitment to drafting an effective forestry law; all point to successive governments operating on a live-for-the-day philosophy, lacking the political will to control the forests, the loggers, and the landholders for the long-term interest of the nation.

Wider policy decisions had implications for both the resource and the people. Devolution, a politically attractive idea, meant handing over forests to authorities lacking the implementation capacity to combat loggers' illegalities, an outcome Prime Minister Mamaloni realised, given Forestry Division's constant advice. The provinces were ill-prepared for the task throughout the 1980s and remained so.[61] The repercussions of this, even at a forest law level, were evident in 1990 with the abandonment of the attempt to draft an effective new law. These were harbingers of the final tragedy. Solomon Islanders when in grievance mode claim their colonial masters did not give them sufficient infrastructure in 1978 to run the national government well;[62] the Mamaloni government seemed determined to do the same thing to the provinces a little more than a decade later. It was as if Mamaloni had set up the provinces for failure in forestry control; an outcome not unwelcome to unscrupulous loggers and local interests who swallowed the bait of devolution. Mamaloni and his ministers had washed their hands of responsibility for the forests and thus those whose future subsistence depended upon them.

— 12 —

Aiding the Forests

> No people can make over another people ... whatever we can do can only be of slight assistance to help [a nation] over its most severe problems ... A nation that comes to rely on gifts and loans from others is too likely to postpone the essential, tough measures necessary for its own salvation.
>
> Robert Taft, Committee on Banking and Currency, United States Senate Report 452, part 2, 'Minority Views', 79th Congress, First Session, 1945, 9.

Decolonisation came relatively late to the Pacific Islands in the 1960s and 1970s, mainly because of their small size and isolation. Following World War Two the colonial powers had poured money into establishing rudimentary national infrastructures; after independence, their contributions continued as aid, along with those of others interested in the region. Throughout its history, the Department/Division, relied considerably on assistance for budgetary support, project implementation and specialist advice. Assistance to forestry has not been confined to the Division. Many agencies, foreign governments, churches, charities, institutions within the country itself, and even loggers have aided the forest.

Reforestation of government land: quantity and quality

Besides attempting to regulate the logging industry, the Forestry Division in the 1980s focused on reforestation. Its replanting target in 1980 had been modified from 3,000–4,000 hectares annually to 1,000–1,500 hectares: even if this had been achieved, it would have been insufficient to guarantee sustainability of commercial log production. Based on an annual cut of 400,000 cubic metres, this cut would mean about 8,000 hectares logged each year, as the natural forest yield per hectare was approximately 50 cubic metres. Since plantation yield would be, on average, 170 cubic metres per hectare,

Aiding the Forests

2,000–2,500 hectares replanted each year would have been required to replace the cut volume.[1]

However, any new replanting programme would start under a handicap. Between 1953 and 1966, before any replanting commenced, 165,292 cubic metres of timber or about 3,306 hectares of natural forest had been logged. From 1967 to 1981, an additional 3,257,285 cubic metres (about 65,146 hectares) had been cut, with an estimated future yield of only 158,170 cubic metres expected from 15,817 hectares replanted, the revised stocking figures for the established line plantings being only 100 cubic metres per hectare. In total then, from 1953 to 1981, 3,422,577 cubic metres were cut and provision was made for a future yield amounting to a mere 4.6 per cent of that extracted, a quantity that mocked the concept of sustainability (see Figure 87).[2]

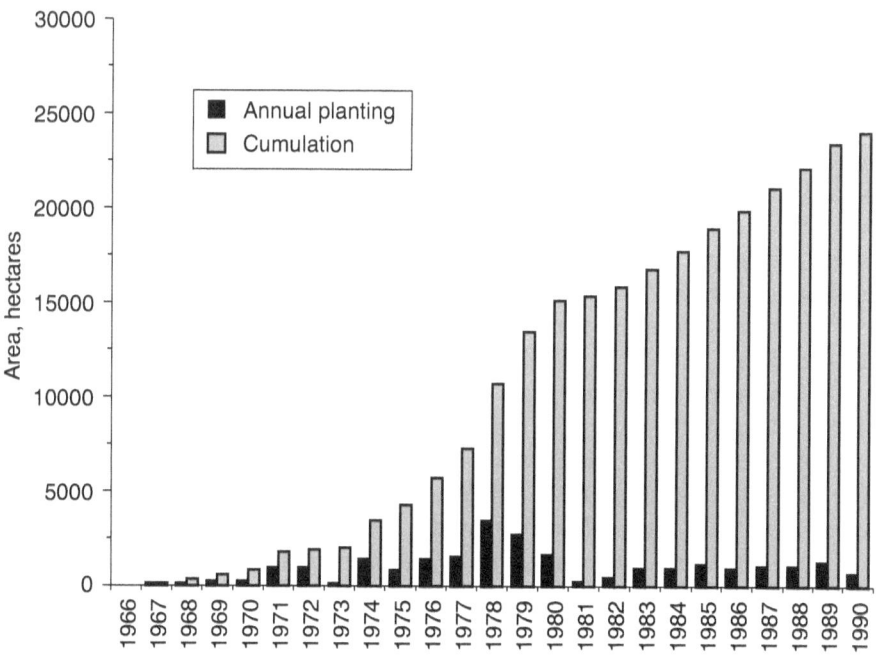

Figure 87. Government land reforested, 1966–1990.

Sources: FDAR 1966–1989; Martin Bennett et al., *Forestry Project Identification Mission, May–June 1991*, 22.

This, of course, does not allow for natural regeneration which could be substantial in areas where early, more size-selective logging had occurred, providing there had been no subsequent disturbance. However, in 1984 when the Division carried out a 'desk study' on regeneration, it found 'a serious lack of knowledge'.[3] Casual observation and a 1988 study of logged land at Barora produced the conclusion that relying on natural regeneration would be uneconomic, as indigenous pioneer species would take at least seven years to re-establish, so great was the damage to the humic mat and soil by heavy equipment and clear-felling.[4] World Bank and FAO mission findings in April 1989 were even more sobering. Logging of trees, including super-smalls on customary land, had been so uncontrolled that it would take the remnant forest at least 80 to 100 years to reproduce the original stocking.[5] The Australian-funded forest inventory project further endorsed this in 1992.[6] Solomon Islanders had mortgaged their heritage for the next three generations.

Even keeping pace with the amount extracted annually proved elusive – never between 1981 and 1991 did annual replanting on government land amount to more than 1,317 hectares, with the average about 980 hectares. Timber cut during that time was a conservative 4,203,024 cubic metres.[7] This represented about 84,000 hectares, but with wastage and under-declaration of cut, was more likely to represent the clearing of 100,000 hectares of mainly customary land.[8] To compensate for the timber cut, at least 28,020 hectares would have had to have been planted, with an ideal yield of 150 cubic metres per hectare. The total planted during that time was about 10,000 hectares, making a grand total of about 24,868 hectares planted since reforestation commenced in the 1960s. The situation worsened, however. From 1992 to 1995 a conservative 2,794,456 cubic metres (representing exports and local consumption of around 40,000 cubic metres annually) or 55,889 hectares has been logged, requiring about 18,630 hectares of reforested land to compensate for the cut, more than that planted in the preceding ten years.[9]

Moreover, significant plantation areas were not effectively stocked due to variable spacing and maintenance over the years, as well as losses to cyclones. To 1980, plantings of the four indigenous species, *Agathis macrophylla*, *Campnosperma brevipetiolata*, *Terminalia brassii* and *Terminalia calamansanai* had made up over 87 per cent. With extensive plantings on Santa Cruz and assisted regeneration on Vanikolo, *Agathis* was dropped in 1983 due to slow growth rate and die-back. The trees had failed to thrive in plantations on Santa Cruz; evidence of the dangers of monocultures. To 1980, exotics had been tried, but on a relatively small scale. From 1981 to 1987, eight major species – *Campnosperma brevipetiolata*, *Cedrela odorata*, *Eucalyptus deglupta*, *Gmelina arborea*, *Swietenia macrophylla*, *Tectona grandis*, *Terminalia brassii* and *Terminalia calamansanai* – and several minor species

Aiding the Forests

were planted, mainly under the close-planting method, in practically every location. Early 'trial and error' methods had got out of hand, dominated more by the search for what would grow well, and less by what would sell. By 1988, *Campnosperma brevipetiolata*, *Eucalyptus deglupta*, and *Terminalia calamansanai* had been dropped because of various species-specific problems in plantations such as poor growth rate and form, difficulties with nursery stock, site-sensitivity, and susceptibility to wind, weeds and pests. Overall, the legacy was too much variety in too many plantation sites, diminishing the value to loggers.[10]

However, monocultures are vulnerable to pests, so the Division's line enrichment planting in the 1980s, with a mixture of two species such as *Terminalia calamansanai* and *Swietenia macrophylla*, promised greater resistance to the *Hypsipyla* borer and *Merremia*. Moreover, debate continued about the potential markets for plantation species, especially *Gmelina arborea*, as well as the inherent risks of large monocultures: for example, on Kolombangara there is a fungus to which cultivars of limited genetic stock are particularly vulnerable.[11]

One positive outcome of such varied stocking has been the accumulation of information on species. This and other data from research plots were assembled in 1990 for a major revision of Marten and Thomson's *Silvics of Species* of 1980. G. E. Chaplin's *Silvicultural Manual for the Solomon Islands* appeared in 1993. Funded by British Overseas Development Administration (ODA), it will be a standard plantation forestry text.

The European Community[12]

Donors continued to supply aid for replanting, though from 1981 Britain's bi-lateral donor involvement decreased. At Independence Britain had sponsored Solomon Islands' membership to the Asia-Caribbean-Pacific group of 'developing' countries associated with the European Community (EC). The EC contributes to a European Development Fund which allocates finance according to guidelines set every five years as well as setting aside finance for Stabilisation of Export Earnings (Stabex) to compensate for losses in export earnings from primary products by developing countries. Stabex funds can be used outside their sector of origin and have aided forestry projects. Complex procedures govern aid allocation, especially since 1992 when the process incorporated a rigorous evaluation component. The EC, in the early 1990s, was the largest donor to Solomons after Australia.

From 1981, the EC funded the reforestation of government land at Santa Cruz and Alu. In 1982 it extended to Viru, New Georgia, taking up where British ODA left off in 1980. The Santa Cruz and Alu projects were completed in 1984. By 1989 the Viru project had replanted 2,139 hectares, providing considerable local employment. It continued to be funded by EC, but a break from 1986–1988 and a switch to Stabex funding meant considerable delays in the programme, so, by 1991, the government was trying to involve the Commonwealth Development Corporation. There was criticism of the Viru project because of lack of early evaluation and regular appraisal, which was primarily the Division's responsibility.[13]

Donors were slowly beginning to realise that the government's increasing earnings from forests were being diverted to other sectors, with aid filling the shortfall in the Division's finance.[14] During 1988–1990, some government plantations were becoming weed-infested and new plantings needed thinning. For the first and last time in 1990, the government of Solomons invested in reforestation, spending $4 million from the $13 million collected in export duties and the levy. Britain's ODA signified a willingness to take on reforestation in 1990–1991, but only if the Solomons would continue funding in the longer term. Mamaloni's government was unforthcoming, so from 1991 there was no direct aid for reforestation and maintenance declined.[15]

In 1992, the EC had premised its aid on the Tropical Forest Action Plan (TFAP), which had a strong reforestation component. It investigated potential for assistance, even though ODA had done a similar study in 1991. The TFAP claimed to assist developing countries to strengthen their capabilities to manage their forests sustainably and required a forestry sector review.[16] Because a review is a costly undertaking, a number of donors normally are involved, one being the 'lead agency'. The EC and British ODA offered, as well as the UN Development Programme. The Japanese offered training in timber utilisation, and the on-going Australia-funded computerised data base, Forest Resources Information System (FRIS), was to be a foundation for planning.[17] The Hilly government of 1993 supported the concept, but it was abandoned when Mamaloni returned to power in late 1994.

Kolombangara: aid business

In 1982, Lever's Pacific Timbers had been interested in replanting on Kolombangara as one inducement offered to the government for the logging option on north New Georgia, but this seemed destined to founder when Lever's withdrew in 1986. Meanwhile, preliminary negotiations began in 1985 between Lever's and the Commonwealth Development Corporation (CDC), involving the appraisal of a joint venture with government participation. The CDC is a British public corporation, a bi-lateral finance agency, engaged in development overseas. It makes long-term loans at competitive interest rates and invests expertise and equity capital, through part-ownership of projects. It gets its funding from the British government's aid programme which it repays with interest, investing any surplus in CDC projects. On its equity investments it receives returns as well. Accountable in a business sense, it seems to get results.[18]

Following Lever's withdrawal, CDC conducted further investigations and concluded that an 11,300 hectare project growing predominantly the exotic, *Gmelina arborea* for woodchip and sawn timber would be a viable operation, with the possibility of an eventual processing plant.[19] Although the government purchased Lever's assets in 1986 in anticipation of the project, it was on a caretaking basis, and deterioration occurred, hindering the Division's work at Poitete. The Mamaloni government was concerned too about financing its share and further reappraisal followed on both sides. Despite attempts to interest the Asian Development Bank, the World Bank and Japanese companies in the SI$140 million dollars (US$70 million) project, when Prime Minister Alebua and CDC signed the agreement in January 1989 to set up Kolombangara Forest Products Limited (KFPL) on a 50–50 basis, they were alone.

In February 1989, Mamaloni, back in power, reneged on the agreement. Wary of past Kolombangara and Enoghae land disputes and the tying-up of government funds from Stabex in a long-term investment, he wanted the original Kolombangara landholders and the Western provincial government to take up the half share in KFPL. Moreover, Mamaloni was privatising several costly government interests, such as the National Fisheries Development and its fleet, including two purse seiners bought with $26.4 million borrowed from an Australian finance corporation, so investment in yet another project seemed risky. Negotiations commenced with the government offering as equity its 8,500 hectares of Kolombangara plantation timber in return for shares to be held by the Investment Corporation of Solomon Islands, for the Western province and the landholders.

CDC liked the idea of incorporating the 8,500 hectares because the actual plantable area of former Levers' land was only 8,000, not 11,330 hectares as thought, and an earlier start to logging and some financial return were appealing. Western province was attracted to a rental from the government land, to which it hoped to get title, but it did not want shares and planned to allocate to landholders part of the rental. But in 1992 Mamaloni's government dealt directly with the landholders, consisting of two contending parties. The government land would pass to the Commissioner of Lands in trust for two years, while the landholders formed the Kolombangara Landowners Trust, which would hold the title and lease the land to KFPL. CDC would put up the cash capital, the land constituted the equity of the landholders who would receive rentals on the land; they would hold 49 per cent equity and the CDC, 51 per cent. The Forestry Division's trees would eventually pass to the trust, but the government would retain its forestry land for the time.[20] On this basis the project resumed with the land, by 1995, under lease from the landholders' trust, providing a unique situation and one that in many ways was a model for any future return of the Forest Estate to original landholders. In having alienated the land from customary use for generations and, in the 1960s, for the Forest Estate, the colonial government, albeit inadvertently, had put the landholders in a position to appreciate a profitable long-term alternative use for already-planted lands. Such was the case on Kolombangara. The question is what other logging organisation, beside KFPL, would be prepared for such long-term commitment with groups of landholders and vice-versa?

Reforestation and loggers

KFPL commenced logging in 1994 as the trees planted by the Division in 1976 reached maturity.[21] It exported 9,500 cubic metres of logs and in 1995 this increased to 32,450. By 2008 it will have 15,000 hectares under trees, producing 306,000 cubic metres annually on a sustainable basis. About a third of this is under exotics, *Gmelina arborea* and the rest *Eucalyptus deglupta*. Despite the retention of maximum biomass and non-burning of timber debris, *Eucalyptus deglupta* is greedy of soil nutrients, and over successive plantings can alter the soil's chemistry. In terms of the native ecology, these exotics will create a 'green desert'.[22] Yet, KFPL is trying to achieve some balance, maintaining over half the area leased under natural forests, providing a reserve for endemics and associated fauna.[23]

KFPL is in for the long haul, investing millions, and has a reputation for involving local communities in enterprises related to the industry, such as sub-contracting tree-felling and supporting the concept of 'outgrower' sources of logs – if communities can overcome the challenges of clan-based tenure. If KFPL can expand its plantings to 50,000 hectares, there is promise of the country's recovery as a producer of logs after the native forest is exhausted.[24]

In 1985, Allardyce considered reforestation on Alu, but the estimated costs of $1,500 a hectare were daunting. Kayuken had shown interest too, if the Japanese Overseas Afforestation Association (JOAA) paid. In February 1983, years before it logged west Kwaio, its director, Lopez, tried to interest JOAA in the land. JOAA wanted vast areas on Malaita and the Guadalcanal grasslands, but shied away from unguaranteed tree ownership on customary land, so the scheme vanished.[25] On Malaita too, Taisol's Suny Tong, applied in 1984 to log from Fauabu to Langalanga, proposing to reforest with teak and cedar. As with Kayuken, this seemed a ploy to win support and escape the moratorium.²⁶ On Choiseul, Eagon offered reforestation of customary land as a feature of its logging deal in 1994. In lieu of paying the reforestation levy to the government, it established nurseries, but landholders remained concerned about ownership of the trees.²⁷

Reforestation on customary land

From the late 1970s there had been increasing concern about reforesting customary land. The Division in 1979 had stated that it planned to replant government land until about 1985. With a target of 20,000 to 22,000 hectares on government land for 1980–1985 it would presumably have replanted a total of 35,000 hectares, a goal never achieved. When a government minister suggested in 1981 that licences should only be approved if the company concerned would reforest, Sam Gaviro raised the perennial objection that customary land tenure provided no security for any company to re-invest between SI $900 and $1,500 a hectare. Although this idea of 'no reforestation, no licence' reappeared periodically until the late 1980s, it appealed less than the alternative of the reforestation levy.²⁸

Nonetheless, the Division secured a post-colonial first in 1982 when it co-operated with Santa Cruz landholders in replanting their 567 hectares. Allardyce had logged much of this when it had wandered onto customary land. The land was first registered in the name of the landholders, with the Division agreeing to plant fast-growing species. The European Development Fund paid for this, as part of its government land reforestation project. Unfortunately, the community was not directly involved, except in allowing its land to be replanted.

When the Division's new extension service tried to encourage such projects elsewhere, it found landholders were suspicious of long-term commitment of land to growing seedlings supplied by the government – the old fear the government might not only claim the trees, but perhaps also the land. Moreover, although the Division did manage to set up some small, one-hectare demonstration plots with villagers' co-operation, shortage of funding and personnel trained to work with communities circumscribed extension work in the early 1980s.²⁹

Extension services and New Zealand aid

Assistance came from New Zealand. The UN Development Advisory team of 1982 had revealed the limited potential of Malaita's forests for the subsistence needs of a growing population. It had recommended a pilot study involving the planting of 1,000 hectares of hardwoods in logged-over and old garden land. After discussions with the Solomon Islands government in 1984, New Zealand opted for 250 hectares of the exotic, *Swietenia macrophylla*, receiving support from the Forestry Division and the province.[30] Two groups, the Anotafa at Fote and the Saenaua, near Auki, agreed to participate, meeting the Division's requirement to register their land in group title to provide security. Project objectives were,

> to stimulate interest ... in the establishment of forestry plantations on customary land; to show that production forestry was a ... worthwhile use of land; to test the technical, social and financial viability of this type of development and the relevance ... to future projects; and to demonstrate the benefits, management operations and skills involved.[31]

New Zealand allocated SI$384,586 for project infrastructure and to fund labour to establish the hardwood plantation.[32] The project originated from the wishes of officials, and not from the people themselves, explaining in part why they felt they should be paid. The Forestry Division agreed to fund two staff members for management. The project commenced in mid-1985 for five years, with the surveying of land and the training of landholders in silviculture techniques. A major review in 1987 found a communication problem, especially between the Division and the landholders. The landholders believed they were to get all the funds and resented Forestry Division's share. They had wanted a link to the main coastal road and that had not eventuated. Landholders resented the attitude of some Division supervisors with no expertise in community-based projects. Confusion over pay rates added to this.

The work was reorganised whereby the community agreed to supply labour on a contract basis, at a higher rate, but not as high as forestry workers elsewhere. The lower scale was more than fair as the landholders would reap the reward when the trees matured.[33] The women finally became involved. Like most aid projects to increase income, New Zealand until the review of 1987, foolishly concentrated on males only. So women end up doing all the 'traditional' household tasks as well as work for cash, a danger to their physical and mental health and that of their children.[34] Moreover, women often have rights to forage and collect firewood in the forest. These 'secondary' rights are increasingly over-ridden by men who can claim 'primary' rights to make gardens in the area which nowadays frequently equate to de facto rights to cut all but known cultivated trees and to plant cash crops and trees.[35] This is particularly so in societies with a patrilineal land inheritance like much of Malaita.[36]

Although both communities planted firewood lots, the women at Saenaua, with the encouragement of a forestry officer, became interested in small-scale agro-forestry, growing gardens under the trees. But although the women worked hard in their market gardens, they did not have as much time for their household gardens and they were not paid like tree workers, so benefited little.[37] By 1990, much had been learned. Local input was needed from the beginning, with the initiative for any such project to come from the people. The role of agro-forestry and firewood lots was appreciated, as was the significance of women and their needs. Though many saw the project as simply a way to earn cash – about SI$180,000 went as wages – some perceived the long-term economic and social benefits to the community.[38] Elsewhere, there was interest in forest-based community projects, so the main aim had been achieved, but not without endorsing the 'handout' attitude.

Partly to remedy this dependency and to respond to a UN Development Programme request, New Zealand assisted the Extension Service's technical capacity, with NZ$2 million to 1997.[39] The Division welcomed this proposal in 1990 as a way to implement the aims of the National Development Plan, 1985–1989 and the 1989 Forestry Policy, to develop community forestry and agroforestry.[40]

Makira, Malaita, Guadalcanal and Western provinces, sites of intensive logging of customary land, were targeted. The expanded Extension Service, advised by Tim Thorpe, took over the Malaita project and placed officers at Poitete, Auki, as well as four in Honiara. Two men each year were sent overseas for training in social forestry; no women were trained.[41] Problems arose with the supply of high quality seedlings and lack of silvicultural knowledge of the native species which villagers valued (it was not until 1982 that species testing had begun). Some landholders, such as those on north New Georgia, wanted to replant thousands of hectares of logged land, which was beyond the capacity of the service, with its six to seven officers. Its maximum project size was 50–100 hectares, but it effectively reforested 10–15 hectares per year, at an annual cost of $0.5 million. The focus was firewood lots, trees to supply subsistence building needs, and agro-forestry, as well as publicity and short courses. One woodlot course at the National Forestry Training Institute, Poitete had representatives of landholding groups, but they complained that the Service did not pay them an allowance for attending! To discourage the embedded 'handout' attitude of many villagers, the Extension Service decided to charge between 50 cents to a dollar for all seedlings, which cost 87 cents to produce. In the loggers' wake, indications are that there will be more demand for commercial re-forestation by landholders. KFPL is interested in the Division's encouraging woodlots of *Gmelina arborea* with a dual purpose, for subsistence needs and for sale to them for a proposed mill, but the Division has been wary of raising expectations for 'outgrowers' until the mill materialises. Outgrowers too have to resolve the problem of rights to planted trees on clanland. Moreover, it is *Gmelina moluccana* (*Arakolo*) that people favour to build their canoes, but it has been an unsuccessful plantation species.

Figure 88. Poster publicity for replanting by the Extension Section of the Forestry division in 1992. Posters were designed for each province.

The expertise gained in project appraisal for these operations could be amplified. The project also had a component for two medium-scale demonstration plantations (100–500 hectares) on customary land and the development of appropriate legal and financial arrangements. To obviate the payment approach used on the original Malaita reforestation project, joint ventures were planned between the central government and landholders whereby profits after logging would be shared. Difficulties in instituting loan arrangements have retarded progress.[42]

Aiding the Forests

The EU-financed Isabel Sustainable Forest Project in 1995 commenced on unlogged land and incorporates conservation and protection functions and community involvement in managing the forest. It has an outlet for its eco-timber through Greenpeace links in New Zealand. Unlike this project, aid-based reforestation has social and financial dangers if done on land logged by commercial companies because, as in the New Zealand project, it takes the responsibility for sustainable logging away from the government and the landholders. Consequences of bad decisions are not faced. Though it was not their intention, relatively small-scale, high-cost projects like New Zealand's indirectly support logging corporations, because they offer assistance for reforestation which should be budgeted into costings either by the national government or the landholders, at the outset of logging.[43]

Australian aid

Bridging the colonial-independence period,[44] Australia had funded a major cattle-under-the trees (CUT) project in 1976 on Kolombangara, utilising British ODA funds for tree-planting, with the cattle project manager supplied by New Zealand. The experiment was unwieldy, as the goals of the forestry project clashed with the cattle project, and it was hampered further by shared responsibilities. Although suitable improved pasture was established on the 1,450 hectares, it did not thrive under the expanding canopy of *Eucalyptus deglupta*. Tinkering with tree density and pruning to increase light transmission only increased costs. Cattle were found to be useful at clearing *Merremia*, but this was nothing new.[45] The 2,300 cattle ate the bark and rubbed the young trees, exposing them to decay and termites. No other tree species proved resistant. Lack of soil fertility was another problem after clear felling. Little thought was given to the project outcomes – in the 1970s, how could a cattle beast be consumed unless there were about 300 hungry people around to eat it before it rotted in the heat?[46] No one asked what effect large, hoofed beasts and introduced weeds would have on the native vegetation and soil. By 1985 the project had earned $234,000 for an outlay of $2.6 million in aid – a failure, it was abandoned eventually in 1987.[47]

In the 1980s, Australia responded to a call for assistance in resuming reforestation at Allardyce, Santa Isabel and for a forest survey for the Division.[48] There was considerable bi-lateral enthusiasm for the project which aimed to reforest 2,000 hectares of degraded forest and to investigate whether enrichment planting was necessary on the 2,500 hectares of government plantations. Australia would give A$2 million yearly for five years, with the likelihood of a five-year extension. AIDAB and the government set terms of reference for the project and AIDAB chose the consultants, Shedden

Agribusiness, based in Australia.[49] The Forestry Division rejected their report in 1990 because it was inaccurate and uninformed. It strayed far from the original terms, recommending a polycyclic logging system using portable saw-mills run by local contractors to feed into a central processing and marketing centre. Reforestation was relegated to a minor place. Shedden and AIDAB seemed to be more influenced by AIDAB's Pacific agenda of increased emphasis on environmental responsibility than the original terms of reference.[50] AIDAB remained wedded to the proposal until 1992.[51] The Shedden project may well have been a prototype for the kind of community and environment-sensitive development that many people wanted, but the Solomon Islands government did not like being dictated to by Australia. Thorpe has calculated the survey and a related one cost AIDAB about A$250,000, a concomitant loss to Australian taxpayers and the Solomons government.[52]

The forest inventory and its FRIS, carried out by consultants headed by Tony Fearnside and to cost Australia A$2.3 million and the Solomons SI$125,000, appear to have been somewhat modified to include sociological and environment surveys and data which had not been part of the original request.[53] The 'add-ons' were mainly at the instigation of AIDAB, although in 1988 the Ministry of Agriculture wanted the study widened to include land evaluation and landholders' views of their resources.[54] The project, began in November 1990 and completed by February 1995, comprised six parts:

> the establishing of the computerised data base (FRIS) ...; the timber values component which estimates the nature and extent of the potentially commercial forest ...; the non-wood values component, concerned with protective and environment values ...; a resources planning component ...; a comprehensive training component ...; and a project management and reporting component.[55]

The project's strength was resource assessment; less thorough were the environmental and social aspects. The Division also criticised the FRIS project for management shortcomings, because, like other interested parties, it wanted more input. Nonetheless, the project produced an enormous amount of data, including thirty reports, manuals and papers. The Division, the industry, and forest users all had a powerful tool at their disposal. It revealed, unequivocally in clear language, a resource at a critical stage, on the edge of being lost for generations unless sustainably managed.[56]

The Timber Control Unit project headed by Hylton Taylor, upset fewer people within the Division, the foundation for the unit's work having been laid before Australia's involvement. The Division believed it could have managed the unit itself, if only it had been given the finance. The goal was to strengthen the Division's capacity rather than to build a new unit.[57] The programme was a three-year one, subsequently extended, with the following aims:

Aiding the Forests

A) To improve the SIG's capacity to monitor logging operations ... to ensure:

   1. that environmental damage is minimised in logging practices;

   2. that wastage is minimised and licensing agreements adhered to;

   3. that information on the utilisation of the resource is regularly and reliably updated;

   4. that information on production and revenue and the invoicing on manifests is correct.

B) To train and assist custom landowners to ensure:

   1. that they negotiate logging contracts on a more informed basis;

   2. that they have a better awareness of the environmental dangers of logging;

   3. that they are able to monitor logging operations on their own land.[58]

It had four components: the timber inspectorate; in-service training which later included saw-mill training and the development of grading rules and standards; research and development of procedures; and landowner extension and advisory services.[59] Five inspectorate field units would be sited near logging operations. As well as expanding the field units to eight in the face of increasing logging operations and the timber grading and saw-mill aspects, a forestry financial policy specialist was added. Units were stationed at Kirakira, Auki, Buala, Gizo and the Shortland Islands, and one planned for Choiseul. A commercial unit also opened in May to provide data on pricing and marketing of exported logs.[60] The units were doing such an effective job by 1993/4 that the logging industry was complaining.[61] The least satisfactory aspects were the courses on saw-milling and grading, which appear to have been cobbled together from disparate sources.[62]

Education and the forests

Sam Gaviro, trained at Bulolo Forestry College in Papua New Guinea, believed that education about the forests was the way to guarantee their future.[63] The Division's extension section plays a key role in educating for community forestry as did the timber control unit to 1995 in teaching people about logging agreements. Australia's aid extended to the education sector, most notably, to Solomon Islands College of Higher Education (SICHE).[64] The School of Natural Resources, established when SICHE opened in 1984, was headed by Dr Rosemary Kinne, a Dominican sister from Australia. Her successor was Frank Wickham from the Western province. Short courses to up-grade unqualified Forestry Division staff began in 1986 and a Certificate in Forestry course the following year. The government and AIDAB paid respectively SI$130,000 and SI$195,000, for the establishment of the Forestry Training Centre (later the National Forestry Institute) at Poitete in 1988. The three-year Certificate of Forestry was designed to educate forest rangers. By 1993, the emphasis shifted to community and industry needs.[65] The school also offers shorter courses. A Forestry Advisory Committee was established in 1985, consisting of representatives from the Division, SICHE, industry, NGOs and others with environmental interests, who ensure that SICHE includes their priorities in the course content.[66] Most graduates from the National Forestry Institute find employment with the Division, KFPL and the logging industry, although groups of chiefs dealing with companies have also sponsored a few.[67]

One of the Division's problems has been the paucity of higher-level trained staff. While strong on the practical side, several key officers have risen through the ranks, but lack formal training. The government seems unwilling to upgrade these. Of 83 members of the Division in 1991, only three had a Bachelor's degree; none had post-graduate qualifications, much the same as in 1984.[68] A significant weakness of the Division and of training is the total lack of women in forestry, a bias that reinforces a perception that the disposition of timber as a commercial commodity is a male prerogative and denigrates the wider importance of forest products to women and for life itself.[69]

Several SICHE graduates have left the Division in frustration at seeing venal politicians undermine its work.[70] Yet, one person can have a significant impact. Myknee Qusa Sirikolo graduated in 1988 and was appointed to the Division's herbarium, located within the botanic gardens at Rove, Honiara. Once a showpiece of Whitmore's collection and Geoff Dennis gardening, both had deteriorated by the early 1980s. The publication of the ODA-funded *A Guide to the Useful Plants of Solomon Islands* by C. P. Henderson and I. R. Hancock in 1988 alerted many to the herbarium's importance. A New Zealand volunteer, David Glenny, worked with Sirikolo, along with a retired Dennis, to rehabilitate the collections in 1990, assisted by a New Zealand grant. Sirikolo continues to collect and care for the trees – a large responsibility, as squatters fell trees and people scrape them for custom medicine. He also assists visiting researchers. This has been mutually beneficial as they often have access to funds and information, which furthers the botanical work.[71]

Like Glenny, many overseas volunteers have contributed their expertise. Australian ecologist Tania Leary was organiser of a national environmental management strategy seminar in 1990, followed in 1992 by a draft plan, prepared for the South Pacific Regional Environment Programme

Aiding the Forests

and the Asian Development Bank and endorsed by the Hilly government in mid-1993.[72] Graham Baines, funded by the Commonwealth Fund for Technical Co-operation to be senior planning officer for Western province in 1983–1985, had a significant input into raising environmental awareness. Western Province's first environment week in 1990 was largely due to the inspiration of lawyer Bridget Nicholls, a New Zealand volunteer. Britain's ODA funded David Bellamy to deliver the keynote speech opening the week. Environment week is now a regular occurrence in Western Province.[73]

On Guadalcanal, American Peace Corps volunteers taught Sosimo Kuki in the mid-1980s how to operate a chain-saw and to work out super-feet, to price and grade timber. Attendance at a World Vision seminar had encouraged Sosimo to try to log for himself, rather than see Hyundai do it. Kuki became a key figure in teaching others to log and mill in a sustainable way at his rural training centre at Komuniboli. Since 1987, organisations, such as the Foundation for the South Pacific and the Seventh Day Adventist Church, have assisted the centre.[74]

NGOs

Excluding the churches, the strongest empowering force in the rural Solomons is the NGOs. Unlike most of their Northern counterparts, they often link human rights and development with environmental issues. The most prominent is the Solomon Islands Development Trust (SIDT). It began in 1982 with funds from the Foundation for the Peoples of the South Pacific and Dr John Roughan as adviser. His philosophy, formed in a Catholic church-based team approach on Makira in the 1970s, has been the guiding principle of the SIDT. The director is the charismatic Abraham Baeanisia, from Malaita.[75] The aim of SIDT is to:

> help villagers realise that development should be seen as an internal growth process, that development projects should be seen as part of an overall village development programme, not an end in themselves.[76]

It receives funding from a range of agencies – overseas churches, AIDAB, the EU, the Canadian government, and Freedom from Hunger, for example. Overseas volunteers have played key roles in training local personnel in theatre, cartooning, the writing and production of colourful publications in Pijin and comic strip form, *Link* and *Mere Save*, and calendars. SIDT has mobile teams and a village theatre group, Sei!. It concentrates on needs as they arise, stimulating awareness on social, political and environmental issues.

Pacific Forest

By permission of SIDT

Figure 89. Cover of Link *magazine published by SIDT to highlight logging issues*

Aiding the Forests

SIDT has conducted many workshops on logging, especially since 1988. At these, loggers and the Division can present their views. SIDT espouses neutrality, but the trust does not support commercial logging as carried out in much of the country, as their publications and plays reveal.[77] It has succeeded in raising villagers' consciousness and pointing them towards alternatives.[78] Recently it has provided some. Its Conservation in Development Unit teaches fibre paper making, ecotourism and ngali nut oil production as alternative ways of utilising the forest resource. In 1994, the Ecoforest Unit began training landholders to cut, mill and sell their own trees.

Another NGO is Soltrust, which also offers an alternative to loggers through its company, Iumi Tugetha Holdings. Sometime parliamentarian, Ezechiel Alebua is trust chairman of the company while Antony Carmel is the executive director. The company promotes 'eco-timber' production and the use of *wokabaot* saw-mills. Originally, the Foundation for the Peoples of the South Pacific imported saw-mills to help rebuilding in the wake of Cyclone Namu in 1986, and then the company started building its own in Honiara. In the early 1990s the trust began selling eco-timber to the Ecological Timber Company of Newcastle-on-Tyne in Britain, but this was temporarily threatened in 1992 when rumours told of shipment deficit being supplemented from a commercial saw-mill.[79] Small donors fund Soltrust, which runs workshops featuring the relative merits of large-versus small-scale logging. The trust also conducts an educational programme.[80] In 1994, it started publishing *Sol Tree Nius* which informs people of developments in sustainable forestry. In a country where the illiteracy rate is around 78 per cent, Soltrust gets its message through variously. It has recently produced a cassette album of songs, called *Tears of the Forest*, for young people.

Soltrust believes that 200–300 small saw-mills could replace the commercial logging industry. This might be so, but the economics of scale and marketing seem against it. Upgrading of saw-mill equipment and skills are needed. Soltrust has attracted a number of former Division staff such as James Iniabu and Mathias Marau, so its expertise is significant in certain fields. In 1995 Soltrust, with British support, set up a monitoring and certification Division under Moses Rouhana, to check that eco-timber had been logged in a sustainable way. Soltrust has not been without its critics. Some see its close relationship with government as a liability. This was not evident when it drew the fire of Mamoloni's government after citing figures in *Sol Tree Nius* and the *Solomons Star* about the extent and value of the Pavuvu forests in mid-1995 as loggers moved in.[81]

The Development Services Exchange (DSE) opened in 1984 as an umbrella organisation for NGOs. Its affiliate, the Nature Conservancy Divi-

From *Solomon Star*, by permission of John Lamani

Figure 90. Former Prime Minister Solomon Mamaloni receives a chest of drawers from John Cunning of R. E. H. Kennedy Ltd (United Kingdom), July 1995. The timber, Calophyllum, *was cut in north-east Guadalcanal by Isaac Thomas and exported through Soltrust's commercial branch, Iumi Tugetha Holdings Ltd. The chest is valued at around $3,500.*

sion, helps to organise workshops on sustainablilty in rural areas. Working closely with inter-church representatives on Malaita, the Conservancy has run workshops to help clergy inform their people of environmental issues through the pulpit. Church leaders have had to face social disharmony in the village caused by logging, but the more fundamentalist churches, notably the South Sea Evangelical Church, seem less questioning of causation.[82] Other NGOs visit periodically or work through local NGOs. The World Wildlife Fund has recently established the Community Resource Conservation and Development Project, involving communities in Marovo Lagoon, Kwara'ae and north-west Isabel.[83]

Since 1993, Greenpeace Solomons has been involved, through founder Philip Pupuka and Lawrence Makili, in logging issues. Unlike SIDT and Soltrust, Greenpeace is more confrontational, resorting to peaceful demonstrations to raise awareness. It has been critical of the government's stand on logging and has sided with the Kokota people against the Isabel Timber Co. in 1994, and with Pavuvu land claimants against the loggers in

1995.[84] This hard-line approach and Greenpeace's international clout have not made it popular with central government.[85] Logging companies are unsympathetic to Greenpeace and other NGOs that seek to empower the people with choice, labelling them as 'outside political and environmentalist pressures causing disruptions'.[86]

Aiding the forest always comes back to aiding the soil, the country's greatest resource. A charity doing this is the 'sup sup' garden, funded mainly by UNICEF. It commenced in 1986 to assist urban dwellers to supplement their diet by means of a garden that would provide the making of a soup or *sup sup* in Pijin. Women have been its main targets and teachers because they are the daily gardeners and responsible for most of the cooking and nutrition of their families. It teaches them to dig trenches and bury vegetable matter to create compost. Crop rotation and mulching also feature, as do simple methods for controlling insect pests without imported chemicals. The project teaches inexpensive ways of fencing, trellising and terracing as well as regularly publishing recipes. SIDT's programme on women's development in 1992 has taken this out into villages, as have the Village Education Programme of the Solomon Islands Christian Association and the Catholic Women's Programme.[87]

This is intensive horticulture, not reliant on the forest fallow. Waste matter formerly consigned to the sea shore or to dumps behind the village can be used to fertilise relatively fixed gardens. This is the first step along the path indicated by colonial agriculturalists and foresters, like Walker. As the forest degenerates in the face of logging, cash cropping, and increasing population, Solomon Islanders will have to move further down that path. A degraded ecosystem needs more energy, and this can come through tillage, mulching, or the addition of compost or more expensive artificial fertiliser. On north Malaita where the land is under stress, the Australian NGO Appropriate Technology for Community and Environment has aided the establishment of *sup sup* gardens in Toambaita, as well as publishing booklets on garden species conservation, *Sapa* by Joini Tutua and *Replanting the Banana Tree* by Robert Waddell.[88]

Planting leguminous cover to combat soil degradation is taught at St Dominic's boys' rural training centre, Kolombangara. *Leucaena* is planted along contours in the sweet potato gardens and the results are good. The Marist Brothers were the only people teaching its use until 1987 when at Fote, the National Agricultural Training Institute introduced leguminous rotations, mulching and contour planting into its courses. The Taiwanese government is interested in soil stabilisation, and has incorporated some of these principles in its Honiara demonstration plot.[89]

In many ways, the NGOs are missionaries for conservation and some of their funders and workers are motivated to save or recreate, not the kingdom of God, but the kingdom of Nature. Just as Solomon Islanders at the turn of the century selected what they wanted from the missionaries and made their own Christianity, so do they now with the NGOs. The NGOs have their own environmental conservation agendas, but their value is to offer choice. All NGOs and a couple of the churches aim to create greater public consciousness of environmental issues which, in many cases, come back to the use and abuse of the forest. Several offer an alternative form of logging – community-based with a market for eco-timber, and are establishing standards for sustainably selective logging. They are answering the call of rural dwellers for empowerment and control of their resources. It was a call to which the colonial government was virtually deaf; it is a call the independent governments have heard, but practically all have chosen to ignore. All NGOs educate, most lobby at home, globally and though the media; and increasingly, a few agitate. In this way they can subvert the hegemony of the logging corporations and their clients among the local elite. They are an emerging political force in rural areas.

Costly aid

Aid to the NGO sector is largely beyond the reach of the Solomons government, hence its revolutionary potential. It is not easy to calculate all the aid given to the forests through NGOs. Overseas churches, NGOs and foreign governments have contributed millions of dollars in cash and personnel. A more exact measure is that foreign governments financed the Forestry Division's budget for the decade from 1982 to a phenomenal 85 to 90 per cent, excluding the costs of visiting specialists and staff attendance at overseas courses.[90] This reflects a minimal commitment of successive governments to the work of the Division. It also left the Division particularly vulnerable to any lessening of aid.

Governmental aid is not always what it seems.[91] A relatively high percentage of it returns to the country or origin. Often big projects create even more environmental and social damage.[92] In Solomons, CUT was a fiasco and degraded soils. AIDAB's Allardyce appraisal done by Shedden in 1990 cost in excess of A$250,000 with zero benefits to Solomons' forests. The FRIS cost about A$2,762,480, of which 80 per cent was spent in consultants' fees and much of the remaining 20 per cent for miscellaneous items such as aerial photography was probably spent in Australia. But this project required a high level of specialist expertise and it is hard to see how much of the grant could not have been expatriated, somewhere, if not Australia.

The EC forestry study of 1992 was extravagant in the light of an earlier ODA one in 1991, and all the more so with EC consultants being paid a fee of SI$1,296 a day, plus a per diem of $263 and fares. The Solomon Island consultant was paid $444 a day.[93] Some consultants rediscover the wheel. Economist Roman Grynberg has been critical of R. C. Duncan's *Melanesian Forestry Sector Study* because it documented material available elsewhere – for a consultancy fee of A$50,000, from the Australian taxpayer.[94] Yet, other than a few consultants' fees and fares for annual visits, British ODA investment in reforestation over the decades left a major asset for the country, much of this aid having reached its target.

From the start, aid to forestry has excluded women's needs and issues, as the New Zealand pilot on Malaita revealed, and even some NGOs and churches fail here. In a society where women are rarely forthcoming in public meetings and in negotiating timber rights, this is an appalling omission for it devalues their often profound knowledge of the entire forest and the worth of it and them to the Solomons' subsistence economy.[95]

Another aid cost is the diversion of public servants.[96] Few in the Division have much time left to read the massive reports a parade of advisers have left, gathering dust at headquarters library. The Division would prefer more direct grants to supplement its budget; yet the scope for abuse and/or continuing dependency is such that most donors will not countenance it. Aid can also tempt would-be patrons. Some of the forestry-related projects Japan offers are of massive scale. The bureaucrat or politician who could tap this for, say, his island could benefit in status, if not in wealth; a temptation public servants avoided to the mid-1990s. In deflecting the bureaucracy from other duties, forestry assistance often suits politicians influenced by the loggers' supporters, while injecting relatively high sums of money into the rural economy, if only for the project's lifetime. New Zealand paid villagers in Malaita to rehabilitate forest and soil fertility while Prime Minister Kenilorea waived the moratorium to let Kayuken profit from the diminishing natural forest.

Thus the morality of aid is to be questioned. Never does a week go by in Solomons without some handing over of money, goods and services by foreign donors. Its pervasiveness has sapped self-sufficiency. Development, the word and the concept, for the bulk of Solomon Islanders now means a handout, usually in cash. Proposal writing has become an industry, not simply for government. Walk inconspicuously through practically any Honiara office: in some corner there is a public servant beavering away at a proposal for his home village to get, for example, a saw-mill from New Zealand, or a 'custom' *kastom* house from Australia.

By the early nineties, against a background of increasing impatience with the Pacific 'welfare lake' and its several mismanaged economies, traditional donors were becoming more wary about aid allocation.[97] Island governments did not have poverty and starvation knocking at their door;

rather, the example of metropolitan lifestyles has created 'rising expectations', reinforced by their own elite.[98] Because most have been shielded from fiscal reality though the decades of assistance, and have failed to address seriously resource sustainability and burgeoning populations, this drying-up of aid has political implications.[99] For Solomons' forests, the return of Mamaloni to power in 1994 and his persistent refusal to attempt to foster sustainablilty saw the dry spell become a drought.

— 13 —

Provincial Ambitions, Community Aspirations: The Western Islands, 1985–1995

> We aim to make money by selling trees and use the hard cash to meet our social needs. This is a business deal. No handouts.
>
> Clement Kengava, *Solomon Star*, 4 May 1990

From their inception in 1981, provincial governments were heavily dependent on the national government for annual grants so tried to tap other sources of finance. Logging companies were obvious targets; and some revenue could be extracted directly through a new mechanism of provincial business licences. Provinces that contained national government lands, especially reforested land, began to claim an interest in these and the so-called timber levy that the government had collected for years, supposedly for re-forestation. Some provinces supported landholders in attempts to oust an existing company to replace it with another that offered more.

Some provincial officials journeyed to South-east Asia to attempt to interest investors in their respective timber, fish and possible mineral resources. The dream of practically every province, and most communities, was a company that would provide employment, roads, wharves, inexpensive sawn timber, clinics, schools, education scholarships, and perhaps even a processing plant and reforestation. Timber companies understandably had only one agenda, though they might sweeten it with a bit of seeming philanthropy. Only those that had to operate more than a few years to exploit a significant contiguous reserve, or saw hopes of further logging concessions elsewhere made more than a token effort to meet these aspirations. In time, communities, wanting more local control and equal or greater financial return than loggers offered, tried small-scale milling. There were sometimes deeply divided opinions as to which was the better course: large-scale logging or small-scale community and family-based logging, ostensibly more sustainable and less damaging to the ecology. While there were commercially-viable forests still standing, provinces, communities, landholders, and the companies continued to seek to maximise the benefit for themselves.

Business licences and government lands

For all Mamaloni's advocacy of devolution, the provinces in 1981 were given only limited revenue raising powers, their major source being central government grants which provided between 70 and 80 per cent of their finance.[1] Provinces could raise head and property taxes, and issue business licences. The Western Province set a licence fee for each logging company at $50,000 annually in the early 1990s, but exempted locally-owned milling companies from such a high fee. Under its ordinances of 1993, the province introduced a charge of SI$10 a cubic metre on logs exported by companies in the province. In late 1994, it raised this to $100. Allardyce Lumber Co. Ltd, Kalena Timber Co. Ltd, Golden Springs International, and Hyundai Timber Co. objected and took the province to court. The ruling in May 1995 was that the charge was unlawful, as it amounted to an export duty, impinging upon international trade, which only the central government could impose. Premier Arthur Unusu, frustrated with this ruling, raised the business licence fee to $150,000 for each separate area of operation of any one company. This huge increase Unusu justified on the basis of declared exports of 404,840 cubic metres from the province in 1994, valued at $158 million. Loggers challenged this through their Solomon Islands Forest Industries Association (SIFIA), which filed a writ with the High Court, claiming the fee was excessive. The court in November 1996 found that the fee was a tax and as such was 'unlawful, unconstitutional and invalid'.[2]

The provinces had claims over alienated lands, both private and government. At independence, the central government had assumed ownership of alienated land. Some foreign interests stayed and entered lease agreements, but much of this land eventually went back to local groups for fairly nominal sums.[3] Government or so-called Crown lands were the targets of the provincial governments, though former landholding groups sometimes signalled their interest too. Various parcels of land went to the provinces for schools and expansion of town centres. In line with his support for devolution, Mamaloni promised that all Crown lands, about 3,658 square kilometres, would be returned in 1990 as soon as the Lands department had carried out the legal steps for the transfers.[4] Nevertheless, throughout the 1980s Mamaloni had no plans to relinquish control of the lands of the 'forest estate', where the Division had its plantations. He had however, moved rapidly in 1992 to rid the government of the land, but not the replanted trees, on Kolombangara. Western province thought it would gain part ownership, but Mamaloni bypassed it in favour of the claimant landholders. Progress on the return of undeveloped government lands languished.[5] Though by 1992, Mamaloni had opened the possibility that Forestry Division lands just might be negotiable, the hopes of both provinces and land-claimants foundered as the privatisation of forestry plantations began in late 1995.

Lobbying for the levy

Since 1970, the government had collected a timber levy to fund reforestation on government lands. After independence, the levy underwent various minor modifications – for example, sawn timber generally remained exempt from the levy in order to encourage local processing. However, on log exports, it remained at 7.5 per cent of f.o.b. value. The levy was always lumped in with export duties because of lack of specific legislation. Throughout the 1980s, the Division wanted this money for its programmes, to make it less dependent on aid donors. The whole issue was an academic one, as this levy went into consolidated revenue and, except in the dying years of the Forestry Division's reforestation programme on government lands, 1991–1992, practically no reforestation was funded from revenue, and virtually all from aid. The Division managed to get the levy by the early 1990s, but only under an accounting head: it played no part in collecting or allocating the funds. The value of the levy was substantial. In 1988, it was worth $2.74 million, alongside export duty of $4.54 million; it 1990 it was estimated to be worth $4.4 million, with export duty at $8.5 million. In the early nineties, the levy revenue *alone* could have covered the Division's recurrent budget of $400,000 and a development budget of $4 million, an outcome Trenaman had anticipated in the 1960s.

The provinces also contested the levy, arguing that, since most logging from 1981 was on customary land, it should go to them, perhaps to allocate to local reforestation projects. Some landholders believed they should have it, as they could then finance re-planting, a motive viewed with scepticism by certain administrators in the light of attitudes revealed in the Malaita reforestation project. Mamaloni, the architect of devolution, was not averse to this provincial revenue in principle – at least in 1986, when he was leader of the opposition. In trying to frame a new forest and timber bill, he recommended that all duties, levies, and licences from logging on customary lands should revert to provincial coffers, less central government service charges, with 50 per cent to go to reforestation in the province concerned. Back in power in 1990, he had changed his mind: the levies and duties remained the prerogative of the central government, and a fully revised timber act is still wanting.[6]

In June 1993, Billy Hilly's National Coalition Partnership (NCP) assumed government and came under pressure to allocate the levy to the provinces. A leading advocate of this was Western Premier Unusu, who perceived his province as having contributed most to the national coffers since the early 1980s, its natural resources having been exploited at an alarming rate. The Honiara budgetary meeting of the provincial premiers in August 1994 echoed similar sentiments.[7] In October Hilly announced that 20 per cent of all duties, including the levy, would go back to landholders. Although the provinces hailed this, they wanted an additional five per cent. When the NCP government collapsed in early November, one of Prime Minister Mamaloni's first moves was to reduce the 20 per cent to the old rate of the 7.5 per cent levy, to be held in trust for landholders and/or area councils' reforestation projects.[8] Although premiers such as Malaita's David Oeta kept reminding the national government of its obligations, the finance did not materialise.[9]

By the early 1990s in the western Solomons, where replanting by the Forestry Division and KPFL had had an enormous demonstration effect, landholders were showing interest in replanting with *Gmelina* for KFPL's planned chipwood mill, but the company, though supportive, wanted an aid-donor to fund an 'outgrower' project.[10] If landholders were to agree to reforest clan land, a case could be made for levy revenue going to fund this – although another view is that these landholders obtained royalty benefits from their initial log sales and will also reap some benefit from the sale of replanted trees, so why should the government or, even more pertinent, an aid-donor subsidise this?

Western Province: Allardyce, Hyundai and Kalena advance

Despite the announcement in early 1995 by Minister of Natural Resources Kemakeza of the devolution of 'certain aspects of forestry management and control to Provincial Governments and Area Councils', the provinces remained sceptical.[11] Premier Unusu claimed that, on revenue collection, Western Province, like others, was subjected to 'financial blackbirding': provincial demands were being brushed aside by the 'political gimmicks' of the central government.[12] In relation to logging, Unusu had hit the mark: between 1990 and late 1994, the logging companies (including Eagon on Choiseul) had cut out more than 30 per cent of the loggable timber in the province, and logging on customary land had been going on since the early 1980s. By late 1995, Choiseul and Western provinces were supplying 76 per cent of the country's revenue from export of round logs.[13]

This had taken a toll on human and natural communities, as the Division and successive governments knew. In Shortlands, for instance,

Provincial Ambitions, Community Aspirations

landholders at Uluma started legal proceedings in 1985 against Allardyce Lumber Company Ltd, as it was logging customary land without having gone through the legal steps. Allardyce was annoyed because the Division seemed to have condoned this practice. When it applied for a Form One to commence moves to regularise this, it came up against the moratorium in 1986. Moreover, the Division waged a constant battle with the company to try to get it to mill 20 per cent of its quota, instead of its actual output of less than ten per cent. Allardyce prevaricated, citing the cost of shipping sawn timber to Honiara as a drawback to marketing a competitive product, and claiming that it only had, even with an extended area, a few years of logging life left on Alu (a life that got shorter the longer it could fob off the Division). The logging life may have been short, but it had been a busy one. The company exceeded its annual quota of 42,000 cubic metres in 1982, 1984 and 1986. It got its extra logging areas on Alu after the moratorium was lifted in early 1987, but by then was looking to Mono and Fauro, which it claimed under the original licence from 1974, and further afield to New Georgia in Koraga and the Kazukuru area.

On Alu, it struck its first major snag when landholders objected to the company extending its log pond at Lofung out to the reef, damaging an area which they held was not part of the original agreement. In 1989, landholders claimed compensation of SI$6 million. There were threats of disruption of company operations, and the company applied for an injunction against landholders. Mediators at ministerial level flew back and forth from Honiara. Eventually the court decided the reef belonged to the state, rather than the local people; an odd ruling in a country where land and reef resources are perceived to be all part of a continuum, but one which suited the government. However, further negotiations brought about a peaceful resolution in 1990.[14]

The dispute over the saw-milling quota dragged on. All the Division could do by way of enforcement was to suspend log exports. But this would have meant local unemployment and no royalties for landholders, including the then-Governor-General of Solomons, Sir George Lepping, an Alu chief. Allardyce also salvage logged about 4500 cubic metres of *Instia bijuga*, used in high value furniture, which was taken off the protected species list after 1989. As the law stood, the timber could only be exported in processed form, but Allardyce had the trees cut and away before it could be stopped. Again, there was no Divisional action on this. Nonetheless, Allardyce was one of the less offensive logging companies in the group, at times had worked closely with the Division and was the only one that paid royalties to landholders at the rate stipulated in the standard logging agreement of 1984; others paid between a third and a half.[15]

Its relatively good reputation was not enough to win unquestioned entrée elsewhere. On New Georgia, following the 1989 court case, Allardyce obtained a licence to log the Kazukuru land, but was soon in dispute with the Voramoli people of Munda. On Vella in 1990, two companies besides Allardyce had tried to get licences: Hyundai and Odalisk (SI) Ltd, an Australian company. Some landholding groups, including Naigao, Vaulula, Makavore and Sarapaita, favoured Hyundai which, by late 1990, appeared to have been through the Form One procedure for their lands, mainly in the south. Disputes followed; a major one being between the Hitukera family and Olive Gina over ownership of the log loading site near Malosova on the south coast. The court found in Olive Gina's favour, so the Hitukeras lost their small coconut plantation and home site.

Conflict was not confined to land disputes. Some objected to the loggers' entire operation. In 1994, Teddy Hitukera led observers into the south Vella hinterland to the Hyundai logging operation, five kilometres inland, to see:

> Swathes of forest 50 to 80 metres wide had been cleared, sparsely dotted with the stumps of rosewood and kwila trees the company was cutting ... The surface of the ground is gouged by the dragging of heavy logs, churned by heavy machinery or compacted into hard, smooth surfaces. Nearly every stream is blocked, diverted or eroded by abandoned logs or banks of bulldozed earth.
>
> Unnecessary damage to other trees and vegetation is immense; on every side are great piles of discarded tree-trunks and broken branches, while bunches of orchid plants torn from falling trees are crushed into the mud by the tracks of the bulldozers.
>
> An almost universal forestry regulation in the tropics is that trees are not cut within 50 metres of a watercourse in order to minimise damage to streams and reduce dangerous erosion that can result from tropical rain falling directly on unprotected soil.
>
> Like most other such ground rules, it is being ignored. Said Hitukera: 'This is no good. They don't build proper bridges, they just cut the trees and throw them into the river. The company has spoiled everything here.'[16]

Like Hitukera, many people did not favour the loggers, but villagers and kinsmen who wanted the coastal road and other benefits the company offered pressured them into acquiescence. Several landholding groups wanted Hyundai out – some not because of any conservation issue, but because they believed they could get a better deal elsewhere. There were irregularities in the Hyundai case: the secretary of the Vella Lavella area council falsified minutes, accepting the Form One application that had been rejected by the council. Although the court found that particular licence void the council was back in court again in 1992 when it refused another application by Hyundai, on the grounds that there was a dispute over land ownership. The Commissioner of Forests questioned this and instructed

Provincial Ambitions, Community Aspirations

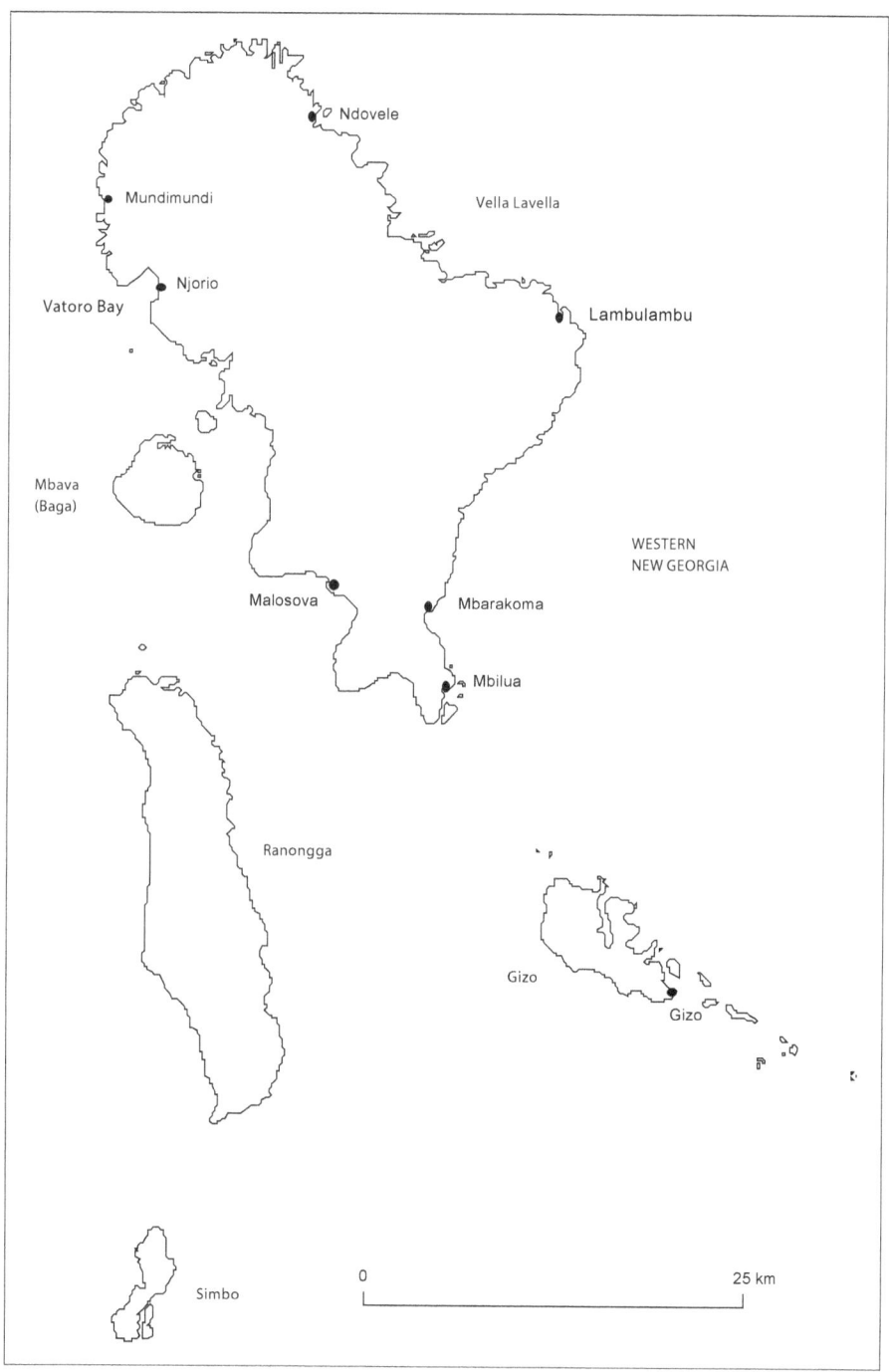

Map 10. Western New Georgia islands.

the council to continue with its determination, despite the dispute. This may well have been a test case for Commissioner Gaviro who, in the light of the advice he had received from the Attorney-General in 1986 and the Forest Resources and Timber Utilisation (Amendment) Act of 1990, may have been seeking clarification as to his powers when faced with obvious failure of the area council to follow procedure, or when its determination was irregular. The High Court held that the Commissioner had no legal basis in requiring the council to continue with its determination.[17] It seems likely that this was just the outcome the Commissioner wanted.

In the Ndovele region in the north, Hyundai tried to get support from landholders, but word of its methods on Guadalcanal and its dealings with the Vella Area Council did not reassure them, so they sent representatives to the Shortlands to examine Allardyce's operations there. Although Allardyce was favoured by the Ndovele people, the company did not in 1992 get the co-operation it needed in adjacent areas.[18]

Allardyce had been a possible successor to Kalena Timber Company in the Saikile area in 1990, when the latter was in arrears with royalty payments to the Saikile Development Association. However, James Boyers for Kalena and the Association came to terms, and the royalties were paid.[19] By 1991, Kalena had logging rights on Rendova at Ughele. The company in 1992 tried to obtain rights inland and further to the south-west, offering some of the landholders gifts of food and very generous payments of allowances at meetings where draft agreements were signed. Advised by landholder Eric Havea who worked for KFPL, they reconsidered the options and refused to sell their rights. At Ughele, too, the landholders were monitoring the logging to see that the company followed the law.[20]

The United Church: human and community development

As the Rendova case illustrates, there was some resistance to large-scale logging. Some of it was based on experience, observation and heightened awareness though various forms of education. Much emanated from the United Church (formerly the Methodist mission), and the work of two of its bishops, Sir Leslie Boseto and his successor, Philemon Riti.

In the early 1980s, the Church had been instrumental in organising a conference to find a better way of settling land disputes than court cases that resulted in winners, losers and fractured communities. The Church was also interested in other social concerns, such as unemployed school leavers and 'drop-outs'. The three issues of finding ways to peace and reconciliation, codifying systems of land tenure on the island through a Land Committee, and seeking practical ways to help the young through the Education and Community Development Programme concerned Boseto (a

Provincial Ambitions, Community Aspirations

Choiseul man) and the Church from that time on. Their basic philosophy combines Christianity with customary ways of dealing with problems. Boseto has no illusions about *kastom*: it was not always life-affirming for all, but the Christian gospel challenges custom, which can be modified and developed.[21] Since the early 1980s, 60–70 conferences and workshops have convened in the west and Boseto was constantly invited to speak and assist with their organisation. Roman Catholics and Seventh Day Adventists attended, although many of the latter were less interested in the future of the forest, believing the second coming of Christ imminent. Moreover, they do not use the forest as a hunting ground, as dietary restrictions on most forms of meat prohibit this.

Boseto has been characterised as having 'actively campaigned against logging'.[22] This was not to the liking of several politicians supportive of the loggers, who told him to get back to his pulpit. His stance was,

> How can I isolate the Gospel from the people and if you are concerned about the people you have to care for the source of life; life is not alone, life is related to nature, to fish, to crops, to trees, you can't separate them. So there is the inter-relationship, interdependence of life. You can't isolate parts that are for the human community, the human community is sustained by environment.[23]

Rural people wanted an alternative to large-scale logging to meet their cash needs. Working closely with Soltrust and through the church's own Integrated Human Development Programme (IHDP), to which the Australasian churches donate substantial finance, the United Church aimed to encourage small-scale logging with *wokabaot* or portable saw-mills to cut trees to sell, but within sustainable limits.

By 1992, Vaghena and Choiseul had their saw-mills. The first shipment of logs by IHDP left the timber yard at Munda, the church's headquarters, in 1994, destined for eco-timber buyers overseas. IHDP was paying, less local shipping and handling, between SI$600–1,200 a cubic metre sawn for premium timbers like *buni*, *gema* (*Calophyllum* spp.), *vasa* (*Vitex cofassus*) and *ringgi*, rosewood (*Pterocarpus indicus*). Commercial loggers paid up to SI$35 a cubic metre for round logs. By felling two trees sustainably for $5,000, a producer can earn the same as if loggers felled 70–100 trees, and the latter would need to give the remnant 100-plus years to regenerate. In July 1994, the Church sponsored for 120 people in Western and Choiseul provinces courses in timber grading and sustainable forest management, provided by SICHE and Soltrust Eco-forestry Programme. It formed an economic development body, Solomon Western Islands Fair Trade Trust (SWIFT), and by 1995 was cutting and shipping timber under conditions of sustainable logging, certified by international monitors, the Forest Stewardship Council.[24] A subsidiary company in the Netherlands buys the timber and sells to those supportive of sustainability. SWIFT's most recent acquisition in 1996 was a landing craft, *Swift*, which can be used to travel around the West and beyond to buy timber from the local producers.

Western Province: wasted heritage or world heritage?

By the late 1980s, Allardyce was meeting resistance to its logging plans on Mono and Fauro, where practically everyone is United Church or Roman Catholic. On Vella, it was the same among the United Church people; in part this was because some landholders now viewed the wider environmental issue as important and in part because by the mid-1990s they had an alternative in IHDP.[25] However, some listen to the churches, to local and visiting environmentalists, to SIDT, and yet still choose to admit the logging companies. They look across the sea to the king of the waters, the cone-shaped Kolombangara, and see it clothed in green where it had been logged by Lever's in the 1960s–1980s, and this cloak is their guarantee all will be well. What is not appreciated always is that Lever's, unlike some more recent companies, did not log above the 400 metre altitude mark, so Kolombangara's green crown remained relatively untouched. Moreover, although KFPL has retained reserves of the natural forest, much of the regrowth below that level is either impoverished and reduced flora or exotics in plantations. The return of the natural 'primary' forest will not occur in their lifetime, or even that of their children.[26]

Allardyce entered north-western Vella Lavella in 1993 amid accusations that there had been irregularities in the area council's timber rights hearings. Disputes erupted in 1994 when loggers moved on to lands at Vatoro, near Njorio. Some landholders wanted Allardyce, others wanted the SWIFT method, the latter group complaining that one of the chiefs had sold out their rights.[27] When logging is an issue, such quarrels are common, and inevitably damage family and community relationships. However, by late 1995 Allardyce's careful logging methods and co-operation with the predominantly Seventh Day Adventist Ndovele people in their development projects had convinced most of the neighbouring Vatoro people to allow them to log their area. In common with many islands distant from Honiara, people on Vella considered themselves to be in a backwater, forgotten in the distribution of national development, so much so that one of the their leaders at Ndovele, Dedili Sasabule, said of Allardyce,

> To us here many people did not know what the word development meant before the arrival of the company. It could just as easily been the brand name for some biscuits.[28]

Provincial Ambitions, Community Aspirations

Vella was not the only island in the Western province where the advent of loggers had created or exacerbated dissension. On north New Georgia, arsonists burned heavy equipment belonging to Golden Springs in March 1994 because of a dispute between the company and the Keru landholders in Nggerasi; and the following month Kalena Timber Company to the south lost bulldozers in the same manner. Police para-military personnel were sent to north New Georgia to keep the peace, much to Premier Unusu's disquiet. The central government, headed by Billy Hilly, had a vested interest in keeping the trees falling, as Golden Springs and Kalena had quotas of 220,000 cubic metres between them, about a third of the total 'allowable' cut, and they were producing close to capacity. For all Hilly's desire to slow the rate of logging, his frail government could not afford a sudden halt to such a significant contribution to government revenue. Despite its wasteful logging methods, Golden Springs was beyond censure by the Forestry Division because of the North New Georgia Act, but the Western provincial government was successful in bringing an injunction against the company in April 1995 on account of its destruction of sacred sites of the Hoavah people in Nggerasi.[29]

The Western Province also brought a case against Silvania Products, which commenced logging in 1992 on Vangunu in the Marovo lagoon. Several environmental groups, including the Maruia Society of New Zealand, had recommended much of the lagoon for world heritage status in 1989. Australian Parks and Wildlife service funded Maruia to do an ecological survey of Solomon Islands. Maruia, assisted by an American trust, has been instrumental in publicising the uniqueness of Marovo overseas. This area holds the largest double barrier reef in the world. For generations, the local people have managed it carefully under a regime of localised conservation and sustainability that reveals a deep knowledge of the local ecology.[30] Eco-tourism here offers an alternative source of income. Yet the loggers had obtained their licence with no great difficulty, as they were to log government land. Their operation was highly destructive. The FRIS team found in August 1993,

> The degree of canopy removal and soil disturbance was the most extensive seen by the authors in any logging operation in tropical rainforest in any country. It appeared more like a clearfelling operation and bore little relation to any attempt at retaining even a token sample of future commercial crop on site.[31]

In early 1994 the Hilly government suspended Sylvania's licence for a time because it had destroyed 'tambu' sites, and so the province could bring a case under its ordinances relating to protection of historical and sacred sites. Silvania's licence was again suspended in March after representatives from the Ministry of Forest, Environment and Conservation inspected the logging areas and found,

> ... the environmental impacts of Silvania's logging operation on Vangunu are among the most serious observed to date in Western Province ... An immediate consequence of the logging operation is deposit of silt in Marovo Lagoon from rivers flowing down from the eastern slopes of Vangunu Island.[32]

In August 1994 Prime Minister Hilly yet again suspended its operations because it had exceeded its quota, was failing to operate a saw-mill and was causing environmental damage.[33] Soil damage here meant waterways and reef damage, as all who had seen the earlier operations of Kalena at Viru and of Golden Springs in north New Georgia knew.[34] The fisheries resources of the lagoon in turn were being effected.

This case on government land, perhaps more than any other, with the possible exception of Pavuvu in the Russell Islands, reveals the direction the Solomons Islands government would have taken if it had had total control over the forest resource. There was little evidence of commitment to custom (*kastom*), and much of a commitment to capitalism.[35]

There may be more to lose. The uninhabited island of Tetepare whose claimants left for nearby islands seven generations ago is the second largest uninhabited island in the world and the home to unique flora and fauna. Most of this island is government land. The Indonesian logging company, Goodwill Company Ltd, one of the Mega group, wants a licence and is supported by the Rendova member of parliament, Danny Philip. Goodwill Company Ltd, like Marving Brothers on Pavuvu, have been linked with a re-settlement scheme (see next chapter). In 1994, the land claimants imposed an injunction to stop any logging, but Philip pursued the matter in the face of their opposition, considerable international criticism, and the fact that Tetepare with 50 per cent of its surface sloping more than 30 degrees, will be extremely susceptible to erosion once the forest cover is removed and soil disturbed. The claimants oppose the logging and have plans for preserving the area's *tambu* sites as a mark of ancestral respect as well as for eco-tourism.[36]

Choiseul: a company province?

In March 1991 Choiseul, once part of Western Province, became separate under Premier Clement Kengava. Eighteen months later, at Choiseul Bay, the administrative centre, the Premier proudly pointed out to the writer a new truck bearing the provincial logo. He said that Eagon had donated this lone provincial vehicle, as indeed they had the timber for the new wharf.

Provincial Ambitions, Community Aspirations

To the premier and his supporters, the presence of Eagon underwrote their province as a political unit. Since colonial times, of all regions in the archipelago, with the exception of Rennell and Bellona, Choiseul was the least touched by capitalist development.[37] Only its churches, the Methodist (United Church) and the Roman Catholic seemed to care about the isolated villagers. Choiseulese perceived themselves as 'neglected people'. The Sirovanga Association or VASA, set up to promote development in the mid-1970s with Kengava as its adviser, supported Eagon because its members believed this was a key to development. In his words, 'We are opening up Choiseul to industrial development. No longer should we look for copra as the only source of income'. The Association, based in west Choiseul, wanted a road from Choiseul Bay to Chirovanga (Sirovanga), clinics, and schools. Supported by the premier of Western Province, Tausinga, they believed they could deal with the company and minimise destructive effects.[38]

Eagon's public relations have improved since 1992, when landholders went to court to force a royalty of $12 per cubic metre instead of the original pittance of $5.[39] Its president, South Korean J. Y. Park, is Christian, a point of contact for most Choiseulese. Eagon had most support in the south-west. Its presence on the opposite mainland overshadows the former significance of the Catholic church, school and clinic on Mole Island. From the air, the company settlement stands as a red scar dotted with buildings and a log-filled waterfront. The neat hospital and school on the island were abandoned with indecent haste in 1992. The company opened the Seoung Min Memorial Hospital, supplied a doctor, a school, and established the Eagon Choiseul Foundation to support secondary and tertiary education of the local youth, all part of its president's philosophy of 'sharing what one has to help the less fortunate brothers'.[40] The company has given to members of the local Catholic church to win support and has plans to localise its work force rapidly.[41] It appeased the Division by setting up a saw-mill in 1991 and logging 19 per cent, not of its quota, but at least of the total volume of the declared cut of 93,000 cubic metres in 1992/3. Saw-mill operatives have been sent to Korea for training. Eagon had plans for a hydro-powered veneer plant at Choiseul Bay. It has set up a tree nursery and planned a reforestation project and assistance with small-scale agricultural ventures. Eagon was a force in isolated Choiseul, as its wages bill in 1995 was around SI$2 million and it employed almost 300 local people.[42]

Yet there were increasing signs of disenchantment. Kengava, like all provincial premiers, wanted to see sharing of revenue from investments in enterprises such as logging and fishing and, for a brief few months under the Hilly government, this seemed a real possibility, in relation to timber. The Eagon reforestation project was to cost $87 million and questions were asked about who would own the trees. Eagon wanted the reforestation levy returned to landholders, to be used as equity in the replanting scheme.[43] Eagon's acquisition in late 1995 of the government plantations at Viru on south New Georgia raised the question of where Eagon will concentrate its reforestation resources: on Choiseul, where it is still trying to win support and where re-planting will be on customary land, or at Viru, where it will have a fixed-term title to the 25,000 hectares of land and a defined, legal right to any replanted trees?

The honeymoon between Kengava's government and Eagon was short. By 1994 the province was growing suspicious of logging export declarations by Eagon, and welcomed plans for the AIDAB-financed forestry inspectorate station.[44] Kengava, once an almost unqualified Eagon supporter, warned his people not to support 'certain elements within the timber industry in Choiseul province' when the Choiseul Bay Association tried to stop the province's environmental protection study of the Sui river waterfall catchment.[45] To control the provincial government, Eagon could use the simple expedient of underwriting the promises of its employees standing as candidates in the provincial elections, in order to stack the assembly with its supporters, an allegation Clement Kengava made late in 1995, not long before he lost office.[46] The tail was beginning to wag the dog.

There are tensions elsewhere, too, as most United Church followers and many Catholics are resistant to the company's attempts to obtain further logging rights. Several people have received support from the United Church and the national government's Provincial Government Unit, which has money from New Zealand for portable saw-mill projects. Within the Catholic community there were differences: in 1992, the Choiseul-born priest then at Mole, Michael Lomiri, referred to by Eagon as a director of the company, was supportive of Eagon, while the Dominican priest then at Chirovanga, Alex Vickers, a New Zealander, favoured the United Church's small-scale sustainable logging.[47] Up to 1995, the Catholic Church's upper echelons remained circumspect in pronouncements on the morality of logging and environmental damage. Without local family ties, the expatriate hierarchy are not subject to the same pressures as local clergy on the logging issue, so do not have the same level of credibility.[48] If they had upset the Mamaloni government, they would not have been the first expatriates to have had their passports seized for their environmental views. However, most bishops actively support programmes that empower the laity – the Catholic Women's Programme in the diocese of Auki being but one example. Through the empowerment of women, the community is better able to articulate all its needs, not just those of the men; in doing this, questions arise about the use of the environment and thus the future of the forests.[49]

Provincial Ambitions, Community Aspirations

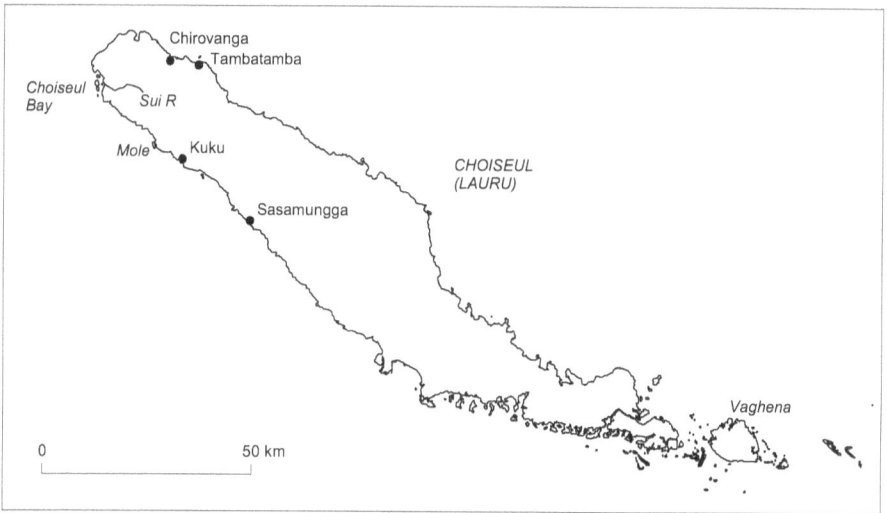

Map 11. Choiseul.

Isabel: a millionaire province?

Each province in Solomons is unique, but Isabel has certain qualities that have made it particularly different. Those qualities were to make post-protectorate logging slow to start and, now that it has come, deeply to divide a formerly united and peaceful people. Isabel's population is almost entirely Church of Melanesia, and it was the first area in Solomons where large-scale conversion to Christianity occurred in the 1890s. The strong chiefly tradition and sense of community evident there stem from the time of Soga, who led the desperate and weakened tribes into a new unity based on the new religion. Hence the church and church authorities have singular influence. The paramount chief is Sir Dudley Tuti, who also was a bishop in the Church of Melanesia, now retired. Even as a young priest in 1959 Tuti had been interested in having a logging company on the island when the Pacific Timber Company visited Allardyce Harbour.[50]

After the 1972 cyclone, Allardyce effectively ceased felling on Isabel, but the six years of logging and the resultant damage to soils appear to have left a strong impression. Moreover, Isabel has a comparatively high level of registered land in local hands, which may well have resulted in more precise awareness of who holds particular rights to land and trees. Being lightly populated has meant that there have been historically fewer land disputes on Isabel than, say, on Malaita. When the 1980s saw the logging rush elsewhere, the provincial government was circumspect; and most schemes up to the mid-decade were rejected or petered out.

In 1985, the diocese of Isabel organised a seminar to educate the clergy, chiefs, politicians, public servants including the chief forestry officer, Sam Gaviro, and the local people on aspects of development. The resulting Jejevo communique was a statement of aims and values pertaining to human and resource development. Regarding logging, it stated:

> The Seminar recognises large scale logging operations should be discouraged and is fully aware of the permanent damages that may [be] caused to land, river and sea. ...small-scale logging operations would be more appropriate for the Ysabel people. The Seminar further notes with concern the attitudes and activities of the multi-national logging Companies that are interested to log in the Solomon Islands.[51]

In 1988, the Isabel Development Authority was established to further economic development and improve the people's quality of life. By the time a major company appeared with an offer of logging in 1989, the SIDT had preceded them. As well, representatives of landholders had visited logging sites on Guadalcanal and Viru. The Forestry Division had shown several groups the film, *Fate of the Forests*.[52] Thus, many people generally had an awareness of the disadvantages and advantages of logging. They were not going to be easily beguiled by loggers or their local advocates and clients.

In 1987, the province sent a delegation to Singapore to discuss logging with Lawrence Tay's South Pacific Forests Ltd. Dudley Tuti, Premier Jason Leguhavi, and the provincial secretary, R. Natowan led the group.[53] The company appears to have lost interest when it discovered that the central government had issued a lease of most of the former R. C. Symes properties (then leased to James Wong) to Jim Boyers, who had just purchased Foxwood Sawmilling near Honiara to operate as Pacific Timbers. The province was not happy about this leasing of valuable lands they wanted, but the issue was resolved eventually when in 1988 the central government handed over the titles to the provincial government. The province confirmed Boyers' rights to lease for 50 years on 13 parcels of coastal lands, although it handed over five small holdings to local communities.[54]

In December 1989, Axiom (initially a Fiji-based company, registered in Jersey, which had connections with the large Monarch company of Hong Kong) proceeded with plans to obtain a licence, but like Walter Jones' Eastern Enterprises they ran up against the short-term moratorium imposed in late 1989. Mamaloni's cabinet granted an exemption as there was no major company on Isabel. Eastern Enterprises appears to have gone steadily through all the required procedures, and was issued with a licence in August 1990 to log 40,000 cubic metres annually in the Suavanao area in the north of the

Provincial Ambitions, Community Aspirations

island. The conflicts involved in the acquisition of the timber rights seem to have been minimal, perhaps because only one land-owning group, the Bulao, has been involved.[55]

Axiom had much bigger plans, and powerful supporters in government and the elite. When it put forward its plans to the province it already controlled Silvania Products on Vangunu; it also owned Integrated Forest Products and Rural Industries companies in Makira.[56] It planned to set up the Isabel Timber Company (ITC) to log 150,000 cubic metres annually over 30 years, mine the nickel reserve on San Jorge, and set up a smelter in Honiara powered by hydro-electricity from the Lungga and Tina rivers. There was to be 30 per cent equity held by the province, local directors, reforestation, jobs for local people, an investment of $400 million by Axiom, with dividends to go to the province after initial capital outlay was recouped. No one indicated just when that might be.[57] There were other unanswered questions: for example, was a hydro scheme feasible, given the intractability of Guadalcanal landholders on this issue?[58] There was no environmental impact study of the effects of mining or processing, a matter of some weight, as asbestos occurs with the nickel. Amid fanciful talk about making perfumes from by-products, there was no credible logging plan in terms of Isabel's ecology, which was fundamental as ITC hoped to log about four-fifths of it.

Although the Foreign Investment Board somehow had approved the concept, Commissioner Gaviro had been unwilling in mid-December 1990 to issue Form One because of lack of any preliminary feasibility study or stand appraisals by Axiom. In the face of delays, advocates of the company appear to have prevailed upon the Acting Minister for Natural Resources, Victor Ngele, who forced a political decision on Gaviro.

Many landholders objected to the plan, several complaints being made of irregularities in attempts by the company's agents to get landholder and area council consents. Old people were woken at night and told to sign papers written in English; young brothers of women landholders and clan heads gave consent to log. Mining also was anathema to many, with the shadow of the Panguna copper mine cast across the Bougainville strait. By 1990–1992, there were two camps. The pro-Axiom one was led by the paramount chief, Dudley Tuti, sometime executive director of ITC; Premier P. Manehatha; two national Isabel parliamentarians, N. Supa and Edmond Andresen who was also a director of the company; and E. Vunagi and Caspar Huhugu, directors also. Their hope was that the company would be the island's economic salvation, that it would cease to be dependent on 'handouts' from aid agencies and the like; in this, they shared much in common with the Sirovanga Association on Choiseul. The anti-Axiom camp, was led mainly by the provincial secretary, J. Vunagi; the Isabel Development Authority; the Church of Melanesia diocesan bishop, Ellison Pogo; the vicar-general of the church, F. Kapu; and the church's Mothers' Union, a very significant body with a huge interest in the outcome as the society is matrilineal in regard to land rights. This group favoured small-scale logging, direct involvement of people in the working of their resources, and the protection of the natural environment.

Pogo enunciated a thoughtful theology of the environment, strongly based on Genesis in the Old Testament. His analysis had much in common with the United Church and Boseto's approach. Accusations, open letters and preaching from the pulpit flowed back and forth, with various respected people trying to mediate. In 1992, the Development Authority was instrumental in running a workshop with government officials and leading figures in church and politics attending. The ecologist, G. Baines, who was married to an Isabel woman, attended for the World Wide Fund for Nature. A plan resulted for 1993-1997, emphasising education, stewardship of the environment and sustainable development, and endorsing the earlier Jejevo communique.

Nonetheless, ITC, with 'legal' rights to negotiate under Form One, continued to seek the logging rights, while Soltrust assisted villagers to obtain *wokabaot* or portable saw-mills to log their own timber with a view to selling some of it, at least, as 'eco-timber' at premium prices. With positive planning, mechanical and ecological/silvicultural education and centralised marketing by government and NGOs it is just possible such a scheme could have been successful, particularly since there was the SWIFT organisation in Western province and within Isabel the small EU financed sustainable forestry project as prototypes. All national governments thus far, despite pious sentiments, have failed to implement what most Solomon Islander rural people have hoped for since the 1970s: their direct involvement in and control of the development of their resources. Mostly, they have had to accept the next best thing, short-term profits which will lead to their ultimate impoverishment, and large-scale exploitation by foreign companies of forests and seas. It was not simply inertia any more, but a matter of those in power getting greater rewards from letting the big companies dominate the logging scene.[59]

In 1994, Greenpeace came into the Isabel picture, with Lawrence Makili being banned from the province by Premier Gigini because of Greenpeace's support for the opponents of the company's activities in Kokota ward in the north, after it had been forced out of Litogahira in the south. By the end of 1995 when the loggers moved on, the people at Goveo village, Kokota, were a divided community, with most of the $4,000 per capita in royalty payments spent on consumables, and their development plan for a 'per-

Provincial Ambitions, Community Aspirations

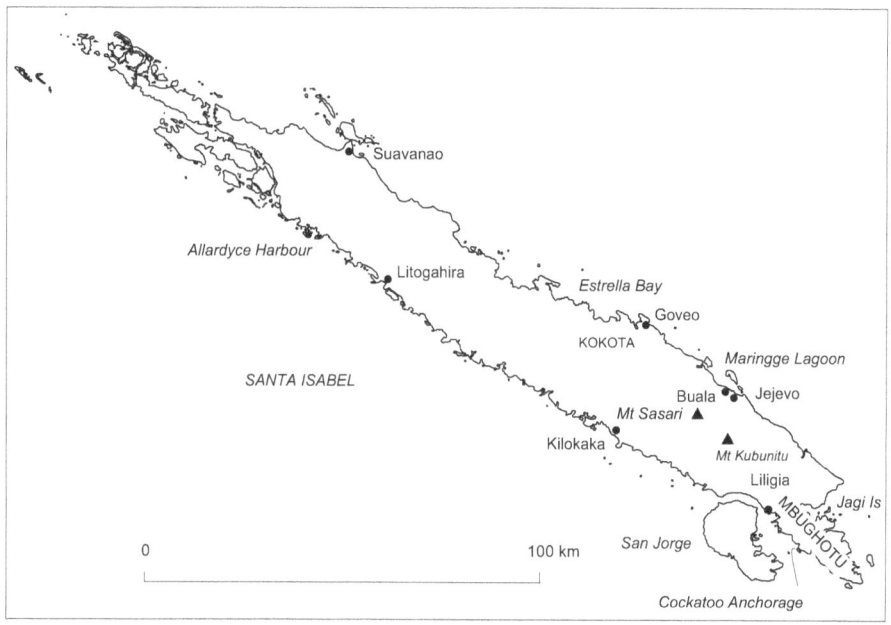

Map 12. Santa Isabel.

manent materials' village and a system of roads, a lost dream.[60]

The companies continued their campaign to obtain more forest. Axiom's successor, Kumpulan Emas Berhad, which controls ITC, ran up against seemingly intractable negotiations with landholders in 1995 when trying to secure adjoining lands for logging: 'True ownership is often difficult to establish and disputes frequently happen'. This hard lesson cost the company time and money.[61] By early 1996, Earthmovers Solomons Limited (controlled by Lee Ling Timber Co., a Sarawak-based company) which had taken over Eastern Development Enterprises as a subsidiary, was trying to convince landholders at Kilokaka, where Soltrust had already established an eco-forest management plan, that the company had more to offer them. A spokesperson for the Kusa clan, Rev. Zephaniah, claimed that landholders were being wooed by 'gifts' of cartons of tobacco, biscuits, corned beef, milk and money and promises of a clinic, school fees and construction of an airfield. The company denied this, but did admit gifts of 'rations' during negotiations.[62] A familiar pattern on most islands, as Jules Makini's poem tells.

GREASED DINNER

Last night I was treated like a king
at a business dinner
Never before had I been treated thus
Along with ten of my wantoks
we who had never entered any hotel
were wined and dined
at the Gizo Hotel.

I sat with a roomful of strangers,
the table-littered with plates,
glasses on thin legs, ashtrays,
two sticks to use instead of fingers,
and handkerchiefs to blow my nose
with.

"Self service, eat as much as you
want."
So I helped myself to many dishes,
very new, different smells, but tasted
great
even had squid, tho't it was strictly for
bait!
I heaped up my plate, as high as
Kolobangara,
asked for a fork cos I couldn't master
the chopsticks.

Pretty girls wearing pink
lingered ready to please,
men in black sulus
eveready with drinks,
fosters was flowing

I tasted it and liked it
I smoked Peter Jackson instead of spear
Our host was too generous
The drinks kept coming,
and I kept going.
Business dinner it was
so we ate, drank and talked

Unfortunately,......I was too far from my
host to fully understand.
The Roviana Bamboo Band
made too much noise,
the waitresses too distracting,
the lights too dim,
and the smoke too hazy.

I heard 'royalty',...'timber rights'...
'roads,...clinics,...schools..'
but through stuffy ears and blurry eyes.
Anyway we'd already talked it over in
the village
Our politician and lawyer were present
so I didn't have to read the 'small print'
I signed.

zzzzzzzzzzzzzzzzzzzzzzzzzzzzzzzzzzzz

©Jules Makini (Jully Sipolo) 1991 Gizo, Western Province, Solomon Islands

By permission of Jules Makini

Figure 91. 'Greased dinner'

Glossary: 'Greased' implies smooth talking, flattery, suasion often associated with bribery. 'Wantoks' includes relatives, friends or speakers of the same language, from the same area. 'Sulus', a wrap-round, about two metres in length, to cover the lower half of the body. 'Fosters' is a brand of Australian beer. 'Peter Jackson' is a brand of tailor-made cigarettes. 'Spear' is a less costly brand of stick tobacco, very strong and tarry, made in the style of the trade tobacco of the nineteenth century.

Provincial Ambitions, Community Aspirations

When the Solomon Islands FRIS report (a publication few have seen) appeared in 1995, it recommended that 75 per cent of Isabel should not be logged because of its 'high ecological and environmental sensitivity'; yet Kumpulan Emas Berhad's ITC alone aims to log about 80 per cent of the total area.[63] In the report the recommended annual cut for Isabel was 38,600 cubic metres, as opposed to ITC's licence for 150,000 and Earthmovers' subsidiary, Eastern Development Enterprises' 40,000.[64] The conclusion that this licensed coup of *five* times that recommended is unsustainable is all too obvious, yet logging proceeds.

Within the Solomons context, the western islands had already 20 years' experience with loggers working within their vicinity, through the operations of Lever's, Allardyce, and Kalena, for example. Provinces realised the west's timber resources had for years subsidised the rest of the country, for limited return. Islands like Vella, and indeed provinces in Choiseul's case, often perceived themselves as forgotten by central government in terms of services such as transport, clinics and schools. In the judgement of most communities or their spokespeople, what the loggers had to offer seemed better than what they had to lose in forests. For some leaders, wealth from the logging industry meant more dignity than begging from donors and even the government. Despite several attempts to capture logging revenues, via a tax on exports and demands for the reforestation levy, the provinces failed to regain from the loggers or the national government what they considered their due, though temporary employment opportunities were significant. Business licences provided some income, but the High Court placed limits on this when it verged on a tax, the constitutional prerogative of the national government. The logging companies seemed to offer opportunities for rapid development and access to cash. For many areas, they were seen as the only such opportunity; at least until bodies such as SWIFT and Soltrust began to offer alternatives. In giving communities a choice these bodies had a powerful political role because they could weaken the links between the patron logging firms and their several client go-betweens, whether area councillor, chief, priest, premier or national politician. In the west, although there was resistance to loggers in the standard form of arson of company equipment, there was little direct confrontation; but practically everywhere the questions of logging and the type of logging created or exacerbated conflicts, not only between the central government and the provinces, but also more destructively within small communities. Experience, in the form of a history of regional logging, did not always equate with wisdom.

— 14 —

Provincial Ambitions, Community Aspirations: Central and Eastern Islands, 1985–1995

> Nothing in this section shall be construed as affecting traditional rights, privileges and usages in respect of land and fisheries in any parts of the Solomon Islands.
>
> The Provincial Government Act, 1981, Solomon Islands.

The central and eastern islands, with the exception of Vanikolo and north Guadalcanal, had far less experience of logging companies than communities of the western islands. This gave the loggers' certain advantages, but left newcomers vulnerable when they assumed that governments, central and provincial, had more power than they actually did. The provinces utilised similar strategies to those employed in the west to obtain some direct return from the timber sector. A few were more inventive. One possibility was for a province to have a share in the company. Some provinces proposed applying for licences from the Division and allocating these to the company which could offer the best deal. Here and elsewhere, provinces were drawn to Mamaloni's idea of devolving the functions of the Forestry Division, but seemed unable to grasp the administrative implications of this.

Malaita: Taisol and Kayuken conflict with Waibona

In February 1986, Malaitan Premier Maomatekwa successfully lobbied for the removal of the moratorium on new logging licences imposed in April 1983. Kayuken Pacific was the main beneficiary, in wards 34 (Sie Sie) and 35 (Waneagu Sulana) on the west coast, although it had failed to get its planned concession of most of central Malaita. One area the company had wanted was to the north, in west Kwara'ae (wards 2, 3 and 4). The first foreign logging company on post-war Malaita, Taisol Investment Corporation

Provincial Ambitions, Community Aspirations

started operating in earnest there in 1983, its licence application apparently having been in the pipeline before the moratorium.[1] Certainly, since 1982, its principal, Suny Tong, in association with the Taiwanese government, had been trying to find a concession in the islands. In 1987, Tong's dealings were such that the Taiwanese interest disassociated itself from Taisol.[2] Toshio Hashimoto, a major shareholder in Dalsol and Integrated Forest Industries, also had shares in Taisol. Taisol came under censure of Ombudsman Qoloni in 1990 because of allegations that two senior government officers, including a lawyer, failed to advise landholders of their legal rights in order to get a licence for Taisol in east Guadalcanal.[3].

Both Taisol and Kayuken were seeking additional areas in Malaita to make their operations viable. They had fallen victims to their own greed and to the endless patience or ineptitude of the Division. Each had obtained large annual quotas – Taisol had 24,000 cubic metres, Kayuken had 50,000 cubic metres. The magnitude of this can be appreciated when compared with the recommended annual quota of 28,000 cubic metres for Malaita, which the 1982 UN Development Advisory team had set for the resource to last 25 years, with 25 per cent reserve.[4] The companies were supposed to mill 20 per cent of their quota. Throughout the years of logging in the 1980s, their declared cut often did not achieve half that quota, because of the scattered nature of the stands and the under-capitalisation of the companies. Periodically, the Forestry Division suspended their licences because of failure to mill 20 per cent, but promises, hired mills, the use of chain-saws for rough milling, and occasionally new machinery would persuade the Division to reinstate the licence.

Logging companies refused to commit substantial capital to processing when they perceived 'a lack of a national forest industry strategy' and they could get higher prices for round logs.[5] Until the early 1990s, neither the government nor the Division seems to have appreciated the extent of the capital outlay involved in establishing first-class mills and the level of expertise required in maintenance, management, and marketing (problems evident in Trenaman's time). British ODA suggestions in 1992 that the Bina saw-mill which had produced export grade timber for New Zealand be studied as a model for a locally-owned and controlled saw-milling industry were ignored. No encouragement for rationalisation of the saw-milling industry issued from government although the Division had recommended it in 1985. Mamaloni, in 1990, as part of his devolution of forestry package, had advocated increased local processing and even a centralised system of milling timber from a network of loggers, but he took no steps to implement this, despite the aspirations of rural people. And why should the government bother if export duties and the levy kept coming to its coffers from the export of whole logs? Each logging company, after a few years of grace to get established, had to run a mill. The timber milled was usually from low-grade logs and inferior saw-mills, so the product did not create much market interest, the entire exercise being wasteful of the resource.[6] It was all a policy farce, a piece of populist political rhetoric, as year after year until 1990, only one company ever achieved the 20 per cent milling requirement.

Ironically, that company was Kayuken, at its Manaba mill in 1989.[7] The year before, Kayuken had been trying to expand the area under licence into west Are'are (wards 31 and 32). Some landholders were in favour; some were not. They were exceptionally well-briefed. There were meetings in early May 1988 led by Premier T. Kauhiona, involving 1,400 people at Hauhui, Kopo and Maru'uru. A representative contingent of 16 chiefs and landholders went to Honiara for a 'look and learn' session and met with Kayuken's Kong Ming Khoo, the Forestry Division, and Are'are men in Honiara who were informed on environmental issues. Many of the educated Are'are Honiara people opposed logging, at least by foreign companies. Following this three-day meeting, the party toured Are'are, meeting at six centres and involving over 600 people, to consider Kayuken's offer of negotiation for a logging contract. At Rohinari, John Roughan, former parish priest and, more recently, SIDT adviser, further explained the pros and cons in October.

As in the 1970s, communities were interested in logging and milling for their own use and sale. In the east, the Bina saw-mill co-operative had been a success, assisted technically by New Zealand church volunteers and also helped by the expanding network of government roads. This created interest in neighbouring areas. The Kwaio-based Waibona Logging and Milling Company had been formed in 1987, with John Gafui and C. Karaori as directors. George Luialamo and Sam Korasimora, one-time provincial Minister of Agriculture and Forestry, were behind the company. They mustered further support following a four-day workshop on the advantages and disadvantages of logging and saw-milling by Waibona, sponsored by SIDT for ward 33 people in November 1988.

For some time, Luialamo had been gathering evidence to appeal against Kayuken's right to log ward 34 (Sie Sie), which it already had under the controversial licence of 1986. Like Kayuken, Luialamo was seeking logging rights in ward 31 (Mareho), Are'are. Kayuken's managing director, Kong Ming Khoo, maintained that Waibona wanted the rights to both wards 32 (Tai) in Are'are and 33 (Kwarekwareo) so that it could sub-contract to Logimex Co. Ltd, a Taiwanese company – a plausible claim, as Waibona lacked investment capital and expertise. Kayuken and supporting landholders continued to pursue a licence for the Are'are wards 31 and 32, but there were certain irregularities in the procedure which, in the light of the ombudsman's criticisms of the scandal surrounding the initial 1986

Provincial Ambitions, Community Aspirations

licence, impelled the Division to proceed carefully. At the same time, in 1988, Kayuken was trying to get a licence to log on Makira, the home island of the leader of the opposition, Mamaloni, who supported the company's bid. By 1990, Kayuken was still waiting on Malaita, despite pressure from Prime Minister Mamaloni to expedite the licence.

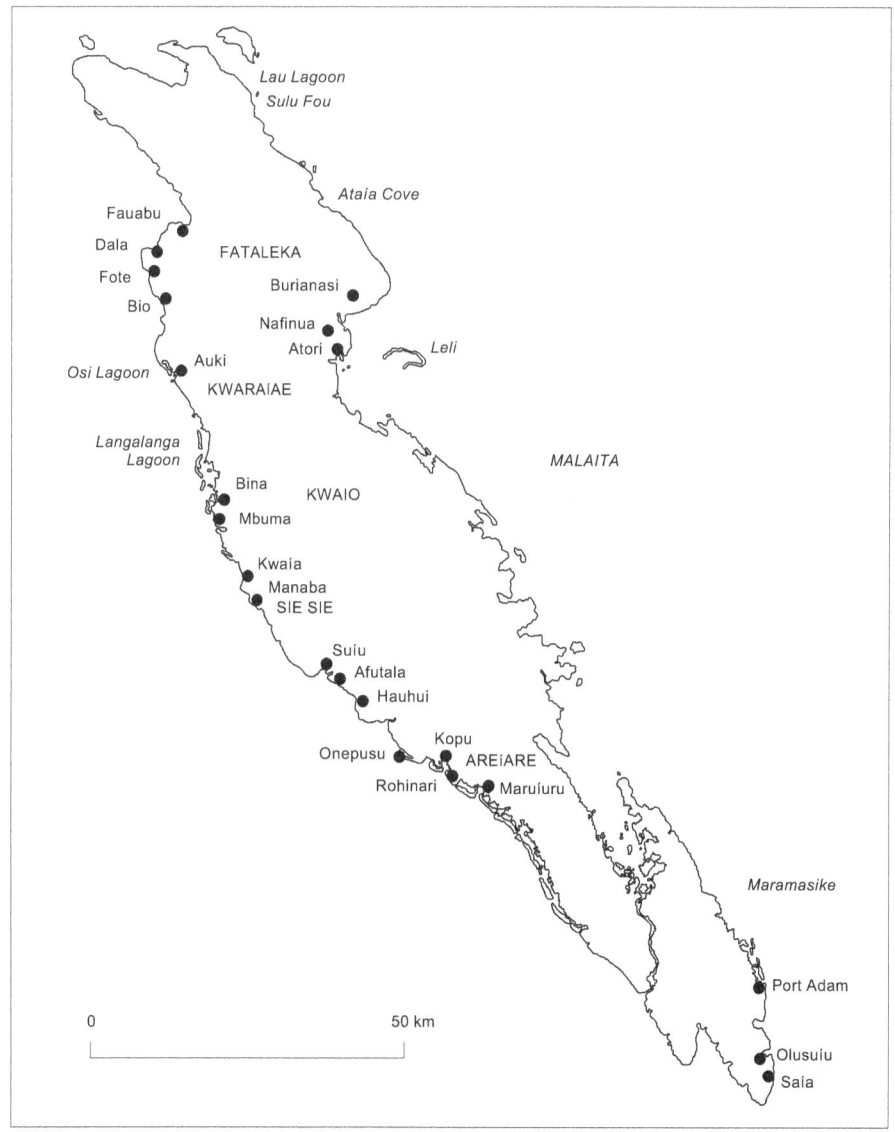

Map 13. Malaita.

George Luialamo was the member for parliament for West Kwaio and also a director and prime mover of Waibona: a pattern of directorship-holding that parliamentarians and provincial politicians were by this time pursuing with increasing alacrity. Conflict of interest never came into the picture, despite pious statements about leadership codes in parliament and in public.[8] In 1990, Kong Ming Khoo maintained that Luialamo's attempt to bring appeals against his company's licence to ward 33 was motivated by the desire to replace Kayuken with Waibona as logger. Khoo believed that Luialamo had been trying for two years to bring in Integrated Forest Industries, Hyundai Timber Co. or nearby Taisol to log the area, and claimed that he had given misleading information to the ombudsman to discredit Kayuken. The ombudsman's reports of 1988–1990 had dealt a severe blow to Kayuken's credibility, incriminating many in high levels of government and politics. A locally-based company headed by Luialamo was likely to appeal to local people, many of whom were disenchanted with Kayuken's record of environmental damage. Khoo was well informed, as Waibona had applied to the province for approval of a partnership with Hyundai early in 1990. Thus there were contending forces at work, with Suny Tong waiting in the wings. Tong's financially-strapped Taisol had already formed a partnership with Marving Bros to log west Fataleka.[9]

From blood money, Golden Springs

Kayuken's plans came to naught with the murder of Suny Tong in February 1991 in his home in Taivo Ridge, Honiara. A single bullet in the head had killed him, upon his return from a social engagement late at night. Earlier, Tong, represented by lawyer Jennifer Corrin, had won a civil case against Kong Ming Khoo and Kayuken Pacific Ltd to recover a debt of thousands of dollars. Corrin's car had been torched in December 1990. Those arrested for the murder were Kong Ming Khoo, Anthony Bara of Guadalcanal, and Keleto Lalani of Kwaio. Bara claimed Khoo paid him SI$40,000, an amount deposited in his bank account from a cheque signed by Khoo. When Lalani discovered this he told police he was offered only $500 by Bara to kill Tong. Lalani was charged with murder and Bara and Khoo held in custody on conspiracy to commit murder. Lalani confessed, was found guilty, and received a life sentence. At the trial, it emerged that Lalani had tried earlier to gain access to Corrin's office to destroy the files relating to the case between Tong and Khoo and Kayuken. He was also found to have set light to Corrin's car at Bara's instigation.

Khoo hired two barristers from Australia. Bara had a local solicitor. The court found Bara guilty of conspiracy to murder and sentenced him

Provincial Ambitions, Community Aspirations

to life imprisonment. Khoo claimed that the $40,000 given to Bara was out of friendship. The court acquitted Khoo on the grounds of circumstantial evidence. However, he was re-arrested on a charge of conspiring to murder Bara. It was alleged that shortly before the trial, Khoo's brother had offered to pay a policeman to murder Bara in prison. The man informed his superiors and, it was alleged, there were police witnesses present when brother Khoo handed over some kind of poison. This brother Khoo was arrested, but, it was alleged, after political intervention, the man was released and left the country for Australia.

For his protection Bara was then moved to Kirakira, Makira while preparations for the trial of Khoo went ahead. When the time for the trial was near, police went by boat to Kirakira to collect Bara, but found plans had changed. Bara was to return to Honiara by plane. The police's chief witness in the trial was inexplicably sent to Kirakira to accompany Bara.

On mountainous south Guadalcanal, heading towards Honiara, the plane crashed with 15 people on board. The pilot was experienced, but had been in the country only a week. The crash occurred in a very isolated area so the government asked Australia to help with search and rescue. The helicopter pilots found everyone dead. Understandably, there was much grief throughout the country. Several versions of events then started to spread in Honiara, suggesting some passengers had survived, but died before the searchers reached them. Post mortem reports did not support this. The Australian army carried out the difficult tasks of gathering and returning the dead to Honiara and investigating, along with the police field force, the cause of the crash. When all the dirty work was done, Prime Minister Mamaloni ordered the Australians out. The report on the cause of the crash has never been published, supposedly because the government refused to pay the typing fees. However, an inquiry did take place with reputable commissioners, headed by Francis Waleila. The cause was found to be pilot error. That the results of the inquiry were never made accessible to the public redounds badly on the government and has provided grounds for more sinister interpretations.

Khoo was released for lack of evidence and flew to Australia. The investigation produced Bara's 'hit list' which included other people who had crossed Kayuken, such as Jennifer Corrin, Luialamo, Qoloni, Commissioner Sam Gaviro, and SIDT's Abraham Baeanisia and John Roughan, as well as the late Suny Tong.[10]

Many blamed Kayuken and Khoo for Lalani's crime and for bringing Kwaio into disrepute. Waibona capitalised on this repugnance, with Luialamo leading the fray. The Forestry Division found that conditions of the Kayuken licence had been breached and so suspended it in mid-1991. Kayuken sold out to Golden Springs, with Kang Wibisono as a director, a transfer endorsed by Minister for Natural Resources Tausinga, who had a close relationship with that company in his home, north New Georgia. A 'landowners' committee', as well as blaming the Division for Kayuken's failings, issued demands for compensation from Golden Springs for Lalani's life sentence, for damage to land, for unsubstantiated labour costs, for the transfer of the rights without the owners' consent, and other claims amounting to over a million dollars. Yet another landowners' committee wanted Kayuken to stay. Negotiations moderated the demands, gaining consent from the landholders who used the opportunity to increase their royalties and to extract $63,000 as compensation.

Golden Springs achieved almost complete control of the remaining timber resources of central Malaita the following year when it bought out Taisol for around $300,000. Soon after, Luialamo, Minister for Lands in the Mamaloni government, sold Waibona to the same company, becoming a director of the new subsidiary bearing the old name.[11] Taisol, still under its old name, ceased logging in ward 4 of West Kwara'ae where the people objected to being paid royalties of about $10 a tree and to the drying up of springs near Fote, and the company concentrated on West Kwaio and West Are'are until late 1991, when it returned to ward 4. Golden Springs bided its time in West Kwara'ae until October 1995 when, just as a seminar for local elders was expressing their wish that no logging be carried out, the company managed to get other elders to sign a timber rights agreement for the Fote region. The company moved its heavy machinery into Bio the same month. Arguments as to just who were the 'true landowners' continued into 1997.[12]

West Kwaio demonstrated similar ambivalence and contest. In September 1991, Soltrust had conducted a three-day seminar at Dala with its representatives, the Forestry Division, the provincial agricultural office, pro-and anti-logging lobbyists, and Peace Corps community workers as speakers, and over 50 participants from the region. As a result, the West Kwaio Producers Cooperative Association Ltd wanted the government to cancel Taisol's licence, so they could log on a smaller scale, using portable saw-mills.[13] Of course, the government could not lightly revoke a licence as this would damage its credibility with overseas investors, something both parties wanted to avoid, so Taisol (Golden Springs or subsidiary Marving Bros Timber Company Ltd.) stayed and the logging industry remained largely in foreign hands.

But not entirely so. In 1989, a Malaitan, Abraham Eke, working at Auki, learned a lot about the logging industry from Taisol's camp production manager, Jimson Tareamu. Already experienced in banking and small business, Eke began saw-milling and came up with the idea of milling on Ndai island. He needed finance, so planned to sell round logs to finance

Provincial Ambitions, Community Aspirations

the mill. Working with the Ndai landholders' company, headed by Clement Diau and Vineat Samo, Eke applied for a logging licence and, with provincial support, obtained a quota of 12,000 cubic metres per year. On Ndai, this could result in the clearing of much of the lowland forest.

Eke tried to get a loan for logging equipment from the Development Bank, but he lacked equity. Morgan Equipment's Alistair Martins sold it to him on the understanding that it would be paid for with the first shipments of logs. Martins' mechanics provided maintenance. An experienced Malaitan workforce, recruited by Tareamu, joined the enterprise, known as Island Logging. By selling his personal assets and raising a commercial bank loan, Eke met his costs and the first shipments went to China and Hong Kong. By June 1993, Eke had paid off the $1.2 million he owed Martins, and the landholders were getting their royalties and were involved in the process of selective logging, taking great care of the land. Eke next opened the saw-mill on Ndai.[14] Operating as A. E. Enterprises, he now concentrates on saw-milling having purchased a Mahoe saw-mill for SI$140,000 from New Zealand in 1994.[15] Eke's business remains an indigenous rarity, a successful mix of skills and business experience with access to finance, a skilled technical workforce, and well-serviced equipment to work a confined, but commercially viable area, free of disputants and unrealistic aspirations. It is ironic that the government's agencies, despite the political rhetoric of 16 years, had no input into this enterprise – except for issuing of a logging licence.

Makira: companies old and new

Attempts to change canoes mid-stream were not confined to Malaita. Neither Integrated Forest Industries (IFI) nor Rural Industries (RI) had escaped criticism for their logging methods on north Makira (Bauro). In 1985, for example, IFI were logging on the western bank of the Wairaha river, including land used for Tikopian resettlement which was subject to dispute as to ownership in the area councils. Land disputes also came to the fore where RI were logging.

Nonetheless, the Makira provincial government wanted to attract more companies, and in 1985 they planned to apply for a licence to sub-contract to an interested company. This was a variation of the practice of a local company obtaining a licence then sub-contracting, entering some form of partnership, or selling an interest to an overseas operator. However, their plan fell foul of the moratorium and the national government's policy against sub-contracting in the light of experience with, for example, west Guadalcanal Solmac and Cape Esperance companies in the early 1980s. Another revenue-earner was the business licence fee, which the province could impose under the Provincial Government Act of 1981, and an enabling ordinance which they passed in 1984. This appears to have netted ten per cent of the value of logs exported or milled.[16] However, although together IRI and RI had an annual quota of 106,000 cubic metres, their declared log exports were only 27,645 cubic metres with no saw-milling being recorded for these companies in 1985, so licence revenue from loggers remained limited. Once the moratorium was lifted in February 1987, the premier of Makira raised the possibility of a joint venture with some foreign company, possibly with Paul Freeman, former principal of IFI in 1982, who had been deported by Prime Minister Mamaloni.[17] Central government financial and forestry advisers were cool towards this, particularly in the light of the scattered nature of much of the Makiran forest.

It was not only the province that was looking for a joint-venture partner. The Hagaparua Sawmilling Association at Star Harbour had been performing badly and so it was hoping for a partner to bail it out. As Ombudsman Qoloni noted, it had a 'non-transferable mill licence, no timber rights, no logging licence and little support from land owners' – hardly an attractive proposition.[18] Surprisingly, an Australian investor was found and Sollumber was born in 1987, with its overseas principals apparently unaware of the legal situation regarding access to trees. A senior Forestry Division officer had been the consultant to the local company in its attempts to interest other companies operating in Solomons, and ended up being a director-shareholder of the joint venture. Accused of giving landholders incorrect advice, he came under the ombudsman's censure in 1989, but by then he had resigned from the public service. By the late 1980s, landholders on Makira, Malaita and elsewhere, though lacking in formal education and literacy, were becoming increasingly well-informed of their rights as well as obligations regarding logging and its consequences, through local workshops, usually jointly conducted by NGOs or church bodies and provincial and national officers involved in forestry. There had been one conducted as part of Makira-Ulawa province's community education in mid-1988 at Kirakira, another at Star Harbour itself, and a repeat at Mwakorukoru in October, each running for three days. Although, in the light of the Star Harbour Sollumber fiasco, these workshops were reactive, rather than pro-active, they appear to have emboldened landholders to complain to provincial and Division authorities, leading to the ombudsman's investigation.

The ombudsman found many irregularities. The milling licence had been unlawfully amended, the company could not obtain actual logging rights, but managed in the interim to convince the Division and the government's Foreign Investment Board to allow it to export 7,000 cubic metres of round logs in 1989. It had only a saw-milling licence and no right to cut trees for logging, although it tried to circumvent the law by supplying equipment so local people could cut and drag trees from their land, a measure that was costly and inefficient. No one was sure who owned the trees as the area councils had not made final ruling. Sollumber failed to get timber rights and faced growing opposition from landholders and the province. Finally, the Division suspended its operation in June 1989. The company ceased logging and it apparently went into receivership, though its Australian subsidiary, Rantex Pty Ltd tried to mill some of the logs, for which no compensation or royalties had been paid. Another Australian company headed by Schmarr and Owen, called the Star Harbour Timber Company, which proposed to take Sollumber's operation, did find support from the Star Harbour Council because it built a new and bigger saw-mill in 1990, but the provincial government was slow to approve the venture. When it got its licence it was for an area of about 45,000 hectares.[19]

Another company on Makira also found its operations suspended in early 1989, in circumstances that involved intervention by Prime Minister Mamaloni. While leader of the opposition in 1988 he had expressed support for both Sollumber and Kayuken on Makira. Kayuken was preferred over existing loggers because of the landholder discontent with IFI and its subsidiary RI. These had both just been taken over by a large Malaysian company, Monarch Leasing Company. Monarch, whose principals included Hii Yi Ging, had tried to take over the licence of RI in 1987, to avoid applying for a licence which would have required the application of the standard logging agreement. The ombudsman took a High Court action against the company, but Monarch won when the court sanctioned RI going into receivership.[20] Under IFIs' banner, in 1989, Monarch was seeking logging rights in the Arosi and Haununu districts where Mamaloni's home village of Macedonia, as well as his electorate of west Makira, were located – hence, Mamaloni's interest. Technically, IFI, when it applied for its licence for Bauro, had started the process of obtaining the logging rights to Haununu in 1982, having completed Forms One and Two, but land disputes had held up operations, until the court settled these in 1988. Kong Ming Khoo wanted logging rights there too, and had given free trips to area council members and landholders to visit Kayuken's operation in West Kwaio, Malaita. Another supporter of Kayuken appeared to be James Morea, the provincial member for Haununu. In 1989, he complained to Mamaloni about alleged bribery by IFI and the discontent of most landholders with the company.[21] There was little doubt considerable environmental damage had been done already in Bauro, as noted at the time:

> The effects of logging around [Kaonasughu, Bauro] can be clearly seen from the air. Rivers with catchments in logging areas dump their muddy discharge far out to sea, covering the once-pristine reefs that surround the island with a fine layer of silt.[22]

An inquiry found Morea's complaints to be overstated, although some Haununu landholders were not happy, so their lands were excluded from logging. Legally, IFI had prior rights over Kayuken, and this prevailed, despite 'toe-pulling' – attempts by provincial officials to win 'favours' from the company representatives in Honiara by playing one off against the other.[23]

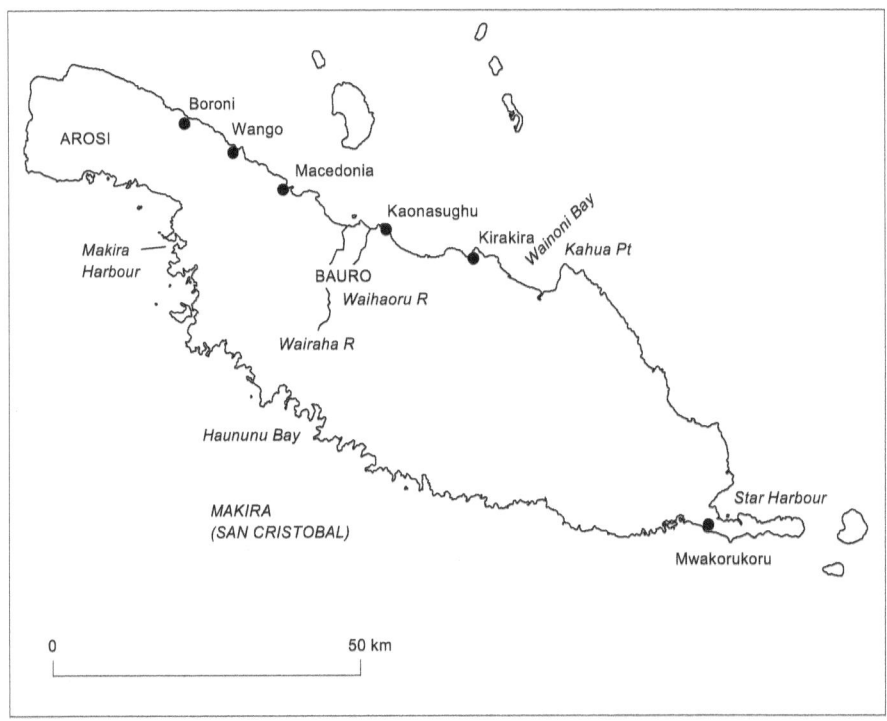

Map 14. Makira.

Berjaya and bribery

Toe-pulling on a grander scale was alleged in 1994 in the wake of the Malaysian-based conglomerate, Berjaya Group (Cayman) Ltd's negotiations for 600,000 hectares of land for logging on Makira and Guadalcanal.[24] Berjaya had courted a Guadalcanal provincial delegation to Malaysia in 1993/4, and their director Tan Sri Dato Vincent Tan Chee Yioun met with Premier Moses in Honiara in April 1994. Tan wanted logging rights to 300,000 hec-

Provincial Ambitions, Community Aspirations

From *Solomon Star* by permission of John Lamani

Figure 92. Tan Sri Dato Tan Chee Yioun of Berjaya, 1994.

tares on Guadalcanal. On Makira, Berjaya was to take over the provincial Makirabelle Fishing Company and the Star Harbour Timber Company. There was talk of a timber processing and plywood factory and even a hotel venture on Guadalcanal. Investment would amount to US$60 million. By June, Berjaya had purchased the Star Harbour operation for US$1.2 million, but central government administrative approval of the logging quota was still pending. The NCP, headed by Prime Minister Francis Billy Hilly, who had replaced Mamaloni in June 1993, applauded the company's plans. Like others before him, Hilly had set a date, 1997, for the cessation of round log exports in favour of local processing, so the processing plant seemed a step in the right direction.[25] The provincial officials were happy with Berjaya, though at least one Makiran, Bincent Bernard Heimaenia, condemned Premier Ramoni's support for 'a multi-millionaire cult', likening it to cargo cult ideas of the early fifties and the hopes aroused by Paul Freeman's logging in Bauro. The press debate was hard-hitting:[26]

> If an Asian investor comes out of the blue and asks you to turn your back-yard into a pig-pen and promises you $100,000 the following week, I know Mr Ramoni would give in.[27]

Money could be used to open other doors. Minister of Commerce, Employment and Trade Joses Tuhanuku claimed that when negotiating with Tony C. T. Yeong, managing director of Berjaya's Star Harbour Timber Co. Ltd, regarding the company's purchase of Star Harbour Timber Company in July, Yeong had offered him SI$10,000 as a token of Berjaya's 'appreciation' for his anticipated support of the company's application then with the Solomon Islands' Foreign Investment Board. Tuhanuku had accepted duty free gifts valued at SI$100 which, Yeong claimed, Tuhanuku returned the following day. Yeong also claimed that he, 'like a good samaritan', had offered his own cash because the minister had complained of money problems. Berjaya's Tan took umbrage at Tuhanuku's allegations of the cash bribe, although he did say that if such an act occurred it was done without company authorisation. Tuhanuku ordered Yeong's deportation on a Sunday, the day after the alleged bribery, before any formal hearing. Aspersions were cast on Tuhanuku, alleging his apparent 'Mr Clean' stand resulted from too small a bribe, rather than his honesty, but Hilly transferred him to the Ministry of Forestry, Conservation and Environment to replace Alebua. The company smoothed its rumpled feathers and proceeded, ostensibly, with the legal requirements. When the requisite approval came in August 1994 it was for the logging of 30,000 cubic metres annually, far less than Berjaya had anticipated.[28]

Provincial Ambitions, Community Aspirations

Guadalcanal: conserving a declining resource?

In the mid-1980s Guadalcanal Province had made several attempts to control loggers. Their pioneering 1985 Protection of Historic Places Ordinance gave the province legal power to protect *tambu* places, such as burial sites, and provided for inspection by their cultural protection officer. Yet there were logistical limitations on what could be done. Cultural officer Victor Totu was responsible in 1992 for enforcing the ordinance, but had no telephone or vehicle to inspect sites. Moreover, he could act only when a complaint was made. Getting allocated funds was not always easy, and the Forestry Division gave little support. However, the cultural focus has enabled aid agencies to assist; thus Australia has funded an oral history project here and in Western province which included the recording of significant sites and the EU also has money for recording and training. Once the sites were recorded, the logging company concerned was given a site list and had to survey and mark the sites clearly, with the landholders paying for their assistance and costs.[29] Action was another thing: though Dalsol was responsible for considerable damage to a site in 1991, the province was slow to prosecute.

Logging, many claim, has affected far more than *tambu* sites. North Guadalcanal had suffered severely from cyclone Namu in 1986. A vast amount of rain fell, making the area between the Lungga and Mbokokimbo rivers a huge 'lake', leaving wrecked schools, gardens, villages and a devastated Honiara. This raised questions about logging exacerbating the impact of the cyclonic floods. The national government denied any link. The truth was somewhere in between. Logs left behind by loggers had come down in the flood and piled up around bridges and then burst, releasing great volumes of water. But entire trees also had washed down from landslides higher up in the foothills than the loggers had worked. However, reduced forest cover certainly diminishes the ability of the earth to absorb water and remain stable. Though civil defence systems were strengthened as a consequence of Namu's impact, little work has been done to study or eliminate possible contributing causes associated with deforestation.[30]

Action was needed elsewhere. Honiara is not only the national capital, but also the provincial centre. It is ideally suited to be a base for saw-milling for local communities, but the province has done little to encourage this. It seemed more interested in large scale industry, such as the Dalsol saw-mill at Ranadi. The addition of a kiln drier was considered so significant that both the premier, Gideon Moses and the new commissioner of forests, Enele Kwainarara assisted in its opening in May 1994.[31]

More recently, Berjaya appeared to be the bright new hope of Guadalcanal province. However, the question was where and what would it log? In the east Hyundai appears to have halted logging in 1992 due mainly to difficulties with landholders, many of whom, supported by their parliamentarian, Hilda Kari, saw an alternative in Sosimo Kuki's rural training centre at Komuniboli, which taught people how to log and run their own chain-saw mills. In 1989 the Zhong Xing Group from the People's Republic of China took over Dalsol, which had acquired B. K. Maurice's concession in the 1980s, and was still logging in the west.[32] These companies, and the Foxwood saw-mill owned by Earthmovers of Malaysia at Red Beach, near Honiara, have logged about 53,130 hectares. Foxwood has been logging inland on the north coast where its various predecessors had felled earlier. This could buoy hopes for those who believe in the natural recovery of logged forest. However, the previous logging was less intensive and did not open up the canopy as much as the loggers are doing in the east and west of the island. Moreover, no assessment has been done of the re-established forest on the north – will it sustain another coup and what will result? The total area of merchantable forest remaining in the early 1990s was about 718 square kilometres or 71,800 hectares and the recommended allowable cut is 45,000 cubic metres a year, including some previously logged areas. Yet Berjaya wanted 300,000 hectares of forest. This is also hard to reconcile with the Dalsol quota which alone is 30,000 cubic metres, to say nothing of the several saw-mills' quota of 150,000 cubic metres a year.[33]

The south coast has never been logged, though Dalsol seemed to be edging closer along the western end of the island. The 'weather' coast is extremely precipitous, with a slope so great that, even where no recent human disturbance has occurred, great landslips scar the hills, evidence of the heavy rainfall and earthquake activity. Once logged, the topsoil of this region, about a third of the island, would simply slide off into the deep sea and with it the Tasi Mauri people's subsistence. Yet in the early 1980s, Hyundai made a bid to get a licence for the eastern south coast to Valasi, and IFI tried to get consent to log from Marau to Moli, but backed off in the face of opposition. The people of this heavily populated area, despite their relative proximity to Honiara, are isolated by a difficult sea, no harbours, and a taxing terrain. They have little means of earning money except by migrating periodically to Honiara and other places of employment. Some have moved permanently to the Aruliho area after devastating earthquakes and floods in the late 1970s. Schools on the weather coast are few and remote. With the right go-betweens, these isolated communities could be easy targets for sweet-tongued loggers. The FRIS put this area in an environmental *tambu* class. There are sites of scientific and ecological significance, most notably Lauvi lagoon, for which plans are in hand for an eco-tourism venture, incorporating a walking track from Aola in the north. The problem for eco-tourism is that the tourism infrastructure in Solomons is verging on the abysmal, despite the country's extraordinary natural beauty and the

Provincial Ambitions, Community Aspirations

friendly people. Moreover, if the forests go, so will the reefs and much of the rarity and richness that attract the tourist.[34]

By October 1995, the spell Berjaya had cast on Guadalcanal was broken. Despite the distribution of $100,000 to supposed 'landowners' in Star Harbour, Makira, in late 1994, the Berjaya subsidiary inherited the land disputes Sollambur and Star Harbour Timber had faced. It decided to pull out of Makira. The loggers had come off second best in terms of loss of venture capital. Any hopes Guadalcanal had of investment in logging vanished. Premier Moses, erstwhile supporter of the Berjaya proposal, announced a moratorium on new logging applications on the grounds that the forests were being destroyed while the province needed time to educate people in resource management and to draft a resource management ordinance. Moses appeared to be reflecting the concerns of his people who had voiced their grievances concerning the existing loggers' activities. There was a growing realisation that the resources of Guadalcanal were finite and needed to be preserved for the future.[35]

Central Province: Pavuvu and the people's resistance

Lever's, under its various subsidiaries, was a symbol of the colonial order. This has less significance than it once did because more than half of today's Solomon Islanders were born since independence; but, to Solomon Mamaloni's generation, Lever's reminded them of much they would rather forget. In 1979, Lever's had relinquished control of undeveloped land in the Russell Islands to the central government.[36] Former landholders early signalled their interest in the 2,000 hectares, but the issue languished until 1992 when the government announced a scheme for re-settlement and development of these lands. Lever's was scaling down its operations in the Russells, mainly because of the depressed state of copra and oil palm prices. In late 1993 it sold out 60 per cent of its shares to James Boyers, of Earthmovers, and 40 per cent to the country's Investment Corporation.[37] The resettlement scheme would have created difficulties enough because the original landholders wanted it, while the government was intending to re-settle unrelated people as well. A more vexing issue was the plan for the Malaysian company Marving Brothers, a subsidiary of Golden Springs, to log the forested land first. Pavuvu claimants' spokesman, Martin Apa, claimed that the licence to log was issued in 1992 without the knowledge of Commissioner Gaviro and without consultation with the commissioner of lands, who expected the unlogged land to go to the people. It was claimed that George Luialamo, as Minister of Lands in the Mamaloni government, granted the logging licence.[38] How he had the power to do this is anyone's guess. The government's plan was that the company would build infrastructure for the resettlement scheme. From the start, there was opposition to the logging from the local people, supported by the NGO, Greenpeace. Although the Hilly government did not cancel the licence, it did put it on hold while surveys were done; that stay lasted only until Mamaloni's return to power in 1994.[39]

To placate the objectors, the government promised the lands to the original landholders, not outsiders. Yet the opposition was more to loggers than outside settlers. The province still held a trump card: it could refuse to issue Marving Brothers with a business licence. After Premier Nelson Ratu did this he was unseated through the machinations of the national government, whose agenda put Marving Brothers first. Greenpeace continued to support the local people and their own development plan for the land which included small-scale logging supported by Greenpeace and the Tropical Timbers group, buyers of timber from sustainable sources.[40] Throughout early 1995, Ministers Luialamo, Orodani, Andresen, and Kemakeza made visits to the Russells to persuade the people to agree to the logging; failing this, the government sent the para-military police field force with their guns to disperse resisters assembled at Marving Brothers' camp in April. The government ignored Opposition Leader Sir Baddley Devesi's call for the suspension of logging. Visits by opposition parliamentarians, Joses Tuhanuku, Patterson Oti, and widely-respected Hilda Kari and various NGOs, including Save the Children Fund, kept up the pressure on the Mamaloni government. Some NGOs had overseas volunteers working for them and SIDT's adviser John Roughan, by birth an American, is a Solomons national. From his retirement, the happy warrior of ecological conservation Geoff Dennis, a Solomon Island citizen, took up his pen to educate, plead, and shame readers into protecting the unique features of the Russells forest in an article in the *Solomons Star* on 19 May 1995. This all enraged Mamaloni: '... foreigners, both citizens and non-citizens are advised not to interfere in internal government development matters of Solomon Islands and to refrain from feeding the media with false information' and he accused NGOs of being the 'same characters who have destabilised Papua New Guinea during the past ten years ...'. The criticism had found its mark among the political elite and hurt.[41]

Ministers again tried to persuade or, as some claimed, to bribe the Pavuvu (Lavukal) people up to the point where, led by Augustine Rose, landholder and lawyer, they marched on parliament in May 1995 with a petition signed by 597 people for the removal of several ministers as well as return of the lands, the cessation of logging, and an environmental assessment of damage done by loggers. The polite words of the Deputy Prime Minister Denis Lulei were seen as delaying tactics, as was an unenforced suspension order issued by Mamaloni; so some protesters in July set fire to Marving's heavy logging equipment, a time-honoured form of resistance in Solomons.[42] The protesters wanted to stop the company from completing its first export of 6,000 cubic metres, valued at more than US$750,000. They were unsuccessful, despite undersized trees making 80 per cent of the cut and logging of watercourse banks. The government received its million Solomons dollars in duties and royalties from exports to Japanese importer, Sumisho Lumber. The Lavukal people's stand was noted within the archipelago and, as significant, abroad. Even more so, the murder of spokesman Martin Apa in late 1995 and the long delay by the police in investigating the matter, was noted on national news broadcasts in New Zealand and the Internet.[43] Augustine Rose claimed the company was trying to exploit disputes among landholders on adjacent customary land in order to extend its operations beyond government's land. So, once again, logging claims divided another local community. The company to 1997 had yet to build any significant infrastructure, other than short-life logging roads.[44]

Mamaloni had made a stand on Pavuvu, putting the rights of the near-bankrupt state, before those of landholders, real or elect. This was more an act of desperation than of perspicacity. Some might argue that the pity of it is that a stand like this had not been taken 20 years before when it would have been less painful and more beneficial to have asserted the rights of the state to control the disposition of the forests: but then, given the calibre of parliamentary leadership in the past 16 years, the question is: would landholders have received even less reward?

Rennell and Bellona and Temotu provinces

No loggers, so far, have obtained licences to log the Polynesian islands of Rennell and Bellona. Rennell is the largest and highest uplifted coral atoll in the world and is a potential World Heritage site. Rennellese have resisted mining companies who wanted their bauxite, but there is some interest in logging, although the eco-tourist potential of the island is potentially far more valuable than the commercial forest. The FRIS recommended an allowable cut of 6,700 cubic metres yearly, a scale more suited to small-scale logging-saw-milling projects than foreign loggers.[45]

In Temotu province, Jones' Eastern Development Enterprises company persisted from 1985 to 1989 with attempts to get a logging licence. However, the province, particularly under Premier Fr John Lapli, resisted. The company was suspected of being a front for another operator, so the province asked it to bring in its heavy equipment as proof of its bona fides. Eastern Enterprises complied, only to find a new executive rejected its application in 1989, so the company retreated to Isabel. Santa Cruz has done well with the EU-aid road building and 'rehabilitation' scheme, which has given it some of the best roads in the country, factors that could make it attractive to loggers. The development of small businesses was an aim of the province which, under the aegis of the Temotu Development Authority, has made good progress on this with the help of loan money from the central government. Some communities in the Nanggu area were interested in selling rights to a commercial logger, but the Development Authority invited the NGO Soltrust to offer an alternative. Soltrust and the province have set up a scheme of sustainable logging management for the extraction and milling of eco-timber. If a success, the returns will be around SI$760 to $1,400 per cubic metre gross instead of $15 royalties most companies were offering in late 1995. When the sale by the central government of its plantation forests was announced in November 1995, Temotu province wanted control of those on Santa Cruz, but the central government refused.[46]

Devolution: butchering the Division

Central to the future of the forest and the ambitions of the provinces was the question of devolution of national government powers. Provinces were interested mainly in the allocation of the revenue from resources, not in the running of Forestry Division, but if they cannot get a fair share of the revenue then they could press for control of the administration. If given this power, they could vary licence conditions and set different logging agreements to suit themselves, in fact, have an open slather. While the argument for a greater share of returns was valid, the rest was folly, in a country where the population and area are so relatively small and resources, particularly human ones, are scarce.

Devolution of the Division's functions was one of the more bizarre ideas of Mamaloni. He saw it as contributing to the wider policy of devolution of government to the provinces as well as to the aims of the 1989 forest policy.[47] By 1992, he was impatient with the Division, which was having enormous difficulty trying to find ways to implement this policy of devolution. Who would be in charge of the Honiara herbarium, who would get which personnel, who would be responsible for personnel training, what about the specialists, where would they go, would every province duplicate services, what would happen to timber control units as loggers finished with one province and moved to another, how would the provinces negotiate aid, and who would represent Solomons on international forestry organisations? Planning for the nation would be a dead letter. There were more questions than answers, and the provinces, many of their existing administrative positions regularly filled by overseas volunteers, had no idea how they could fund these people or even accommodate them. The Division cobbled together a timetable. which went on hold with the fall of the Mamaloni government in 1993.

Mamaloni should have listened less to the logging lobby and more to public servants in the Division who thought the policy was a mistake. He could also have looked across the Coral Sea to the chaos created in Papua New Guinea when that state did exactly the same thing. Papua New Guinea had set up provincial government in 1976, just after independence, and their Forestry Department was carved up more or less on a 'where is, as is' basis. Any ability the national government might have had to control the supervision of forestry operations collapsed and the industry soon became the shame and scandal of the Pacific for its exploitation of the resource, the environment, and the landholders. Judge Tos Barnett's Commission of Enquiry into Forestry in 1987–1989 revealed horrific corruption and abuses at every level of the industry and politics and the Judge almost lost his life to some he exposed. In relation to the devolution in Papua New Guinea, the Commission described it vividly, despite the muddled imagery,

> as if a mad butcher had chopped up the carcass of the national forestry service, cutting out the limbs and putting them in the provinces and leaving the head in Port Moresby with no limbs to carry out its commands or to feed it with information to enable it to make plans.[48]

Among the Commission's recommendations was the establishment of a single forestry unit under national control and this came into law in 1991/2, just as Mamaloni was ordering the dividing up of the Division in Solomons.[49] The question has to be asked: to whose benefit? Certainly not the landholders, and ultimately not the provinces.

The fall of Mamaloni's government in mid-1993 postponed the issue; but the new policy of his 1995 government put the Division's future in question under the pretext it has been too much influenced by 'foreign' concepts – a classic Mamaloni diversionary tactic.[50] Provincial governments and resource owners in theory were to have greater involvement in the forestry sector, but the policy also states specifically that 'certain aspects of forestry management and control' would be devolved to the provinces.[51]

To 1995, the provinces had shown no greater aptitude for conserving the forests for future generations than the national government. Provincial officials were often better placed to manipulate community affiliations to advance the loggers' aims. Some, like Kengava in Choiseul and Ratu in the Russells, thought they could control the companies to benefit their province; but when environmental and social costs mounted to an unconscionable level, the companies simply provided support for pliable client candidates for provincial office. This has suited the national government because it has facilitated the companies' access to the remaining forests and has been a plausible means of shifting accountability from the centre to the region, albeit sugar-coated with the rhetoric of devolution.

— 15 —

Profits and Loss, c. 1982–1996

> Bless us with your wisdom Father. May we, who often laugh at our ancestors for selling their lands for bottles, beads, knives, axes and guns, not be guilty of the same mistakes. Will the next generation not laugh at and blame us for trading our resources for outboard engines, trucks and under-the-table payments?
>
> Archbishop Adrian Smith, at Independence commemoration, Honiara, July 1995, *The Australian*, 21 Sept. 1995.

Central to understanding the state of the forests in Solomon Islands is the part the timber industry plays in the country's macroeconomics and the less easily quantifiable, but far more valuable, part the forest plays in underwriting the livelihood of most of the population. There are contradictions that require examination: a government that needed revenue to pay for services to its people, yet has squandered one of its few resources; and resource-holders who accepted cash payments worth far less than the subsistence value of the forest. Both produced short-term gains, but have resulted in immeasurable loss of the nation's natural capital.

Money grows on trees?

From a macroeconomic perspective, the primary reason for the over-exploitation of the forest lies in the Solomons' cash economy and the timber industry's contribution to it, as well as Asian demand. For the decade prior to 1990, the economy was dominated by export-oriented production, involving tree-crops of copra, cacao, and oil-palm, commercial fishing, and logging. These contributed more than 93 per cent to export earnings. The manufacturing sector was tiny, at no more than four per cent of GDP. Government services consisted of non-export items, so contributed little to export earnings. A large public service sector accounted for over a third of all full-time employment. Yet, the main source of livelihood for 80–90 per cent of the population was subsistence agriculture, generating little or no surplus. So there was a fairly narrow base of economic activity to provide domestic savings and investment, with less than 20 per cent of the population in the formal sector. Thus, 'economic performance [was] strongly influenced by changes in the external terms of trade and the availability of foreign exchange'.[1] This was demonstrated by two exogenous recessions in 1982/3 and 1986/7 that decreased earnings from primary industries and led the government to put greater demands on resources to obtain foreign exchange. Smallholders producing copra and cacao suffered considerably as commodity prices fell from the mid-1980s and the price of imports rose. In the 1970s, agricultural exports made up more than 40 per cent of the total value of exports; in the 1980s, under 30 per cent, falling to about 20 per cent in the mid 1990s.[2]

Although the country is relatively well-endowed with natural resources, it has failed to achieve a consistent annual growth rate higher than its 3.5 per cent annual increase in population. GDP grew only by about 2.8 per cent per annum during the 1980s and around 3.5 per cent in the early 1990s and was, in cash, over SI$200 million in 1990. Higher growth of 5.2 per cent recorded for 1994 and seven per cent in 1995 was partly attributable to a construction boom, financed mainly by bank credit and aid. GDP fell by an estimated three to four per cent in 1996.[3]

In 1986, cyclone Namu severely affected parts of Malaita and north Guadalcanal. Relief aid counteracted the potential dramatic rise in the deficit, but the government was not able to control current expenditure after aid receded to the previous level of over 20 per cent of the country's budget. The deficit in 1990 was 20 per cent of GDP. By then, the pattern was of increasingly large fiscal deficit, rising current expenditure and declining public investment. With an increase in external debt, more export earnings were needed to service it. Most loans had low interest rates, so although the external debt in 1995 was SI$340 million, the servicing was only $21 million. The imported capital was largely consumed rather than invested, thus devaluing the currency and fuelling inflation. Domestic public debt also escalated and with it, interest: from 1979 to 1988 debt reached $76 million; from 1989 to 1991 it increased to $97 million. By 1996, it was $353 million. Servicing this cost the government 20 per cent of expenditure, $83 million.[4] Around 30 per cent of GDP serviced debt in the early 1990s, increasing to 36 per cent in 1996.[5]

The economy's fragility was reflected in its dollar. Between 1984 and 1990, it depreciated with respect to a basket of currencies (Australia, United States, United Kingdom and Japan) by 50 per cent and it has continued to fall to 1996. The country's rate of inflation in the late 1980s was 10–12 per cent and by the 1990s, real inflation was probably around 12–15 per cent, significantly higher than most of its trading partners and the main cause of falling real incomes. External reserves dropped steadily throughout the late 1980s and early 1990s to a critical level, providing only one month of import cover in 1994. Although increased by 1996 to two months' cover, this was mainly because the government simply was not servicing its external debt.[6]

Rural dwellers have been the more disadvantaged in the declining economic situation. About 75 per cent were involved in some form of commercial agriculture as well as subsistence.[7] As Ian Frazer has shown, this has occurred at a time when the public service sector expanded, boosted in part by the establishment of the provincial tier in 1981. Between 1977 and 1992 private sector jobs increased by 49 per cent compared to 82 per cent for government jobs. The private sector in 1992 employed 62 per cent of the work-force of 26,842, and the public service the remaining 38 per cent.[8] Public service numbers, however, decreased by 1996, as did those in the wage sector.[9] Public service unions have won regular increased wage awards from the mid-1980s on, particularly in 1989, 1991, 1995 and late 1996, forging well ahead of private sector wage scales. Frazer points out that from the early 1980s to 1990s, the average local buying price for a tonne of copra fell by 18 per cent. As the average amount produced yearly per household was about 0.8 tonnes, depending on location, a household would have earned, when copra rose again in 1992 to SI$500 a tonne, about $400. Of course, the rural cultivator also would have produced commodities for domestic consumption. In terms of cash-in-hand, however, the rural dwellers have less to spend compared with a novice on the lowest rung of the public service receiving $4,000 yearly.

Holders of jobs at higher levels were rewarded far more – contract officers, consultants, holders of constitutional posts, parliamentarians, earned in the range of SI$787–5,361 a month. Often these people had access to other earnings – as directors and board members on statutory bodies, state enterprises, and private companies. They had free or subsidised housing, the use of government vehicles, overseas travel and so on. A government survey in 1990 reflected the emerging economic and social stratification in Honiara at least: among indigenous households, 48 per cent were earning less than SI$750 a month, 85 per cent earned less then $1.500 a month. One third earned 75 per cent of total household income, with the top 0.3 per cent earning 34.6 per cent. Frazer shows that these highly paid civil servants and politicians invested significantly in property, agriculture, transport and other businesses. They had the most to lose if the government were to retrench public service, constitutional and political office numbers, or reduce salaries, so they were committed to the status quo.[10] As much as any rural landholder, they were part of a system that relied heavily on over-exploitation of timber as well as fisheries resources.

Timber was and remains a major contributor to the economy. Throughout the decade 1980–1990, it averaged almost 27 per cent of export earnings, growing in 1991 to 31.7 per cent, the largest source of foreign exchange. Since 1992, it has constituted over half of the country's export earnings. From 1985 to 1991, log exports produced four to eight per cent of actual government revenue. More significant, for the period 1993–1995, this has risen by an annual average of 23.5 per cent, and was up by 50 per cent in 1996, even with tax exemptions for 'local' companies. Timber's contribution to GDP in 1990/1 was around eight per cent, the same as fisheries. In 1990 the government obtained from the industry, through its combined reforestation levy and export duty, 25 per cent of the f.o.b. value of timber exports, higher than any scaled on other exports, and this rose to 32 per cent in 1994 (see Table 4). However, from early 1995, the effective rate of log taxation has fallen, because of exemptions for local companies. That year, $34 million of log taxation was lost to government in this way, rising to an estimated $53 million in 1996. Whether measured in 1996 as a gain for

	Duty rate % FOB			
	< SI$175 per cu.m	> SI$175 per cu.m	< SI$250 per cu.m	> SI$250 per cu.m
1988	17.5	20.0		
1989 (a)	20.0	22.5		
1990 (b)	22.5	25.0		
1991	22.5	25.0		
1992 (c)	25.0	30.0		
1993	25.0	30.0		
1994 (d)	27.0	32.0		
1994 (e)			35.0	65.0
1994 (f)			32.0	35.0
1995 (g)			35.0	38.0
1996			35.0	38.0

Table 4. Rates of duty on round logs 1988–1994.

(a) From February 1989
(b) From July 1990
(c) From February 1992
(d) From January 1994
(e) From July 1994
(f) From November 1994
(g) From June 1995
Sources: Price Waterhouse, *Forestry Taxation and Domestic Processing Study*, 12; CBAR, 1995, 1996.

Profits and Loss, c. 1982–1996

government, contributing around $65 million or 50 per cent of its revenue, or as a loss of $53 million in exemptions, unsustainable logging of timber underwrites much of the economy.[11]

The subsistence economy

Utilisation of the forest resource was not only a question of logging sustainably or unsustainably. Logging affected the subsistence economy that still largely supports most Solomon Islanders. Orthodox economists, focused on the industrial society model, have given little consideration to the valuation of subsistence activity. More recently, some have come up with estimates of the value to GDP of the 'non-monetary' sector: 'the production of households in the form of food, housing, canoes, and land improvements'.[12] From 1989 to 1991, this sector was estimated at $41.3 million to $44.3 million of a real GDP in the range $229.8 million to $ 243.0 million – about 18 per cent of GDP.[13]

Before World War Two, administrators realised the importance of the subsistence economy, and monitored various factors, such as birth rates, related to the people's ability to sustain themselves from their forests, lands, and seas. The subsistence economy subsidised the plantation economy and taxation and underwrote the colonial state and its peace.[14] Agricultural officers, though they spoke deprecatingly of the 'primitive' character of the 'bush fallow' method, admitted its value in preventing soil erosion.[15] Some foresters recognised the forest's pivotal role:

> It is a big departure from usual practice to regard 'bush fallow' as one of the forest resources. Shifting cultivation has claim to exceptional regard in the Protectorate because of its unusually widespread incidence and firmly established position in native life ... Agriculturally the practice is effective ... To exclude so very real a value of forest vegetation from a list of forest products may leave the escapist impression that the practice deserves little but derogatory consideration, except in so far as it may be used for forestry purposes ...[16]

This approach was sidelined with the establishment of the Forestry Department in 1952. In the early 1970s, Trenaman saw soil conservation as resting 'largely, perhaps mainly, with the Agricultural service'.[17] By 1978, this compartmentalisation seemed less rigid as the Division was co-operating with the agriculture department in tree replanting on north Malaita and on the CUT project on Kolombangara. Some in the Forestry Division sounded a warning:

> By the year 2000 the population is estimated to be 500,000 people ... who will need 1 million hectares of land ... for shifting cultivation. If steep mountains are excluded probably between 0.75 and 1.00 million hectares of land will be left for

all cash cropping, cattle plus food production for all the extra people in future generations. Even now customary land tenure means some small groups of people have plenty of land; others ... do not have enough land for development ... in the future ... land will also be urgently needed for production of post, poles and firewood for local use.[18]

However, once logging began on customary land, the Division was so overstretched that its purview extended little beyond commercial concerns. Though the UN Malaitan survey in 1982 had recommended tree planting as part of the bush fallow cycle, the Division's endorsement of plans to increase village wood supplies was mainly in order to capture a proposed New Zealand aid package.[19] The Ministry of Agriculture and Lands (MAL), rather than Forestry Division, worked on systems of agro-forestry that would meet the need to strengthen both the bush-fallow system and cash-cropping with an economic tree cover.[20] When MAL suggested in 1989 an assessment of the economic value of non-timber forest products, the Division's response was that the government's pressing need for foreign earnings could not be compromised by such long-term projects.[21]

Although reduced to platitudes by subsequent Mamaloni governments, the *Forest Policy Statement* of 1989 at least asserted the forest's value to the bush-fallow system and that policy should safeguard the 'large, unmeasured benefits to the people': 'providing the population with essential food, water, fuel, building materials, traditional herbal medicines, and recreation from the forests'.[22]

Unmeasured benefits

In 1991/2, a forester named Ross Cassells tried to measure these on Choiseul by a form of cost-benefit analysis that attempted to incorporate wider social and environmental effects.[23] Cassells concentrated on two comparisons: first, logging royalties versus the loss in subsistence production for a single year, and secondly, cumulative royalties versus the long-term value of sustainable subsistence production.

Using the 1986 census data, Cassells calculated the annual value of the forest produce to a seven-person household at SI$10,512.15. Kuku village's losses resulting from logging of 41 hectares were the destruction of garden land, nut trees, betel nut palms, sago palm and trees for food, firewood, housing, canoes, miscellaneous produce and medicine on common land, amounting to $176,613.13. This was far in excess of the total royalties due from Eagon of $18,162.18 for logging the same area. The uncompensated loss in subsistence production was thus $158,450.95 (less $2,164.50 which Eagon later paid for loss of food trees and betel palms). For some households, the situation was even more unfair as losses were not evenly distributed.

When Cassells did his study, not all the Kuku-owned land had been logged. Future royalties on the remaining 535 hectares would have amounted to $236,994, which, added to the $18,162.18 to give a total of $255,156.18, would have completely offset the loss of $176,613. However, the royalty would be a single payment per hectare logged, while the villagers' loss of subsistence would continue over many years – and logging of the remaining 535 hectares would add further to this loss.

Cassells next compared the long-term benefits of sustainable household subsistence production with those the household might gain from logging. Using discounting procedures, the overall value per household for forest subsistence production was $95,565, far more than the estimated total of $12,150.31 per household the people could expect from logging. This simple comparison assumes that logging would totally destroy the Kuku people's ability to practise any kind of subsistence production on the land, whereas in fact they could still plant gardens and useful trees in the logged areas.[24] But when the depreciation in subsistence value of the degraded land was taken into account in simple, conservative terms, the ratio of losses to gains per household in cash terms was still of the order of 7 to 1.

Supporters of tax exemptions for local logging companies could argue that landholders involved get a far greater return than that provided by Eagon, so that the comparative loss is less; but analysis based on the concept of economic rent indicates that the contracted foreign logging companies capture between 53 per cent and 77 per cent of the foregone tax revenue and the landholders get a mere 23 per cent to 47 per cent.[25] The loss of forest resource still remains significant if Cassell's calculations are applied.

Cassells' price measure is only indicative of the many values inherent in the forest which are too diffuse, long-term, or uncertain to be fully measured in market terms. There are values, for example, attached to conservation uses. Pivotal ecological functions of the forest are watershed protection, nutrient cycling (particularly for the soils), carbon storage, and microclimatic regulation. Recreation values exist for rural Solomon Islanders because the forest provides much for their way of life. Foreigners place a value on recreation through tourism, and several NGOs and even the government in 1995 presented this as a commercial option for forest use.[26] Other option values of the use of forests relate to their biodiversity. There is also a value 'derived from the desire of people to pay for the very existence of these resources'.[27] If a forest and its ecology were reserved exclusively for traditional use, this would involve compensating the landholders and thus attaching a direct economic value to such uses.[28] These conservation and largely non-market uses of the forest are not easily expressed in monetary terms, so no government, logging company, or timber merchant factors these values into their costing when trees are logged.[29]

Cassells' research, however, highlights two things that can have a monetary value attached. First, a seven-person rural household produces goods from the forest worth about $10,512 yearly and this sustains them. If households are calculated at a conservative seven each, then there are 50,000, less ten per cent which do not rely on the subsistence economy, making 45,000.[30] Thus non-monetary sector production is about $473 million, at least ten times the loading given by the Central Bank for 1989–1991. If these calculations are accurate, economists have underestimated the value of the non-monetary sector and thus the potential cost, likely to be exponential, of its loss, which pivots on forested land. Second, logging does not pay the resource holders commensurate returns. This includes no costing for social disharmony induced by company logging. Some things are beyond price.

Compounding this loss is the annual population growth of 3.5 per cent. Demographers maintain that with increased per capita income, population begins to level out. 'Demographic transition' though, seems more a characteristic of urbanised industrialised societies and it is a pattern, not a law;[31] it could just as easily be a 'demographic trap'.[32] Moreover, increased per capita income almost certainly will mean increased demand on the environment. Population increase will not change significantly until well after the year 2000 and by then most of the commercial forest on customary land will have been logged. More people need land for gardens, cash crops, with trees for forest products and perhaps cash. A recent calculation sets the amount of land needed for gardens per household of 6.7 persons at 0.25 hectares yearly. The average active life of food gardens is 1.35 years. If all used garden land were allowed a full cycle of 30 years to return to secondary forest then over this period a household would use 7.5 hectares averaged over the 1.35 years of the life span of the garden; that is, 5.5 hectares. It has been calculated that for an estimated increased population of 53,700 rural households in 1996, the amount of land needed would thus be almost 300,000 hectares or ten per cent of the country's land. All else being equal, this is the limit sustainable for shifting cultivation as the remaining 90 per cent of arable land is in some part of the bush fallow cycle (see Chapter 2).[33]

It seems that the population in the early nineteenth century was no greater than 150,000–200,000, well within the number sustainable by shifting cultivation when the *only* demand on the forests was subsistence.[34] Even with the European alienation of five per cent of lowlands, not all was lost to subsistence as only about half was planted before World War Two.[35] The current population of 367,400 is making greater demands on the forest than even a comparable number, let alone only 200,000, would have done

100 years ago. Moreover, the opportunities of the modern world – health care, education, as well as imported technology, clothing, and foodstuffs – must be bought with cash. In 1986, about 27,000 hectares were being used for smallholder cash-cropping, excluding significant areas of commercial company plantations. Smallholder areas are likely to remain fairly stable because of their mainly coastal location. However, with increasing population, garden fallows will decrease in length and the land will eventually degrade, unless it is otherwise fertilised, or the people will have to cultivate more primary forest areas, particularly inland. At present, between 1,200 and 1,800 hectares of primary forest are currently being cleared yearly for food gardens.[36]

In the Solomons, there were over 367,400 people in 1996, almost 13 people per square kilometre of all land, compared to half that in the early 1970s and a conservative four people per square kilometre before World War Two. Deducting cleared and/or degraded land, amounting to 608.7 square kilometres in 1993, there were 14 people per square kilometre. This is a very crude measure, as many areas, because of altitude, slope, isolation and even soils, are not suitable for gardening. Moreover, access to good gardening areas is not evenly distributed across the provinces.[37]

Thus from subsistence needs, cash cropping and logging, the primary forest disappears at an alarming rate. Its ability to sustain the bush fallow system is near breaking point in several areas. No longer the bush fallow of 7–20 years of 20–50 years ago; now it is closer to 3–4 years and in some places less than six months. These trends began to emerge in the 1970s, becoming evident between 1980 and 1990.[38] Little has been done to limit population growth except as by-product of the AIDS pandemic.[39] Even if widely practised, declining human fertility is unlikely to relieve pressure on the forests for at least 40 years.[40] The major causes of deforestation – gardening, cash cropping and forestry – have not been systematically addressed.

There is another issue which could be significant. Cash cropping means individuals are often asserting long-term claims to land, thus reducing the amount that can be returned to the clan 'commons'. If such individuals can register the land, clan access will be undermined. Land could be sold among Solomon Islanders, thereby creating the basis for a landed, possible *rentier* or landlord class and its obverse, a landless proletariat or tenantry. In towns, Solomon Islanders have bought former government land, so there is a precedent. A growing 'middle' class implies an underclass: beggars appeared in Honiara's streets in 1992.[41]

If the forest is left alone, even in its disturbed state it will eventually recover, but time is the issue. Estimates vary from 50 to 100 years.[42] Where soils are good and access easy, available land is likely to be converted to

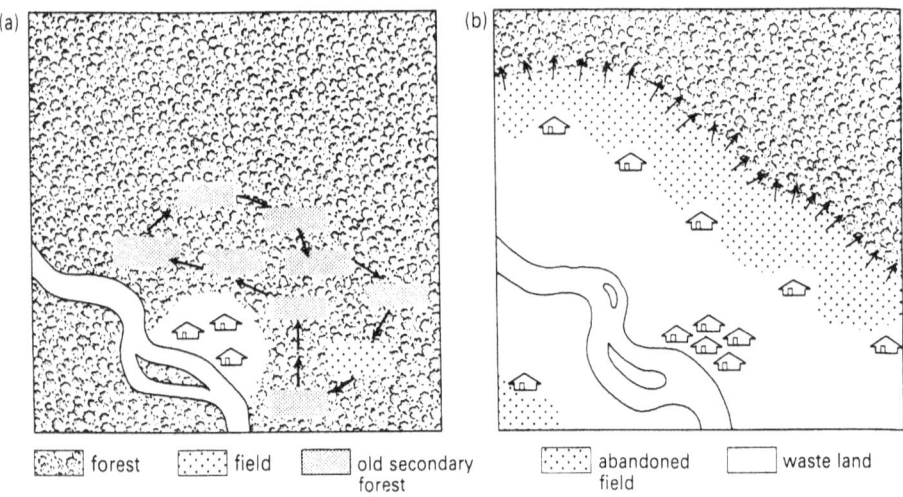

Figure 93. Sustainable agriculture compared to unsustainable agriculture. Until recent times, (a) was the common pattern throughout most of Solomon Islands. The second pattern could be created by the combination of clear-felling for logging, cash cropping, and subsistence agriculture.

cash cropping. As population increases, some parts of the forest will remain ecologically changed as gardens follow a shortened rotation. If a second cycle of logging gets underway before, say 20 years, the forest probably will undergo progressive species simplification. Even the 'new' forest of 50–100 years will be different from the old in structure and composition. The lifetime of a forest is of a different order to that of its human inhabitants. Human needs almost always demand satisfaction within a human time frame. Manifested within a Melanesian cultural and societal context, these needs have left resource-holders susceptible to the suasions of the loggers.

Resource-holders' susceptibilities: land tenure

Many 'resource owners', to use the suggestive 1980s' terminology, wanted commercial logging; some did not. What created situations where those wanting logging almost always prevailed was the nature of land tenure, designed to accommodate subsistence activities. In pre-capitalist times, although land wars were rare, disputes over land were common.[43] Particularly on populous islands, such as Malaita, 'land disputes have always been part of the dominant discourse'.[44] However, there were mechanisms for

Profits and Loss, c. 1982–1996

resolving them, usually based on the knowledge of elders whose decisions were accepted by consensus, by *force majeure* or, failing all, by emigration.

Societies recognised a range of use-rights to land and sea resources of clan members and affines. The apical founding ancestor usually had obtained land by being the first to clear it; in time, the lineage grew and sub-lineages developed, all related through descent. The territory claimed expanded as numbers increased leading to the situation as shown in Figure 5. The sub-lineages A, B, and C all have use rights for hunting and foraging in area ABC of primary forest, unused for gardens within living memory. Each can do the same in their respective sub-lineage land, which has been used within memory as garden land. Unless mutually agreed, members of group A, for example, would not be allowed to garden in areas B and C, but in many societies, sub-lineage members usually could hunt and forage in the lands of related sub-lineages without obtaining permission or paying compensation. In some societies, one sub-lineage, say, A could have genealogical seniority and could invoke this as a rationale for decisions where inter-lineage conflict arises.[45] Lineage members who have moved away from their land still have claims to use it, but those residing on the land generally have stronger claims. Relatives of the first settlers, those who did not seek a new home, also have some rights, but these are subordinate to those of the first settlers' descendants.[46] Tenure of land depends on the use context and the social identity of the potential users. Although this model focuses on land-based resources, similar principles apply to those societies with marine resources.[47]

Since European contact, the situation has become more complex. Solomon Islanders began to produce copra and shell that they exchanged for trade goods or, later, cash. Particularly under the stimulus of the head tax of 1921/2, individuals began to plant coconuts to sell, though this was not a dominant trend before World War Two because existing groves supplied a surplus. After the war, especially from the 1950s, the government encouraged cash cropping. In the 1960s this had various subsidies attached which attracted potential growers. Cattle projects were the vogue in the 1970s.[48] These were both signs of, and means to 'development'. Most took up these ventures to improve their standard of living. Like the pre-war Europeans, they too believed that transformation of the landscape could make it more productive.

Along with intensification of demand for accessible coastal land came population increase, a function of peace and basic health care. Individuals or small groups took garden and other subsistence-use forest out of the lineage 'commons'. Tree crops like cacao and coconuts removed land from circulation for several decades. The trend of relocation of inland peoples, fostered by government and missions from the late nineteenth century on, put coastal land under pressure, just as it did the concept of permissive residency, granted to relatives. Clans were angered to find lands they had offered relatives or co-religionists for subsistence turned to capitalist use. All this meant increased land disputes. However, villagers no longer had to come to terms directly with their neighbours regarding land claims. The courts and introduced legal system of a foreign culture became the resort of the discontented.[49]

Selectively and lightly utilised for subsistence, timber trees afforded a new resource, usually beyond the prized coastal lowlands. However, like land alienation before 1912, the shift from subsistence to sale stressed tenure systems.[50] Reverting to the diagram, loggers would negotiate with some members of sub-lineage A; but their forest was not particularly viable by itself, so some would pressure relatives in B and C for their sub-lineage land, as well as the common primary forest land. Perhaps some would agree, particularly if A had genealogical seniority. Often A would manufacture so-called rights to B and C, based on the common land holding of 'primary' forest and sign away the timber rights. The only legitimate agreement would be one in which all the adult members of groups A, B and C agreed. This would take long, careful negotiating, because of the diverse use and property rights involved, as well as the absence of some members in paid employment.

Resource-holders' susceptibilities: conflicting perspectives

Yet sometimes these absent, urban-based clan members have been the point of contact for the logging company and they have persuaded their elders, although there have also been cases of elders wanting material benefits, like imported foodstuffs, before they die, while younger people in their late twenties and thirties were less willing to sell off resources needed to raise families. In more recent years, urban-based groups often have opposed logging on their lineage land: some, because of the environmental and social consequences; others, because they see loggers as destroyers of the old way of life. Their lives do not depend intimately on the resources of their lands and they are able to see these forests in a broader context. For some of these people, their motive is environmental concern. For others, it is more nostalgia. The rural idyll is valued by those who come home once a year to visit and celebrate or to retire on superannuation and the reciprocity due for having assisted 'country cousins' over the years. These urbanites are not the householders who trudge up muddy hillsides to gardens after days of rain and who have to beg a neighbour for a light if their cooking fires die. These are not the parents who grieve the departure of young children as they go far away to school, year after year, or who agonise as they try to find

Profits and Loss, c. 1982–1996

a way to transport a malaria-racked child across 120 kilometres of open sea to a hospital. So rural people often ignore the arguments of their urban kin, perceiving them as having, through employment, access to cash, to shops and their comparatively cheaper imported goods, schools and health facilities. Scarce in many isolated areas, these commodities became more remote in an economic sense when world prices fell in the 1980s and early 1990s for the staple cash crops of cacao and copra, just as government services declined in the provinces. To such people with no alternatives, loggers' royalties, wages, and promised infrastructure were virtually irresistible.

Within this gender plays a role. Tree-felling traditionally is men's work; colonial foresters and loggers and their successors were men; so now logging is men's business, even though logging has an enormous impact on women and children and even though most rural women probably have a greater knowledge of the garden, forest and foreshore ecology than most men. The impact of logging on the ecology, particularly in relation to pollution of water supplies and inshore waters and lagoons, has added further to women's heavy work load. Good land is appropriated by men for cash crops, forcing women to locate gardens at increasing distances from their village, as has happened on Malaita. Even in matrilineal societies women's concerns are not always considered with the same care as men's, but as ever, strong women do have a significant voice, though it is not often a public one outside the descent group. Between and within landholding groups there is no neat pattern of age, residence or education characterising those who welcome the loggers and those who do not, but women remain more sceptical.[51]

Resource-holders' susceptibilities: loggers' stratagems

Particularly since 1990, as people have been educated though NGOs and churches, those opposed to logging have initiated legal action when they feel that their rights as 'owners' have been violated. Disputes based on trespass or conversion, improper granting of licences, royalty entitlement between clans, and breaches of the SLA are the most common.[52] If these actions reached court, the judge has usually found for the plaintiffs. Thus far, the Solomons' judiciary appears to be beyond the reach of those who would subvert by bribes.

However, many cases never get to court. The most common reason for this comes back to tenure. A purported landholder, say from group B, might claim damages in trespass because a company has logged his clan land without his consent. An interim injunction is sought and usually granted, particularly if the plaintiff has the support of the Public Solicitor or an expensive private lawyer. However, the companies play a delaying game that means long and expensive trips by complainants to appear at court. The cost incurred is too much for some and the action is dropped. Another ploy is to intercept the would-be plaintiff, representing his family or clan, before the application for the injunction is heard and spirit him away to Honiara, ply him with 'gifts'; perhaps even get him to agree to accept the company's lawyer to represent him in signing a consent order to discontinue proceedings. Sometimes, the pay-off is a shared royalty agreement with another compliant claimant, say, someone from group A, whose right to the common land is dubious anyway. Family and clan members do not know what to do next and often give up to preserve family harmony in the face of the company's massive power. And this is all the more so when few have much understanding of legal procedures and the law. Moreover, the Public Solicitor's Office, which provides some legal aid, is under-funded by a government that has no incentive to change this, as loggers might suffer and revenue and other 'benefits' fall.[53]

This applies to the land of one descent group, but loggers usually want extensive areas owned by several clans.[54] The opportunities for mistake, coercion, and fraud expand accordingly. Among timber-rights holders, individual and collective greed along with fear that another clan might profit by the promised development, are as much motivating factors as the desire for royalties, employment, services and clearance of land for cash-cropping. It is often clan leaders or village 'big-men' who precipitate such feelings.

Resource-holders' susceptibilities: big-men and societal values

The nature of leadership still sees the 'big-men' or *sifs* (chiefs), as major players. Constrained by Western law and rivals created by new opportunities, they cannot bind their communities as did the strongest of their pre-Protectorate predecessors. Their local power commonly hinges on their appearing to be able to deal with outsiders as much as it does on patronage: attracting followers, mobilising community resources and support, and dispensing community and individual satisfaction. Knowledgeable about their own community, their position as negotiators with outsiders enables them to gain access to benefits that enhance their role as patron and dispenser of good things. Thus they are pivotal players in the nexus between loggers and politicians, on the one hand, and resource-holders on the other, and are instrumental in any company's strategy to win logging rights. However, few big-men have the knowledge to weigh up the strength of a foreign logging company and the legal and financial resources at its disposal. Usually, such leaders end up as clients of the loggers, as they do not willingly admit their limitations and are prey to their own ignorance or to the suasions of kinsmen from town. These big-men are the recipients of gifts and promises of development; if they can make the promises reality then their position is assured – the more susceptible and vain even go as far as demonstrating their 'big-manship' to foreigners by agreeing to assign rights that are not theirs to dispense. Within this nexus of vulnerability too lies the attraction of 'local' logging companies fronting for overseas contractors – a 'big-man', say a director of a company with tax exemption, can earn commissions, royalties and 'bonuses' and then dispense this largesse to the local community. This comes nowhere near reflecting the economic rent, let alone the non-monetary value of the forest, but it is more socially satisfying and immediate than services that never materialise from a faltering, distant government.

Big-men, politicians, government personnel and even the resource holders who grew up to adulthood in the colonial era often resented the restraints the foreign government imposed; many believed that independence, almost magically, would somehow removed these, that that which had been the government's would become the people's. Throwing off colonial power and paternalism often has meant throwing off caution and asserting control by rejecting what appeared to be arbitrary rules and conventions. 'It is ours, we can do what we like with it' was a common attitude; but careless management has the same consequences to a state and a forest, no matter the colour, nationality, or race of the manager. In asserting an independent identity often the price has been profligacy and neglect of the future.

There are other values and motives operating. Walk behind a Solomon Islander on a path through the forest; he swings his bush knife or machete to clear away the vegetation. The trees and undergrowth seem endless, are always on the move, and have to be controlled. Watch a group of young men chatting to each other resting by the bush path or under the banyans in Honiara. The machetes chip away at the tree bark; down go small branches, even entire saplings. Adolescent energy doodling on treescape; the graffiti of the unlettered. Look around the town of Honiara: young and old shelter gratefully under the trees out of the glare; but they swarm even more gleefully for firewood from trees felled without warning to make way for urban 'development'. Blessed with a beautiful sea front and rising foothills, Honiara is the ugliest capital in Melanesia,[55] not only because of uneven footpaths and litter-filled drains, but also because its remaining trees, survivors of an unperturbed colonial township, are scarred isolates, no longer able to filter dust, noise and sunlight. People fell trees inside the boundaries of the botanic garden near the prison, itself a reforested area

From *Solomon Star* by permission of John Lamani

Figure 94. Three trees felled within the boundary of the Honiara botanical gardens and water catchment. Myknee Sirokolo and assistant Patty Dofu survey the damage, September 1994.

after most of the original forest was cut for firewood in the 1950s. Consider too the logo of the Solomon Islands Development Trust: a machete and axe stand in heraldic affirmation of what signifies and brings forth development.

To Solomon Islanders, trees have, at best, a neutral value unless turned to socially-meaningful uses. The forest still represents a challenge to many. The forested land, once cleared, might bring forth crops and spawn schools, clinics and towns. It might empower its inhabitants. As of old, it evokes ambivalent emotions: on the one hand, there is the comforting feeling it will always be there or will rapidly regrow, just as it has after cyclones or gardening; on the other, there is control asserted in cutting down the *dak bus*, dark bush. This act of opening the forest is especially appealing to young men who flash chain-saws with bravado and drive the bulldozers up the hills through the forest undergrowth. Like the warrior chiefs of old they are leaving a mark on the land and doing a man's job; the 'virgin' forest is no more. Over half the population is below 20 years of age. Like most of their parents, the young rarely have the same respect for the forest as their pre-Christian ancestors; they may recognise old sacred sites, but no longer clear them for use as sacrificial altars. For some at this stage of their life, it is not only the ancestors who are dead, but their spirits as well, and all the more so as logging confuses familiar vegetative and even physical topography, the markers of boundaries and pivots of clan history. For these, the forest is simply dollars, not life. It is the present, not the past and future. However, the spirits have ways of reminding their descendants and others of their enduring presence.

Old beliefs have been modified and the new want a Melanesian conceptualisation. In using the forest for gardens, most older Melanesians and their ancestors planned for a few years, perhaps a decade. The soil regained its fertility without human intervention, except through invocations to the ancestors. In Christian times, Divine Providence took a similar role. The world view this encourages can be very short-sighted in relation to forest conservation, where a tree's life cycle may be longer than a human being's. Christianity is so recent that it has only just begun to address the question of human responsibility for, and relationship to the environment: concepts of dominion and stewardship for it have yet to be reconciled. Although Christianity and its parent Judaism are world religions, they originated in semi-desert places. They were not religions of tropical islands, but of continents. Christianity was expansionist and had few strictures on resource use, encouraging population growth through sanctions on abortion, infanticide and often on artificial means of birth control. A Christian theology of the Melanesian environment awaits precise articulation, though Bishops Boseto and Pogo have opened the debate on the degradation of Creation.

Resource-holders' susceptibilities: an unlimited development resource

Solomon Islanders' experience of large-scale clearing of their forests has been extremely recent. Although indigenous gardening practices had changed the structure and composition of the forests, this was over hundreds of years and imperceptible to any one generation. The small clearings perhaps became a little bigger with metal tools, but then population declined in the late nineteenth and early twentieth centuries, increasing the amount of land under secondary forests, just as the few Europeans began seeking plantations along the coasts. Before World War Two, the European planters clear-felled, but they were quick to plant coconuts; and colonial alteration of the natural environment, though rapid, was of the littoral and scattered. Until chain-saws became common in the late 1970s and 1980s, the clearing of customary land was driven by subsistence plus some cash cropping. Except where population was great, most patches of clearing soon reverted to bush. Prior to a few years before independence, large-scale clearing was almost always the result of cyclones. For all their initial ferocity, they do little permanent damage because they do not damage soils and leave remnant forest to spring to life. So these were the dominant models of forest clearance impressed upon the minds of rural people for generations. The forest heals, it will always be there and there is plenty of it: an illusion of an infinite resource.

Thus, most could not imagine the damage heavy machinery and clear-felling could inflict. Some old people must have witnessed this during the war, but, it too was localised and mainly coastal. Anyway, it passed quickly; the forest grew for another twenty or thirty years before the loggers came in numbers. Graham Baines believes that even when rural people see what has happened on someone else's land they cannot imagine it happening on theirs, perhaps because each sees his/her own land as unique instead as part of a greater system.[56] Paradoxically, except for an isolated island, the part is sometimes perceived as having greater resilience than it has because it exists against a backdrop of larger adjacent forest. No villager had attached comparative values to forest products like Cassells and few had the education to do so. There had been no existing indigenous economy focused on a timber trade, before or after the Europeans came.[57] The loss of natural and cultural resources afforded by the forest was not often realised until it happened.[58] Many rural dwellers know much about the ecological characteristics of their localised environment; what they did not know was how that environment would respond to a new kind of clearing. Their ancestors, as they moved into the Solomons, learned, through trial and error, how to manage it so it would sustain them, using a stone age technology.

Profits and Loss, c. 1982–1996

They learned it was not infinitely productive. Though occupying the same physical space, islanders since the 1960s have been in a seemingly infinite forested world of new opportunities revealed by modern technology and new markets. They are still learning the limits of what its soils, rivers and reefs can tolerate once the forest canopy is reduced or lost.

Trickery and greed apart, those who allowed logging on their land did it believing that it would bring them development and a place in the modern world. Both the colonial and independent governments had raised aspirations in the hope of stimulating enterprise. Both were full of plans and talk. In the eyes of the majority, they both failed, by and large, and in doing so left the rural people feeling deprived and thus open to those who seemed to offer more. The colonial government did not envisage the rate and extent of logging of the 1980s; by the time the damage was evident to the independent government in the early 1990s it seems too much had been underwritten by logging – political and public service careers, foreign capital, the formal economy, local self-respect, even national hubris – to slow the juggernaut. Most landholders arrived at their position more out of ignorance and lack of alternatives than cupidity, though there were many communities that had options and outcomes of logging clearly presented to them and still chose the royalties' road. Knowledge alone is no guarantee that individuals or groups will make decisions that will be ecologically and, in the long term, economically the wisest for the country.

The politicians and public service advisers suffered from another kind of ignorance. They and often their parents were products of formal education, often provided away from home and by outsiders, offering little instruction in the cycles of nature within Solomon Islands.[59] Along with deforestation of the land there has been a 'devegetation of the mind' regarding the environment, what it affords and how to use this sustainably.[60] The new learning may have provided an understanding of the wider world, but it has not generally produced a conservationist and environmental ethic; rather a desire to share the profits of resource 'development' to obtain what the capitalist world offers; a view little different from that of early European settlers and certainly mirroring that of Asian loggers and buyers.

Thus politicians, well-travelled, relatively well-educated in a Western sense, and certainly well-advised, were far more culpable when they facilitated the loggers. Many implemented policies that in the widest sense were economically inefficient, but were politically rational, because they served to keep their initiators in power. Their public servants, wittingly or not, were paid in part by logging revenue. To different degrees, the landholders, politicians and public servants alike have been so lacking in imagination that they have been unable to deny the profits of the present to prevent loss for the future. In logging areas, experience, education and exhortation are beginning to hit home – at least, to landholders. The extent of loss is being realised. For most, the hiatus between realisation and cessation will be too late, since there are massive differences in time scales between social and natural phenomena. Unless another resource can be 'mined' like the forest, or some hapless donor milked for the nation, repair and rehabilitation, rather than sustainability, will be perforce the dominant discourse in relation to forests for coming generations.

— 16 —

Contested State: Contest of States, 1993–1997

> If this government is a people's government put in by the people, then it should be listening to the people. But if the government is a loggers' government, then it should be listening to the loggers.
>
> Philip Pupuka, Greenpeace Solomon Islands, *Solomon Star*, 13 Jan. 1995.
>
> We are too small to change the conditions tied to foreign aid.
>
> Prime Minister Bart Ulufa'alu, *Solomon Star*, 29 Aug. 1997.

By the early 1990s, the collapse of the Solomons formal economy loomed. Its chief resources, timber and, to a lesser degree, fish, were being exported at an unsustainable rate.[1] The logging of the forests, vital to the weakening economy and to its main beneficiaries, became highly political. The future, not only of the forests, but also of the Solomons' state was contested at several levels. In 1993 and in 1997 lines were drawn between loose national political coalitions – one wanting to keep the old order and one wanting massive fiscal overhaul. Within the public service there were university-educated professionals particularly in finance, economics, taxation, and forestry, who contested the fiscal and forestry policies that politicians increasingly dictated without consultation. Working for change, NGOs, like the chorus in a Greek tragedy, warned of environmental cost. Churches, formerly silent, found a voice. Western donor countries pressed for sustainable logging and for structural reform, as did agencies such as the World Bank. For rural people, the contest was between a state that largely failed to deliver and political patrons and companies that might, and sometimes did.

Local and international pressures

For all its worth to the economy, there is little evidence that the government had used logging revenue wisely. In 1989, the International Monetary Fund (IMF) urged structural adjustment through restraint in expenditure, reduction of government involvement in the economy, and diversification in the private sector. Mamaloni's People's Alliance Party (PAP) espoused this in its 1989 platform, but achieved little once in government. The public service was cut by 17 per cent in 1992 and by a further seven per cent in 1993, but most of the posts cut had existed on paper only, so few real savings were achieved. Although the government sold its holdings in several major companies, the wage increases given to the public service in the late 1980s and early 1990s flowed on to the private sector, reducing competitiveness. In turn, this partly forced up interest rates, as did local borrowing. This reduced credit to the private sector, impeding its growth, and thus defeating the intentions of IMF's recommendations.[2]

The Central Bank, under its Governor, R. N. Houenipwela, was the most outspoken local critic of this fiscal recklessness. Mamaloni's Finance Minister, Christopher Abe, had suspended the Bank's board in January 1993.[3] Public dissatisfaction with aspects of the logging industry was but one of the many reasons Mamaloni lost power in mid-1993. However, the composition of the parliament did not alter dramatically, partly because of the tactical use by incumbents of special funds pre-election and the splitting of votes amid large numbers of candidates standing in several electorates.[4]

During this time, several high court injunctions had been sought by dissatisfied landholders, particularly in Choiseul and Western provinces, against companies for not following procedures according to law. Moreover, the Australian aid-funded Timber Control Unit (TCU) of Forestry Division had begun exposing abuses by logging companies. Under-pricing continued to be common through the mechanism of transfer-pricing, worth at least $15 million in 1993.[5] In April that year, Kalena Timber Company had to pay the government a total of SI$900,000 in understated duties and royalties. The TCU also had bases at Kirakira, Auki, Buala, Gizo and the Shortland Islands, and was becoming so effective that, 'some companies are now approaching politicians to get rid of the unit'.[6]

In the mid-1993 elections, Francis Billy Hilly became Prime Minister heading the National Coalition Party (NCP), but his one seat majority was tenuous (see Appendix 6). Following his defeat, the leader of the opposition, Mamaloni, was lobbying among Hilly's supporters to win them over – an easy task in a system where factions are fluid and party loyalties evanescent: all the more so with the NCP, an amalgam of seven parties. While Prime Minister from 1990 to 1993, Mamaloni had become the darling of the loggers, as he and 11 of his 15 cabinet ministers were connected to the industry.[7] Hilly, however, was to become their demon.

Since the annual licenced extraction rate was four times the 325,000 cubic metres considered to be sustainable, Hilly was determined to reform the timber industry.[8] In July 1993, the NCP Minister of Natural Resources, Ezekial Alebua, announced plans for increased local processing and the phasing-out of large-scale logging in favour of eco-forest logging.[9] Consultants of Solomon Islands Forest Industries Association (SIFIA) estimated there would be an annual monetary loss of SI$100 million in foreign exchange, $30 million in tax and levies, and $10 million in royalties to landowners along with the loss of 1,000 jobs as a result of government plans to phase out log exports by 1997. This could spell political suicide in a country where logging provided more than half the foreign exchange earnings and 20 per cent of the government's revenue.[10] By January 1995, loggers would have to process 50 per cent of their timber quota, instead of 20 per cent. This setting of target dates was nothing new, so loggers were not too worried.[11] Even a moratorium introduced to stop foreign companies getting new licences in September 1993 was of little immediate concern to the existing companies in the archipelago; it simply gave them more potential scope.[12]

Externally, the IMF, the World Bank and aid donors, such as Australia, had been pressuring the Solomons government to make wiser use of natural resources, particularly timber.[13] The World Bank, for example, in 1993 had set sustainable extraction at 300,000 cubic metres a year. By early 1994, the implication was clear: without urgent action, aid and soft loans could dry up. The Hilly government was on the move, reinstating the board of the Central Bank and planning a structural adjustment program prepared by IMF.[14]

Structural adjustment found international support; the Australians, for example, indicated they would assist with a fishing boat debt incurred by Mamaloni in the 1980s, that was costing the Solomons government around US$ 600,000 a year. As part of the overhaul, Hilly set up a Commission of Inquiry into corruption headed by Moses Pitakaka. Dubious land sales in Honiara were slated for investigation. 'Operation Scorpion' started to round up all the government vehicles that were being used for the personal convenience of leading public servants. Low public service rents came under scrutiny. All these planned changes did not win Hilly friends among those who had benefited under the old order.[15]

The Hilly government started to act on forestry. Interim data gathered by the Australian-funded Forest Resources Information System (FRIS) were presented to interested parties. In August 1993 in a seminar opened by Minister Alebua, the state of the forests was discussed. Information about the role of the FRIS and the Timber Control Unit was publicly disseminated.

The NCP government announced detailed policy late in March 1994. Timber control further intensified, with a new inspectorate station scheduled for Choiseul. The commercial unit within Forestry Division was set up in May 1994, as part of the TCU, to provide accurate data on log and processed timber pricing, marketing, and exports. More disturbing to the logging companies was a plan to re-negotiate their quotas to a sustainable level, depending on the results of the forest inventory, due in late 1994. Hilly's government in July raised the combined duty on logs to 35 per cent on the first SI$250 of the f.o.b. price per cubic metre and 65 per cent on any balance of the price above $250 per cubic metre because, so the Finance Minister, Andrew Nori argued, the average export price of Solomons timber had risen from $170 per cubic metre in 1991 to $400 in early 1994, giving the companies a gross operating margin of 25 per cent of the f.o.b. price after all costs, taxes and royalties. SIFIA registered their protests and claimed they could not absorb this increase.[16]

The duty increase resulted in the logging companies' stockpiling logs in August, decreasing revenue when the government was under pressure to increase public service wages. Amid defections of some of Hilly's ministers to Mamaloni and pressure from SIFIA, the coalition announced in October 1994 a reduction of the 65 per cent export duty component to 50 per cent on round logs. More radical was the allocation of 20 per cent of all government logging revenue, including the 'levy', for return to the landowners. The same month, the new minister, Tuhanuku, launched a public forum, preliminary to the development of a National Forestry Action Plan and revised forestry legislation, shelved by the Mamaloni government since 1990. In late October 1994, representatives of landowners, provincial governments, village councils, public servants and citizens attended; and the consensus was a total ban on new licences, reduction of the quota, 50 per cent duty on logs for the government of which 20 per cent should go to landowners and five per cent to provincial governments. A significant start was made on revising the legislation pertaining to forestry, including a logging code of practice, drawn up largely by the TCU.[17]

However, Hilly had shot himself in the foot. In order to encourage greater local processing and participation, his government had increased the number of local saw-milling licences, permitting the extraction of 1.2 million cubic metres of logs with an export quota of up to two shipments in order to raise capital, similar to the arrangement Abraham Eke had on Ndai.[18] However, men like Eke were rare and the new licences ended in most cases as fronts for foreign companies or logging contractors. As Hilly admitted, 'the foreign logging companies and our own local licence holders seem to have become extremely good bed-fellows with the one aim in mind

Contested State: Contest of States, 1993–1997

From *Solomon Star* by permission of John Lamani

Figure 95. Former Prime Minister Francis Billy Hilly, 1994.

to cut down trees for money …'.[19] For example, the Malaysian timber-logging consultants, Mega Corporation and a subsidiary Goodwill Industries, headed by Hii Yi Ging, handled the logging for two local licence holders, Kololeana (15,000 hectares) and Somma Ltd (45,000 hectares in Arosi, west Makira), the latter having Solomon Mamaloni and his son, Audie Murphy Mamaloni, as directors.[20] The Marovo lagoon people of Tamaneke logged an area of their barrier islands and the logs went out on Asian ships, further damaging the integrity of an area slated for World Heritage status.[21] Even SIFIA, representing some of the companies that wanted to stay in the Solomons, had supported the moratorium on increasing export of logs, favouring 'responsible restructuring' for sustainability.[22] Overall, the demands on the forest had increased rather than decreased. This resulted in bad publicity for Hilly's government through an Australian Broadcasting Corporation (ABC) television documentary in July.[23]

A foreign scapegoat

In late July–August 1994, Hilly attended the 25th annual meeting of the South Pacific Forum, hosted by the Australian government in Brisbane, where unsustainable logging was on the agenda.[24] From there, he played into his opponents' hands by announcing peremptorily the suspension of the licence of the Malaysian Kumpulan Emas Berhad's subsidiary, Silvania Products, for Vangunu in Marovo lagoon. Australia was portrayed as the instigator of this suspension, though it was the third in a series of such actions against the company.[25] Hilly agreed to an arrangement whereby the Australian government would give a 'debt for nature swap' worth about A$2 million (US$1.5 million) to preserve the Marovo area, in return for payment of some of Solomons' debt.[26] A task force funded by Australia would investigate Silvania's practices and facilitate the introduction of the independent Societé Générale de Surveillance to control the export and inspection of timber. Hilly extended the moratorium on new licences for foreign companies to local businesses (effective as of 29 July 1994), blocking the gap he had opened for foreign logging contractors.[27]

Given the timing of these measures during the Forum meetings, Opposition Leader Mamaloni had a field day. He portrayed Hilly as an Australian puppet, usurping the rights of local landholders. Mamaloni implied that Australian Prime Minister Keating's information about forest resource exploitation had been leaked via Australian aid personnel in the Forestry Division, where the data had been 'fabricated'. Keating had criticised Malaysian companies for the environmental damage they caused in Solomons.[28] The influential *Australian Financial Review* asked why Australia and New Zealand should continue to pour aid into Solomons and other Pacific Islands when these countries were allowing 'logging and fishing rip-offs' to go on unabated. The *Review* claimed that the Solomons government lost A$13.2 million to Malaysian loggers in 1993, enough to build 1,100 clinics.[29] Mamaloni told Keating to 'shut up and stop interfering with the domestic affairs of Solomon Islands', defended resource holders' rights, and fired a broadside at Australia's record of exploitation of the Aborigines' resources.[30] Such a stance invariably won Mamaloni political points in the perception battle among Solomon Islanders. Most saw Australia with its strong military capability, as a neo-colonial 'big brother' and indirect instigator of the Bougainville crisis that had generated diplomatic, financial, and moral problems for Solomon Islands.[31]

Arthur Unusu, the Premier of Western Province, which was preparing a court case against Silvania Products, gave the issue some perspective when he posed the nationalists' dilemma,

> While I agree with Solomon Mamaloni's views and comments, I also wonder whether we should do the same with these foreign logging companies. So the question is, are we to bow down to Australia for the sake of our environment but with very minimal help in the way of protecting our environment or should we bow down to the whims of these foreign logging companies?[32]

Hilly's reforms ended when his government fell in early November 1994, amid a constitutional crisis. Mamaloni, throughout late 1994, had been successful in attracting several of Hilly's supporters – including Dennis Lulei, Eric Seri, Francis Orodani, Alfred Maetia, Oliver Zapo, John Musuota, Edmund Andresen, Allan Paul, and the party-peripatetic Allan Qurusu – most of whom were subsequently rewarded with ministerial portfolios.[33] Tuhanuku's involvement in the Berjaya bribery allegations had sown doubts about the Hilly government, which were reinforced when Finance Minister Nori came under investigation for alleged improper financial dealings and was forced to resign. Parliament had not met since May 1994. High profile government projects had ground to a halt. As well, heavy government borrowing from local institutions without the consent of parliament provided grist to Mamaloni's mill, while the loggers had no reason to support the government of the NCP.[34]

On 7 November, the Mamaloni-led Solomon Island National Unity, Reconciliation and Progressive Party (SINURP) became the government. Simultaneously, a series of articles appeared in the Malaysian *New Straits Times*, deploring Hilly's policies and extolling the Malaysian loggers and Mamaloni. Praised too was 'logging consultant', George Luialamo of Waibona reputation, who subsequently was to become the Minister for Commerce, Trade and Employment in the new government.[35] Many in the opposition, along with Tuhanuku, blamed the logging lobby for Hilly's demise. In an interview in Australia he accused an ethnic Malaysian, Robert Goh, of being the go-between in paying off those who defected from the NCP.[36] Whatever the truth of these allegations, Hilly had made enemies and had also failed to reduce logging, with declared production for 1994 hitting a new high of 750,000 cubic metres.

The loggers' government

With Mamaloni back in power it was business as usual. Structural adjustment was sidelined. In 1994, recurrent expenditure was $170 million higher than 1990. The 1995 budget of $360.4 million was the most expansionary ever, with a deficit of $78.7 million. Yet government services declined and unbudgeted items absorbed expenditure. By mid-year, the Finance Minister directed the Central Bank to extend the legal borrowing ceiling from 30 per cent to 40 per cent of the average annual ordinary revenue. From July 1994 to June 1995, the government had borrowed so heavily by means of government bonds and treasury bills that the Central Bank announced it would no longer pay any interest on these, as the government had no funds. Throughout, the Bank advised reduced public spending, logging sustainably, and the abolition of tax remissions to 'local' logging companies.[37]

SINURP revoked the moratorium on local licences and reduced the duty on export round logs. From the modified rate in October of 35 per cent on the first $250 f.o.b. per cubic metre and 50 per cent on the balance, the government reduced it to 32 per cent and 35 per cent respectively. The 20 per cent of all duties to go to landowners introduced by Hilly returned to the old 7.5 per cent timber levy for landowners via approved development projects, as of January 1995. This immediately released the 'log jam' imposed by the companies and declared exports jumped from around 53,000–62,000 cubic metres per month during January–October 1994 to 82,000 in November, levelling out at 63,000 in December.[38]

A more startling, if not sinister, innovation in terms of governmental responsibility and post-logging consequences was a remission of five per cent of the export duty to companies which provided community projects and services.[39] Companies were now in the business of development, but development for whom and audited by whom? The government increased exemptions from various taxes, including export duties on whole logs, for local companies, several of which were fronts for Asian-controlled businesses or logging contractors. Some established companies also had exemptions (see Appendix 7). Several were closely connected to politicians. A major beneficiary was Somma, which profited by an estimated SI$4.3 million in exemptions during 1995. Somma's directors included Mamaloni. Thus although export duty (or tax) went up by three per cent to 35 per cent for logs up to SI$250 per cubic metre and 38 per cent for logs above that price in June 1995, log tax revenue fell. Moreover, for 1995, though there was an increase in whole log exports of 12 per cent over 1994, log export earnings fell from 51 per cent of all export earnings to 46 per cent, while the effective rate of log taxation fell from 30 per cent of all revenue to 23 per cent, reflecting the change in the export duty regime and the several exemptions (see Table 5). Income lost from exemptions alone amounted to SI$24 million in 1995. As the courageous Central Bank Governor, Houenipwela, put it:

> By the end of 1995 the effective rate of log taxation was little over two thirds of the announced rate. This is the same as saying that for every two ships that leave Solomon Islands paying the full rate of export taxes, another full shipment went tax free.[40]

The vigilance of the TCUs had dramatically reduced transfer pricing margins and pricing variations of species, producing increased foreign exchange receipts. The glaring discrepancies in prices companies achieved for their logs, however, suggests that forms of transfer-pricing still persisted. By an interesting coincidence, the declared prices of those companies with duty exemptions were considerably higher than those without. This applied also to companies, such as Somma, working in Makira where logs are smaller, fewer relative to area, and of worse quality than logs say from New Georgia. Finance Ministry officials estimate $50 million was lost in this way in 1995. The nation was getting less income, but was permitting the extraction of more trees.[41]

	1991 SI$	1992 SI$	1993 SI$	1994 SI$	1995 SI$
Value of round log exports*	51,093,000	104,020,000	221,564,000	249,471,000	266,143,124
Logs as a percentage of total export earnings	22%	34%	54%	51%	46%
Average price declared per cubic metre	169	189	375	401	366
Direct government revenue from logging duties	11,930,000	26,061,205	61,156,374	82,937,629	70,295,243
Logging duties as a percentage of total government revenue	8%	13%	28%	30%	23%
Percentage of sales value captured by logging duty	23%	25%	28%	33%	26%

Table 5. Log exports, key financial indicators.

* Round log export figures exclude plantation logs from KFPL
Source: *1995 Forestry Review*, Honiara, 1996.

Government versus the administration

Regarding the forestry policy of the new Mamaloni government, the Minister, Allan Kemakeza, issued a statement in late February 1995: the government would compile new forestry legislation; phase out the export of round logs by 1999; devolve aspects of forestry management to provincial governments and area councils; ensure greater benefits to resource owners and the community; 'provide a wider and integral scope of development and investment to members of the Solomon Islands Timber Association which will commensurate with the development plans and activities of the National and Provincial governments' (sic); reduce wastage; diversify wood by-products; specify and identify plant species and reserves; diversify research and investment in the biodiversity sector; maintain the current moratorium until a review of all licences is done; and privatise all government forestry plantations.[42]

This was an edited version of a paper of the previous December which issued from the Prime Minister's policy evaluation unit without any consultation with Forestry Division. In part, the December version was ill-informed, with nonsensical plans such as

> ... markets should be found for the sale of bottled tropical forest oxygen, an activity that can be very beneficial to local people. The other resource is clean pure water from the forest which can also be bottled and exported to overseas markets. For crying out loud, will Australia buy these two commodities? The other most valuable resource of our forests is herbal medicine which have been used by succeeding generations to cure and prevent human ailments such as VD, TB, cancer, skin disease, stomach problems, broken bones, etc....[43]

The Mamaloni government's increasing distrust of the public service and sensitivity to Australia were well illustrated in this document and its genesis. Lack of procedure increased public service insecurity and emasculated its critical role. When the Forestry Division presented constructive comment on how it might assist in implementing these policies and detailing errors of fact, the policy evaluation unit's response was to blame the Division for all past failures. The Finance Ministry was another victim, as it had its role in budget-making replaced by the whims of competing interest groups, including private businesses.[44]

In line with professed forest policy, the government in March 1995 commissioned a study of forestry taxation options and domestic processing possibilities, financed by the EU. Price Waterhouse started work in June, and assembled an impressive body of data in a surprisingly short time. They were assisted by the Division and by the Ministry of Finance, which was preparing its own damning report on the forestry sector. The Minister of Finance, Abe, suddenly cancelled the Price Waterhouse study in July, but the resultant final draft report, *Forestry Taxation and Domestic Processing Study*, revealed the perilous state of forest fiscal and tax management as well as offering remedies for the situation.[45]

Everywhere there were indications that the politicians wanted to administer, not simply govern. Their fear was that the public service would do its job and carry out enunciated parliamentary policy, not unstated 'policy' practised covertly to win pivotal supporters in their electorates and to feather the nests of several in government, the private and corporate sectors. To this end too, Mamaloni had dispensed with permanent secretaries of ministries under public service conditions, hiring and firing his own under short-term contracts; a standard ploy to prevent the growth of power loci within the administration. Several were his special advisers, allotted particular tasks and duties set by him. Their loyalties were to him personally, not to the nation.

The politicians' discretionary fund introduced by Mamaloni during his 1989–1993 term was part of this pattern. Justified by recourse to 'the Melanesian way' or *kastom*, the fund was supposed to take the pressure off politicians whose wontoks or constituents solicited their help and who felt obliged to assist.[46] In 1993 this Constituency Development Fund increased to SI$100,000, half of which was doled out just before the elections. Instead of schools, books, water supplies, wharfs, and clinics being allocated in a rational way where need was greatest, and according to set administrative criteria, parliamentarians personally could dispense largesse to communities and individuals in their electorates to the tune of an increased $200,000 a year in 1995. Or they could spend it on themselves – no one audited their annual spending of $9.4 million. In doing selective favours to the groups most disadvantaged by the policies of his governments, Mamaloni and his ministers disorganised the opposition and diffused public criticism – a relatively simple task as the party system is weak. The Melanesian big-man, a patron winning a local followership, was alive and well.[47]

FRIS and public information

Just before Minister Kemakeza's forest policy hit the press in February 1995, the AIDAB forest resources inventory and the Forest Resources Information System (FRIS) were finally presented to the government. At the same time, the Permanent Secretary of the Ministry of Natural Resources indicated that the future of the TCU was in doubt, 'threatened with closure by the P[rime] M[inister]'.[48] The government, Forestry Division, and SIFIA were all tight-lipped about the data that the FRIS and the province-by-province reports revealed and the implications that a continued excessive level of logging would have socially, ecologically and economically. The *Solomon Star* summed up the situation,

> The government has disregarded reports and recommendations that forestry experts and advisers have supplied them.[sic] ... How foolish we are. Should we kill the messenger just because he brought the bad news?[49]

In the national press there was no more than the most cursory mentions of this significant study throughout 1995 and early 1996, but then this too had been the fate of the Price Waterhouse report.[50]

This reticence says a lot. The inventory report endorsed 325,000 cubic metres sustainable annual level of extraction, assuming a 45-year cutting cycle, revealing the potential timber resource at 13.3 million cubic metres for logging. When the last official set of Division figures was released for 1989 the total licensed quota was 664,000 cubic metres for the year; in 1994 the quota had reached 3,300,000 cubic metres – *ten times* that considered to be sustainable.[51] If all the companies did extract their quota, the resource would be exhausted by 2000. Real extraction rates continued to mount (see Figure 96). Another eloquent silence was the failure of the Division to publish annual reports since 1989, though the 1990 one was ready for publication in 1992. It was widely believed within the Division that this was because the public would have then had relatively easy access to the quota information and, more importantly the data TCU had collated, revealing the extent of abuses by the logging industry. The Mamaloni governments since 1990 had been anxious to keep this information within the realm of administrative communications, though some was available through the Statistics Branch and Customs which confirmed the existence of transfer-pricing.

Prime Minister Mamaloni was sick of foreign advice on resource use.[52] Yet no aid donor answerable ultimately to its citizens and its own agenda of national interest would be willing to pass an unsupervised purse across to the Solomons government. Advisers, technical or supervisory, are part of the package, the reality of the aid culture that has helped sustain and subsidise the Solomons government since 1978 to the tune of SI$2.9 billion.[53] In 1995/6 Australian aid alone amounted to SI$11.2 million a year, and Australia was not the only donor, although a substantial one to the forest sector. The reports and FRIS made no difference: the first quarter of 1995 registered a monthly average extraction of 95,000 cubic metres, three times the sustainable rate. The total harvest for the year was 846,425 cubic metres, of which about 738,425 went out in round logs. Furthermore, FRIS had provided enough computerised information to point loggers in the right direction, a fear early expressed by an Australian consultant.[54]

Contested State: Contest of States, 1993–1997

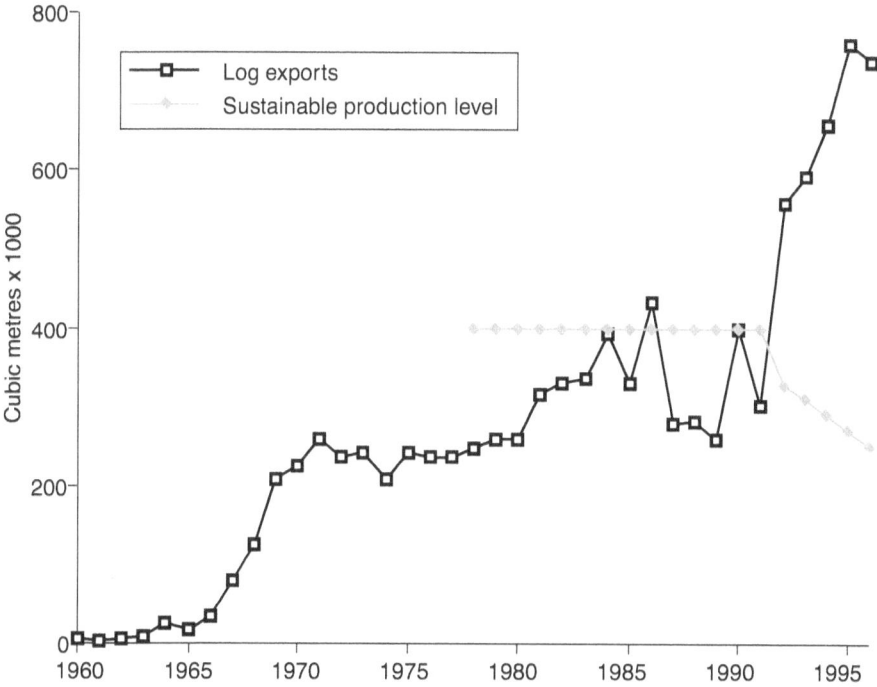

Figure 96. *Log exports and sustainable rate of logging.*

Source: The exports for 1996 exclude 54,000 cubic metres from KFPL plantations. *1995 Forestry Review;* FDAR, 1960-1989; *CBAR, 1996.*

The 'aid war'[55]

Being brought to book by international financial organisations and donors infuriated the government. The World Bank in mid-1995 again warned the Pacific islands, notably Solomon Islands, against large-scale logging.[56] At the South Pacific Forum meeting in Madang in 1995, the Solomons' government, along with Papua New Guinea, whose record on forestry is notorious, 'endorsed' but refused to sign a regional code of practice for forest sustainability to which they had agreed in Brisbane. Before the Madang meeting in September 1995, it was widely speculated that Mamaloni would close down the TCU because of alleged 'inaccurate and erroneous reports on the real state of the country's forests', but that he was biding his time until the contracts of the Australians expired in March 1996.⁵⁷ The Australian Minister for Pacific Islands Affairs, Gordon Bilney pulled no punches:

> Corruption is a cancerous thing and there is a developing crisis of government in some parts of Melanesia. If the leaders of a country do not have the best interests of their citizenry at heart and don't act in those best interests, then no amount of aid will save them. Indeed to continue to give aid in those circumstances is to condone and encourage malfeasance and poor government.⁵⁸

There was increasing awareness in Australia, with an ABC programme on television on 10 September 1995 revealing that the Solomons forests were being cut out at a great rate, while the country was getting closer to bankruptcy. Australian government officials also admonished Solomons. Soon after, the Solomons government informed Australia that it no longer wished to receive technical assistance in forestry matters.⁵⁹

This was part of a pattern.⁶⁰ British ODA had allocated funds during the time of the NCP government for forestry policy development such as Tuhanuku's public meeting in October 1994, involving the wider community. Britain sent a planner, Keith Campbell, as part of the aid package. The overall aim, agreed during the Mamaloni government of 1990–1993, was to implement the Tropical Forestry Action Plan, to strengthen the capabilities of Solomon Islands to manage their forests sustainably. The Mamaloni government abandoned the resultant plan in late 1994. The British adviser, using EU funds, commissioned an inventory and valuation of the government's plantations, a move recommended in 1992 by the forest resource inventory project. Now, the valuation was to be a step in the privitisation of the Forest Estate. By selling the plantations at least 400 workers were to be made redundant. This was the first major 'adjustment' of the public service, notably though of the lowest paid. The asset, purchased by the colonial government and funded by overseas aid until 1991/2, was to be capitalised.

The logical course would have been first their valuation, then their competitive tender. But public tendering had been increasingly abandoned by the government. The government terminated the study two weeks before it was complete and the ODA adviser Campbell was sent home. In early November 1995, Eagon agreed to pay SI$23 million for Viru's 25,000 hectares, of which about 11,000 had been replanted. When the purchase agreement was drawn up the government failed to consult the Division, and did not place a caveat on the forestry research plots that Eagon subsequently logged. Eagon could log the portion of the land that was not under plantations, without any of the steps under the Act to obtain the consent of the Division or the province. The former landholding *butubutu* had no chance to assert claims, though several perceived the land as still being theirs. Moreover, there has never been available to the public a valuation of this plantation, said to be

Contested State: Contest of States, 1993–1997

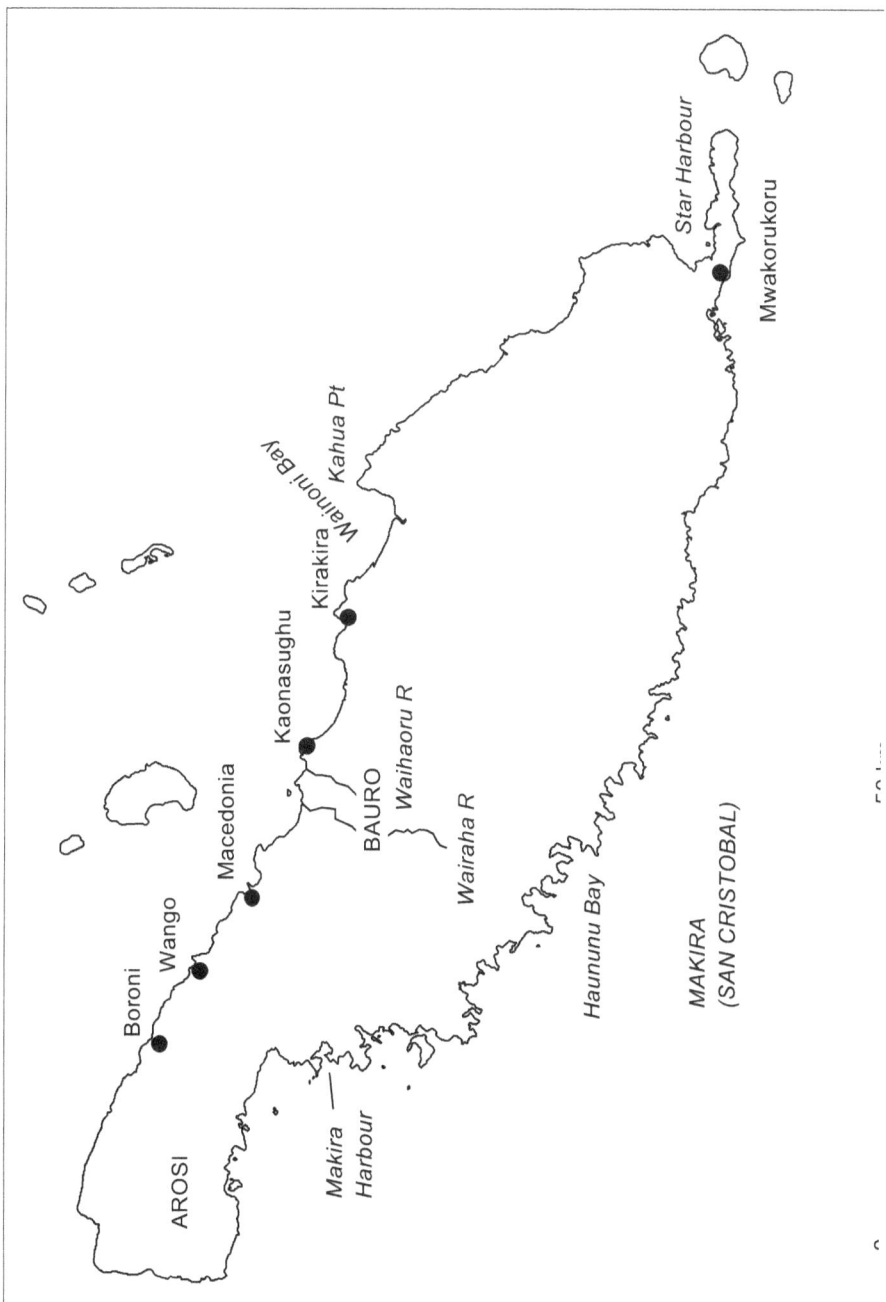

Map 15. Solomon Islands forest estate, 1987, on land owned by the government.

worth around SI$80 million. This and the government's cancellation of the Price Waterhouse study caused the EU to stop funding all national forestry projects, except its small Isabel Sustainable Forest Project.[61]

The ethics of government continued to plummet. In November 1995, several ministers, Luialamo, Abe, and Supa were said by the newspaper *Solomon Star* to be recipients of SI$7 million in bribes provided by Integrated Forest Industries, a subsidiary of Kumpulan Emas Berhad. The ministers threatened to sue, but backed down. Questionable relationships with commercial operators were not confined to the logging industry. It was public knowledge that members of Mamaloni's caucus were 'supported' by certain Honiara businesses and several were anxious to obtain interests in government agencies that were to be privatised. Partiality was manifest. Failing to find a donor to extend the Rove Prison in late 1995, the government gave Chan Wing Ltd duty remission on imported beer for two years in exchange for $20 million for the prison's construction, an estimated saving to Chan Wing of $6 million. There was no public tender for the construction.[62] Such stories and practices sullied the reputation of the government and sapped the morale of the public service.

In the wake of the scandalous loss to the government coffers of SI$24 million in exemptions and remissions, as calculated by the TCU for 1995, and the continuing over-cutting of the resource, Australia finally followed through – with the repercussions foreshadowed in 1993. In early January 1996, Bilney, Minister in the Labor government, reduced aid to the forestry sector by SI$2.2 million, of the $11.2 million grant.[63] For Mamaloni, it provided the justification for the gutting of the TCU, the *bête noir* of those who spirited away illegal shipments, protected species, and undersized logs and who paid politicians and others bribes to look the other way. Minister Kemakeza stated the aid would not be missed, and Mamaloni held that most of it found its way back to Australia anyway.[64] Nonetheless, MP and advisor to Mamaloni, Sam Alasia, during a visit to Canberra, stated that he did not blame Australia 'because the heart of the problem is with the Solomon Islands Government'.[65]

Although the New Zealand government, much to Australia's annoyance, continued its customary land reforestation project, the major forestry donors saw Mamaloni's policies in much the same light as the Australians.[66] British Minister of State, Foreign and Commonwealth Office, Jeremy Hanley stated on 22 January 1996 that ODA was willing to continue to aid the Pacific states if recipients were 'able to convince us that it will contribute to sustainable development'. Forestry in Solomons received mention, when Hanley noted that the government had 'repeatedly spurned' British, EU and Australian efforts to aid the sustainable use.[67] And this was within a

Contested State: Contest of States, 1993-1997

context of overall aid reduction from 22 per cent to ten per cent of Solomons GNP between 1990 and 1996, because of the diminishing ability of the government to present viable projects and its refusal 'to enter into policy dialogue on sector specific policy issues'.[68]

Nonetheless, it was Australia that aroused the greatest ire, partly because Bilney's speech was re-broadcast on Solomon Islands radio in early February. Bilney's stance was portrayed by Solomons commentators as a hypocritical election year ploy to sway Australian voters. More realistically, the secretary of SIFIA, Eric Kes, defended the logging industry against Bilney's 'one-sided' allegations, pointing out that it was finalising a binding code of practice for all logging operations, devised in close co-operation with the TCU. The only contentious point was the preservation of 25 per cent of the canopy at a logging site. SIFIA had plans for an industrial training programme with the government and 'possibly outside assistance'.[69] Like many a supposedly binding regulation of the government, such as the Standard Logging Agreement, the gap between theory and practice was evident. For example, the fundamental difficulty was the emasculation of the TCU, the body that could most effectively police the code and the law. Kes claimed that the industry was not clear-felling on customary land, as Bilney believed, but that selective felling was normal. Allardyce, for example, claimed they took only about six big trees per hectare on their Ndovele operations on Vella Lavella. If that were so for all the members of SIFIA, why the industry's reluctance to agree to the 25 per cent canopy cover retention – surely the amount left after 'selective logging' was close to that anyway? And no matter how careful its methods, it is hard to imagine how it could force a reduction of the four million cubic metres annual quota that loggers had acquired. They were unlikely voluntarily to reduce to the sustainable level recommended of 270,539 cubic metres. And the question must be asked, why should SIFIA expect 'outside assistance' (read 'aid donors'), when the $24 million exemptions and remissions to local companies in 1995 would pay for it all and more?

Although some long-established companies, such as Allardyce, appeared to be following the code, there were no sanctions on those who did not, as SIFIA was to admit when it eventually drew up its own version of proposed legislation in 1997.[70] The Isabel Timber Company claims to operate under the code as well as the SLA, yet its methods include logging of buffer zones on banks of waterways, destruction of residual stands and food trees and felling above 400 metres. Abundant soil erosion from so-called roads and skid tracks has done truly terrible damage to the land-sea environment of Kokota ward on the north coast. If SIFIA was and is genuine about sustainablilty, then it can only succeed with the government's backing and this was lacking; in Mamaloni's time on Isabel for example, where the lone TCU officer had to beg petrol from ITC and had no telephone.[71]

The failure of the Labor Party in Australia's elections in February 1995 elicited a certain amount of satisfaction in Solomons, particularly when Bilney lost his seat. Though the Solomons government welcomed a new era in its relations with Australia, Australia's government continued to voice concern with financial mismanagement in several Pacific states.[72] This concern was reinforced by the release of the *1995 Forestry Review*, not by the Division, but by the Finance Department in April 1996. It highlighted the several anomalies in the tax on loggers, the unsustainable logging of the resource, the spiralling decline in the sustainable rate, and the unconscionable national loss.[73]

The Mamaloni government: death rattle or hiccup?

The country lurched further into crisis in 1996, with loss of confidence by the banking sector and international bodies. After the Central Bank in August 1995 had barred Mamaloni's attempts to borrow in excess of the statutory limits, he had threatened to change the legislation governing it and set up a tribunal to investigate the actions of Governor Houenipwela and his deputy, John Kaitu. In the face of banking and commercial sector opposition, Mamaloni backed down, and the banks carried the country by accepting delayed repayments of interest on loans of A$3 million, for a time. In April 1996, the banks tired of waiting for their money. They confronted Mamaloni with the possibility of legal action and the collapse of the economy.

May was Mamaloni's month. Eagon's deposit of SI$20 million for the Viru plantation along with substantial yet decreased revenues from logging, rescued the government from insolvency. Mamaloni even went so far as to revoke monies for the Constituency Development Fund – at least until July when, with the 1997 elections looming, politicians pressured for funds to sweeten their constituents. Mamaloni also revived a scheme for selling citizenship to 5,000 Hong Kong Chinese in May, but shelved it in the face of widespread opposition. Another bright hope was for the finalisation of negotiations for the Gold Ridge mine with Ross Mining of Australia. Not at the end of the rainbow, but in Tulagi, Mamaloni sent his police field force on a futile search for a horde of gold supposedly hidden during the war. It was also in May that Mamaloni announced that he had 'discovered' SI$35 million (US$2 million) of government funds had been misappropriated during 1995/6, involving 15 ministries; though the media had raised this issue as early as February.[74]

Contested State: Contest of States, 1993–1997

'Solo', as Mamaloni was known, is an astute politician and survivor. A master of the diversionary tactic, he tried to re-establish his reputation by drawing attention away from the systematic loss of revenue from the timber industry, brazening out the continuing admonitions of the Central Bank and the patent conflicts of interest that were commonplace among politicians who supported logging.[75] In response to the commercial banks' ultimatum, Mamaloni even revoked the remissions and exemptions to loggers briefly, but he reneged on this in June. Though there were plans to trim the fat off government bodies such as the post office, at the end of 1996 Mamaloni had done nothing of substance and was blaming the commercial banks for failing to solve his government's 'cash flow' problems.[76]

Over the last seven years, Solomon Islanders have become uneasy with the image their country and thus they themselves have in the world. A small emerging middle class wanted predictablity and a stable economy. There have been a series of 'get rich quick' schemes by Mamaloni and his Ministers of Finance in the last couple of years. The list includes a timber-for-oil deal with Kuwait, a loans deal, duty waiver for patrons to build public buildings, as well as the sale of passports to Hong Kong Chinese. These just might have been passed over, for such pipe-dreams have visited other Pacific states. However, Mamaloni made a exhibition of himself and his country through his inept behaviour at the Commonwealth Heads of Government meeting in Auckland, New Zealand in 1995, when Nelson Mandela was present. He seemed to decline to support a motion condemning gross human rights violations in Nigeria, and jumped the queue at a wreath-laying ceremony.[77] At home, several of his ministers have made drunken fools of themselves on official occasions or by trying to intimidate pilots in flight on the national airline. The regional and local press have reported all this with derision, most notable in the monthly *Islands Business*, whose editorial of May 1996 dwelt on the 'internationally infamous' story of what is happening to the Solomons' natural rain forests.[78]

Mamaloni had contained both critics and potential rivals by the expedients of sending them overseas into the diplomatic service or for tertiary study. Another method had been the dismissal of ministers when they began to coalesce around another possible leader, as seen in the September 1996 sacking of six ministers during the run-up to the 1997 elections. Mamaloni increasingly faced public censure, however, and he was not able to bridle the media entirely. A young countryman, Tarcisius Tara, ridiculed Mamaloni's 'cargo cult' gold hunt in the May 1996 issue of *Islands Business*.[79] Tara was also instrumental in an exposé of corruption in high places in the ABC's *Foreign Correspondent* in August 1996, which opened another verbal battle with Australia.

Television journalists from the ABC had wanted to investigate the Pavuvu conflict, to which Greenpeace and other NGOs had drawn media attention. Mamaloni refused because 'the issue had already been heavily covered'. A couple of Australian journalists in July 1996 came into Solomon Islands as tourists. With a small but effective video camera they recorded scenes of dissatisfaction with Kalena's logging on Rendova. A brave Rendova woman, Mary Bea, described just how a few 'go-betweens' in the villages made money, while the bulk of the resource holders saw little benefit from the loggers, while losing their forest and clean water. One of these 'go-betweens' openly admitted his part and the personal benefits. Nonetheless, the journalists openly contacted high level supporters of the government. Eric Kes of SIFIA told the journalists that any people with complaints about loggers had the right to take legal action, but seemed unable to defend the Rendova case. Tara, a university lecturer, and Duran Angiki, a journalist, as well as Joses Tuhanuku, MP, were critical of the government's policies and corruption. Mamaloni's connection with Somma was highlighted, and also the fact that the company, being on paper a local one, paid no tax. Finance Minster Abe's appearance justifying the Prime Minister's directorship in Somma convinced few that there was no fundamental conflict of interests. After the video screened in Australia, Somma's manager, Kaipua Tohi, accused the Australian government of a racist 'vendetta' against the Solomons government and Mamaloni.

The programme had little to say of any good to come from more responsible logging practices, such as on north Vella Lavella. This one-sidedness, and the secretive way in which the television crew entered the country, hit an already sensitive spot of Solomon Islands' nationalism, in a country still coming to terms with the role of the media in politics. Somma representatives circulated a booklet defending their activities. They stated that under Solomon Islands law, the fact that a prime minister could be a major beneficiary of a company was considered acceptable as long as it was known to the Leadership Code Commission. Moreover, the 'development' the company was bringing to the area was more positive than the activities of earlier 'Australian' companies (presumably Sollumber and Integrated Forest Industries when it had some Australian principals in the 1980s). Mamaloni protested to the Australian High Commissioner. In the eyes of his compatriots as well as supporters in west Makira, who saw the incoming royalties as big-man Mamaloni's doing, 'Solo' became for a time a martyr to the foreign media, an apparently mild but determined David personifying hurt national sensibility against Goliath, Australia.[80] Another victory in the perception battle.

But these several smoke screens could no longer hide the reality of social distress or the country's accounts. The Catholic Archbishop Adrian

Smith during 1995–1996 criticised the unfair distribution of wealth paid for from the resources of the people; and the Church of Melanesia on Malaita condemned logging because it was making it harder for the population to survive by subsistence.[81] In the wider economy, although some forestry workers had been sacked, by early 1996 there had been no systematic rationalisation of the public service. Reductions were not significant, reflecting the more general decline in wage employment. Instead, the Finance Minister planned a goods and service tax, and introduced a debit tax of two to three per cent on all withdrawals from banks and lending institutions. Resistance to the debit tax was widespread, culminating in a High Court ruling of March 1996 that it was illegal on overseas transactions and beyond the minister's powers. An unexpected lift in revenue from domestic sources in 1996 was a chance to begin to balance the books, but the government used it to defer the new tax regime and radical reform.[82] A law to remove provincial governments and devolve power to area assemblies was in part a belated attempt to control expenditure, but one which met with popular resistance in 1996.[83]

Through early 1997, the public service was near collapse – wages were not paid, maintenance on public buildings ceased, roads deteriorated, shipping declined, SICHE was forced to run a bazaar to find funds, tertiary student allowances went unpaid, the government high school in Honiara, the once proud King George VI, was a decaying shambles, local businesses refused supplies to government, and the government could not pay even its telephone bills. Parliament often was suspended for lack of a quorum, because parliamentarians were largely operating outside parliament and were busy with business interests, including logging. So ineffectual had the government become that SIFIA took it upon itself to draft a new forest industries bill; rightly or wrongly, a conflict of interest was suspected. The newspapers, even the circumspect *Solomons Voice*, reflected criticism of Mamaloni's government. In May, the Finance Minister, Maena, finally revealed the state of the government's accounts, an open secret in Honiara.[84] In the following August election, the Alliance for Change led by Bart Ulufa'alu gained power, though not without Mamaloni trying to get him into his SINURP.[85]

There was a sense of relief when Ulufa'alu came to power, though many perceived him to be part of the old guard, as a member of the first 1978 parliament. But relief was tempered by caution. Like Hilly in 1993, he leads a shaky coalition. 'Politician' is a term of opprobrium among the young while the old have learned to be sceptical. When Ulufa'alu was involved in a drunken incident at a nightclub in October 1997, public censure was rapid, more so than under the old regime. It seems the Solomons' people are weary of sloppy leaders who make fools of themselves and their country. There have even been calls to recruit expatriates for key positions to re-introduce an attitude of 'service' in administration, reinforced by the outstanding work of the new English Police Commissioner, Frank Short.

Having little choice, Ulufa'alu, in 1997, is committed to structural reform, albeit 'home-grown' to suit local needs, and he and Finance Minister Manasseh Sogavare have had extensive discussions with IMF, the World Bank and donors like the EU and Australia.[86] In November, the forestry sector was back on the agenda, with Australia promising Minister Hilda Kari to send reviewers to assess areas of need for assistance. Ulufa'alu has reduced the number of government ministers and they will not be missed, but moves to cut the public service by ten per cent have met union opposition. More importantly for the forests, plans are afoot to revitalise the TCU, review forest legislation, and reforest customary land.[87] However, the government's cancellation of all remissions and exemptions has seen the loggers stockpiling logs again, just as they did when Hilly increased duty.[88] Moreover, the uncertain economic future of South-east Asia began to depress the market. Donor countries, now called 'development partners', are to be asked for grants for recurrent expenditure as well as special programmes; but no longer can the Solomons government so easily capture aid to subsidise wasteful practices as was the case in forestry. Sovereignty bends to solvency. It will take a determined government and powerful funding agencies to break old habits and the stranglehold of the overseas loggers – and they still have friends at court, as well as in the villages.

Conclusion: The Forest, Contested and Uncontested

> ... one might very well cast the history of ecology as a struggle between rival views of the relationship between humans and nature: one view devoted to the discovery of intrinsic value and its preservation, the other to the creation of an instumentalized world and its exploitation.
>
> Donald Worster, *Nature's Economy: A History of Ecological Ideas* (Cambridge, 1994), xi.

> We need to cut down trees for making gardens and to take fish and shells from the sea. These things live together with us. Now, these white men who come and talk about conservation want to protect nature, they say. But what is this thing called 'natural beauty,' that they always talk about, anyway? As I see it, what they call 'nature' is simply the place where we live.
>
> Unidentified leading Marovo spokesman, cited in Hviding, *Guardians of Marovo Lagoon*, 56.

Until the 1930s, forest conservation was not an issue to the colonial power. All the lessons learned and policies developed in the British Empire, and indeed in other colonial empires, remained inactive in the Protectorate. The reasons for this lie, not in a forgotten imperial history or absence of contemporaneous awareness, but in lack of demand on the resource. Subsistence gardeners did not trade in trees or processed timber, other than the occasional canoe. There were no timber-hungry colonial ventures such as underground mining or railways. Demand was minimal due to the distance of the Protectorate from possible markets for its timber and more particularly because of the relatively small population scattered over a large land mass – 3.5 people to the square kilometre in the 1930s. Even then, many administrators believed, just as Charles Woodford had earlier, that the population was declining still. The early administrators saw the future of the land, if not the diminishing population, in 'development' mainly of copra plantations. Although several Europeans recognised the potential value of the forests as timber, the clearance of lowland areas was primarily to make way for plantations and the panoply of the colonial establishment – towns, mission stations and plantation buildings.

The capitalist reality was that timber was worth only what the market would pay; the demand for tropical timber before and immediately after World War Two was met by suppliers closer to the markets and in less remote regions who produced timbers, such as mahogany and teak, that had become known over decades, if not centuries, for certain attractive characteristics. Conditions in Solomons, with its relatively constant rainfall and scattered islands along with the typical mixed and relatively poorly stocked stands, made the extraction of even high-value timbers such as kauri a costly undertaking, fraught with risk as the history of the Vanikolo operation demonstrated. Local, small scale mission mills succeeded because they were geared to local consumption and were in reality subsidised by the expertise of dedicated, low-paid overseas mission personnel and by donations from mother churches. Their incoming logs, too, sometimes represented a subsidy because Solomon Islander Christians often gave all or part of the log to the mission. For them and the foreign missionaries, the work of the mission in teaching, education and healing was a community effort and returned material as well as spiritual and psychological benefits in a period of great change. Moreover, the products of the mills and schools – timber, boats, and basic furniture were often utilised by mission people in their daily lives. This kind of local involvement from the selection and transportation of the log to the utilisation of the timber was one model that, by and large, both the colonial government and the governments of independence have seen fit not to follow or foster on any significant scale.

As late as 1946, the forester Walker was still convinced of population decline, but then the Pacific war had caused a temporary drop in the birthrate, at least in the western and central areas, where entire populations 'went bush' to survive for a year or more and had their gardens raided by retreating Japanese. Once the Pacific war was over, the forest in battle zones began to recover, as did the human population, while the flurry of overseas interest in the timber industry almost completely faded.

Though the colonial administrators prior to the early 1930s had seen little need to 'conserve' the forest or even replant it, for it was not under threat, in the isolated areas where commercial logging occurred, they did make sure that the means of subsistence would not be undermined by logging. This was not difficult. It is hard to imagine, for example, the 60–80 inhabitants of Vanikolo ever being bereft of sufficient garden and foraging

Conclusion

land on the 72 square miles that made up the island; all the more so given the irregular pattern of logging practised by the Kauri Timber Company.

Vanikolo, as well as the Fijian experience, changed this. It alerted the Colonial Office to the dangers to both government revenue and the sustainability of the commercial forest where there were no forestry officers to supervise loggers. These administrations in 1933/4, unable to finance additional staff, ceased to encourage loggers. For the Solomons, the Colonial Office effectively imposed a moratorium on new licences on customary land (non-leased or unalienated land) that lasted over 30 years – with the exception of war emergency in 1942–1945 – until adequate control could be provided, despite the post-war need for increased revenue.

When the conservation issue came to be reconsidered in the 1950s and 1960s, a time of increasing demand for tropical hardwoods, it was in relation to the creation of the Forest Estate, with a nod to the protection of water catchments which, in a little-urbanised country, was still not a major problem. Although recognised as a possibility in the distant future, population pressure and the needs of subsistence agriculture were not yet major parameters, and certainly not characteristic of the areas Keith Trenaman aimed to designate as lands for the forest estate. His purpose was to create a long-term asset for the state which, one day, would be an independent one, in every sense of the word. It was premised on the principle echoing Britain's political designation of the country – protection: protection, at the time, of a near-illiterate and naïve people against modern resource raiders, the sophisticated overseas loggers whose guile might just prove to be as great as their technological expertise, as well as protection of the forests from logging without replanting and from unsustainable logging methods. Yet there was an implicit paradox. For the Protectorate, the planned estate was to be the largest extension of state control over land contemplated since the Waste Land regulations of the early 1900s. It was potentially a form of enclosure because clan commons would be withdrawn from future use by original rights' holders. The government may have believed this could go back to the local people if the demand arose, but if valuable timber trees were to be planted, the probability of this was remote. The hope of the government for 'vacant' land proved futile, then it moved to purchase its forest estate lands from local land-holding groups. This was a protracted exercise hampered by the complexities of customary tenure. As tracts of forested land became state land, private capital in the form of logging companies arrived in response to rising, if sometimes erratic, market demand and, in a sense, thwarted the government's plan, thus complicating in the long-term the relationship of logging companies with local communities. The 'public good' seemed a government subterfuge to acquire land. The Solomon Islanders began to realise they could make more money by selling logs or timber rights to the loggers, while still retaining their lands. To prevent forested land from slipping completely out of its control, the government declared the forest areas in 1968. This met a wall of resistance from the people, and aroused ineradicable suspicion.

In terms of colonial forestry practice elsewhere in the humid tropics, in India, Africa and South-east Asia, the British government had simply left its run too late in the Solomon Islands, by almost 100 years. The time was past when colonial coercion and violence could be marshalled to enforce policy. Its finale in the late 1940s against Ma'asina Rule achieved less than the High Commissioner's subsequent negotiations and accommodation. If there were to be sustainable forestry on government land, negotiation, persuasion, purchase, and lease remained the colonial administration's only means. Post World War opinion, voiced by the United Nations, was condemnatory of colonies; any attempts by the colonial government to contest the issue of the control of land in the Protectorate were beyond serious contemplation. Somewhat ironically, this time lag meant that, unlike the other colonies, forest protection was designed as a future asset for an independent state, not a plantation to feed Britain's seemingly endless colonial occupation.

Trenaman's obsession with establishing the forest estate on government or Crown land had other ramifications. With the notable exception of the Vanikolo native council, there was no attempt to involve Solomon Islanders in the management and sustainable commercial exploitation of their own forest, other than on a centralised state level; although High Commissioner Grantham and Forester Logie contemplated involving the local councils during the more leisurely late 1940s and early 1950s, when the creation of a viable modern economy, although desirable, seemed extremely far off. In their empires, British and other Western administrations in tropical colonies usually had appropriated more power over their native subjects than any government in their own homelands. The ideology of the nation-state and its pre-emptive power was hundreds of years in the making in Britain and Europe. Even there, provincial and regional mini-nationalisms had not been completely extinguished. To import concepts prefaced on a system evolved over centuries, painfully and often in bloodshed, thousands of miles away was unwise – especially in relation to land and the property thereupon. Perhaps had there been more time, another hundred years, before independence, the concept might have taken root in Solomons, but this remains speculation. In the event, so focused had Trenaman been on a government-owned forest estate, that before independence, no structure had been set up to deal with the commercial logging of customary land in an efficient and sustainable manner that maximisd local participation and control.

Conclusion

The independent government was no better: in fact its record is even more reprehensible because it purports to understand and reflect its own people, a claim no colonial government could legitimately make. The post-colonial governments have increasingly wanted fast revenue and soon abdicated any direct responsibility for the difficult task of fostering 'tree-roots' involvement in logging, beyond that of Solomon Islanders being vendors of the logs and employees of the foreign companies or their puppet contractors. Solomon Islanders wanted far more, as demonstrated since the late 1980s in their burgeoning support, albeit often in the loggers' wake, of alternative systems of small-scale logging and milling offered by the NGOs. Most NGOs aim to empower local communities to control their own 'development' of forest and other related resources by means of systems many consider economically inefficient. Still captured by the capitalistic colonial economic model of unprocessed primary products feeding into a metropolitan centre, few in government or big business have sought to assist and perhaps rationalise these systems through centralised marketing of milled timber and systematic training in production of value-added timber products. The essential radical restructuring of the industry would mean the loss of mass extraction and productivity would fall for a time, depriving those who profit from large-scale logging of their considerable benefits. It is not easy to weigh up this loss of forest earnings against the wider economic, social and political benefits at the national level. It is easier, however, to weigh the loss of forest against its total economic value to subsistence – yet still fewer in government have calculated this along the lines indicated by Ross Cassells. This is not simple a question of equity, participation, and social justice, but also of choosing an alternative that at least has a chance of achieving sustainability, as opposed to large-scale, commercial logging as practised since about 1980 which, with a few exceptions, has simply raided and 'mined' the resource and moved on.[1]

Commercial logging has not been the economic bonanza rural Solomon Islanders first envisioned. The colonial government did an excellent job as publicist for all the blessings logging was supposed to bring in the mid-1960s. Fluctuating markets, relatively low stocks of mixed and untried species, and general establishment problems for the early logging companies on government land saw a gulf between promise and performance. Added to this, some villagers perceived the mediation of government, via land acquisition, as draining potential profits away, so the pressure was on for direct dealing. Since at least the early 1980s, the negative effects of logging appear to outweigh its benefits, though not all of this should be laid at the door of the logging companies. They do what the people and government directly or indirectly permit them to do. As Asian loggers have said privately and publicly, many resource holders have frittered away their royalty payments. The beer barons in Australia have done more business than local banks. The visions of jobs, schools, clinics, shops, roads and even townships have sometimes materialised in whole or part; but usually only for the brief life-time of the loggers' operation in any particular area – a pattern the same world-wide where natural forests are logged. Loggers are there for the short-term; though some which claim otherwise do so only to gain entry into yet another adjacent area of forest or perhaps, for the one or two with superior logging techniques, to return in ten or twenty years to cut out the immature trees left from their first coupe. Excepting the successful plantation-based logging of KFPL – based on an ethic not far from Cecil Rhodes' 'philanthropy plus five per cent' that is, aid as business, not as handout – the protracted research and processes involved in replanting logged land with timber trees far exceed the time horizon of the highly mobile logging companies.

The government has failed to maximise the market value of logs because of its tardiness since 1980 with funding the timber and tax control capacities of the Forestry Division and the Ministry of Finance; millions of dollars of potential revenue have sailed out to sea in undeclared shipments, as mislabelled species or via the mechanisms of transfer-pricing. Remissions and exemptions in the early 1990s profited loggers far more than landholders; directors and big-men far more than ordinary villagers; and foreigners more than citizens. The level of commitment by recent governments of Solomon Islands can be discerned in Prime Minister Solomon Mamaloni's ousting of the Australian-funded TCU – for the Unit knew too much about these illegal tricks. By failing to act as the guardian of the resource, the government has further reduced the likelihood of schools and clinics, the yardsticks of development for rural communities. Not only did the government fail to extract its due and to protect the country's human and non-human resources, but also its leaders profited from and facilitated the plundering. The greatest losses, however, which have still fully to be appreciated, are the country's loss of subsistence potential (by the old method of bush fallow or swidden horticulture) in logged land and the extensive damage to land and sea habitats for flora and fauna. Sadly, it is only as the loggers move in and then on that that realisation sinks in. Together with the diminishing availability of forested land because of population pressure and the need for cash crop production, logging has been more of a disaster than a bonanza, a calamity rather than an opportunity.

There have been other shattered visions. The independent government from 1978 made little attempt to extend government lands for the forest estate; and indeed began to divest itself of other land formerly classified as alienated, espousing an ideology of returning land wherever possible to

Conclusion

traditional claimants or at least for community purposes, such as provincial centres and high schools. Original rights' holders and/or claimants began regaining the commons that had been government property. The Forest and Timber Ordinance of 1977 was, in the main, the creation of the independent government, for it did nothing of significance subsequently to rescind the provisions for direct dealing between the forest rights-holders and logging companies; in fact, with time, it has weakened central bureaucratic control of logging agreements by choosing for 20 years not to revise the act and its several amendments to clarify its presumable original intent. Particularly in the period 1980–1987, the Solomons state, or at least Forestry Division, has faced opposition to preservation of customary land for future supposedly sustainable commercial extraction. Like its predecessor, but with less justification, the independent government made no consistent attempt to address the challenges of systematically regulating the demands of local timber rights holders and loggers. Ad hoc solutions were tried, but each failed and indeed further complicated any sort of enforceable regulation. The new state buckled under this pressure, abetted most enthusiastically by Asian loggers who filled the coffers of government, the bellies of rights-holders, and the deep pockets of venal politicians.

Another major source of resistance to national control has been the wish of the provinces to get a share in the forest estate in their regions; in other words, to replace central control with regional. However, this prospect is disappearing amid increasing financial problems because the state has chosen to divest itself of its assets with unseemly and uneconomic haste, as the sale of Viru to Eagon demonstrates. Likewise, on Pavuvu, state-owned land with forest cover was offered to the loggers first, before resettlement by land-hungry local people and people from other islands. The government's commitment to restoring traditional 'owners' to their ancestral lands has been well and truly compromised by its financial difficulties. Mamaloni who had wanted no central government control of the KFPL project in 1987, and who had handed over potentially valuable shares to the local Kolombangara land claimants in 1992, clearly was less concerned with the claims of the Pavuvu people in 1995/6 and did not consult the former Viru landholders who still perceive it as 'their' land.[2]

Neither the loggers nor the government have fulfilled the hopes of most the rights' claimants. The Pavuvu case called forth the usual incendiarism of company equipment, a pattern of resistance set with the Enoghae incident in 1982. However, resistance to logging companies has been autonomous and localised. Thus far there has been no country-wide mobilisation of this resistance by any significant elite or political group, although some of the NGOs and churches (as well as the EU project on Isabel) are providing alternatives to foreign logging companies' activities – in the form of small-scale logging, under (ideally) a sustainable extraction regime, with export to eco-timber dealers. This community-oriented process has been given mainly lip-service by those in political power, since most profited far more and more quickly through their connections of patronage with the Asian logging companies and international capital. At least until late 1997, the state has been locked into 'mining' the forests, because the revenue earned – although only part of the worth of the timber cut – underpins the indebted government and, of course, the careers of both public servants and politicians. The logic of economics is not always the logic of politics.

The government's commitment to the notion of provincial government has also weakened in the light of budgetary realities. The costs of a three-tiered structure of government – area councils, provincial governments and the national government – have proved too much. However, Mamaloni's attempts to disempower the provinces was a political gamble that failed. Where the former roles of Forestry Division will fit, relative to the provincial governments, still remained uncertain in 1997; but then, there is less and less commercial timber left on customary land.

For the bureaucracy, as the elected government's dependency on log revenues increased, the power of the Forestry Division decreased, a pattern common in many newly independent countries. Its high dependency for 85 per cent plus of its funding on overseas aid speaks volumes for the government's lack of interest in it and for its vulnerability once donors got wise to the folly of funding wastage. The temptation to capitalise the forest resource base coupled with other temptations from loggers were too great. To 1997, the Division is now but a shadow of its once-prosperous self. It has very limited power over licences – in reality it can delay the process, but seldom prevent it completely; its estate, a word redolent with colonial overtones, is being sold off to loggers who just may decide to replant, but with oil palms rather than timber trees. Its once rich treasure of aid has halted and its timber control became nominal once the aid from Australia ceased. Many of its former staff members work in industry or for NGOs and its former head, Sam Gaviro, became the Permanent Secretary of the Ministry of Natural Resources and thus directly answerable to Prime Minister Mamaloni until mid-1997. Yet the Division has proved it can develop effective extension programmes. This co-operation with and involvement of the landholders in management and planning of forest use seems the only way that rehabilitation of the forests will succeed, unless land accrues into large holdings owned by the few, an eventuality that would appeal to Japan's JOAA. Fortunately, there remains a substantial database on silviculture, though much remains to be done on valuable native species.

Conclusion

Trenaman's dream of an enduring forest estate for which he fought valiantly for almost 20 years did not long survive the glare of dawning political independence. But there was another issue that not even the colonial government, with all its experience, had contested. The significance of ecological diversity in relation to the forests and, through land-maritime linkages, the sea, received scant attention. Trenaman, though aware of its role, appeared more concerned with the lessons it taught about how to manage plantation forestry; but he did not foresee the massive scale that logging was permitted to reach during the late 1980s and 1990s. By the 1990s there was increasing world consciousness of the ecological characteristics of the closed tropical rainforest – its stability of flora and fauna form and function that remain in balance over time and despite natural events such as cyclones. Its co-adaptation with the 'bush fallow' system as practiced in the pre-World War Two Solomons created a system apparently sustainable for itself and its inhabitants, both human and non-human. Now all that has changed and continues to do so as the human population becomes greater in number as well as more involved in capitalism or its euphemism, the global economy, which in turn pressures the forest and its ecology.

It is a truism today that the tropical rainforest is a genetic storehouse. At least 24 of the world's major food crops originated from wild forms found in such forests. Certainly, most of these came from South America or South-east Asia. Melanesia has not contributed markedly to this store. To that, we should add the qualification, 'yet'. Melanesia has entered world perception relatively recently; much is yet to be learned about its flora and fauna. It may be, for example, that the varieties of canarium almond, banana, or pandanus could be a crop to help feed an expanding world population. Cures for disease may lie in the unique chemical composition of forest plants and fungi.[3] The diversity of the Melanesian forest and the potential uses of its store are imperfectly described and more often than not simply unknown – to destroy before one even knows what one is destroying must be the nadir of ignorance and arrogance. The timber industry recently has claimed that much of the forested land of Solomons remains and will remain untouched by logging – up to 50 per cent of it being land with little stock of commercial trees or located above the regulation 400 metre level declared to be out of bounds to loggers; but even if this regulation is adhered to, and there is no guarantee a government will not change it, much of the forest below the 400 metre level has been or will be logged or otherwise modified by clearing, thus disturbing the lowland ecology, and all the more so as much of this has been logged, as the locals say, 'Solomons style'.[4] Loggers forget that the forest is not just trees and certainly not just commercial trees. The lowland ecology has a series of interconnections with both the upland and coastal systems. Conservationists would like to see the ecological diversity of all regions sustained.

Solomon Islanders could counter with the question, why they should conserve habitat and species so that the rest of the world might benefit while they live in what they perceive as relative deprivation? This is a valid argument in part. If the world cares about genetic diversity of the tropical rainforest and its potential economic value, some costing for this must go into the pricing for timber, which will involve not only the resource owners, but also the purchasers of the finished product, governments and the timber industry. With higher prices, the returns to the owners should be the same, but with less demand on the resource and some breathing space for the development of a sustainable regime for both the timber and the entire indigenous forest eco-system.[5]

Aid packages could provide part of the compensation for what Solomon Islanders' might conceive as loss of opportunity costs in conserving forest habitat. However, the acceptance and monitoring of 'debt for nature' or 'aid for nature' raise many complicated issues, one of which is the spectre of ecological colonialism by which the North bullies the South to preserve a tropical reserve of genetic and ecological diversity. More recently, the concept of the tropics providing a 'carbon sink' has been mooted: trading in greenhouse gas 'credits' was discussed by economists at the Kyoto convention on global warming in December 1997. Carbon sequestration appears to be a promising area because in the private sector, electricity suppliers of the North are seeking ways to mitigate their production of greenhouses gases from the burning of the fossil fuels of coal and oil and so avoid being charged predicted environmental taxes by their governments. The tropics are perceived as ideal places for carbon sinks because of the growth rates of heavy density tree species which absorb and fix large amounts of CO_2, but only while they live. Degraded areas can be rehabilitated for this purpose and, in logged-over land, enrichment plantings can be carried out. Related to this are forestry approaches to conserving existing forest, including improved logging techniques which significantly reduce damage to residual stands. Under a sustainable regime, any such mature timber logged could be marketed as sustainably produced if wood certification schemes like those in which Soltrust is involved give their approval. If a tropical country such as the Solomons takes this seriously and its government can enter into agreements with its people and the outside agency, the way is open for transfer of funds from the North as a commercial transaction rather than an aid handout. The companies on the receiving end of the carbon credits are likely to be alert monitors of whether or not the arrangement is working as they have so much at stake.[6]

Conclusion

Such international deals are difficult, as Prime Ministers Keating and Hilly found, and global ones even more daunting. Any kind of protected area management or sustainable forestry regime can only succeed if it is genuinely desired not only by the world community, but also by the local community, which must always have ultimate control. Solomon Islanders will not do this unless it preserves and/or enhances their lifestyle. Western ideas of environmental preservation easily become confused with conservation. Westerners would be advised to remember when they tout their early models of national parks and various nature reserves of the late nineteenth and early twentieth centuries, set apart from human settlement, that, at the time, these often were of limited apparent economic value – to the second wave settlers. Their scenic and aesthetic qualities, and their remoteness gave them a worth mainly as tourist attractions. The value of this wilderness for tourism was such that in the case of Yellowstone National Park set up by the United States government in 1872, the resident Soshone Indians were evicted, some 300 being killed in subsequent clashes. In colonies of white settlement – and perhaps in all imperial possessions – the 'wilderness' that second wave settlers, including conservationists, to-day argue over remained because the indigenous people were taken out of it or were extirpated by diseases and colonial violence. For indigenous people, wilderness, as Westerners understand it, does not exist; the land- and seascape is a cultural artefact.[7] Most rural Solomon Islanders do not talk about nature or see it apart from themselves. They live in it, on it, with it and from it; so to set it aside from their use which affords them a good life is an alien concept. It is both folly and unfair to demonise the forest farmers for inhabiting their forest.

However, because Solomon Islanders, as human beings, are part of the habitat, when it is degraded it will not only be the rest of the world that is poorer, but also they themselves, as emerging nutritional, economic and social patterns in Solomons to-day would seem to confirm. The loss of opportunity costs adhere far more to the real value of the forest for subsistence, a value that royalty payments presently simply do not recompense. Loggers may claim that they are not getting their logs cheaply when world prices come into play; but Solomon Islanders are paying dearly when they sell at such prices because, for the most part, they are selling the foundation of their subsistence for decades to come. Foreign loggers at least have the option of going back to Asia where they came from, to contemplate their own altered landscapes.

As yet, within the Solomon Islands, there are few voices contesting this instrumental view of ecology and of nature. To the outsider, it would seem that most Solomon Islanders lack the concept that an ecological system should be preserved 'because it is there', for its own sake or as part of Creation and not simply because of its usefulness to human beings. Indeed, it is rare enough within other societies with longer remembered histories of ecological change, destruction and loss. Refraining from disturbing or destroying an ecological system perversely may be a result of knowing one has the power to do so as well as an understanding of the linkages between the elements of any such system. Such power was much circumscribed in relation to the forests when the ancestors of Solomon Islanders had only their stone tools and fire to modify relatively small sections of the forests at any one time. This is not the case any more. Local chiefs and big men are no longer the main manipulators of the forest to produce wealth – practically one young man bearing a chain-saw can do the work of a hundred stone-age fellers in less time. New technology makes clearance of forests easier not only for logging, but also for cash cropping and gardening. The limits of this multiple usage of the forests are coming closer each year under the increasing population and that population's ever-rising aspirations, which have meshed nicely with the Japanese and Korean demand for timber to meet their own and the world's hunger for light hardwoods and prized furniture timbers. Thus the preservation of forest ecology for both its utilitarian value and for its own sake paradoxically becomes both more urgent and less likely.

However, in relation to valuing the intrinsic worth of the forest-sea ecology, as the 'deep ecologists' would, it seems more a case of outsiders having asked the wrong questions, naturally couching these in their own cultural experience, idiom, and, above all, perceptions.[8] In former times and still in the minds of most rural people in Solomons, there are *tambu* places, redolent of history. These commonly were or are sites of religious significance: burial sites, sacrificial altars, certain stretches of forest used as pathways for the spirits, places near the shore and in lagoons where ancestral heroes began or finished their voyaging, places of unusual natural features formed by ancestors or creator spirits, and places where spirits dwelt such as high mountains. These may be conceptualised differently because they are envisioned differently, but like the sites of Westerners who wish to conserve for their intrinsic worth, there is an implicit recognition of the *genius loci*. A unique value exists in the elements that make up a forest valley, a seashore, an island or a mountain top, a value where the sum is more than the parts, yet every part is interlinked with the other; although by this, total understanding of ecological connections is not implied. Westerners, especially those urban dwellers in settler societies, want to see 'remote' (that is, from them) places preserved; they seek what their societies have so often destroyed through the 'development' process – wilderness. Yet there is no wilderness in the thinking of Solomon Islanders about their own

Conclusion

lands – though especially in pre-European times there were areas held as dangerous and unknown because they were the territory of a stranger or enemy group. The wild is tamed and the separation between human beings and nature bridged when through human intelligence and spiritual power the wild is used for human societal needs; but taming is domestication, not destruction. Most Westerners perceive the world as a dichotomy – nature and culture – but this is an alien concept to most Melanesians, as indeed it is to most indigenous peoples still *on* and *of* their ancestral lands. The forest, the foreshore, the lagoons, the reefs and seas are all part of a whole and they are part of it through their mythic and historic connections and relationships with the human living and the dead.[9] What many a Western preservationist would see as the *genius loci* or the spiritual significance of nature, already exists as part of the world view of the Melanesian conservationist, but not as something apart from the everyday world of getting a living and sustaining the life of the community. *Genius loci*, even in the Western tradition, is an intellectual construct and would not exist conceptually without the human mind and imagination. Rural Melanesians, to many an unobservant foreigner, live in a forest wilderness; to Melanesians themselves, they live in their home.

Yet there is no doubt that many Solomon Islanders have contributed to the impoverishment of their home. Through consenting to logging and commodifying the trees, many sought to stretch their use of the forest to improve their quality of life; some were tricked or persuaded by relatives and neighbours; others were greedy and became the paid tricksters for loggers. In the emic sense, most rural Solomon Islanders knew, to varying degrees, far more about the localised ecology than any logger – yet in the final analysis it was often those closest to the land who assented to the opening of their forest to loggers. Guile, bribery, and human weakness aside, the explanation for this complicity of the people in the degradation of their own natural and cultural environment seems to lie in the long history of land use in this archipelago.

Large-scale logging of the forest as a possibility had not been part of the land-learning experience of the archipelago's first settlers and the vast majority of their descendants throughout much of the colonial era. Even the European planters, as their name proclaims, replanted the cleared land with a type of tree, albeit the coconut palm. Logging posed a choice that was totally novel. From such a unprecedented perspective, most Solomon Islanders saw the forests which had 'always' been there, as an apparent unlimited resource and, all the more so, when they recalled how the forest heals after cyclones and, as some old people knew, even after the destruction of modern warfare. In pre-Christian times the limitations the natural world placed on human beings and their activities produced, if not enforced, an inherent respect and care for the environment. That spiritual ecology did not weaken simply because of the advent of Christianity and Genesis concepts of human beings bringing the natural world into submission, or even because of the demonstration effect of the few white settlers and missionaries, despite debilitating endemic malaria, so determinedly transforming 'their' new environment. Imperial Adam, seeking to domesticate the natural world for his and his family's benefit, was alive and well in Melanesia long before Christianity and the European colonists touched the shores of the archipelago: the main factors responsible for the stability of so much of the forest-marine ecology was the limited technology of the stone age and the relatively small human population, itself probably kept in check primarily by the presence of malaria and by deliberate and incidental methods of fertility control.

Resilient as the Solomons forest is, it now cannot be left largely to its own devices to recover, as it was with classic shifting cultivation or even war damage – unless, of course, all logging, cash cropping and gardening in areas of diminishing fallow were suddenly to lessen substantially or cease. Solomon Islanders' traditional knowledge of the forest and maritime ecology, though based on the observations of generations of first-nation scientists, has not been and is not sufficient in itself to protect and salvage the forests of the archipelago. Moreover, there are many in the younger generation who know little of it. Where, for example, logging companies have been irresponsible in their methods of extraction, where the trees have been virtually clear-felled and the canopy reduced dramatically, the 'result will be a plethora of vines and secondary species of no commercial value'.[10] This may support the bush fallow system if no other demands are made on the land, but it will not recreate a forest of commercial trees. Conscious education, planning, implementation, and management on both a local and national scale are now needed to rehabilitate the forest if it is to sustain in the widest sense its human and non-human population. If this kind of pressure on the land continues it will not sustain the predicted growth in population and its attendant needs. It would seem an emerging scenario for Malthusian forces to come into play. The alternatives are several and dependent on many variables. Some new industry could create employment and move people off the land into larger settlements, but this seems unlikely in the near future. Social or community forestry could provide some income and conserve soil for subsistence needs: large-scale plantations like KFPL could provide economic stability and sustainability, but need some kind of long-term land tenure security and commitment from all parties. In many ways, KFPL shows what Trenaman's vision of the Forest Estate, with appropriate management structures and the means to come to terms with local landholders, might have been.

Conclusion

Yet, in the short duration, the Solomons experience is not novel. Since the 1960s in the Malesian forests, from South-east Asia, Papua New Guinea through to Vanuatu and indeed in other tropical areas, loggers have found out the institutional and cultural susceptibilities of the people and 'mined' the forest. Governments and greedy individuals opt for short-term development capital and personal gain rather than the wise use of a permanent forest asset. Like their European counterparts of the North, these people of the tropical forests will have to address the hard realities of declining forest resources by replanting and conservation because, unlike most of the timber-hungry west Europeans from the late fifteenth century to the early twentieth century, they will not find any more 'new' lands with abundant forests to fuel their development.

Viewed through a pre-historic as well as this historic lens, the Solomon Islands is not the first group of Pacific Islands to feel the pincer grip of population pressure (and/or the need for increased social production) and the shrinkage of forest, the pivot of successful swidden or bush fallow agriculture and hunting of game. It is also possible that anthropogenic change from climate or tectonic forces helped to create or exacerbated these pressures on the environment.[11] Before the Europeans arrived, in parts of Melanesia, such as New Caledonia and the highlands of New Guinea, similar developments occurred. Further research in prehistory may well show more precisely similar situations in ancient times in the Solomon Islands. This has been the common experience of people on most of the high islands of Polynesia where the available land area is much smaller than in most of Melanesia. The least successful Polynesian adaptation was isolated Rapanui (Easter Island). A combination of increased socio-religious competition and rising population appears to have put enormous demands on relatively limited resources, particularly the forests and soil. Climate change through a fall in temperature in about 1300 AD played a part, too: an example of the several great and enduring forces of nature that elude the puny efforts of humanity. Soon after the forest disappeared into grassland, food production appears to have declined and the formerly expanding population collapsed from an estimated 10,000 in 1600 AD to 2,000, who were marooned on their treeless island. This may have become the ecologically modified island's maximum carrying capacity. We will never know, for European contact exposed the remnant to 'continental' diseases and brought the 2,000 to the point of extinction.[12] Similarly, the Maori population of New Zealand's South Island fell in numbers along with the contraction of the forests in the sixteenth century. Though not an agriculture-based society, because they lacked food crops adapted to the cold, the land could not carry the former large population as the forest was burned off, in part at least, to drive out the staple moa, a large, meaty, flightless bird. The moa numbers shrank to extinction due to a combination of loss of their habitat and human hunting practices.[13] On the high islands of Mangareva and Pitcairn, swidden gardening over the centuries before the Europeans came reduced forest and induced erosion. With reduced productive capacity and a dearth of suitable trees, the people could no longer make their regular long-distance voyages between their islands and the atoll of Henderson Island, whose population subsequently disappeared, cut off from vital supplies that supplemented those of their own marginal habitat.[14]

Elsewhere in the pre-European Pacific, agricultural and other food-producing methods changed with a happier outcome, though not always free from conflicts. What can be seen in the short term as land degradation often became, in the mid-term, land enhancement. In some areas with upland erosion river valleys silted up to create fertile lowlands. Reefs may have been lost, but coastal flats grew. As the forests retreated under decreasing fallow and soil loss, shifting cultivation, though still possible in pockets at the remnant forest up the steeper valleys, such as on Futuna and Upolu, was replaced by more intensified cultivation.[15] These 'semi-permanent forms of agriculture' evolved according to local conditions.[16] This involved weeding and mulching in dry-field systems such as in parts of Hawai'i and New Zealand and parts of the New Guinea highlands; or irrigation and drainage in wet-field systems, such as in lowland Futuna. Limited field rotation combined with arboriculture is another form of intensification. Within the borders of the modern Solomons, Tikopia illustrates the last method, while Anuta utilised a form of dryfield cultivation, involving mulching, weeding and rotating taro and cassava with a grass fallow of up to three years.[17]

This transformation was costly: although several variables influence specific quantification, in the main, more labour was required for survival. On Anuta, Douglas Yen calculated the amount of work required to produce the same output from dryfield agriculture was double that needed for shifting agriculture – about 7,000 person-hours per hectare per year.[18] For irrigated taro the person-hours needed range from about 4,800 to 5,900, depending on terrain, at least in New Caledonia.[19] So here is another index of the worth of the forests of the Solomons being shipped off-shore. In Solomons, there could be a re-discovery or, in some places, a revival of intensified cultivation of disease-resistant taro though large systems of irrigation and drainage.[20] The pressure of population is already seeing coastal peoples, often descendants of 'bush' people who came down to the coast in the late nineteenth and twentieth centuries, returning to some inland areas, at least for gardening and cash crop purposes. Compared with the typical sweet potato- and, since about the 1930s, the cassava-dominated plantings of the reduced fallow gardens close to coastal villages, these bush gardens are often more productively diverse and represent agroforestry.[21]

Conclusion

This is yet another phase in the ebb and flow of populations from coast to mountain crests over thousands of years, patterns emerging in the history of the long duration.[22] The recent re-opening of old lands has been facilitated in some areas by the creation of logging roads, but these dissolve or are smothered in weeds when the loggers depart. Without easy access to the coast, it seems unlikely that bush populations in the immediate future will expand dramatically unless all other possibilities to gain subsistence are closed or some beneficent aid donor constructs and maintains roading throughout the archipelago.

In the towns and increasingly in the villages, women of the Solomons are already exploring one technique of intensified agriculture – the *sup-sup* garden, incorporating compost, crop rotation and mulching. This method, introduced more to provide variety in the diet and improved nutrition in semi-urban areas than to replace the typical gardening of the Solomons, points to possibilities, as bush fallow rotations become shorter. All else being equal, the sharp spur of necessity will force Solomon Islanders to more widely adopt this and other forms of intensified agriculture, sooner or later.

Solomon Islanders, so long dependent on their lands for their physical and spiritual life, have a fierce attachment to them and, as both the colonial and independent governments found, to the trees thereupon. No government has ever been able to drive significant numbers of people off their lands or deprive them entirely of their birthright, though the colonial government did make local movement onto the 'Waste Lands' of Kolombangara difficult. Unlike many other post-colonial states, in south Asia and Africa for example, no village people have suffered forced re-location to make way for a government forest or a national park, reserve or wildlife corridor. In the case of the Queen Elizabeth II Park near Honiara, few ordinary people have heard of it, including the hundreds of 'squatters' living there in 1997. There was little precedent for this during colonial times, other than the iniquitous Waste Land regulations and these were aimed at areas of minimal if not nil resident population density.

Ever suspicious of governments, Solomon Islanders in the rural areas still contest the claims of the state to rule on land matters. Although some, both inside and outside the archipelago, might have wished for strong government to successfully control the direct dealing between loggers and rights holders and thus assist the sustainability of timber extraction, the government's weakness here has been the nation's strength. Though appalling mistakes have been made in regard to the exploitation of the forest resource, the ordinary people still hold in their hands the means to their subsistence, though probably not for Western-style affluence. Given the record of recent Mamaloni governments in enforcing the rights the state has over lands which are legally its own, such as Pavuvu and Viru, and its neglect in enforcing stringent logging methods on its own lands, it may well have been that a wider state power over lands and forests, so desired but eventually abandoned by the colonial government, would have meant, at least to 1997, an even more ruthless reduction of the forest with far less return to the inhabitants of the land. Apart from the mounting degradation of the forest cover and soils, the major present threat to the loss of the means of a more general subsistence for the people would be the registration of lands for the purpose, not of keeping the land in the hands of its rightful holders for their use (perhaps under, say, clan incorporation and leasing schemes for large projects), but for freehold consolidation under a few individuals, as some orthodox economists, including the ubiquitous IMF, have urged on Pacific states for years.[23] Paradoxically, if the economy continues to deteriorate the state may still have its day and introduce significant land tenure change along these lines, propped up by international funding bodies and governments. Under the global schemes of Tropical Forest Action Plan large-scale replanting of logged areas, implementation and monitoring would be simpler, at least in the short term, if a government had control of major land areas.

As the logging industry on customary land neared collapse a tired Mamaloni stepped aside for Ulu'afalu in the elections of mid-1997. This may signal an end to the frontier phase in independent Solomons: a confident generation, unpreoccupied with grievance against colonial paternalism, is coming of age and looks to find its own path. Many an aging resource robber now wants respectability for his educated children. A growing middle class seeks economic and political stability. This change in the political and economic landscape could make reform of the broad environmental, financial and legal issues associated with logging more politically feasible. The remaining stakes are high. The losses and missed opportunities of almost 20 years could still take their toll. The forests and other natural resources are not and have never been unlimited. The burden of continuing economic problems and burgeoning population may fuel social unrest. Most Solomon Islanders and indeed their neighbours have put their hopes in Ulufa'alu, but all are more guarded than ever before about political promises. Aid donors are poised to assist on several fronts, including the forests. His stepping aside may not mean Mamaloni or his like will not return to power. For those powerful interests who would control the disposal of the remnant commercial natural forest, this new regime may be just a minor blip in the downward curve of resource exploitation of this Pacific archipelago, the once 'Happy Isles'.

For most historians, including this one, the past of Solomons is elusive enough; the future is for others to make their own.

Afterword

Historians who have the temerity to discuss recent events are always vulnerable. Recency strips their armour of hindsight, fine focus, and seeming closure possible with the analysis of the more distant past. When the events are still fresh there can never be a satisfactory assessment because new things continue to happen, variously and often unpredictably connected to their immediate historical antecedents. Since I submitted this manuscript, however, significant changes have emerged that warrant mention.

One of the first actions of the new government was the suspension of the area councils in early 1998, putting a brake on new logging applications. The government accepted the Australian offer of aid. This enabled the drafting of new legislation resulting in the Forests Act 1999 and subordinate acts. Despite widespread scepticism and even opposition, the Forest Act became law in July 1999. The three-year contract that the Solomons government signed with Australia provided SI$16.5 million for a bilateral Forest Management Programme, the implementation of the legislation being a prime component. As the Acting Commissioner, Enele Kwainarara went into parliament, the government selected Peter Sheehan, Associate Professor of Forestry at Melbourne University, as the new Commissioner of Forests.

Part of the new Act was a revised and mandatory form of the optional Code of Forestry Practice, initiated by the Timber Control Unit in 1994–1996. The Act seeks to address the changing circumstances that began to emerge after 1969, but most particularly since 1977/8, once logging of customary land became common. It eliminates myriad anomalies such as the old sawmill licence that had become a blind for unregulated tree felling, selling, and exporting by local owners.

The Act enunciates guiding principles for the Minister and the Commissioner: sustainability of resource utilisation, the rights of customary owners, application of the precautionary principle to management decisions, the balancing of economic and ecological objectives, the protection of biodiversity, consistency with international treaties and obligations, and consistency with the national policies for forest resource conservation and timber industry development.

The Act remedies the major flaws in earlier legislation as well as introducing new measures. It recognises the intricate legal and ecological relationship among land, soils, and forest. First, it removes the confusing separation between ownership of trees and ownership of land. A break with old Melanesian tradition, it acknowledges the significant changes in tenure concepts of the past 30 years as trees came to have a monetary, rather than a use value. Since owners of the land are synonymous with owners of the trees, the recording of tree and land ownership with concomitant rough mapping could well mean, on customary land, a form of registration. In turn, this may lead to the sale and concentration of land in the hands of fewer owners.

Secondly, a radical, but potentially soil-saving requirement, is in the resource management clauses. The Forestry Division is to assess all land on which its owners plan some kind of 'development', be it logging or clearing of any kind, except for garden sites. Terminology echoes the emphasis on conservation: 'logging' is now 'timber harvesting'. Trees are a crop, not a non-renewable resource. To preserve water catchments as well as soil, the Minister retains the power to declare Forest Reserves. There is no specific power to declare areas reserved because of special conservation value, though if the government ratifies international agreements – relating, say, to World Heritage sites – then the Minister, it would seem, is bound to consider this. State land may be used for plantation forestry and/or to protect significant conservation values.

The Act specifies procedures for the identification and notification of all landowners regarding any timber harvesting application. Provincial governments have a role to play, but must report on how they fulfil their new responsibilities. The councils of chiefs, local courts or, if necessary, the customary land appeal court are to settle disputes. Groups of identified landowners may negotiate through representatives, but, like the provincial governments, their responsibilities are explicit. As the North New Georgia Act has demonstrated, getting mutually agreed representatives may not be so easy. The Commissioner is responsible for seeing that landowners understand the nature of any harvesting agreement and that it is fair. If the provincial government then gives its approval, the licence to harvest may be issued. The mandatory Code of Practice entails heavy fines for abuses and/or gaol sentences and gives legal protection to landowners. Stringent conditions apply to the sale, marketing, and export of timber, with prices set by the Commissioner.

Another innovation is a Forestry Board with members from concerned ministries and interests including the Central Bank, commercial loggers,

Afterword

SIFIA, the Development Services Exchange, customary owners, along with the Commissioner and Secretary of the Ministry as ex-officio members. This Board must meet at least three times a year to advise the minister.

The Minister, in consultation with the Board, is to allocate funds for reforestation. These funds seem to be for tree planting on both customary and State land. A new Forest Trust receives money for this from a forest development levy, along with 50 per cent of fines for infringements of the Act, 50 per cent of licence fees, as well as any other monies allocated by parliament or by donors. No longer is the levy, in a concept dating back to Trenaman's time, destined to be lost in consolidated revenue. The government recognises the industry should pay to replace what it has removed and should be seen to be doing so. This will convince donors also of Solomon Islands government's commitment to sustainability and could attract more forest aid.

Overall, the Act demonstrates a far more active role by both central and provincial governments in the relations between growers and loggers. It recognises the imbalance of the former relationship and protects resource owners as well as the resources of forest and soil. A clear and predictable legal context ensures security for both foreign and domestic logging companies. Through this Act, a passage to the open sea of international commerce seems to have been discovered beyond the reefs of the Independence-induced licence and the shoals of colonial British paternalism.

For all that promise, however, the ship of state is listing badly. Recent events may negate the Act. In late 1998, the premier of Guadalcanal Province, and one-time prime minister, E. Alebua, let the genie out of the bottle by declaring that the national government should pay the province compensation for having the capital there and that Malaitan settlers should show more respect to their Guadalcanal hosts. This crystallised resentment against Malaitans that had been brewing since the mid-1980s. Certainly, from the 1950s, Malaitans have bought, rented, and squatted upon land on north Guadalcanal, a situation permitted by many older Guadalcanal people, mainly males. These have orchestrated negotiations and often over-ruled their womenfolk who traditionally inherit the land. Many of their children, now adults, resent this dispossession and blame the Malaitans. Their extended families augmenting their numbers, Malaitans often flaunted their customs, offending the sensibilities of Guadalcanal (Isatabu) people. Add to this the discontent of the landholders of the CDC and government-leased land on north Guadalcanal over what they consider a poor return from the oil-palm company, and land control becomes even more contested. Moreover, because Malaitans are hard-working people they were sought after as employees. As the economy weakened over the years, unemployed Isatabu people, particularly from the isolated, services-poor south coast, resented this, because they considered any jobs their prerogative. The example of

Bougainville's success in driving out mainland Papua New Guineans, and its defeat of the army, recounted by Bougainvilleans in Honiara, was not lost on the Isatabu malcontents. For their part, Malaitans believe that their labour over decades has helped the country. Like the people of several other islands, they feel the colonial government neglected their island's development in favour of Guadalcanal with its national capital. They see the successor government as no better, so many give it no loyalty.

Rioting in 1999 shut down the huge oil-palm operation on north Guadalcanal. Guadalcanal men attacked and even killed Malaitan settlers there, causing about 24,000 to return to Malaita and others to flee also. The destruction called forth a Malaitan demand for compensation from the Isatabu people. The government, however, failed to act decisively, and prominent foreigners failed to bring the parties to negotiation. Though by the end of 1999, most Malaitans had retreated home, their characteristic assertiveness soon resurfaced, fed by supporters of the Mamaloni way of government who were losing power and money as Prime Minister Ulufa'alu doggedly implemented fiscal, structural, and forest reform.

Each side attracted its militia of young men, mainly unemployed school dropouts, armed with World War Two ammunition, home-made guns, and stolen police arms. In June 2000, the Malaitan 'Eagle Force' took over Honiara and, for a time, captured Prime Minister Ulufa'alu. Headed by the disaffected Andrew Nori, a lawyer and former minister of Mamaloni, the militia forced Ulufa'alu's resignation after some of his parliamentary supporters deserted him. Most of the expatriate population left and, with them, their contribution to the economy. Manasseh Sogavare now heads a largely ineffectual national government, with an impotent police force and a stalled judicial system. The Eagle Force 'protected' Honiara, but soon degenerated into thuggery and theft, ferrying dozens of vehicles and other loot back to Malaita. Both sides have carried out 'pay-back' killings and torture, even on non-militia personnel. About 200 people have been killed. Ulufa'alu and others called on Australia and New Zealand to send in military support for the government. Until August, when this book went to press, they have declined, but have supported and financed peace negotiations. After Solomon Islanders themselves negotiate peace, they have promised assistance with peacekeeping. Though the government and militia have made several so-called agreements, peace and reconciliation are elusive, as rogue elements continue the conflict. One hope is the increasing public desire for peace, expressed strongly by women, the churches, and certain NGOs.

This coup could be portrayed as arrogant, expansionist Malaitans making trouble and resentful, hot-headed Guadalcanal men resisting them. At the root, however, is the irresponsible performance of past governments, particularly those of the now-deceased Mamaloni, which crippled the coun-

Afterword

try's chances of addressing its problems. In the forestry sector, these either squandered the forest capital or simply failed to collect the timber revenues that should have benefited the whole country. They set a poor example of the use of power and undermined the qualified confidence the people had in the state. Parliamentarians, often with no more than ten percent of their electorate vote, have bought their minute following with favours. Lacking legitimacy, such men are not trusted or respected.

Their economic mismanagement has meant fewer employment opportunities for a still-increasing and youthful population of about 410,000, and jealousy about who should have jobs, education, and resources. More-over, rightly or wrongly, the former British presence had forced only a seeming unity upon disparate Melanesian societies. Until now, when economic bad times exaggerate cultural sensitivities and differences, Solomon Islanders have not had to systematically address the issue of how to make their islands function as a viable entity.

The inter-island conflict has exponential impacts that affect the forest. Malaita's soils and forests, long stressed from subsistence gardening, received a further shock as the returning Malaitans made gardens to feed their families. The resident population of 120,000 is three times what it was in 1931. In the Marovo lagoon, NGO and New Zealand government aid to eco-tourism has encouraged heavy investment of time and effort by local people in visitor facilities. As most visitors come via Honiara, tourism is at a standstill. Unless tourist numbers increase, eco-tourism will founder and the people involved will seek other ways of getting cash. Projects such as the Vangunu oil-palm of Kumpulan Emas, which aims to clear more forest beyond its own concession, then could seem financially attractive for a government since the closure of the Guadalcanal oil-palm industry. More significant, in 1999, the country had been getting over 50 per cent of its export earnings from logs with about 2,700 employed in forest industries. The Forests Act was to be a prelude to a national forest policy that would have probably reduced extraction rates to conserve the forest for sustainable production. Such a policy, in the short term, could only intensify the country's deepening economic crisis. Already, Tommy Chan, the new Minister for Forests, Environment and Conservation, is welcoming the return of the expatriate loggers, the 'businessmen', as potential saviours of a cash-strapped government.

Given that political uncertainty will continue, with or without peace-keepers, it seems even more unlikely now than it was in late 1997, that the forest will be given time to recover from the extraction regime of the last twenty years, in spite of the comprehensive Act. It, along with the projected forest policy, may become just another set of forlorn hopes for this much-contested Pacific forest.

Appendices

Appendix 1.
Annual Property Schedule 1925

date	cost	area (acres)	value (1925)	site/property
Roviana Circuit				
1902				Roviana – Mission House,
				Roviana – Sisters' House, Store
1912				Roviana No. 2 House
1921				Roviana – Electric light, plant and furniture
1923				Marovo – Native House
1913–1970				Institutional Building
1915				2 boys' houses
1910				Teachers' House
1917				Store
1902	£15,200 freehold		£4,920	Kokenggolo – land
1918				Cattle
1902		1? acre		Island of Nusuzonga
1920				Sisters' boat
1902	£1	1000 acres	£21,500	Island of Mbanga
1912–1914				Mbanga – building
1907	£10	25 acres	£40	Simbo land
1910		3/4 leasehold		Gizo Store and Site
1910			£1,400	'Tandanya'
1921				'Saga'
				'Hilda'

Total £48,963

continued …

Appendices

date	cost	area (acres)	value (1925)	site/property
Mbilua Circuit				
1907				Mbilua – Mission House
1919				Electric light plant and furniture
1921			[sub-total £1890]	Sisters' house
1905	£50	1000 freehold	£11,105	Land
1918–1920				Cattle
1907	£10	160 freehold	£200	Njorio – land
1907	£5	15 freehold	£40	Ranonnga – land
1912				Launch
Total £13,035				
Choiseul Circuit				
1905 (rebuilt in 1917)				Bambatana – Mission House
1913–1914				Sisters' House
1920				Additions and furniture
1912				Store
1922			[sub-total £1394]	Cattle
1911	£40	40 freehold	£1,250	Land
1921				Senga - seminative house
1911	gift	2 freehold	£50	Tambatamba Island
1921				Whaleboat
				'Te Karere' launch
Total £2,649				

Sub-totals not recorded
Source: Box White A107, Methodist Mission Archives, Auckland

Appendix 2.
Plants in Woodford's experimental garden at Tulagi, 1898, as described in his own words.

INDUSTRIAL FOOD AND FRUIT PLANTS	
Bermuda arrowroot	Lemon
Sugar cane	Lime
Papau (Caries papags)	Pumpkin
Orange (Mandarin)	Water melon
Orange (Paramatta)	Tomato
Egg fruit (Solanum)	Yam (2 varieties)
Granadilla (passiflora)	Cassava
China beans	Chili pepper
Piper betel	Capsicum
Piper methisticum	Custard apple
Areca palm	Pineapple
Coco-nut palm	Coffee (4 varieties)
Sago palm	Cocoa
Banana (4 varieties)	Ginger (Jamaica)
Taro (Caladium esculentum)	Cinnamon
Tamarind	Mangosteen
VanillaSweet potato	
Turmeric (2 varieties)	Ceara rubber
Agava rigida	Ficus elastica
Foureroya gigantes	Tobacco

ORNAMENTAL	
Coleus	Orchids – cont.
Dracoens (varieties)	Local species (15)
Bougainvillea	Frangipani
Alamanda (2 varieties)	Gardenia florida
Eucharis amazonica	Celosia
Amaryllis (varieties)	Hibiscus (5 varieties)
Canna (4 varieties)	Phlox (3 varieties)
Caladium (2 varieties)	Taxonia manicata
A malvacious shrub?	Ixora
Lagerstroemia ovata	Anthurium crystallinum
Clerodendron	Crinum moorei
Adiantum farleyense	Crotons (various)
Lygodium (2 species)	Mignonette
Orchids:-	Palms (5 species)
Cypripedium barbatum	Poinciana regis
Dendrobium spectabile	etc.

Source: AR-BSIP 1898/9

Appendices

Appendix 3.
Public Notice of Forest Areas

WESTERN PACIFIC HIGH COMMISSION,
HONIARA,
BRITISH SOLOMON ISLANDS PROTECTORATE.
PUBLIC NOTICE 77/68
F.327/2/18   21st June, 1968.

Orders were made on 28th May, 1968 under provision of the Forests Ordinance declaring 17 separate tracts of land to be Forest Areas. These occur on Guadalcanal (4), Isabel (4), New Georgia (4), Vangunu (2), Vella Lavella (1) and Shortland Island (2). The effect of the declarations is that "no trees may be felled other than for domestic purposes" within the tracts "except in accordance with the terms of a licence or permit issued by the Conservator of Forests" (section 24 of the Forests Ordinance). Cultivation and settlement is thus prohibited without a licence or permit, and so is the cutting of trees for sale, or for sale of timber from them.

2. Every effort has been made however to exclude from the Forest Areas land already used for agricultural purposes, or likely to be needed for this, at least in the near future. Furthermore Forest Areas are not intended to imply permanent land-use allocation, so that should need arise in future years for more land for cultivation or other purposes, particular Forest Areas may be readily amended or even revoked.

3. Rights to hunt and fish and to take any produce of the forest for personal and domestic use, such as poles, leaf, or timber for houses, canoe trees firewood, medicines nuts and fruits are not affected by the Forest Area declarations.

4. Forest Areas were declared in 1967 over parts of Kolombangara and Shortland Island. Licences and permits to authorise timber working in these areas have subsequently been issued.

5. The 1968 Forest Area declarations are based on the results of forest surveys done mainly in the period 1965–1967 and it is thought that the most of the important timber tracts in the Protectorate have now been located. The purpose of these declarations is to bring under clear control that part of the Protectorate's timber resources not yet committed to timber working, so that the whole resource may be used and developed to the best long-term advantage. In particular it will be possible to control the rate at which the country's natural timber stands are cut out, and to ensure that regeneration takes place in tracts which are suitable for productive forestry use on a permanent basis.

6. Regeneration, in the suitable tracts, will normally involve acquisition of land by the Government. It is stressed however that the Forest Area declarations do not affect or alter the ownership of the land and have nothing to do with land acquisition. Any land purchase proposals by the Government will be made quite separately. As in the past for the tracts bought at Allardyce, Vangunu and the Viru-Kalena region of New Georgia, it is the Government's policy to purchase land needed for forestry use by negotiation and agreement with the owners concerned.

7. A large part of the country's forest resources has already been committed to timber-working and it is now necessary to conserve supplies for the future needs of the timber industry. It is not anticipated therefore that any of the newly declared Forest Areas will be opened to timber working in the near future, except insofar as necessary to meet urgent needs of existing timber companies.

8. Maps to show the location of the Forest Areas declared on 28th May, 1968 may be inspected by members of the public at Forestry Offices, at Honiara, Gizo and Munda.

for *acting Chief Secretary.*

Appendices

Appendix 4.
Forest Areas and People's Protection Party

Rere Plantation,
via Honiara, B.S.I.P.
16th July, 1968.

T. Russell, Esq.,
Chairman,
Legco. Special Select Committee on Forestry Areas.
Honiara

Sir,

1. On 19th June 1968 it was proclaimed that certain areas had been declared Forest Areas under the Forestry Regulations.

2. A brief reference to this was made some days later in the S.I.B.S. broadcast. Subsequently, the Hon. Baddeley Devesi forwarded a copy of two declarations affecting the area from Bokokimbo River to the Kau Kau district.

3. As a result of some publicity being given to this matter, a very large meeting of headmen and others from all parts of the area affected was held at Rere on 30th June.

4. Mr Devesi had sent a copy of the Forestry Regulations and the effect of these declarations was explained to the meeting. A long discussion took place and it was decided to form a Protection Society with a view to defending what these people consider to be their inherited rights. A Committee was appointed consisting of two members from each and every section of the district. The meeting decided to oppose to the utmost any attempt to implement Part 4, Section 23 of the Regulations.

5. In this matter they consider that no one possibly could entertain the slightest suspicion that Government is acting from purely altruistic and unselfish motives. Government is at least consistent in its policy of opportunism. The people concerned believe that the application of this Section 23 is unjust and discriminatory, and that it contravenes the Pacific Orders in Council of 1875 and 1893.

6. The basic guarantees and promises made under these Orders in Council, in the opinion of the people of this area, preclude Government from taking any action of this nature. A mere intention of goodwill and expressions of good faith on the part of Government is not sufficient to justify arbitrary confiscation and permanent control of the peoples' property. The sovereignty of chiefs and headmen was guaranteed by those early Orders in Council, and the people of this area are determined to take the matter to the highest British Court of appeal, and, if necessary to the relevant committees of United Nations.

7. This "forestry area" item first saw the light of day in 1960. For eight years this misbegotten progeny of a Government, in which the native people of these islands at that time had no direct representation, has been allowed to remain in limbo. Only a self-centred Government, entirely concerned with its own welfare, ever would have allowed the monstrosity to come crawling out of its hole in the ground. No academical questions of legality or politics should be allowed to deny simple justice.

8. As no one in the area had noticed in the last few years (at least it had not been reported) that survey teams had been on the job of demarcation, it was rather puzzling how the "lines" had been covered. It was assumed, therefore, that the demarcation lines had been obtained from aerial photographs. Did this include precision dropping of marker pegs from the air by hand, or was it done by a bow-and-arrow or other mechanical device? Imagine a native who wishes to avoid a heavy fine and/or a severe sentence trying to elucidate "thence by grid bearing <u>approximately</u> 34 deg. to the point of intersection of the middle thread of an unnamed stream with the middle thread of the Kolosahata River at grid reference <u>approximately</u>..." and so on ad infinitum ad absurdum. Keep in mind, or course, there are no out lines on the ground or marker pegs and imagine the difficulties of the most competent surveyors dealing with so many "approximations".

Appendices

9. Since the first meeting at Rere on 30th June, other meetings have been held up and down the coast and the people affected have unanimously agreed to oppose and resist this iniquitous measure :

    (a)   By appeal to the Government appointed Committee.

    (b)   By recourse to the local Courts.

    (c)   By appeal against any adverse judgments of the local Courts.

    (d)   By such further appeals as are available to at law.

    (e)   By appeal to the United Nations Colonial Committee.

They wish to appear before the appointed Committee on August 6th and have appealed to me for assistance and to act as their adviser.

10. They realise, of course, that to make any approaches to a "stacked" committee of six Civil Servants, and three native members of Legco, is akin to spitting against the wind, but, nevertheless, are prepared to submit to orthodox methods of procedure. They are confronted with a fait accompli and are well aware that any representations to such a committee cannot possibly succeed.

11. Compare the method of acquisition of timber lands in Papua/New Guinea. Recently in New Guinea, after lengthy discussions, a timber area in the mountains was acquired for $160,000. The native owners asked for, and obtained, this amount in silver. There was no question of acquiring title to the land or re-planting without the approval of the native owners. The natives sold only the timber and their title to their land remains untouched. (Authority: Pacific Island Monthly)

12. Quite recently, the S.I.B.S. released an interesting item of New Guinea news. The United Nations Special Committee visiting New Guinea strongly recommended that the New Guinea people should take a greater financial interest in timber development in that country both as regards working their own timber and as partners in timber operating companies.

13. The questions the natives ask are these: "Do we own the land? If so, don't we own the timber?" "Why can Government steal our timber and our land in this manner?" "Government has told us that they are not confiscating our land, but Section 25 contains a provision that the Chief Forestry Officer may order the people who work the timber on our land to re-plant. This could go on for ever. Isn't this the same as stealing our land?" "Why didn't Government consult with us first before making use of Section 23?" "What moneys have Government ever spent in the past on the areas now being expropriated?" "Can we trust the Government to spend the royalty money for our benefit?" "Why can't we make our own contracts for the sale of timber in what is unquestionably our forest country? Is it because Government thinks we are not fit and proper people to make such contracts?"

14. Such expropriatory legislation as this does not exist in any other country of the "free" world. This is confiscation without any mention of compensation. The native people regard themselves as trustees of their own land and as trustees for its timber. While they live they have the right of using this land with the permission of their chiefs and headmen. The protection of this right was promised to them by Queen Victoria in Pacific Orders in Council.

15. Recently in Legco, Mr. Twomey said that no modern country permitted private ownership of the wealth beneath the surface of the soil. (I may have got this wrong as I heard it over the air). I think Mr. Twomey is wrong in this matter. The U.S.A. has always recognised the right of the owner of the surface to the minerals beneath it. In Great Britain, the royalties of the Duke of Northumberland and the Marquis of Bute on the coal extracted from their estates have been enormous. Whether these have since been compounded or not I cannot say. In Australia, many old titles still give ownership of what is beneath the surface to the proprietor; the Australian Agricultural Company obtained a million acres of this type of title.

16. Can these local landholders afford to take the risk of allowing their valuable timbers to be grabbed and their land tied up for an indefinite number of years in return for some vague intimation by Government that the spending of the proceeds from timber exploitation may benefit them eventually? These people, whose property is in jeopardy, are not prepared to concede the divine omniscience and self-sacrificing intentions of Government.

Appendices

17. It is said that the first missionaries who went to Hawaii owned the bible and the native people owned the land, but that as time went on the native people owned the bible and the missionaries owned the land. In this instance, when Government first arrived here, the native people owned the land and Government the law. With the application of this type of legislation it shouldn't be long before Government owns both the land and the law.

18. Because of the nature of the Legco, Select Committee appointed to deal with this matter, it seems that any protest, in order to be effective, must go outside the perimeter of the B.S.I. Government's authority. Any organised protest, in the first place, would have to exhaust the legal remedies provided in the Protectorate, and to make the necessary appeals against any judgment of the local Courts. The question may then arise that, this being a Protectorate, to what extent would Government be prepared to assist them in this matter. Their point is that any expense they would be compelled to incur would be, of course, as a direct result of Government's action, and it would seem only fair that Government should cover their legal expenses in such a matter.

19. An 1866 United States Federal Statute says: "all citizens of the U.S. shall have the same right, in every state and territory, as is enjoyed by white citizens thereof, to inherit, purchase, lease, sell, hold and convey, real and personal property". The present Forestry Regulations of B.S.I.P. have a flavor more of 1768 than 1968.

20. A year or so ago, the S.I.B.S. announcer made the remark that with the removal of the restrictions on natives buying "hard" liquor, the last piece of discriminatory legislation had come to an end. How is this reconciled with the fact that Section 23 of the Forestry Regulations is aimed at confiscation of native property? I know of no civilized country which expropriates the property of its own people without reasonable and adequate compensation.

21. In the early part of last century and until quite late in the century, the happenings in the Indian country of U.S.A., amongst the Maoris of New Zealand, and then again in South Africa give a very clear indication of what the reactions of native people are against what they consider the theft of their tribal lands. Most British people never own any land during their lifetimes, but, to these native people, the land is a part of themselves and they are a part of the land. The application of Section 23, in the form in which I have seen it, indicates that Government may even remove timber from the "tabu" areas which are sacred places to the natives.

22. Has the British Government in England, since the last enclosure acts of the early 19th Century, without paying full compensation, ever applied expropriatory legislation of the nature of Section 23 of the B.S.I. Forestry Regulations.

23. This is not good Government, nor is it likely to promote peace and order in the country. It is economic rape under a very thin, and, I believe, untenable veneer of legality.

Yours truly,
PEOPLES PROTECTION PARTY
Committee Members

I. Tagoa	Charles Mau
Timon Taria	Andson Cook N.
P.M. Kaomaue	John Golagola
J. Mali	Paul Pirua
J. Bosa	Tago
Reuben Mostyn Bula	Ronouania

This letter was prepared at the request of the Committee of the Peoples' Protection Party. It has been read to them and to a General Meeting of the Party and has been adopted unanimously as their letter.

(Sgd) J. Bryan.

Appendices

Questions posed by the People's Protection Party for the Special Select Committee on Forestry. (Handed to Committee on 6.8.68.)

Q.1. Government claims that it is not expropriating or confiscating our land. Nevertheless, the Regulation provides that the Forestry Officer may order any person or company exploiting the timber thereon to replant. Under such circumstances, which may be repeated again and again, would not this amount to a permanent deprivation of the use of our land?

Q.2. The Forestry Area Regulation prohibits the owners of the land from using it for cultivation or the making of a settlement. How can these prohibitions be reconciled with standard methods of native cultivation or with the requirements of a fast growing population?

Q.3. How much rent per acre is Government willing to pay for this land while the people who own it are deprived of its use?

Q.4. After reaching agreement on the question of rental per acre per year to be paid for the land, how much per 100 super feet of timber taken away from the land is Government willing to pay to the native owners of the timber?

Q.5. We think this Forestry Area law would be a good law for wild pigs. Does Government recognise us as people or wild pigs?

Q.6. Do you know of any country in the free world where the Government has taken the power to confiscate the timber belonging to the people or the land belonging to the people without paying full and proper compensation? We believe this sort of thing went out of style over a hundred years ago and the one or two revivals since then, i.e. Hitler and a few others, proved unpopular and did not last too long.

Q.7. This Forest Area regulation has been in the B.S.I. law book since 1960. Why did Government wait until 19th June 1968 before first applying it, and why did Government not talk to the people before then to find out their ideas of it?

Q.8. Fiji is a Crown Colony answerable to the British Government through its High Commissioner. Why has there never been a similar provision in the Forestry Regulations of Fiji?

Q.9. How many of these Forestry Area regulations orders have been served on the island of Malaita?

Q.10. Is there any similar law to this Forestry Area regulation in Britain, Canada, Australia, New Zealand, or other British colony or within the British Commonwealth?

Q.11. Why did Government not introduce the same Forestry Regulations here as they have in Britain?

Q.12. Under what Order in Council was this Forestry Regulation made, and under what particular clause of such Order in Council?

Q.13. As this Forestry Area has not been surveyed, how are we to know when we are trespassing? No boundary lines have been cut and no pegs placed in the ground.

Q.14. Has this Special Legco Committee any power to cancel these orders, and, if not, what is the purpose of this Special Committee?

Q.15. Assuming this Forestry Area is granted to an operating company, would the money received by Government as royalties be handed over to the owners of the land?

Q.16. How does Government reconcile part 3 of Clause 24 "...nothwithstanding the terms of any forestry concession or agreement made.....before the date of the making of such order" with common law guarantees of freedom to contract and where does Government's authority derive from to put such a clause in the Forestry Regulation?

Q.17. Part 2 of Public Notice 77/68 of 21st June 1968 says every effort has been made to exclude from the Forest Areas land already used for agricultural purposes, or likely to be needed for this at least in the near future. Can any member of this Committee show me where any effort has been made in the Forestry Area regulation ensuring this?

Appendices

Q.18. Part (a) of Clause 30 Forestry Area Regulation says: "Whoever falls any tree". How does Government, in the absence of any definition in the Forest Area regulation of a "tree", reconcile part 3 of Public Notice 77/68 with the Regulation? For instance, when does a tree become a "pole" and how long and thick is it?

Q.19. Para. 4. of Public Notice 77/68 says parts of Kolombangara and Shortland Islands were declared in 1967. Were these orders applied to private properties which were at the time being worked for timber or were the orders applied to customary land? If the orders were applied to private properties what protests, if any, were made by the owners?

Q.20. Para. 5 of Public Notice 77/68 refers to the timber resources (being) "used and developed to the best long-term advantage". The question is "to whose best long-term advantage"? So far as the Forest Area regulation itself is concerned, can any member of this Committee show me any part of it designed to ensure the "long-term advantage" to the native owners of the land?

Q.21. Para. 6 of P/N 77/68 says the Forest Area declarations "do not affect or alter the ownership of the land and have nothing to do with land acquisition". Does Government really expect the native owners to believe this? Land ownership implies full rights of usage, physical possession, and enjoyment, of the soil and its products, and the right to lease, sell, assign, or otherwise dispose of the land. The deprivation of the owner of any part of his privileges as owner must affect his "ownership" and to take away his full rights of ownership or a part of his rights is, for the authority doing so, to acquire those rights while the deprivation continues. The cold language of the Forestry Area regulation is LAW. Public Notice 77/68 is just a piece of bureaucratic sophistry apparently for no other purpose than to make the regulation somewhat more palatable or to obscure the main issues.

Q.22. Para. 7 appears to say "it is not anticipated (expected?) that any of the new Forest Areas will be opened up for working except to meet the needs of existing timber companies. Is this supposed to be conciliatory? The Regulation gives power to Government to dispose of the timber cutting rights to anyone at any time Government pleases.

Q.23. To what extent is P/N 77/68 valid at law as against the express provisions of the Regulation?

Document Three. People's Protection Party

(Received 13th August, 1968)

T. Russell Esq.,O.B.E.,M.L.C.
Chairman of the Legislative Council Select Committee on Forestry,
Honiara.

Sir,

FORESTRY AREA REGULATION.

1. My Committee, supported unanimously by the members of this association under the name People's Protection Party, have instructed me to write to you and to your Committee. This letter is intended to make clear our very strong objections to the implementing of the Forestry Area Regulations.

2. When the British Government first took an interest in this country many years ago our own people had clearly established ownership through having been settled in the country for an unknown number of generations. By cultivating the soil and establishing our homes here we became a unity of families and clans with a past and with the promise of a long future to come. The Government did not bring us the Law. We had a well defined legal structure before Government came.

3. The British Government proclaimed a Protectorate over these islands and promised to guard the sovereignty of our Chiefs and Headmen.

Appendices

4. Now, by means of this Forestry Area Regulation, Government seeks to deprive us of the use of our land and of the timber growing on our land. Against this we strongly maintain that the timber on our lands and the land itself is ours and ours alone by right of ownership and that any curtailment or restriction placed upon us in this regard is an act of confiscation, expropriation, and suppression of our rights.

5. We claim also that part 3 of Clause 24 of the Forestry Regulations "...nothwithstanding the terms of any forestry concession or agreement made...before the date of the making of such an order" is an interference with our common law right of making legal contracts.

6. We are not opposed in principle to Government giving power to the Forestry Department which will ensure a continuity of timber operations in this Protectorate, but we challenge the right of Government to confiscate our timber and/or our land to enable Forestry Department to do this. So far as Forestry Department is concerned we regard it as quite free to institute a policy of perpetual forests on Government land, but not on land owned by the people unless the people are fully and clearly agreed on permitting the Department to do so. The fact that Forestry Department wishes to do so and considers it in the best interests of the country (or of Forestry Department?) is, in itself, not sufficient justification for this action.

7. We cannot regard Public Notice 77/68 of 21st June, 1968 as a responsible Government document. The Forestry Area regulation, as set forth in the Law book, to us is the LAW. The Public Notice is not law in our eyes. There are too many variations between what the Law book says and what the Public Notice says.

8. Part (a) of Clause 30 of Forestry Regulations says "Whoever falls _any_ tree". The definition of a tree in the Regulation includes saplings, brushwood &c.. The Public Notice says that poles and building timber may be cut. It appears to us that the Public Notice is merely an attempt by Government to put a sugar coating on the pill.

9. Para. 5 of P/N 77/68 refers to the timber resources (being) "used and developed to the best long-term advantage". The Forestry Regulation, however, states that the Chief Forester may order the re-planting of areas in which the original stands have been cut out. Presumably, this could be done over and over again. Meanwhile the owners of the land are deprived of the advantages and benefits of ownership with no question of being awarded adequate compensation for the timber or an adequate rental for the land.

10. The Forestry Area section of the regulation, ever since it appeared in the Law book, has been a threat held by Government against us native people making our own contracts for the exploitation of timber on our lands. Government cannot claim that we native people have not shown any initiative in the matter. That particular part of the Regulation was in itself a veto placed on us native owners and any person or company which wished to make a contract with us for the purchase of our timber.

11. Government has claimed many times that it is actuated by a desire to develop the country for the general benefit of the country, but the application of these Forestry Area regulations is discriminatory. The "Robin Hood" policy never gained a firm footing at law, nor the policy attributed to the Jesuits: "the ends justify the means". A benevolent burglar cannot expect to receive much leniency from his judge on that score.

12. We people claim that we ourselves should have the sole right to determine whether worked out timber lands should be re-planted or diverted to agriculture or pastoral purposes. We believe it to be foolishness to think of re-planting with trees good agricultural land giving an annual crop, or providing permanent pasture for cattle, in favour of a crop which would take fifty years or more to grow and then give a crop which measured in dollars per acre per annum would be woefully low.

13. My Committee wish me to emphasise that the failure of us native people to develop our lands in the years gone by has been due not to a lack of energy, initiative, or even technical ability, but to an entire lack of capital. For a long time now we have been looking forward to timber royalties providing us with the capital to carry out the development of our own lands. We realise only too well the advantages of tractors hauling ploughs and harrows over digging our gardens with pointed sticks.

Appendices

14. That as a prelude to self-government and political independence, experience in industrial management and in the development of our own lands would be desirable and rewarding. The United Nations Committee which visited New Guinea recently appeared to be firmly of this opinion.

15. We believe this section of the Forestry Regulation to be ultra vires the promises contained in the 1893 Pacific Order in Council and that the "peace, order, and good government" blanket clause would not override the specific promise to protect the sovereignty of chiefs and headmen contained in the same Order in Council.

16. It is the strong belief of all our people that any attempt to implement the Forestry Area Regulation is bound to cause deep discontent and resentment.

17. We people are a part of the land and our unity is tied up with our ownership and physical possession of our land. We do not wish to be placed in the position of other peoples whose lands were over-run by outsiders. Our history books tell us that annexation of weaker countries went out of fashion over a hundred years ago, though the filching of native lands by means of false promises of government and others still continues in many places.

18. We are confident that given the capital which timber royalties paid to us would provide, we could initiate a programme of agricultural and pastoral development that would put to shame any Government projects of this nature over all the years gone by. In this regard, we do not consider ourselves inferior to our brother Melanesians in New Guinea, and their performance is something to be proud of.

19. We wish to take a hand now in building the framework of our own economic future as participants and not as mere spectators or hired hands. We need to obtain experience in the management of our own agricultural and pastoral projects and also in the management of the capital which we hope to obtain from selling our timber.

20. We pray in our Churches and homes each night that Government may be moved to repeal these regulations which are so obnoxious and repugnant to us. We ask Our Lord to change the heart of Government so that we may continue to hold our land. The Bible tells us that we were given this land by God to possess, to use, and to inherit. (Deut. 19:40)

    for the Members and Committee of the
    PEOPLE'S PROTECTION PARTY
    and written under their instructions.
    (sgd.) James Bosa
    President.

Appendices

Appendix 5.
Forest and timber ordinances and acts

AN ORDINANCE
NO 9 OF 1969
TO CONSOLIDATE AND AMEND THE LAW RELATING TO FORESTS AND TO CONTROL AND REGULATE THE TIMBER INDUSTRY AND FOR MATTERS INCIDENTAL THERETO AND CONNECTED THEREWITH
ENACTED by the High Commissioner for the Western Pacific with the advice and consent of the Legislative Council of the British Solomon Islands Protectorate, as follows:–
PART I – PRELIMINARY
1. This Ordinance may be cited as the Forests and Timber Ordinance, 1969, and shall come into operation on such date as the High Commissioner may by notice appoint.
2. (1) In this Ordinance except where the context otherwise required -
"Conservator" means the Conservator of Forests appointed under section 3;
"controlled forest" means a controlled forest declared under section 13;
"forest offence" means an offence punishable under this Ordinance;
"forest officer" means a forest officer appointed under section 3 and includes the Conservator;
"forest produce" means trees, timber, poles, branches, charcoal, wood ashes, palms, bamboos, canes, climbers, creepers, grass, moss, fungus, lichens, plants or parts thereof, leaves, flowers, fruit seeds, roots, fibres, bark, bark extracts, wood extracts, gums, oils, resins, pitch, sap, latex, rubber, tar, wax, honey, products of animals, litter, humus, earth, sand and stone found in or brought from a state forest or controlled forest and such other things as the High Commissioner may by notice declare to be forest produce;
"livestock" includes asses, bulls, cows, geldings, goats, horses, stallions, mares, mules, oxen, pigs, sheep and steers and the young thereof;
"mill" means a sawmill and includes any mechanically powered plant, machinery or equipment for converting unmilled timber into milled timber, but does not include any plant, machinery or equipment which the High Commissioner may by notice declare not to be a mill for the purpose of this Ordinance;
"milled timber" includes sawn timber, wood particles, wood pulp and veneer, produced in or by a mill;
"state forest" means a state forest declared under section 9;
"timber" includes trees when they have fallen or been felled, and all wood whether cut up or fashioned or hollowed out for any purpose or not;
"timber levy" means a timber levy imposed under section 8;
"tree" includes any root, stump, stem, branch, brushwood, young tree or sapling
"unmilled timber" means timber that has not been converted into milled timber.
(2) In this Ordinance, except where the context otherwise requires, the expressions "customary land", "freehold interest in land", "land register", "leasehold interest in land", "public land", "registration" and "Solomon Islander" shall have the respective meanings ascribed to them in the Land and Titles Ordinance, 1968.
3. The High Commissioner acting in his discretion shall appoint a Conservator of Forests and such number of forest officers and other officers as he may consider necessary for carrying into effect the provisions of this Ordinance.
PART II – FELLING OF TREES FOR SALE
4. (1) Any person who fells any tree or removes any timber from any land for the purpose of sale thereof or of the products thereof otherwise than -
    (a)   for use within the Protectorate as firewood or unmilled timber;
    (b)   for supplying logs for milling to a mill licensed under section 7, from within the area that mill is by its licence authorised to draw unmilled timber;

(c) for such other purpose declared by the High Commissioner by notice to be exempt from the provisions of this section; or

(d) under and in accordance with the terms and conditions of a valid permit issued under section 5,

shall be guilty of an offence and liable to a fine of four hundred dollars or to imprisonment for one year or to both such fine and such imprisonment.

(2) Any person who fells a tree or removes timber from any land shall, until the contrary is proved, be presumed to have felled that tree or removed that timber for the purpose of sale.

(3) The onus of proving that a tree has been felled or timber has been removed for any of the purposes specified in paragraphs (a) or (b), or under paragraph (c) of subsection (1), shall lie on him who so alleges.

5. (1) Upon application therefor and payment of the prescribed fee (if any) the Conservator may, after consultation with the Commissioner of Lands and subject to the approval of the High Commissioner, issue a licence authorising, subject to such terms and conditions as he may therein specify, the felling of trees upon and the removal of timber from any public land, land in which the Government holds a freehold interest in land or leasehold interest in land, land leased by or on behalf of the Government and any land contiguous or island adjacent to such land.

(2) The Conservator may, subject to any general or special directions that may be given by the High Commissioner, at any time alter or amend a licence issued under subsection (1) to include or exclude any land contiguous or island adjacent to any public land leased by or on behalf of the Government comprised in the licence.

PART III – LICENSING OF MILLS

6. Any person who installs or operates a mill otherwise than under and in accordance with the terms and conditions of a valid licence issued udner section 7 shall be guilt of an offence and liable to a fine of four hundred dollars or to imprisonment for one year or to both such fine and such imprisonment:

Provided that this section shall not apply to any mill or class or description of mills declared by the High Commissioner by notice to be exempt from the provisions of this section.

7. (1) Upon application therefor and payment of the prescribed fee (if any) and subject to any general or special directions that may be given by the High Commissioner, the Conservator may issue a licence to install and operate a mill subject to such terms and conditions as he may therein specify and may, at any time, with the agreement of the licensee, alter or amend the licence.

(2) Every licence issued under subsection (1) shall specify -

(a) the area or areas from which unmilled timber to be milled at the mill may be drawn; and

(b) the maximum quantity of unmilled timber that may be acquired or milled or the maximum quantity of milled timber that may be produced, during any specified period, and, without prejudice to the power to specify terms and conditions under subsection (1), every licence may specify the maximum quantities to be drawn or acquired from any specified area during any specified period.

(3) Any person who is aggrieved by any of the terms or conditions of a licence issued to him under subsection (1), or by the refusal of the Conservator to issue him a licence under sub-section (1), may, within two months of being notified of such term, condition or refusal, appeal in writing to the High Commisioner whose decision thereon shall be final.

PART IV – TIMBER LEVY

8. The High Commissioner may by order -

(a) impose a levy on unmilled timber exported from the Protectorate and on milled timber exported from the Protectorate or sold in the Protectorate or milled for the purpose of such export or sale;

(b) impose different levies in respect of different licensees, species of trees, products, grades, places or other circumstances; and

(c) provide for the levy to be assessed on the quantity or value of timber or milled timber, or otherwise howsoever.

PART V – STATE FORESTS

9. (1) The High Commissioner may at any time by notice declare any land that is public land, land in which the Government holds a freehold interest in land or a leasehold interest in land, or land leased by or on behalf of the Government, to be a state forest.

(2) Upon such land or part thereof ceasing to be such land as aforesaid, that land or that part shall cease to be a state forest.

10. The Commissioner of Lands shall not grant any interest or licence in any land comprised in a state forest without the prior written consent of the Conservator, and such restriction shall be noted on the land register and no such grant shall be registered until the consent of the Conservator has been produced to the Registrar of Titles.

11. Any person who within a state forest otherwise than under and in accordance with a valid permit issued under section 12 or in pursuance of any right which existed prior to tghe declaration of that state forest (the onus of proving which shall lie on such person) -

(a) fells, cuts, taps, damages, burns, removes, works or sells any tree;

(b) causes any damage therein by negligence in felling any tree, dragging any timber, lighting any fire or otherwise howsoever;

(c) clears or breaks up any land for cultivation or any other purposes;

(d) resides or erects any building, shelter or structure;

(e) grazes or permits to be grazed any livestock;

(f) has in his possession any machinery, equipment or implement for cutting, taking, working or removing any forest produce, without being able to show that such machinery, equipment or implement is in his possession for a lawful purpose; or

(g) constructs or re-opens any road, saw-pit or workplace,

shall be guilty of an offence and liable to a fine of two hundred dollars or to imprisonment for six months or to both such fine and such imprisonment.

12. Upon application therefor and payment of the prescribed fee (if any), the Conservator, and subject to the directions of the Conservator, a forest officer, may issue a permit authorising any of the acts mentioned in section 11 upon such terms and conditions as the Conservator or forest officer may therein specify.

PART VI – CONTROLLED FORESTS

13. Where the High Commissioner is satisfied that for the purpose of conserving water resources within the Protectorate it is necessary or desirable to protect the forest or other vegetation in any rainfall catchment area, he may, subject to the provisions of this Part, by notice declare such area or part thereof to be a controlled forest, and shall in the same notice specify what rights and the exztent to which such rights may be exercised in the controlled forest.

14. Before declaring any area to be a controlled forest the High Commissioner shall -

(a) cause to be published in such manner as he may in his discretion consider to be adequate or most effective for the purpose of bringing it to the attention of all persons likely to be thereby affected, notice of the intention to do so;

(b) cause to be made such enquiries as he may, in his discretion, deem fit for the purpose of ascertaining -

(i) what rights exist in that area and to what extent such rights may be expected to be exercised in the absence of a declaration as aforesaid;

(ii) the extent to which the exercise of such rights could be permitted without prejudice to the purposes of a controlled forest;

(iii) in respect of the extent to which the exercise of such rights could not be so permitted, what reasonable alternative arrangements could be made or what compensation would be appropriate, and shall, in respect of such rights which cannot be permitted to be exercised, cause such arrangments as aforesaid to be made or such compensation as aforesaid to be paid, within one month of the making of the declaration under section 13.

15. (1) Any person not permitted to exercise any of his rights in a controlled forest who is aggrieved by the amount of the compensation paid or offered or the alternative arrangements made or offered to be made under section 14, may within three months of the declaration of the controlled forest, appeal to a Magistrate's Court, which may make such order as it considers just.

(2) Any person who is aggrieved by the order or decision of a Magistrate's Court under subsection (1) and desires to question it on the ground that it is erroneous in point of law may, within two months of the date of the order or decision, appeal to the High Court.

(3) The High Court may, if satisfied that the order or decision is erroneous in point of law, make such order as it considers just.

(4) The order or decision of the High Court and, subject to the provisions of this section, the order or decision of a Magistrate's Court under this section, shall be final and conclusive and shall not be questioned in any proceedings whatsoever.

16. (1) Any person who within a controlled forest otherwise than under and in accordance with a valid permit issued under section 17 -

    (a)   fells, cuts or removes any forest produce otherwise than for his own personal or domestic use;

    (b)   clears or breaks up any land for cultivation or any other purpose;

    (c)   resides or erects any building, shelter or structure; or

    (d)   grazes or permits to be grazed any livestock,

shall be guilty of an offence and liable to a fine of one hundred dollars or imprisonment for three months or to both such fine and such imprisonment.

(2) Subsection (1) shall not apply to any person acting in exercise of any right specified under section 13, but any person charged with an offence under this section shall, until the contrary is proved, be presumed not to have been a person acting as aforesaid.

17. Upon application therefor and payment of the prescribed fee (if any), the Conservator and, subject to the directions of the Conservator, a forest officer, may issue a permit authorising any of the acts mentioned in section 16(1) upon such terms and conditions as the Conservator or forest officer may therein specify.

PART VII – PROCEDURE AND PENALTIES

18. Any person who -

    (a)   knowingly counterfeits upon any tree or timber, or has in his possession any implements for counterfeiting, any mark used by forest officers to indicate that such tree or timber may lawfully be felled or removed by some person; or

    (b)   unlawfully or fraudulently affixes to any tree or timber any mark used by forest officers; or

    (c)   alters, defaces or obliterates any such mark placed on any tree or timber by or under the authority of a forest officer, shall be guilty of an offence and liable to a fine of four hundred dollars or to imprisonment for one year or to both such fine and such imprisonment.

19. Any person who receives any forest produce knowing or having reasonable cause to believe it to have been obtained in contravention of this Ordinance shall be guilty of an offence and liable to a fine of two hundred dollars or to imprisonment for six months or to be such fine and such imprisonment.

20. (1) When any person is convictged of erecting any unauthorised building, shelter or structure or of planting any unauthorised crops in a state forest or controlled forest, the court may in addition to any penalty it may impose, order such building, shelter, structure or crops to be removed and the land restored to its previous condition within such time as it shall fix.

(2) Any person who fails to obey an order made under subsection (1) within the time fixed shall, unless he satisfies the court that he has used all diligence to carry out such order, be liable to a fine not exceeding ten dollars for every day during which the default continues.

Appendices

(3) Notwithstanding the provisions of subsection (2), when an order has been made under subsection (1) and not carried out within the time fixed, the Conservator may cause the order to be carried out and may recover the expenses of so doing as a civil debt from the person convicted.

21. (1) Any forest officer or police officer may without a warrant -

(a) demand from any person the production of any authority or licence for any act done or committeed by such person in any state forest or controlled forest or in relation to any forest produce for which a licence or permit is required under this Ordinance;

(b) require any person found within any state forest or controlled forest or in the vicinity of such forest, and who has in his possession any forest produce, to give an account of the manner in which such person became possessed of such produce, and may arrest that person if he fails to give a satisfactory account;

(c) arrest any person reasonably suspected of being guilty of a forest offence or of being in possession of any forest produce in respect of which an offence has been committed: Provided that no person shall be arrested under this subsection unless such person refuses to give his name and address or gives a name and address which there is reasonable cause to believe is false or there is reasonable cause to believe that he will abscond;

(d) seize and detain any livestock found trespassing or found without any person in charge of them in any state forest or controlled forest;

(e) enter any timber yard or mill by day to inspect forest produce therein.

(2) Any forest officer making an arrest under this section shall without unnecessary delay take or send the person arrested to a police officer, and any police officer making an arrest under this section or to whom a person arrested under this section is taken, shall deal with such person in accordance with the provisions of the Criminal Procedure Code Ordinance relating to persons arrested without a warrant.

22. (1) Where there is reason to believe that a forest offence has been committed in respect of any forest produce, such produce, together with all tools, machinery, equipment, boats, conveyances and livestock reasonably suspected to have been used in the commission of such offence, may be seized by any forest officer or police officer.

(2) Every officer seizing any property under this section shall place on such property, or on the receptacle, if any, in which it is contained, a mark indicating that the same has been so seized and shall, so soon as may be, make a report of such seizure to a Magistrate.

(3) In any proceedings in respect of a forest offence alleged to have been committed in respect of any forest produce, the averment that any substance is forest produce shall be sufficient without proof of such fact unless the person charged prove the contrary.

23. (1) When any person is convicted of a forest offence, all forest produce in respect of which such offence has been committed, and all tools, machinery, equipment, boats, conveyances and livestock used in the commission of such offence, shall be liable to be forfeited by order of the court recording the conviction.

(2) Such forfeiture may be in addition to any other penalty prescribed for such offence.

24. (1) Where there is reason to believe that a forest offence has been committed by a person who is unknown or cannot be found, any property seized in respect of such suspected offence under section 22 shall be taken possession of and may be disposed of by or under the direction of the Conservator, but no such property shall be sold or otherwise disposed of until the expiration of one month from the date of the service or publication of the notice given under subsection (2), or without hearing the person, if any, claiming any right thereto and the evidence, if any, which he may produce in support of his claim within such period of one month, or until after the determination of any appeal under section 26.

(2) When possession is taken of any property under sub-section (1), the Conservator or, subject to the directions of the Conservator, a forest officer, shall cause notice thereof to be served upon any person whom he has reason to suspect to be interested in the property or may publish such notice in such manner as he deems fit.

25. A Magistrate or a forest officer may, notwithstanding section 24, direct the sale of any property seized under section 22 and which is subject to speedy and natural decay, and may deal with the proceeds of such sale as he might have dealt with the property had it not been sold.

26. Any person claiming to be interested in any property seized under section 22 may, within one month from the service or publication of a notice in respect of such property under section 24(2), appeal to a Magistrate against the taking into possession of such property.

27. When possession has been taken of any property under section 22, and after the expiration of the time limited for appealing under section 26 or the determination of any such appeal in favour of the Conservator, such property or the proceeds of sale thereof if sold under section 25 shall vest in the Crown absolutely.

PART VIII – MISCELLANEOUS

28. (1) Where the Conservator is satisfied that the holder of a licence or permit issued under this Ordinance has contravened any of the provisions of this Ordinance, or that there has been a contravention of any of the terms or conditions of such a licence or permit, he may by notice in writing cancel or suspend the licence or permit.

(2) Any person who is aggrieved by the cancellation or suspension of his licence or permit under subsection (1) may within two months thereof appeal in writing to the High Commissioner whose decision thereon shall be final.

29. No licence or permit issued under this Ordinance shall convey or be construed to convey any right which the Government does not have and in particular no such licence shall convey nor be construed to convey any right or authority to enter any private land nor take any action with respect to any thing without the authority of the owner of that land or thing.

30. No action shall lie against any person in respect of any act done by him in good faith in the execution or intended or purported execution of his duties or powers under this Ordinance.

31. (1) The Conservator shall furnish to the Registrar of Titles every declaration of a state forest and of a controlled forest and every amendment and cancellation thereof, authenticated in such manner as the Registrar may require, and the Registrar shall note in the land register, in such manner as he thinks fit, every such declaration, amendment and cancellation affecting registered land.

(2) Without prejudice to the power of the Registrar of Titles to require further information, a copy of the declaration purporting to have been made under this Ordinance shall be sufficient evidence to support a note in the land register as aforesaid.

(3) The Registrar of Titles shall not be concerned to note any licence or permit issued by the Conservator or any forest officer under this Ordinance in respect of state forests or controlled forests nor to file in the land registry, nor to furnish certified copies thereof, nor to provide for inspection, any instrument embodying such licence or permit.

32. Nothing contained in section 221 of the Lands and Titles Ordinance, 1968, shall prohibit or invalidate the acquisition by a person other than a Solomon Islander of any right to cut and remove any trees growing on customary land, or of any right of access to or over customary land for the purpose of cutting or removing trees growing on customary land.

33. The High Commissioner may make regulations for the better carrying out of the provisions and purposes of this Ordinance, and in particular, without prejudice to the generality of the foregoing power, such regulations may -

(a) provide for the forms of licences and permits to be issued under this Ordinance, the procedure for the issue of such licences and the terms and the conditions to which they may be subject;

(b) prescribe the fees to be paid on the application for a grant of any licence or permit under this Ordinance;

(c) prescribe the form and manner in which returns of timber felled, sawn, or otherwise processed or milled, shipped, or sold shall be made and may require the holder of any such licence to submit for inspection, books, records and accounts relating to any transactions in respect of forest produce or timber;

Appendices

(d) prescribe the procedure for payment of timber levies and provide for the effective recovery thereof;

(e) prescribe rates of royalty to be paid in respect of forest produce obtained from state forest or from other land owned or leased by the Government;

(f) prohibit or regulate the taking of any specified kind of forest produce from any state forest;

(g) provide for the survey and demarcation of state forests and controlled forests and for rights of entry upon any land for the purpose of carrying out any such survey;

(h) regulate the marking of timber and the manufacture, use and possession of timber marking instruments;

(i) provide for the grading of timber for export and sale and for the prohibition of the export of timber not of good merchantable quality;

(j) provide for the protection of timber against insect and fungus attack.

(k) regulate the lighting of fires within a state forest or a controlled forest and prescribe the precautions to be taken to prevent the spreading of fires;

(l) regulate the entry of persons into state forests or controlled forests and the periods during and the conditions under which persons may remain therein;

(m) provide for the closing of roads and paths in any state forest or controlled forest to persons or traffic or such persons or traffic as may be specified;

(n) prohibit or regulate the hunting of birds or animals or fishing in any state forest or controlled forest.

(o) provide for the impounding of stray livestock found in any state forest or controlled forest and of any livestock found in any such forest in contravention of this Ordinance, for the cost and fees to be paid by the owner of such livestock and for the disposal of such livestock as are not claimed or for which the costs and fees are not paid;

(p) provide for the regulation of tree felling and timber milling operations;

(q) provide for the safety of persons employed in the timber industry;

(r) prescribe anything by this Ordinance required or authorised to be prescribed.

34. Section 11(1) of the Mining Ordinance, 1968, is hereby amended by substituting for paragraph (h) the following new paragraph:-

"(h) any state forest or controlled forest within the meaning of the Forests and Timber Ordinance, 1969, except with the consent of the Conservator of Forests and subject to such conditions as the Conservator might impose."

35. The Forests Ordinance is hereby repealed.

36. The Registrar of Titles may cancel the registration in the land register of any order constituting a Forest Reserve under section 15, or declaring a Forest Area under section 23 of the Forests Ordinance (repealed by section 35).

37. (1) Nothing in sections 4, 11 and 16 shall apply to any person in respect of anything done by such person under and in accordance with any agreement providing for the felling of trees and removal of timber, entered into by the Government prior to the commencement of this Ordinance, or under and in accordance with a valid licence or permit issued under section 25 of the Forests Ordinance (repealed by section 35).

(2) It shall be deemed to be a term and condition of every agreement providing for the felling of trees and removal of timber, entered into by the Government prior to the commencement of this Ordinance, that the other party thereto shall pay all timber levies as may be payable and when due under this Ordinance, and such timber levies shall be deemed to be payments due under such an agreement; and any reference in such an agreement to export duty shall be deemed to include a reference to timber levy.

(3) For the purposes of subsection (1) a licence or permit as aforesaid which is valid upon the commencement of this Ordinance shall, subject to this subsection, continue, to be valid according to its terms and conditions, and in addition and without prejudice to its terms and conditions relating to its limitation, suspension or revocation, the Conservator may cancel or suspend it for the failure of its holder to pay timber levies in accordance with this Ordinance.

Appendices

Passed by the British Solomon Islands Legislative Council this twenty-sixth day of June, one thousand nine hundred and sixty-nine.

This printed impression has been carefully compared by me with the Bill passed by the Legislative Council and found by me to be a true and correct copy of the said Bill.

S. MAMALONI,
    Assistant Clerk to the Legislative Council.

Published and exhibited at the Public Office of the High Commissioner for the Western Pacific this ninth day of July, one thousand nine hundred and sixty-nine.

L. M. DAVIES,
    Chief Secretary to the Western Pacific High Commission

AN ORDINANCE

NO. 16 OF 1977

TO AMEND THE FORESTS AND TIMBER ORDINANCE

(13th January 1978)

ENACTED by the Governor of the Solomon Islands with the advice and consent of the Legislative Assembly of the Solomon Islands, as follows:-

Forests and Timber (Amendment)

Ordinance 1977 – No. 16 of 1977

1. This Ordinance may be cited as the Forests and Timber (Amendment) Ordinance 1977.

2. Section 5 of the Forests and Timber Ordinance (hereinafter referred to as the principal Ordinance) is hereby amended by inserting, immediately following subsection (2) thereof, the following new subsection.

    "(3) Upon application therefor and payment of the prescribed fee, if any, the Conservator may issue a licence authorising, subject to such terms and conditions as he may specify therein, the felling of trees upon, and the removal of timber from -

    (a)   any land, not being customary land or land to which subsection (1) applies; and

    (b)   when such felling and removal are the subject of rights granted under an agreement duly approved by the Minister under Part IIA; any customary land."

3. The principal Ordinance is hereby amended by adding, immediately following Part II thereof, the following new Part -

"PART IIA – APPROVED TIMBER AGREEMENT AFFECTING CUSTOMARY LAND"

5A. In this Part, unless the context otherwise requires

    "act" includes omission;

    "approved agreement" means an agreement approved under the provisions of this Part;

    "area committee" means, in relation to any land, the committee appointed under section 33 of the Local Government Ordinance within the area of authority in which such land is situated or, if no such committee has been appointed, or is competent, the Council within whose area of authority such land is situated or a sub-committee of such Council; Forests and Timber (Amendment) Ordinance 1977 – No. 16 of 1977

    "cause" bears the meaning ascribed to such expression in section 2 of the Magistrates' Courts Ordinance:

    "council" bears the meaning ascribed to such expression in section 2 of the Local Government Ordinance;

    "matter" bears the meaning ascribed to such expression in section 2 of the Magistrates' Courts Ordinance;

    "timber right" includes a right to -

    (a)   inspect, survey, enumerate, mark and map any area or tree;

    (b)   cut, trim, lop, top and crop any tree;

    (c)   plant and cultivate any seed, seedling or tree;

    (d)   take any measure whatsoever for the healthy growth and protection of any tree;

Appendices

(e) have access to and extract (whether with or without any agents, servants, workmen, animals, vehicles and machines, or all or any of them) any timber or tree;

(f) take any timber into possession or ownership;

(g) for the purpose of any of the things specified in paragraphs (a) to (f) construct and maintain -
  (i) any works (including railways, tramways, roads, waterways, slipways, harbours, port areas, wharves, jetties, bridges, dams, pipelines, aerodromes, yards and camps);
  (ii) any buildings and structures (including warehouses, sheds, mills, kilns, offices, houses and fences); and
  (iii) any cables, power supplies, lines or other means for the distribution of power, cranes, weighbridges, sawbenches or other things;

(h) for any of the purposes specified in this subsection -
  (i) subject to the provisions of the River Waters Ordinance, to take and use water; and
  (ii) to quarry, extract, move and use any stone, earth or other roadmaking or building material.

5B. (1) Any person carrying on business in the Solomon Islands as a timber exporter or sawmiller who wishes to enter into an agreement whereunder he acquires timber rights on customary land shall first obtain the consent of the Conservator to carry on negotiations and thereafter make application in that behalf in the prescribed form and manner to the Conservator.

(2) Upon receipt of an application under subsection (1), the Conservator shall forward a copy thereof to the appropriate area committee.

5C (1) After receiving a copy of an application forwarded to it under section 5B, an area committee whose membership shall include persons having particular knowledge of customary land rights in the area affected by the application shall –

(a) fix a place within the area of its authority and a day, not being a day less than two months, or more than three months, after the day on which such copy is so received, for a meeting of the area committee to consider such application and to determine in regard to it the matters specified in sub-section (4); and

(b) forthwith give in such manner as it shall consider most adequate and effective to the public within the area of its authority and, in particular, to persons who reside within such area and appear to it to have an interest in the land, trees or timber in question, notice of -
  (i) such application;
  (ii) the parties to, and terms of, the proposed agreement; and
  (iii) the time and place fixed for the relevant meeting under paragraph (a).

(2) Any notice given under subsection (1)(b) shall require any person who has reason to believe that the persons intending to grant timber rights under the proposed agreement are not the persons, or all the persons, as the case may be, lawfully able and entitled to grant such rights to attend the meeting referred to in the notice and at such meeting to state to the area committee the particulars of such belief and the reasons for it.

(3) At the time and place referred to in any notice under subsection (1)(b) the area committee shall meet and consider the application to which the notice relates. In considering the application, the area committee shall hear any representations made to it in response to the requirement provided for in subsection (2) and shall take into account those representations and all other matters relevant to the application known or believed by the area committee to be true.

(4) Upon the conclusion of its considerations under subsection (3), an area committee shall issue a certificate in the prescribed form setting out its determination as to whether -

(a) the persons proposing to grant the timber rights in question are the persons, and all the persons, lawfully able and entitled to grant such rights and, if not, who such persons are; and

(b) whether such timber rights, or such rights in any modified form, may be granted, giving particulars of such modification, if any.

(5) After giving such certificate, the area committee shall give notice thereof in the prescribed form and in a similar manner to that in which it gave notice of the relevant application under subsection (1).

(6) The Clerk to an area committee shall cause any certificate issued by it under this section forthwith to be forwarded to the Conservator.

5D. (1) Any person who is aggrieved by any act or determination of an area committee under section 5C. may, within one month from the date of the determination, appeal to the customary land appeal court having jurisdiction for the area in which the customary land concerned is situated and such court shall hear and determine the appeal.

(2) Any provision in any other law to the contrary notwithstanding, the order or decision of a customary land appeal court on any appeal entertained by it under subsection (1) shall be final and conclusive and shall not be questioned in any proceedings whatsoever.

(3) It shall be the duty of the clerk to any customary land appeal court forthwith to notify the Conservator of the lodging in his court of an appeal under this section and, when such appeal is finally determined, to inform the Conservator of such determination and the terms thereof.

5E. When the Conservator has received a certificate issued under section 5C. and has satisfied himself that -

(a) at least one month has elapsed since such certificate was issued; and

(b) no appeal under section 5D. has been lodged against the issuing of such certificate or, if an appeal has been lodged, it has been finally disposed of; and

(c) an agreement for the granting of the timber rights referred to in such certificate has been duly completed in the prescribed form and manner and that the parties to, and the terms and provisions of, such agreement accord with such certificate or, where there has been an appeal under section 5D. relating thereto, the order of the court determining such appeal, the Conservator shall recommend to the Minister that approval under this Part be given to such agreement.

5F. (1) Upon receipt of a recommendation made under section 5E. and the relevant agreement, duly stamped, the Minister may complete a certificate in the prescribed form approving the agreement.

(2) The Conservator shall within 14 days of completion by the Minister of a certificate under subsection (1) notify the parties to the agreement of such completion.

5G. The Minister may make regulations for the better carrying out of the provisions and purposes of this Part, and in particular, without prejudice to the generality of the foregoing power, such regulations may provide for -

(a) the form of applications for approval of agreements under this Part;

(b) the form of agreements which may be approved under this Part and the manner in which they are to be executed;

(c) the form of certificates to be issued by area committees under section 5C;

(d) the form of the certificate of the Minister under section 5F; and

(e) the fees, if any, to be paid for any act or thing done under the provisions of this Part.

5H. Any person proposing to negotiate the grant of any timber rights affecting customary land or to make representations under section 5C. shall be entitled to seek and to be given, in relation to any question touching on those rights, the advice of the Conservator or any public officer nominated by him.

5I. Any provision of any other law to the contrary notwithstanding, original jurisdiction to hear and determine any cause or matter arising out of, or relating to, an approved agreement shall be exercised only by the High Court.

Passed by the Legislative Assembly this twenty-third day of November one thousand nine hundred and seventy-seven.

This printed impression has been carefully compared by me with the Bill passed by the Legislative Assembly and found by me to be a true and correct copy of the said Bill.

(J.A. JONES)
Clerk to the Legislative Assembly

Appendix 6.
Solomons governments since Independence

Year	Prime Minister	Minister responsible for forests
1978	P. Kenilorea	P. Tovua
1980, Aug.	P. Kenilorea	
1981, Sept.	S. Mamaloni	P. Salaka
1981		P. Tovua
1983		P. Salaka
1984		A. Tropa
1984, Oct.	P. Kenilorea	D. Sande
1987	E. Alebua	P. Kenilorea
1988, Apr.		D. Phillip
1988, June		R. Bera
1989, July	S. Mamaloni	A. Paul
1990, Oct.		J. D. Tausinga
1993, June	F. Billy Hilly	E. Alebua
1994, July		J. Tuhanuku
1994, Nov.	S. Mamaloni	A. Kemakeza
1997, June	B. Ula'afalu	H. Kari

Appendix 7.
Duty remission granted to landowner companies/individuals[a] (as at June 1995)

1.	Soma Company	Pays only 10% on 50,000m^3.	Timber levy exempted.
2.	RAFEA	100% duty remission on 40,000m^3.	Timber levy exempted.
3.	Kololeana	50% remission on 40,000m^3.	Timber levy payable.
4.	Ofongia Association (Mr Silas Milikada)	100% remission on 12,000m^3.	Timber levy also exempted on 12,000m^3.
5.	Golden Fountain (Mr T Chan)	100% duty remission on 20,000m^3 and 50% remission on the next 30,000m^3.	Timber levy exempted on the 20,000m^3.
6.	Dai Sawmill	100% remission on 10,000m^3. Concession given on the basis that proceeds be used by the community to build water supplies, a clinic and school.	Timber levy exempted on 10,000m^3.
7.	SE Sawmill Ltd	100% remission on 12,000m^3. 50% remission on 23,000m^3.	Timber levy exempted on the first 12,000m^3.
8.	Bola Timber Ltd	100% remission on 12,000m^3. 50% remission on 23,000m^3.	Timber levy exempted on the first 12,000m^3.
9.	Tropical Forest/ Makira	100% remission on 12,000m^3.	Timber levy exempted on the first 12,000m^3.
10.	Basakana Sawmill (P Fulaburi)	100% remission on 12,000m^3.	Timber levy remitted as well.
11.	Tamaneke Community (Mahlon Ali)	100% remission on 10,000m^3.	Timber levy exempted.
12.	Daido SI Ltd	100% remission on 12,000m^3.	Timber levy exempted.
13.	SLH Timber Corp (Mr D Damikura)	100% remission on 12,000m^3.	Timber levy exempted.
14.	Mubu Sawmill	100% remission on 12,000m^3.	Timber levy exempted.
15.	VR & JS Lumber Co	100% remission on 10,000m^3.	Timber levy exempted.
16.	Afeala Milling Ltd	100% remission on 3,000m^3.	Timber levy exempted.
17.	Paripao Timber FNT	100% remission on 12,000m^3. 50% remission on 46,000m^3.	
18.	Success Co Limited (HHB Dettke)	100% remission on 80,000m^3.	

a. In addition to the above exemptions, there are other special exemptions that have been granted to individual companies. For example, Eagon resources has a total timber levy exemption in recognition of its reforestation activities on Choiseul. Isabel Timber, Kalena and Silvania all have partial exemptions from the timber levy.

Source: Comptroller of Customs and Excise, cited in Price Waterhouse, Economics Studies and Strategies Unit, *Forestry Taxation and Domestic Processing Study*, 1995, 154–155.

— Notes —

Introduction

1. Donald Worster, 'Transformations of the Earth: Toward an Agroecological Perspective in History', *Journal of American History*, 76, no.4 (1990), 1087–106.
2. Michael Williams, *Americans and their Forests: A Historical Geography* (New York, 1989); John Dargavel, *Fashioning Australia's Forests* (Melbourne, 1995); Michael Roche, *History of New Zealand Forestry* (Wellington, 1990); Peter Dauverge, *Shadows in the Forest: Japan and the Politics of Timber in Southeast Asia* (Cambridge, Mass., 1997); Peter Dauverge, 'Corporate Power in the Forests of Solomon Islands', *Pacific Affairs* 71, no.4 (Winter 1998/9), 524–46; Colin Filer with Nikhil Sekhran, *Loggers, Donors and Resource Owners* (Port Moresby and London, 1998), 52–60.

Chapter 1

1. T. C. Whitmore, *An Introduction to Tropical Rainforest* (Oxford, 1990), 9.
2. Patrick D. Nunn, *Environmental Change in the Pacific Basin* (Chichester, England, 1999), 43–6.
3. For a detailed study of the geology, see D. A. Falvey et al., 'Petroleum Prospectivity of Pacific Islands Arcs: Solomon Islands and Vanuatu', *The Apea Journal* 31, pt 1 (1991), 191–212. A general summary is in John R. Schenk, *Forest Ecology and Biogeography in the Solomons* (Honiara, 1994), 3–5.
4. AIDAB and Ministry of Natural Resources, *Solomon Islands National Forest Resources Inventory: Forests of the Solomon Islands. National Overview and Methods* (Honiara, 1995), vol. 2, i, 15, 23–4 (hereinafter *SINFRI*).
5. Jared Diamond cited in *SINFRI* vol.2, 23.
6. T. C. Whitmore, 'Land Flora: Geography of Flowering Plants', *Phil. Transactions of Royal Society* 255 (1969), 549–66; Whitmore, *Introduction*, 92–4.
7. Schenk, *Forest Ecology*, 23–30.
8. Ibid., 24; Whitmore, *Introduction*, 9–12; T. C. Whitmore, *Guide to the Forests of the British Solomon Islands* (London, 1966), 1–2.
9. F. S. Walker, *The Forests of the British Solomon Islands Protectorate* (London, 1948), 14.
10. Schenk, *Forest Ecology*, 18–19, 31–2; Geoffrey Irwin, *The Prehistoric Exploration and Colonisation of the Pacific*, (Cambridge, 1992), 18–19.
11. Schenk, *Forest Ecology*, 55.

12. Ibid., 55. New Caledonia is a special case. It was part of the ancient continent of Gondwanaland about 83 million years ago, and has several species of flora originating in what is now Australia. There are dozens of genera of flowering plants, including *Araceae, Guettarda, Heliconia, Macaranga, Saurauria, Sloanea,* and *Spathiphyllum,* which are represented in Solomons (and South-east Asia) and tropical America. These 'amphiPacific tropical disjuncts' are believed to have migrated northwards around the Pacific rim across the Bering Straits before two million years ago when the climate in the north was warmer.(Whitmore, *Introduction*, 92–4). Food species from South America, such as the sweet potato, pineapple, and paw-paw (papaya) were introduced in the historic period.

13. C. P. Henderson and I. R. Hancock, *A Guide to the Useful Plants of Solomon Islands* (Honiara, 1988), 328.

14. For example, Walker, *Forests*, 11–14; T. C. Whitmore, 'The Vegetation of the Solomon Islands', *Philosophical Transactions of the Royal Society Bulletin* (1969), 255–70; J. R. F. Hansell and J. R. D. Wall, *Land Resources of the Solomon Islands* (Surbiton, 1976); L. Dahl, *Regional Ecosystem Survey of the South Pacific Area* (Noumea, 1980); Foreign Investment Board (hereinafter FIB), *Forestry and Forests Investments: Forest Resources of the Solomon Islands* (Honiara, 1984), 6.

15. So-called Western botanical classification owes much to systematised indigenous knowledge of plants in south-west India and Malabar collected by the Portuguese, Garcia da Orta and the Dutchman, van Reede in the sixteenth and seventeenth centuries. See Richard Grove, *Green Imperialism: Colonial Expansion, Tropical Island Edens and the Origins of Environmentalism, 1600–1860* (Cambridge, 1995), 77–90.

16. Whitmore, *Introduction*, 13.

17. *SINFRI*, vol. 1, 28–33.

18. Whitmore, *Guide*, 4; Whitmore, *Introduction*, 146–7.

19. H. C. Brookfield with Doreen Hart, *Melanesia: A Geographical Interpretation of an Island World* (London, 1971), xliv—xlv.

20. Whitmore, *Guide*, 1.

21. Whitmore, *Introduction*, 24–5, 99–132.

22. Irwin, *Prehistoric Exploration*, 12; *SINFRI*, vol. 1, appendix 3.

23. T. C. Whitmore, *Change with Time and the Role of Cyclones in Tropic Rain Forest on Kolombangara, Solomon Islands* (Oxford, 1974).

24. D. A. Radford and R. J. Blong, *Natural Disasters in Solomon Islands* (Sydney, 1992) gives a useful notional coverage of cyclones and other disasters, especially for more recent decades, but there are several omissions for the Santa Cruz district alone, including this cyclone and one in January 1952 which affected much of the area from Vanikolo to Tikopia.

25. Western Pacific High Commission, British Solomon Islands Protectorate Series (hereinafter WPHC, BSIP) 1/111, F20/7: Report of Hurricane, Dec. 1935.

26. WPHC, Inwards Correspondence 1053/36: Annual Report (hereinafter AR) Guadalcanal, 1935.

27. *SINFRI*, vol. 1, appendix 3.

28. Ibid., 15, 36; K. E. Lee, 'Some Soils of the British Solomons Protectorate', *Philosophical Transactions of the Royal Society Bulletin* 255 (1969), 211–257; Whitmore, *Introduction*, 51–2, 136–44.

Chapter 2

1. J. Allen, et al., 'Pleistocene Dates for the Human Occupation of New Ireland, Northern Melanesia', *Nature* 331 (1988), 707–9; S. Wickler and M. Spriggs, 'Pleistocene Occupation of the Solomon Islands, Melanesia', *Antiquity* 62 (1988), 703–6; Irwin, *Prehistoric Exploration*, 18–19.

2. G. F. Dennis, 'Nature Notes on the Solomons', *The Journal of the Solomon Islands Museum Association* 1 (1972), 31–42; Les Groube, 'The Taming of the Rainforests: A Model for Late Pleistocene Forest Exploitation in New Guinea', in *Foraging and Farming*, ed. David R. Harris and Gordon C. Hillman (London, 1989), 292–303; Matthew Spriggs, 'Pleistocene Agriculture in the Pacific: Why Not?' in *Sahul in Review*, ed. M. A. Smith, M. Spriggs, and B. Fankhauser (Canberra, 1993), 137–43.

3. D. E. Yen, 'The Domestication of Environment', in *Foraging and Farming*, ed. David R. Harris and Gordon C. Hillman (London, 1989), 55–75
4. Edgar Anderson, *Plants, Man, and Life* (Berkeley, 1952), 1–15.
5. Ross M. Cassells, 'The Valuation of Subsistence Use of Tropical Rainforest on the Island of Choiseul, Solomon Islands' (M. Ph. thesis, Massey University, 1992), 41–82; Ian Heath, 'Land Policy in Solomon Islands' (Ph. D. diss., La Trobe University, 1979), 9–10; H. B. Guppy, *The Solomon Islands and their Natives* (London, 1887), 66, 85–7, 280–1; Lindsay Wall, 'The Stone Carvings of Nggatokae', *Journal of the Solomon Islands Museum Association* 2 (1974), 35–6; Edvard Hviding, *Guardians of Marovo Lagoon: Practice, Place, and Politics in Maritime Melanesia* (Honolulu, 1996), 67; WPHC 434/18: Knibbs to Barnett, 5 Jan. 1918.
6. Judith A. Bennett, *Wealth of the Solomons: A History of a Pacific Archipelago* (Honolulu, 1987), 2–14.
7. Leon F. Peters and William J. Neuenschwander, *Slash and Burn: Farming in the Third World Forest* (Moscow, Idaho: 1988), 1–3; H. C. Conklin, *The Study of Shifting Cultivation* (Washington, DC, 1963), 1.
8. D. E. Yen, 'Agricultural Systems and Prehistory in the Solomon Islands', in *Southeast Solomon Islands Cultural History: A Preliminary Survey*, ed. R. C. Green and M. M. Cresswell (Wellington, 1976), 67–9; Lennox Barrow, Outlying Interlude: an Account of BSIP, 1942–1947, PMB 517.
9. Lord Amherst of Hackney and Basil Thomson, *The Discovery of The Solomon Islands by Alvaro de Mendana in 1568* (London, 1901), vol. 1, 308.
10. Ibid., 331.
11. Walter G. Ivens, *The Island Builders of the Pacific: How and why the People of Mala Construct their Artificial Islands: the Antiquity and Doubtful Origin of the Practice, with a Description of the Social Organization, Magic and Religion of their Inhabitants* (London, 1930), 266–75.
12. J. G. Marwick, comp., *The Adventures of John Renton* (Kirkwall, 1935), 30.
13. WPHC BSIP Series F 46/35: Kuper to District Oficer (hereinafter DO), Eastern Solomons, 2 Oct. 1933; M. M. and J. L. O. Tedder, 'Yam Cultivation on Guadalcanal', [1970?], 8–9, 15–17; Kaj Birket-Smith, *An Ethnological Sketch of Rennell Island: A Polynesian Outlier in Melanesia* (Copenhagen, 1956), 56; Axel Steensberg, *New Guinea Gardens: A Study of Husbandry with Parallels in Prehistoric Europe* (London, 1988), 60; Harold Ross, *Baegu: Social and Ecological Organization in Malaita, Solomon Islands* (Urbana, 1973), 80–1; Cassells, 'Valuation', 192; Henderson and Hancock, *Guide*, 3, 156. See also Harold W. Scheffler, *Choiseul Island Social Structure* (Berkeley and Los Angeles, 1956), 11.
14. Rickie Burman, 'Time and Socioeconomic Change on Simbo, Solomon Islands', *Man* 16, no. 2 (June 1981): 257; Ivens, *The Island Builders*, 355–64; WPHC BSIP Series F 46/35: Kuper to DO, Eastern Solomons, 2 Oct. 1933. Gardeners in the Philippines also prefer to cultivate secondary forest. See Harold C. Conklin, 'An ethnographic approach to shifting agriculture', in *Environment and Cultural Behaviour: Ecological Studies in Cultural Anthropology*, ed. Andrew P. Vayda (New York, 1969), 225.
15. WPHC BSIP Series F 46/35: Kuper to DO, Eastern Solomons, 2 Oct. 1933.
16. H. I. Manner, 'Buma Village, West Kwara'ae. Malaita, the Solomon Islands', in *Agroforestry in the Pacific Islands: Systems for Sustainability*, ed. W. C. Clarke and R. R. Thaman (Tokyo, 1993), 44.
17. Ivens, *Island Builders*, 355–64; Peters and Neuenschwander, *Slash and Burn*, 15–16, 19–20, 35–7; Whitmore, *Introduction*, 133–8.
18. Tedder, 'Yam Cultivation', 15–17; WPHC 133/99: *AR BSIP, 1898–9*; Samson Tahuniara, 'Posi: A Unique Yam Cultivation Practice in Guadalcanal', in *Land Use and Agriculture: Science of Pacific Peoples*, ed. John Morrison, Paul Geraghty, and Linda Crowl (Suva, 1994), vol. 2, 181–90; R. G. Rainbow and S. Teteha, *Shifting Cultivation in Solomon Islands* (Honiara, 1983).
19. Tedder, 'Yam Cultivation', 25, 34; D. E. Yen, 'Environment, Agriculture and the Colonisation of the Pacific', in *Pacific Production Systems: Approaches to Economic Prehistory*, ed. D. E. Yen and J. M. J. Mummery (Canberra, 1990), 267; WPHC BSIP Series F 46/35: Kuper to DO, Eastern Solomons, 2 Oct. 1933

20. Leonard P. Maenu'u, *Bib-Kami na Ano: Land and Land Problems in Kwara'ae* (Honiara, 1981), 19–20; W. C. Clarke and R. R. Thaman, eds, *Agroforestry in the Pacific Islands: Systems for Sustainability* (Tokyo, 1993), 218–9; Yen 'Agricultural Systems and Prehistory', 66; Margaret M. Tedder, 'Old Kusaghe', *Journal of the Cultural Association of Solomon Islands* 4, (1976): 41–95. Cf. William Balee, 'Indigenous History and Amazonian Biodiversity', in *Changing Tropical Forests*, ed. Harold K. Steen and Richard P. Tucker (Durham, NC, 1992), 186–7.

21. Raymond Firth, *Social Change in Tikopia: A Restudy of a Polynesian Community after a Generation* (London, 1959), 32; Patrick Vinton Kirch and D. E. Yen, *Tikopia: The Prehistory and Ecology of a Polynesian Outlier* (Honolulu, 1982), 37.

22. Walker, *Forests*, 33; Judith A. Bennett, 'Cross-cultural influences on Village Relocation on the Weather Coast of Guadalcanal, c. 1850–1950' (Master's thesis, University of Hawai'i, 1974), 35–40.

23. L. Verguet, 'Arossi et ses Habitants', *Revue d'Ethnologie* 4 (1885), 197.

24. Henderson and Hancock, *Guide*, 20–32, 34–7, 45; C. E. Fox, *The Threshold of the Pacific* (London, 1924), 280.

25. Ivens, *Island Builders*, 266.

26. Size of gardens varied with local circumstances. About 0.2 to 0.3 acre (0.08 to 0.12 hectare) in rainforest was needed per head yearly. A family of four would need about 0.8 to 1.2 acres (0.32 to 0.49 hectare) which is a medium to large gap. Jaques Barrau, *Subsistence Agriculture in Polynesia and Micronesia* (Honolulu, 1961), 25.

27. Leonard P. Maenu'u, 'Traditional Farming in the Solomon Islands', in *The Melanesian Environment: Papers presented at the Ninth Waigani Seminar held at Port Moresby*, ed. John H. Winslow (Canberra, 1977), 141–2; Henderson and Hancock, *A Guide*, 322–3; Ian Hogbin, *A Guadalcanal Society: The Kaoka Speakers* (New York, 1964), 40–2; Tedder, 'Yam Cultivation', 8–9; Ross, *Baegu*, 81; W.G. Ivens, *Melanesians of the South-east Solomon Islands* (London, 1927), 356; Ben Burt and Kwa'ioloa, 'The Forest of Kwara'ae' (draft manuscript, 1998); John Roughan, pers. comm., Dec. 1998.

28. Barrau, *Subsistence*, 19–24; R. C. Pendleton, 'The Rain Shadow Effect on the Plant Formations of Guadalcanal', *Ecological Monographs* 19, no. 1 (1949): 72–93; K. D. Marten, *Solomon Island Tropical Rainforest Ecology* (Honiara, 1979); G. E. Chaplin and A. J. Neumann, *Afforestation on the Guadalcanal Grasslands* (Munda, July 1987).

29. John Selwyn, 'The Islands of the Western Pacific', *Proceedings of the Royal Colonial Institute* 25 (1893–94), 366–7.

30. Whitmore, *Introduction*, 134, 155.

31. Les. M. Groube, 'Contradictions and Malaria in Melanesian and Australian Prehistory', in *A Community of Culture: The People and Prehistory of the Pacific*, ed. Matthew Spriggs et al. (Canberra, 1993), 164–86; Tim Bayliss-Smith, 'Melanesian Interaction at the Regional Scale: Spatial Relationships in a Fluid Landscape', in *Migrations and Transformations: Regional Perspectives on New Guinea*, ed. Andrew J. Strathern and Gabriele Stuzenhofecker (Pittsburgh and London, 1994), 295–331.

32. Parkinson claims that, under chief Gorai in the mid-nineteenth century, tribute was collected from west Choiseul. Trade did exist between the Mono-Alu and Choiseul and the former raided the latter periodically to obtain human sacrifices for canoe-launching, but regular tribute is not confirmed by Wheeler. (R. Parkinson, 'Ethnography of Ontong Java and Tasman Islands with Remarks re: The Marqueen and Abgarris Islands', *Pacific Studies* 9, no. 3 (July 1986), 8, 9; G. C. Wheeler, Mono Alu Notes, MS 184245, pagination inconsistent).

33. Yen, 'Environment, Agriculture', 267; Amherst and Thomson, *Discovery*, 303–38, 370–81; WPHC 1290/30: AR Santa Cruz, 1929 enc.

34. Lindsay Wall, 'Guadalcanal Stone Adzes', *Journal of the Solomon Islands Museum Association* 2 (1974), 38–9.

35. W. G. Ivens, 'Flints in the South-East Solomon Islands', *Journal of the Royal Anthropological institute of Great Britain and Ireland* 61 (1931), 421–4.

36. Ivens, *Melanesians*, 356; George Bogesi, 'Santa Isabel', *Oceania* 18, no. 3 (1948), 217; WPHC 1540/34: DO to Resident Commissioner (hereinafter RC), AR, Malaita District, 1933; Hviding, *Guardians*, 126–7, 165–6, 233–40. Much the same applies to lagoons and their islets (Hviding, *Guardians*, 127).

Notes to pp. 26–29

37. This is only a model. Most lineages were augmented by spouses and adoptees, including captives and purchased persons.
38. Maenu'u, *Bib-Kami*, 13; Gideon Zoloveke, 'Traditional Ownership and Land Policy', in *Land in Solomon Islands*, ed. Peter Larmour (Suva, 1979), 1; I. Q. Lasaqa, *Melanesians' Choice: Tadhimboko Participation in the Solomon Islands Cash Economy* (Canberra, 1972), 87, 100–101,109–13; Groube, 'Contradictions and Malaria'.
39. Ivens, *Island Builders*, 266–7; WPHC 1540/34: AR Malaita, 1933; Maenu'u, 'Traditional Farming', 141–2. In Santa Ana and Santa Catalina the fallow period was about five years, even in a time of decreased population pressure (WPHC BSIP Series F 46/35: Kuper to DO, Eastern Solomons, 2 Oct. 1933).
40. See for example, 'The "Generation" of the People of Oau', in *Pacific Protest: The Maasina Rule Movement, Solomon Islands, 1944–1952*, ed. Hugh Laracy (Suva, 1983), 58–64.
41. Guppy, *Solomon Islands,* 280–1. See Cassells, 'Valuation', 48 for similar remarks.
42. Henderson and Hancock, *Guide,* 276.
43. Ivens, *Melanesians,* 355; Henderson and Hancock, Ibid., 276.
44. Tedder, 'Yam Cultivation', 3, 60–72.
45. Hviding, *Guardians,* 191–7.
46. Cf. the Western tradition, Peter J. Bowler, *The Fontana History of the Environmental Sciences* (London 1992), 2.
47. Roger M. Keesing, *Kwaio Religion: The Living and the Dead in a Solomon Islands Society* (New York, 1982), 70–3; Ross, *Baegu:,* 125–7; Hviding, *Guardians,* 127.
48. *Australasian Record,* 27 Aug. 1927, 4.
49. Ross, *Baegu,* 195.
50. Ibid., 194–6; Keesing, *Kwaio,* 75–91.
51. *BSIP Newsheet,* 4/61, June 1961; Colin H. Allan, *Solomons Safari, Part II* (Christchurch, 1990), 129. Many, it is said, die as a consequence.
52. Bogesi, 'Santa Isabel', 212.
53. Interviews on Temotu, 1992.
54. Keesing, *Kwaio,* 46–49.
55. WPHC 1634/38: AR Malaita, 1937.
56. In the 1980s a small island was created in this way off Lauvi lagoon, south Guadalcanal.
57. WPHC 84/98: Woodford to O'Brien, 27 Jan. 1898.
58. Santa Ana and Santa Catalina, east of Makira, were hit by a severe drought in about 1850 (WPHC BSIP Series F 46/35: Kuper to D. O., Eastern Solomons, 2 Oct. 1933). Savo experienced 'a great drought' in about 1877 and, along with nearby north-west Guadalcanal, in 1915 (Woodford Papers (hereinafter WP): C. M. Woodford Diary, 14 Apr. 1887; *Southern Cross Log,* 1 Apr. 1916, 28, 33–4).
59. WPHC 1103/31: AR Guadalcanal, 1930.
60. Ivens, *Melanesians*, 181.
61. Ivens, *Island Builders of the Pacific,* 133
62. Matthew Cooper, 'Langalanga Religion', *Oceania* 43, no. 2 (1972), 120.
63. Christine May Dureau, 'Mixed Blessings: Christianity and History in Women's lives on Simbo, Western Solomons' (Ph. D. diss., Macquarie University, 1994), 73–4.
64. Bennett, 'Cross-cultural influences', 10–12.
65. A. I. Hopkins, *In the Isles of King Solomon, An Account of Twenty-five Years Spent amongst the Primitive Solomon Islanders* (London, 1928), 178–80; Charles Morris Woodford, *A Naturalist among the Headhunters* (London, 1980), 198.
66. Groube, 'Contradictions and Malaria', 179.

67. Bulmer notes a similar dichotomy in Papua New Guinea. R. N. H. Bulmer, 'Traditional Conservation Practices in Papua New Guinea' in *Traditional Conservation in Papua New Guinea: Implications for To-day*, ed. Louise Morauta, John Pernatta, and William Heaney (Port Moresby, 1982), 68–9. Cf. Geoff Park, 'The Polynesian Forest: Customs and Conservation of Biological Diversity', in *Science of Pacific Island Peoples*, ed. John Morrison, Paul Geraghty, and Linda Crowl (Suva, 1994), vol. 2, 132–54.

68. For Papua New Guinea, see Bulmer, 'Traditional Conservation', 72.

69. J. A. R. Miles, *Infectious Diseases in the Pre-European Pacific*. In Press, University of Otago, Dunedin.

70. Matthew Spriggs, 'Landscape, Land Use, and Political Transformation in Southern Melanesia', in *Island Societies: Archaeological Approaches to Evolution and Transformation*, ed. Patrick Vinton Kirch (Cambridge, 1986), 6–19; Anelle Stevenson and John R. Dobson, 'Palaeoenvironmental Evidence for Human Settlement of New Caledonia', *Archaeology in Oceania* 30, no. 1 (Apr. 1995), 36–42.

71. Amherst and Thomson, *Discovery*, 303–38, 370–81.

72. Richard Walter, pers. comm., Mar. 1999.

73. Patrick Vinton Kirch and D. E. Yen, *Tikopia: The Prehistory and Ecology of a Polynesian Outlier* (Honolulu, 1982), 310, 346–9.

74. *SINFRI*, 1, 8.

75. WPHC 84/98: Woodford to O'Brien, 27 Jan. 1898.

76. WPHC 1290/30: AR Gela and Savo, 1929.

77. Walker, *Forests*, 33.

78. Amherst and Thomson, *Discovery*, 302; WPHC 84/98: Woodford to O'Brien, 27 Jan. 1898; Pendleton, 'Rain Shadow'; Marten, *Solomon Island Tropical Rainforest*; P. D. Nunn, 'Recent Environmental Changes on Pacific Islands', *Geographical Journal* 156 (1990), 125–40.

79. Amherst and Thomson, *Discovery*, 291–2, 303–9; Henderson and Hancock, *Guide*, 318–9; WPHC CF 164/2/9: Dennis to Chief Sec., 6 June 1959. For El Niño generally, see Timothy Flannery, *The Future Eaters: An Ecological History of the Australasian Lands and People* (Chatswood, NSW, 1994), 81–5; Nunn, *Environmental Change*, 288–90. For the labour trade, see Ralph Shlomowitz, 'Mortality and Workers', in *Labour in the South Pacific*, ed. Clive Moore, Jacqueline Leckie and Doug Munro (Townsville, Australia, 1990), 124–7.

80. Whitmore, *Guide*, 4; Whitmore, *Change with Time*, 57; Schenk, *Forest Ecology*, 22; Bennett, *Wealth*, 112.

81. John M. Aldrick, *The Susceptibility of Lands to Deterioration in the Solomon Islands* (Honiara, 1993), 36.

82. Methodist Church, New Zealand (hereinafter MCNZ) and PMB 928: Bensley to Scriven, 8 Oct. 1933.

83. M. D. Chapman, 'Environmental Influences on the Development of Traditional Conservation in the South Pacific Region', *Environmental Conservation* 12 (1985), 217–30. The most notable exception is Easter Island which may have suffered demographic and social collapse because of human induced environmental degradation, though a deterioration in climate in AD 1300 was probably a contributing factor. See Paul Bahn and John Flenley, *Easter Island, Earth Island: A Message from our Planet* (London, 1992); G. McCall, 'Little Ice Age: Some Speculations for Rapanui', *Rapa Nui Journal* 7 (1993), 65–70.

84. Yen, 'Agricultural Systems and Prehistory', 61–74. See also Kirch and Yen, *Tikopia*, 28–30.

85. Henderson and Hancock, *Guide*, 156, 163.

86. WPHC 2777/32: RC to DO, 27 May 1932; Raymond Firth, 'Anuta and Tikopia: Symbiotic Elements in Social Organization', *Journal of the Polynesian Society* 63 (1954), 109.

87. Yen, 'Environment, Agriculture and the Colonisation', 270.

88. Fox, *Threshold*, 278–279.

89. Ivens, *Melanesians*, 178; Keesing, *Kwaio*, 119; Tedder, 'Yam Cultivation', 51–2.

90. Ivens, *Melanesians*, 160; J. G. B. Nerdrum, 'Indtryk og oplevelser under et 7 aars ophold paa Salomon-oerne', *Norske Geog. Selskabs Aarbog,* 13 (1901/02), 9–10; Burman, 'Time', 260.

91. Arthur M. Hocart Papers: Chieftainship, Foreign Trade; Scheffler, *Choiseul Island*, 186–7; Guppy, *The Solomon Islands,* 32, 45, 81; Hugh Laracy, *Marists and Melanesians: A History of Catholic Missions in the Solomon Islands* (Canberra, 1976), 52.

92. W. C. Clarke, 'Agro-deforestation in Melanesia', *Cultural Survival Quarterly* 15, no. 2 (1991), 45–8; Patrick V. Kirch, *The Evolution of the Polynesian Chiefdoms* (Cambridge, 1984), 123–51; H. J. Enright and C. Cosden, 'Unstable Archipelagos: South-west Pacific Environment and Prehistory since 30,000 BP', in *The Naive Lands: Prehistory and Environmental Change in Australia and the Southwest Pacific,* ed. John Dobson (Melbourne, 1992), 190–2.

93. Ivens, *Melanesians,* 402.

94. Clarke and Thaman, eds, *Agroforestry,* 12.

95. E. K. Fisk, 'The Economic Structure', in *New Guinea on the Threshold: Aspects of Social, Political and Economic Development,* ed. E. K. Fisk, (London, 1966), 23–43.

96. See, in relation to the American Indians, Kent H. Radford, 'The Ecologically Noble Savage', *Cultural Survival Quarterly* 15, no. 1 (1991), 46–8.

97. Ester Boserup, 'Environment, Population and Technology' in *The Ends of the Earth: Perspectives on Modern Environmental History,* ed. Donald Worster (Cambridge, 1988), 33; Bennett, *Wealth,* 1–19. For Papua New Guinea see Morauta, Pernatta, and Heaney, eds, *Traditional Conservation.*

98. Maenu'u, *Bib-Kami,* 9. This can also be translated as 'A person is hard to obtain' (Ben Burt, pers. comm., 16 March 1998).

99. Life expectancy here includes infant deaths. Some people reached old age, but the majority did not. Bennett, *Wealth,* 1–19; Hviding, *Guardians,* 96.

Chapter 3

1. Admiralty Library, London: John MacGillivray, Journal, Dec. 1852; Bennett, *Wealth,* 83; Codrington, *Melanesians,* 314; R. F. Salisbury, *From Stone to Steel: Economic Consequences of a Technological Change in New Guinea* (Melbourne and Berkeley, 1962), 145–9; Steensberg, *New Guinea Gardens,* 6, 12, 25–34, 37–40; WPHC BSIP Series F 46/35: Kuper to DO, Eastern Solomons, 2 Oct. 1933.

2. Marwick, comp., *Adventures of John Renton,* 29–30.

3. Ivens, *Melanesians,* 48–9.

4. Guppy, *Solomon Islands,* 127–8.

5. J. F. Goldie, 'The People of New Georgia, their Manners and Customs and Religious Beliefs', *Royal Society of Queensland, Proceedings* 22 (1909), 26–7; Walter Coote, *The Western Pacific* (New York, 1969), 144. Even a staged re-enactment of this by some former participants for film-makers almost forty years later makes this clear in the film, *The Transformed Isle: Barbarism to Christianity* (1920). For north Malaita, see Ivens, *Island Builders,* 186–7.

6. Ivens, *Island Builders,* 263, 273; WPHC BSIP Series F 46/35: Kuper to DO, Eastern Solomons, 2 Oct. 1933.

7. For the Roviana district, see Andrew Cheyne, *Trading Voyages of Andrew Cheyne, 1841–1844,* ed. Dorothy Shineberg (Canberra, 1971), 396; for Makira, see John Webster, *The Last Cruise of the Wanderer* (Sydney, c. 1863), 94.

8. Bennett, *Wealth,* 78–102, 113; WPHC BSIP Series F 46/35: Kuper to DO, Eastern Solomons, 2 Oct. 1933; WPHC 83/1932: Garvey, The Depopulation of Vanikoro, 1932. See also Ivens, *Island Builders,* 267.

9. Hopkins, *In the Isles,* 40, 180; Ivens, *Island Builders,* 191; Goldie, 'People of New Georgia', 25. There is evidence that the practice of fortification preceded the advent of firearms; however, the practice may have been extended with the greater penetrative power of bullets and musket balls (Ivens, *Island Builders,* 53, 191–3).

Notes to pp. 38–44

10. See Peter Corris, *Passage, Port and Plantation: A History of Solomon Islands Labour Migration, 1870–1914* (Melbourne, 1973), 140–1; Ivens, *Island Builders*, 268–9; Bennett, 'Cross-cultural influences'; WPHC 481/17: Bell to H. C., 16 Dec. 1916.

11. Ivens, *Island Builders*, 269.

12. WP: B. B. Footaboory, P. Ambuover, J. Gwoefoon, H. Rumsalla, H. Nuferia to C. Woodford, 17 Oct. 1912.

13. Colonial Office, Public Record Office, Kew (hereinafter CO) 225/80: Footaboory and Peter Ambuaffer(sic.) to King Edward VII, 2 May 1907. I have followed the writers' spelling.

14. Arthur I. Hopkins, *Southern Cross Log*, 6 July 1906, 19.

15. CO 225/80: B. B. Footaboory and P. Ambuaffer to King Edward VII, 2 May 1907.

16. CO 225/80: Minutes on file enclosing B. B. Footaboory and P. Ambuaffer to King Edward VII, 2 May 1907.

17. Bennett, *Wealth*, 125–66.

18. WPHC 282/05: AR 1903–05; Ivens, *Island Builders*, 268–9.

19. WPHC 477/96: AR BSIP, 1896; WP: Woodford Diary, 14 Oct. 1888; CO 225/89: Woodford to Johnson, Sept. 1909.

20. WP: Woodford, Diary, 24 June 1886.

21. Woodford, *Naturalist*, 188.

22. WPHC 477/96: AR BSIP, 1896; WP: Woodford, Diary, 14 Oct. 1888. See also Pacific Islands Company Papers, Unilever: Grant to Denson, n.d. [May 1900] and AR BSIP, 1913; Bennett, *Wealth*, 126–9.

23. Bennett, *Wealth*, 138–9; Harold Hamel Smith and Fred A. G. Pape, *Coconuts: The Consols of the East* (London, 1912).

24. Peter Sack, *Land between Two Laws* (Canberra, 1973), 20; Pacific Islands Co. Papers: Langdale to Denson, n.d., May 1900, enc.

25. Bennett, *Wealth*, 125–9.

26. Ian Heath, 'Land Policy in Solomon Islands' (Ph.D. diss., La Trobe University, Melbourne, 1979), 112–9; AR 1912–1913. See also *British and Australian Trade in the South Pacific*, 1916, vol. 5, Report of Interstate Commission, Melbourne: Commonwealth of Australia, Australian Parliamentary Papers, 1917–1919, 68.

27. Heath, 'Land Policy', 115.

28. Bennett, *Wealth*, 137–8, 146–9; WPHC 3386/27: List of Lands, Mar. 1927, enc.

29. References are too numerous to detail. For a sample, see: WPHC files 2431/99, enc. cutting from *Sydney Herald*, n.d., c. 1900; WPHC 40/02, 830/08, 1803/12, 1505/13, 1779/16 AR BSIP, 2582/18,2352/25; *The Sun*, 6 Aug. 1910; *Sydney Mail*, 12 Aug. 1912; S. G. C. Knibbs, *The Savage Solomons as they Were and Are* (London, 1929), 264–76; Corres. Relating to Solomon Islands, Mf 175, National Library of New Zealand: Giblin to Lord Stanmore, 20 Mar. 1905; Pacific Material, Unilever Archives: Meek to Lord Leverhulme, 2 June, 1922; *The Planters' Gazette*, 14 Feb., 7 Dec. 1922.

30. Michael Havinden and David Meredith, *Colonialism and Development: Britain and its Tropical Colonies, 1850–1960* (London, 1993), 4, 16.

31. Bennett, *Wealth*, 177; For concern about the depopulation question, see W. H. R. Rivers, *Essays on the Depopulation of Melanesia* (Cambridge, 1922).

32. WPHC 1196/11: Memo on the Duties of District Magistrates, in the Gilbert and Ellice Islands and the Solomon Islands, n.d., c. 1910.

33. WPHC 63/13: Vernon, Notes on the Solomon Islands Protectorate, 7 Dec. 1912.

34. L. M. Raucaz, *In the Savage South Solomons* (Lyons, 1928), 57.

35. Laracy, *Marists and Melanesians*, 69, 72–4.

36. Ibid., 41.

37. Ibid., 93–5, 98, 106.

38. Ivens, *Island Builders*, 45–6; Hopkins, *In the Isles*, 224

39. *Southern Cross Log*, 20 Oct. 1910, 76.
40. Ibid., 75.
41. Ibid., 12 Oct. 1911, 238; 1 April 1916, 33,
42. Florence S. H. Young, *Pearls from the Pacific* (London, 1925), 168, 176–7, 194, 202.
43. Bennett, *Wealth*, 196–7; WPHC 63/13: Vernon, Notes, 7 Dec. 1912; WPHC 830/08: Mahaffy to Sec. of State, 21 Dec. 1908; WPHC 2015/13: Woodford to Escott, 21 Aug. 1913; WPHC 1991/13: Bell to Woodford, 6 Sept. 1913. The Catholics were also critical of the use of 'school attendants' for work on the Malayta Company's plantations (WPHC 1151/15: Bertreux to HC, 24 Mar. 1915).
44. Young, *Pearls*, 241–5.
45. Methodist Church, *Report of the Australasian Wesleyan Methodist Society* (Sydney, 1905), xxiii, cxiv; *Report* (Sydney,1907), 27–8.
46. Methodist Overseas Mission (hereinafter MOM), Mitchell Library: Goldie to Danks, 11 May 1910.
47. MOM: Nicholson, Vella Lavella Circuit Report, 1916, 3.
48. MOM: W. A. Sinclair and J. W. Court, Report of the Representative to the SI Mission District, 1920, 12–13. See also WPHC 1184/25: Goldie, Deputation to HC of Representatives of Residents, 29 Sept. 1925.
49. PMB 68: John R. Metcalfe, 6 May, 1938, General Letters.
50. Dennis Steley, 'Juapa Rane: The Seventh Day Adventist Mission in the Solomon Islands, 1914–1942' (Master's thesis, University of Auckland, 1983), 33, 52, 89–94, 120–6.
51. Ibid., 40–1. In the existing records I examined at the Ellen White Research Centre, no reference was made to the commercial 'development' of Solomon Islanders, and all involvement in commerce by teachers or missionaries was forbidden. Seventh Day Adventist archives, Cooranbong, (hereinafter SDA): Solomon Island Mission Minutes, 187th Meeting, 8 July 1938.
52. For example, see Barnabas Pana, Letter, *Australasian Record*, 4 Mar. 1931; Steley, 'Juapa Rane', 62, 207–8; Hviding, *Guardians*, 92–8. Hviding seems to imply that there is an association between their religion and the modern Marovo people's commercial entrepreneurship (Hviding, *Guardians*, 317). If this is the case, it was not the stated aim of the early Australian missionaries.
53. Cited in Steley, 'Juapa Rane', 263.
54. K. B. Jackson, 'Tie Hokara, Tie Vaka, Black Man White Man: A study of the New Georgia Group to 1925' (Ph. D., Australian National University, Canberra, 1978), 286–93.
55. Clifford W. Collinson, *Life and Laughter 'midst the Cannibals* (London, 1926), 118–93.
56. James E. Brook, *Jim of the Seven Seas: A true Story of Personal Adventure* (London, 1940), 192; CO 225/87–89: Minutes, Johnson, Sept. 1909.
57. WPHC 3343/26: Barley to RC, 9 Aug. 1926.
58. WPHC 281/11: Mahaffy to Sec. of State, 21 Dec. 1908; *Sydney Morning Herald*, 26 Feb. 1913; Rivers, *Essays*; P. Hermant and R. W. Cilento, *Report of the Mission Entrusted with a Survey on Health Conditions in the Pacific Islands* (Geneva, 1929), 78; Deland Papers, National Library of Australia (hereinafter DP): Deland to Everyone, 12 June 1928; WPHC 3808/33: Garvey's comments, 15 Jan. 1934.
59. AR BSIP 1912–13; WPHC 2929a /20: Rodwell to Workman 3 Oct. 1920; WPHC 1184/25: Goldie, Deputation to HC of Representatives of Residents, 29 Sept. 1925. See, for examples, DP: Deland to Everyone, 12 June 1928; Rev. J. W. Burton, *The Fiji of To-day* (London, 1910).
60. WPHC 2288/16: Acting RC to the Chief and people of Alu, 13 July 1916.
61. *The Australasian Methodist Review*, 4 Sept. 1917, 4.
62. WPHC 2352/25: Eyre Hudson to Sec. of State for Colonies, 26 Oct. 1925.
63. CO 225/95–96: May to Sec of State, 1 Mar. 1911.
64. Burns Philp Archives, Noel Butlin Archives, Australian National University, Canberra (hereinafter BPA): W. H. Lucas, Notes on the Western Pacific and Australian Interests therein: Actual and Potential, March 1917, South Seas Misc. file.

Notes to pp. 46–50

65. Boyle T. Somerville, 'Ethnographical Notes on New Georgia, Solomon Islands', *Journal of the Royal Anthropological Institute* 26 (1897), 411; See also Beatrice Grimshaw, *In the Strange South Seas* (London, [1908]), 56.
66. WPHC 2352/25: Eyre Hutson to Sec. of State, 26 Oct. 1925.
67. WPHC 977/09: Woodford to Im Thurn, 14 Feb. 1910.
68. WPHC 1184/25: Deputation to HC of Representatives of Residents, 29 Sept. 1925. See also Pacific Material, Unilever Archives: Meek to Lord Leverhulme, 2 June 1922.
69. WPHC 977/09: Woodford to Im Thurn, 14 Feb. 1910.
70. BPA: Lucas to Managing Director, 28 July 1908.
71. F. J. Hickie, 'Coconut Planting in the British Solomons', *The Planters Gazette*, Dec. 1922, 12.
72. Fairley, Rigby and Co. Papers, Melbourne University Archives (hereinafter FRP): Rigby to Fairley, 14 Dec. 1912.
73. FRP: Rigby to Fairley, 21 Nov. 1912.
74. F. J. Hickie, 'Coconut Planting in the British Solomons', *The Planters Gazette*, May 1923, 8.
75. Hickie, 'Coconut Planting', Dec. 1922, 12.
76. FRP: Rigby to Fairley, 7 Feb. 1913.
77. BPA: Lucas to Managing Director, 28 July 1908; AR BSIP, 1912–13.
78. FRP: Rigby to Fairley, 14 Dec. 1912.
79. *The Planters' Gazette*, Aug. 1922, 19; Hickie, 'Coconut Planting', Dec. 1922, 12, May 1923, 8.
80. *Australasian Record*, 19 Oct. 1914, 3.
81. Hickie, 'Coconut Planting', Dec. 1922, 12.
82. Hickie, 'Coconut Planting', May 1923, 8.
83. WPHC 1121/24: LDAR 1923; Hickie, 'Coconut Planting', Dec. 1922, 12.
84. Gilchrist Alexander, *From the Middle Temple to the South Seas* (London, 1927), 261.
85. *Australasian Record*, 19 Oct. 1914, 3.
86. WPHC 830/08: Mahaffy to Sec. of State, 21 Dec. 1908.
87. *The Planters' Gazette*, Aug. 1922, 3.
88. Alexander, *Middle Temple*, 258,
89. WPHC 1779/16; WPHC 67/18; WPHC 92/18; WPHC 1522/33; AR BSIP 1912–1913; *BSIP Agricultural Gazette* I, no. 1 (1931–1932), 3–4.
90. J. von Mauler and William Kesslitz, 'The Scientific mission of SM *Albatros*, 1895–1898', *Report from the K. K. Geographical Society in Vienna*, vol. 1, ed. A. B. E. von Böhmerscheim, trans. V. C. Wasem (Vienna, 1899), 64; Roy Struben, *Coral and Colour of Gold* (London, 1961), 85; Margaret Clarence, *Yield Not to the Wind* (Sydney, 1982), 40; WPHC 42/98.
91. WPHC 1290/30: AR Santa Cruz (hereinafter SC); WPHC 1540/34: AR Shortlands, 1933; AR BSIP 1912–1913.
92. Hector MacQuarrie, *Vouza and the Solomon Islands* (Sydney, 1946), 21–2. See also Steley, 'Juapa Rane', 141–2, for the view that a missionary brought the clover in from the New Hebrides in 1917.
93. See for example WPHC 350/98; AR BSIP 1898–99; WPHC 1584/12; WPHC 254/24; WPHC 3726/31.
94. Graeme A. Golden, *The Early European Settlers of the Solomon Islands* (Melbourne, 1993) photographs, 162, 166; Judith A. Bennett, 'Wealth of the Solomons: A History of Trade, Plantations and Society, Solomon Islands, c. 1800–1942' (Ph.D. diss., Australian National University, Canberra, 1979), photographs, 256, 291; WPHC 3726/31: Journal, 30 Dec. 1929–12 Jan. 1930; BSIP 9/111/3: QRSC, Mar. 1929, enc. See also Vanikoro Kauri Timber Co. records, University of Melbourne Archives (hereinafter VR): Corres. to and from, 1932–1936: Wilson, Report of Timber Survey, 21 Dec. 1933 and enc.

95. MOM: George Brown, Letterbook, 24 May 1902. Nusuzonga was purchased from the traders, Kelly, Williams and Woodhouse (C. T. J. Luxton, *Isles of King Solomon: A Tale of Missionary Adventure* (Auckland, 1956), 30).

96. MOM: George Brown, Letterbook, 24 May–2 June 1902.

97. Billings Papers, Macmillan Brown Centre, University of Canterbury: Billings, c. 1945. The mission inherited a title to this property from S. Edmunds Prat who claimed to have purchased it in 1893 from Lepe, Kendo, Tuita, Koomba, Mea, Veto and Enbeenbo. Nonetheless, the mission also purchased the same land for 15 pounds from Ingova, Mea and Gumi in 1902. (Luxton, *Isles of King Solomon*, 31–33; Lands Titles Office, Honiara,(hereinafter LTO): LR 120–001–1, No. 23/64 Lot 2 of LR 142, Kokeqolo.)

98. MCNZ : Effie Harkness to Scriven, 15 Jan. 1938. The soils here are mainly Mollisols (Rendolls), and usually the limestone is within 50 cm of the surface (Hansell and Wall, *Land Resources*, 44 and Map 4d).

99. Luxton, *Isles of King Solomon*,. 32.

100. *Open Door*, 1, 2 (Sept. 1922), 12.

101. *The Australasian Methodist Missionary Review*, 5 Aug. 1907, 16.

102. Luxton, *Isles of Solomon*, 59, 119, 124.

103. Laracy, *Marists and Melanesians*; Raucaz, *In the Savage Solomon Islands*, 82, 96–100, 118–9, 161–4, 174–7. For similar work by the Melanesian Mission, see *Southern Cross Log*, 15 Feb. 1901, 129, 131; 1 Aug. 1903, 44; 12 Oct. 1911, 239; 20 Dec. 1912; 2 Oct. 1916, 741.

104. *Australasian Record*, 9 Sept. 1918, 3.

105. *Australasian Record*, 13 Oct. 1919, 3.

106. PMB 68: Metcalfe, 5 Aug. 1927.

107. WPHC 929/25: Kane to HC, 4 April 1925.

108. Verguet, 'Arossi', 194; *Southern Cross Log*, 13 Aug. 1910, 40.

109. WPHC 830/08: Mahaffy to Sec. of State, 21 Dec. 1908.

110. Campbell, Instructions to Native District Headman, 1918, copy kindly supplied by Hugh Laracy.

111. WPHC 1679/22: Kane to HC, 10 May, 1922; WPHC 287/27: Bell to HC, 24 June 1927.

112. WPHC 929/25: Kane to HC, 4 April 1925; WPHC 2768/22: KR, No. 17 of 1922, enc; WPHC 939/25: Eyre Hutson to Kane, 14 May 1925.

113. WPHC 1679/22: Kane to HC, 10 May 1922.

114. WPHC 1951/23: DO to RC, extract, n.d. [c. 1923], enc.

115. WPHC 2932/26: Filose, AR Guadalcanal, 1925.

116. Martin Johnson, *Through the South Seas with Jack London* (London, 1913), 304.

117. Superior methods of cultivating kumara and possibly superior varieties were introduced by Polynesian missionaries attached to the Methodist mission in the western Solomons (Sione Latukefu, 'Pacific Islander Missionaries', in *The Covenant Makers: Islander Missionaries in the Pacific*, ed. Doug Munro and Andrew Thornley (Suva, 1996), 34).

118. Bennett, *Wealth*, 193–4, 112–4; WPHC 1897/30: Crawfurd to Sec. to the Government (hereinafter Gov.), 8 May 1930, and encls; WPHC F 2/25: Report of DO, 23 Oct. 1940; T. Russell, 'The Culture of Marovo, British Solomon Islands', *Journal of the Polynesian Society* 57 (1948), 318; Hviding, *Guardians*, 60–61. This was probably Gisu Piko, a Seventh Day Adventist and a Native Medical Practitioner who studied in Fiji. Cassava was a common crop on Santa Isabel by the 1940s and in the Russell islands by the early 1960s.(George Bogesi, 'Santa Isabel', 223; Robert H. Black, 'The Russell Islanders of the British Solomon Islands Protectorate' (Dip. Anthrop., University of Sydney, 1963), 5).

119. AR BSIP, 1897–98.

120. AR BSIP, 1901–02.

121. AR BSIP, 1898–99; 1902–03.

122. WPHC 297/97: Woodford to Collet, 3 May 1897.

Notes to pp. 56–64

123. Knibbs, *Savage Solomons*, 264–6; Ted Ashby, *Blackie, a Story of the Old Time Bushmen* (Wellington, 1978), 69–70.
124. CO 225/89: Woodford to Im Thurn, 31 May 1909, enc.; AR BSIP 1898–1899.
125. Knibbs, *Savage Solomons*, 266.
126. WPHC 1184/25: Deputation to HC of Representatives of Residents, 29 Sept. 1925 and encls.
127. See, for example Barnabas Pana's Letter, 4 Mar. 1931, *Australasian Record*.
128. PMB 925: Goldie to RC, 29 Dec. 1931, MM Correspondence.
129. WPHC 297/97: Woodford to Collet, 3 May 1897.
130. Caption on photograph, *The Australian Methodist Missionary Review*, 6 Dec. 1902.
131. PMB 557: Arthur I. Hopkins, Autobiography, 29.
132. Solomon Mamaloni, 'Rural Development Then and Now: Legacy of Colonial Miscalculation', in *The Road Out: Rural Development in Solomon Islands*, ed. Stephen Oxenham (Suva, 1981), 2–4.
133. Cited in *Southern Cross Log*, 15 Oct. 1900, 71.

Chapter 4

1. Amherst and Thomson, *Discovery*, vol. 1, 128.
2. Ibid., vol. 2, 208. Mendaña's brigantine may not have been the first European-built boat made in the Solomons, as there is a tradition of an earlier party coming in a small boat to Wainoni Bay, building a large boat and sailing away with some local people. This might be an out-of-sequence account of the Spanish settlement found near Pamua, whom some believe to be the survivors of the lost ship *Almiranta* which was one of the ships to take settlers led by Mendaña and Quiros to Santa Cruz in 1595 (BSIP 9/11/1, Folder B: AR San Cristoval,1947).
3. Peter Dillon, *Narrative of La Perouse's Expedition*, vol. 2 (London, 1829), 112–315; FRP: Report of the Lands Commissioner, Native Claims 38 and 39, Vanikoro 10 Dec. 1923, 16.
4. WPHC 2531/29: Garvey, log of *Tulagi*, 4 Feb. 1929.
5. PMB 572: Journal of *Alfred Gibbs*, 1856; PMB 680: Journal of *Belle*, 1856; PMB 284: Journal of *Two Brothers*, 1860; PMB 390: Journal of *Mohawk*, 1861; Bennett, *Wealth*, 21–44.
6. Bennett, *Wealth*, 390.
7. Karl Scherzer, *Narrative of the Circumnavigation of the Globe by the Austrian Frigate Novara*, vol. 2 (London, 1862), 611, 619.
8. Otto Finsch, 'Uber Naturprodukte der westichen Sudsee, besonders der deutschen Schutzgebiete', *Deutsche Kolonialzeitung. Organ des deutschen Kolonialvereins* 4 (1887), 17.
9. R. Gerard Ward, 'The Pacific Bêche-de-Mer Trade with Special Reference to Fiji', in *Man in the Pacific Islands*, ed. R. Gerald Ward (Oxford, 1972), 114–15.
10. Scherzer, *Narrative*, 613.
11. Bennett, *Wealth*, 45–102; WPHC 22528/28: Miller to Sec. to Gov., 5 April 1929.
12. This was the third mission vessel of this name. (Selwyn, 'The Islands of the Western Pacific', 363.
13. Cheyne, *Trading Voyages*.
14. Dorothy Shineberg, *They Came for Sandalwood: A Study of the Sandalwood Trade in the South-West Pacific, 1830–1865* (Melbourne, 1967), 140–1.
15. Scherzer, *Narrative*, 609–10.
16. H. E. Raabe, *Cannibal Nights: The Reminiscences of a Free Lance Trader* (New York, 1927), 124, 169.
17. AR BSIP, 1901–1904.
18. Called kwila in Australia and New Guinea. This tree is technically *Intsia bijuga*.
19. AR BSIP 1913–1914.
20. Unilever: Report on *Rob Roy* expedition: Grant to Denson, May, 1900.

Notes to pp. 64–69

21. BPA: Lucas to General Manager, 28 July 1908.
22. WPHC 428/12: Woodford to May, 21 Mar. 1912.
23. BPA: Lucas to Managing Director, 25 Nov. 1907.
24. Bennett, *Wealth*, 145.
25. BPA: Lucas to Managing Director, 28 July 1908.
26. BPA: Lucas to Forsayth, 3 Nov. 1908; WPHC 2081/22: Forsayth to Chairman, SID Co, 32 Oct. 1911.
27. *Protectorate of British Solomon Islands, Statistics to 31 March, 1909* (Sydney, 1909), 18.
28. WPHC 428/12: Woodford to May, 21 Mar. 1912.
29. WPHC 428/12: Woodford to May, 21 Mar. 1912; BPA: Henry Fenn, Offer of a sawmill proposition in Solomon Islands, 8 Sept. 1913.
30. BPA: Lucas, Reports on Solomon Island Plantation Properties, 1 Oct. 1910.
31. LTO: Land Title 019–001–7 Application for first registration, 14/67, LR 146 Konokai; Bennett, *Wealth*, 179, 203–205; PMB 1021: Monckton to Uncle Parry, 13 Oct. 1909, 27 May 1910.
32. PMB 1021: Monckton to Uncle Parry, 30 Oct. 1910.
33. BPA: Lucas, Reports on Solomon Island Plantation Properties, 1 Oct. 1910.
34. AR BSIP 1912–13, 1913–14. Woodford identifies the species as *Calophyllum inophyllum*, but this may not be so, as 'Beach Calophyllum' is mostly a poorly shaped tree. However, it can be used in furniture-making and would have been the most accessible and numerous species, as it is found in beach forest and Alu has limited forest zones, being raised coral and having a maximum altitude of 613 feet. Walker, *Forests*, 41, 122–3; Henderson and Hancock, *Guide*, 220; G. J. Pleydell, *Timbers of the British Solomon Islands* (London, 1970), 7.
35. WPHC 428/12: Monckton to HC, 29 Feb. 1912.
36. Bennett, *Wealth*, 98-9, 114, 177.
37. WPHC 38/13: Copy of Monckton's lease, 3 Dec. 1912; WPHC 1789/12: Sec. of State to Sweet-Escott, 7 Aug. 1912, and encls; WPHC BSIP File 5: Ashley to HC, Timber Concessions in Solomon Islands, n.d. c.1938; WPHC 245/33: R. A. Sykes, Report on Tour of Inspection of the Forests of the BSIP, 1932.
38. WPHC 3845/32: Tench to Sec. of Gov., 4 Jan. 1941; WPHC 428/12: Woodford to Major, 26 June 1912.
39. WPHC 3845/32: Tench to Sec. of Gov., 4 Jan. 1941
40. BSIP NA, F 21/5 Part 1: QRSI, 30 June 1938,.
41. Kimale (Kimmele) was a grandson of Ware by a secondary wife; his father was Mamatau (WPHC 3845/32: Tench to Sec. of Gov., 4 Jan. 1941).
42. BSIP NA, F 21/5 Part 1: QRSI, 30 June 1938.
43. WPHC 228/25: Report on Santa Cruz, Tikopia, Vanikoro; Jacob Piopio of Pirumeri, Shortland Islands; interview, 1976.
44. WPHC 6541/23: Middenway to RC, 13 Jan. 1922; WPHC, 1779/39: AR Shortlands, 1938; Burns Philp Archives (hereafter BPA): Burns Philp (South Seas) (hereafter BPSS) Inspection Reports, Faisi Branch, 3 Aug. 1930, April 1932; VR: Boye to Sec. 14 Nov. 1939, Sec. to Manager, 12 April 1940, Corres. to and from Vanikoro, Sept. 1939–Aug. 1949; Sec. to Manager, 30 Mar. 1939, Corres. to and from Vanikoro, Sept. 1937–Aug. 1939; WPHC 1522/33: AR Shortlands, 1932, enc.; Joseph Odofia of Mari Mari, Malaita, interview, 1976.
45. VR: Sec., Vanikoro Timber Co. (hereafter VTC) to Gullett, Dept of Trade and Customs, Aust. 4 Feb. 1932, Comptroller- General of Customs to Sec. 23 Dec. 1932, Box Gov. Depts to and from, 1932–34; AR BSIP 1933; BPA: BPSS Inspection Reports, Faisi, Aug. 1930.
46. BPA: BPSS Inspection Reports. 1931, Faisi.
47. VR: Collector of Customs to Director, 27 Feb. 1933, Box Gov. Depts to and from, 1932–1934.
48. WPHC 1359/34: Shortlands DO's Diary, 23 Mar. 1934; WPHC 1588/35: Quarterly return of Leases, 30 June 1935; BSIP NA F 21/5 Part 1: QRSI, 30 June 1938.

Notes to pp. 69–71

49. LTO: Adjudication Officer's Report, Kokonai (LR 146), 1968; Justus Malalifu of Maoa; Joseph Odofia of Mari Mari, Malaita; Mikelo Ebinuwi of Nuhu; Joseph Nikolas of Samanago; Joseph Normani; Jacob Piopio of Pirumeri, Shortland Islands, interviews 1976.

50. WPHC 1522/33: AR Shortlands, 1932; WPHC 1540/34: AR Shortlands, 1933; WPHC 1589/35: AR Shortlands, 1934; WPHC 1634/38: AR Shortlands, 1937; WPHC 1779/39: AR Shortlands, 1938.

51. WPHC 1779/39: AR Shortlands, 1938.

52. WPHC 1587/35: Shortlands DO's Diary, 4 Feb. 1936. By 1939 this plant was a major pest in the Shortlands. The government unsuccessfully tried to control it by introducing a parasitic thrip (WPHC 1203/39: Agricultural Committee Minutes, Mar., Nov. 1939).

53. WPHC 3386/27: List of Lands, Mar. 1927; WPHC 1522/33: AR Shortlands, 1932; WPHC 1540/34: AR Shortlands, 1933; WPHC 1589/35: AR Shortlands, 1934; WPHC 1634/38: AR Shortlands, 1937; WPHC 1779/39: AR Shortlands, 1938; Justus Malalifu of Maoa; Joseph Odofia of Mari Mari, Malaita; Mikelo Ebinuwi of Nuhu; Joseph Nikolas of Samanago; Joseph Normani; Jacob Piopio of Pirumeri, Shortland Islands, interviews 1976; *Pacific Islands Monthly* (hereafter *PIM*), Sept. 1953, 133, 135; BPA: BPSS Inspection Reports, Faisi Branch, 22 Dec. 1933, 17 July 1939; VR: Monckton to Manager, Vanikoro Kauri Timber Co. 30 June, Oct. 1937, File 2, Box Vanikoro General Corres., 1935–1941.

54. WPHC 1587/35: Shortlands DO's Diary, 7 Feb., 21 Oct. 1936.

55. Superficial feet or feet square in Hoppus measure is calculated as follows: quarter girth in inches squared, multiplied by the length in feet, divided by 12 = superfeet content. One superfoot of log equals a twelfth of a Hoppus cubic foot or 0.106 of a true cubic foot. Conversion from superfeet to cubic metres is 100 superfeet Hoppus measure = 0.301 cubic metres. In sawn timber 100 superfeet = 0.236 cubic metres.

56. BSIP, NA F14/19: AR Shortlands, 1940

57. WPHC 1587/35: Shortlands DO's Diary, 28 Feb. 1937.

58. WPHC 1587/35: Shortlands DO's Diary, 21 Oct. 1936; Walker, *Forests*, 41, 104.

59. Walker, *Forests*, 41; WPHC BSIP F45/23: Logie, Activities of the Forestry Dept. 1953.

60. VR: Kelly to Sec., 18 Apr. 1935, Employees' Corres. 1932–1953; WPHC 64/24: Meeting of Third Advisory Council, 20 Nov. 1923; WPHC 243/24: MacCrimmon and others to Acting RC, 21 Dec. 1923.

61. WPHC 3386/27: List of Lands, Mar. 1927.

62. WPHC 1422/29: AR Malaita 1928; WPHC 2151/25: Bell, Log of *Auki*, 30 June, 21 Dec. 1925.

63. WPHC 1832/28: AR Malaita, 1927; BSIP F 28/18: RC, Notes on discussion with N. Willis, 2 Oct. 1950, Tenaru Timbers Pty Ltd; VR : Curtis to Chairman, 31 Aug., 1 Nov. 1927, Vanikoro Corres.1927–1931; Peter Plowman, interview, Sept. 1975.

64. WPHC 1832/28: AR Malaita, 1927.

65. WPHC 1422/29: AR Malaita, 1928.

66. WPHC 3726/31: RC, Tour, 30 Dec, 1929–13 Jan. 1930; WPHC 1214/32: AR Malaita, 1931; VR: Gov. Depts to and from, 1932–34: Sec, VTC to Gullett, Dept of Trade and Customs, Aust. 4 Feb. 1932, Comptroller-General of Customs to Sec. 23 Dec. 1932, Collector of Customs to Director, 27 Feb. 1933; AR BSIP 1933).

67. WPHC 3726/31; WPHC 1214/32: AR Malaita, 1931; VR: Director's Minute Book, 7 July, 19 Nov. 1930. Cheetham had commenced negotiations for a timber lease in the vicinity of Su'u in September 1930 (WPHC 827/30: Log of *Auki*, 24 Sept. 1930).

68. LTO: Title 72/69, LR 271 Olisu'u; VR: Robinson to VTC, 11 Dec. 1930 and enclosures, Corres. Gen. 1928–31. It appears that N. MacCrimmon held some interest in the Mala Development Company at least until 1931.

69. Judith Bennett, 'Oscar Svensen: a Solomons Trader among "The Few"', *Journal of Pacific History* 16, no. 4 (1981), 185; WPHC 3116/37: AR Mamara Plantations Ltd., 1939; Knibbs, *Savage Solomons*, 106, 135; *The Sun*, 6 Aug. 1911; AR BSIP, 1912–1913; WPHC 63/13: Vernon, Notes on the SI Protectorate, 9 Jan. 1913.

70. WPHC 2894/27: AR Guadalcanal, 1926; WPHC 1422/29: AR Guadalcanal 1928.

Notes to pp. 71–83

71. Judith Bennett, *Wealth,* 55, 382, 391.
72. WPHC 1479/13: Woodford to HC, and encls 14 July 1913; WPHC 203/85: Statement of Otto Ashe, mate of *Douro,* n.d. c. June 1885; WPHC 140/87: Musgrave to Mitchell, 21 Mar. 1887, and enc. See also WPHC 4/87; WPHC 283/87; WPHC 43/88; WPHC 145/88.
73. WPHC 1479/13: Woodford to HC, and encls 14 July 1913; See also WPHC 3/98, 94/98; WPHC 458/96: Tarves to Woodford, 20 Oct. 1896, and enc.; WPHC 203/85: Statement of Otto Ashe, ca. June 1885; WPHC 1641/12: Sec., Dept. of External Affairs to HC, 30 July 1912; WPHC 1886/12: Woodford to Sec., Dept. of External Affairs, 2 Sep. 1912, and encls.; Leslie to HC, 17 Feb. 1892, enc 39/92; WPHC 1886/12: Woodford to Sec. Dept of External Affairs, Australia, 2 Sep. 1912, and encls; WPHC 1479/13: Woodford to HC, and encls, 14 July 1913; WPHC 1972/13: Law to HC, 2 Oct. 1913.
74. WPHC 1837/25: Schachne to Stanley Baldwin, 24 Mar. 1925, and encls. Woodford seems to have been unaware of, or chose to ignore, an earlier reference to this sale (see WPHC 302/90). The concept of improving upon nature for the tourist pound or dollar was current at this time. For example, in New Zealand, the government supported plans to introduce not only red deer in 1896, but also grouse and heather in 1922, to its first national park, Tongariro, in order to attract gentlemen game shooters from Britain. Craig Potton, *Tongariro: A Sacred Gift* (Auckland, 1987), 142.
75. WPHC 202/30.
76. WPHC 209/00: Woodford to O'Brien, 11 Sept. 1900, and encls; Henderson and Hancock, *A Guide,* 225–7; Walker, *Forests,* 138. The vegetation on San Jorge was relatively stunted, due to the high metal content of the soil. Coote may have known more than he admitted as the island has a nickel deposit. The only other 'ebony' wood likely is *Diospyros* spp. found in coastal and lowland forest on coral (Walker, *Forests,* 113–14).
77. MOM: Goldie to Danks, 5 Dec. 1908; 5 July 1909; 22 July 1909.
78. MOM: Box 174–190, Synod Minutes, Solomon Islands (hereafter SI), 1914; Goldie, Roviana circuit Report, 1915, 15. A saw pit is used to convert heavy timbers to smaller sections. The double-handed saw is operated vertically by one man in the pit and another above.
79. *The Australasian Methodist Missionary Review,* 4 Sept. 1917, 4.
80. MOM: Sinclair and Court, Report of the Representatives to the SI Mission, 1920.
81. George Carter, *A Family Affair: A Brief Survey of New Zealand Methodism's Involvement in Mission Overseas 1922–1972* (Auckland, 1972), 173–4; *New Zealand Methodist Times,* 6 Dec. 1924, 15; PMB 946: Lina Jones, letter 28 Sept. 1924.
82. PMB 925: Goldie to Scriven, 19, 29 Mar. 1934; Methodist Archives, Christchurch: E. C. Leadley, Diary, 8 Nov. 1936.
83. Steley, 'Juapa Rane', 128–9; *Australasian Record,* 8 Oct., 10 Dec. 1923; SDA: SI Mission Minutes, 83rd Meeting, 3 Aug. 1931; 132nd Meeting, 12 June, 1934; 135th Meeting, 25 July 1934; 140th Meeting, 14 Jan. 1935; 142nd Meeting, 20 Mar. 1935; 186th Meeting, 7 July 1938.
84. SDA: SI Mission Minutes, 83rd Meeting, 3 Aug. 1931.
85. *PIM,* 20 Dec. 1935, p. 41.
86. SDA: Dated photograph at Telina, 1922.
87. *Australasian Record,* 8 Oct. 1923; Steley, 'Juapa Rane', 209.
88. SDA: SI Mission Minutes, 113th Meeting, 26 June 1933; Steley, 'Juapa Rane', 165.
89. Pana's letter, *Australasian Record,* 4 Mar. 193, 8.
90. SDA: SI Mission Minutes, 132nd Meeting, 12 June, 1934; 135th Meeting, 25 July 1934.
91. BSIP FO 5/45: Walker, Draft, Forest Resources Survey, 15 Feb. 1946
92. Dennis Steley, 'Juapa Rane', 205, f/n.
93. Ibid., 198–205.
94. SDA: SI Mission Minutes, 186th Meeting, 7 July, 1938; 192nd Meeting, 6 Nov. 1938; 199th Meeting, 1 Mar. 1939; 202nd Meeting, 6 July, 1939; 206th Meeting, 29 Sept. 1939; 209th Meeting, 13 Nov. 1939; WPHC 512/20: Leases for quarter ending 31 Dec. 1918; WPHC 1588/35: Leases for quarter ending 31 Mar., 1935; BSIP NA F 28/13: Clemens to Sec. to Gov., 14 Oct. 1944.

95. Golden, *Early European Settlers*, 28; Hugh Laracy, *Marists and Melanesians*, 38–40; VR: Bladier to Burnell, 10 April 1938, Kauri, Job Applications Box. Hugh Laracy believes that the Buma saw-mill opened in between 1938 and 1940 (Pers. comm., Mar., Nov. 1996).

96. Collinson, *Life and Laughter*, 156; Jack McLaren, *My Odyssey* (London, 1923), 215.

97. See, for example, WPHC 241/99, WPHC 208/00, WPHC 1137/13; Dorothy Shineberg, pers. comm. Nov. 1995.

98. See WPHC 22/00; WPHC 1150/13.

99. Dillon, *Narrative*, 273.

100. J. W. Davidson, *Peter Dillon of Vanikoro: Chevalier of the South Seas* (Melbourne, 1975), 29–31, 80, 81, 90–1, 100–1, 102–5.

101. See also Lesson, 'Vanikoro', 22.

102. *La Caledonie*, 11 May 1898, reference kindly provided by Dr Dorothy Shineberg. English translation: 'A huge forest of kauri.'

103. Knibbs, *Savage Solomons*, 28-9; WPHC 2024/24: Kidson to HC, 23 July 1924; David Hilliard, *God's Gentlemen: A History of the Melanesian Mission, 1949–1942* (Brisbane, 1978), 184–7, 230; WPHC 801/14, 698/15: AR Labour Inspector 1913/14.

104. WPHC 1137/13: Woodford to Sweet-Escott, 3 May 1913.

105. WPHC 1150/13: Woodford to Sweet-Escott, 20 May 1913; FRP: Rigby to Fairley, 12 May, 14 June 1913.

106. WPHC 1137/13: Woodford to Sweet-Escott, 3 May 1913.

107. Documents cited in Joint Committee of Public Accounts, *Interim Report on the Transactions of the War Service Homes Commissioner with Mr J. T. Caldwell*, 5 April 1921, Australia Parliamentary Papers, 1920–21. Also enclosed in WPHC 2854/20.

108. WPHC 1150/13: Woodford to Sweet-Escott, 20 May 1913; WPHC 1641/16: Bonar Law to HC, 6 May 1916; WPHC 2854/20: Greene's notes on file Milner to Rodwell, 7 Dec. 1920; LTO: Application for registration 64/70, LR 299, Waimasi; Joint Committee of Public Accounts, *Interim Report*. Also enclosed in WPHC 2854/20.

109. Joint Committee of Public Accounts, *Interim Report*. Also enclosed in WPHC 2854/20.

110. WPHC 1958/21: Kane to HC, 30 Sept. 1921.

111. WPHC 3312/22: Acting RC Francis to HC, 10 Nov. 1922.

112. This is the only existing record of kauri on Utupua, as the loggers, botanists and foresters who visited subsequent to Caldwell found none. It seems likely that kauri was once on Utupua, but may have died out because of successive cyclones; a recent extirpation due largely to natural processes.

113. WPHC 3312/22: Caldwell to RC, Oct. 1922, encl.

114. WPHC 2462/23: Bennett to Dept of Overseas Trade, (England) 14 Apr. 1922.

115. Before the San Cristoval Estates obtained the concession, the company had to sign a deed of indemnity, freeing the High Commission from all possible liability arising out of the French claims (WPHC 882/25: Notes on file and enclosures; see also WPHC 1473/22).

116. VR: Phillips, Report of the Lands Commissioner, In the Matter of Native Claims Nos 38 and 39, respecting lands at Vanikoro, Santa Cruz, claimed by the Societe Francaise des Nouvelles Hebrides, 10 Dec. 1923, 13–14, 17; Also at WPHC 1843/24; WPHC 1584/25: Eyre Hutson, Memo., 16 July 1925. A legatee of John Higginson, head of the successor company, revived the French claim to Tevai in 1939, but, by 1945 in the wake of World War Two, the claim was dropped (see WPHC F 48/25).

117. WPHC 1023/22: Middenway to RC, 24 Feb. 1922.

118. WPHC 184/23: Kane to HC, 28 Dec 1942

119. WPHC 184/23: Copy of Agreement, 4 Nov. 1922, and encls Re Tevai. Tua seems to have agreed to lease Tevai where he was the chief, but then changed his mind, so the lease was never finalised.

120. VR: Memo of Agreement, 30 May 1940. The people sold a small area of land at Paeu to the government for the site of the government station and the company's main buildings (WPHC 882/25: Kane to HC, 22 June 1925).

Notes to pp. 86–92

121. WPHC 245/33; WPHC 572/29: Sykes, Report on Tour of Inspection of the Forests of the BSIP, 1932.
122. WPHC 1244/21, Kane to HC, 22 Aug. 1921; CO 225/205/40401: Minutes on file, c. Aug. 1925; WPHC 882/25: Amery to HC, 17 Mar. 1925 and minutes, c. July 1925; WPHC 1244/21: Kane to HC, 22 Aug. 1921.
123. WPHC 1244/21: Rodwell to Kane, 12 May 1921; Kane to Rodwell, 22 Aug. 1921.
124. CO 225/205/40401: Minutes on file, c. Aug. 1925; See also WPHC 2677/25: Amery to Hutson, 7 Oct. 1925 and enclosures.
125. WPHC 1244/21: Kane to Rodwell, 22 Aug. 1921.
126. WPHC 1843/24: Phillips to HC, 12 Sept. 1924. The British authorities in the High Commission ran the risk that the French may have pushed for international arbitration and, had the British lost, they would have had to pay compensation for the concession given to the San Cristoval Estates. When this company was subsumed in the VKT Company the High Commission stipulated that it would approve the transfer of the concession only if the company would not claim compensation from it should the French succeed, and, if the French did and wanted compensation, then the VKT Company would have to pay it. In the event, the French did not pursue the claim in the international arena (WPHC 1584/25: Eyre Hutson, Memo, 16 July 1925).
127. WPHC 184/23: Copy of Agreement, 4 Nov. 1922, and encls; Phillips, Report of the Lands Commissioner, In the Matter of Native Claims Nos 38 and 39, 10 Dec. 1923, 17.
128. VR: Land Claim no. 540, Noumea, 28 June 1887, enc., Phillips, Report of the Lands Commissioner (hereafter C) 10 Dec. 1923, 13–14, 17.
129. The population of Vanikolo and Tevai was around 1200–1500 in 1828 when Dumont D'Urville visited. He predicted with remarkable accuracy its fate: '… within a century this island will not contain more than a few families scattered over its whole area'. In 1922 the population was 83; in 1929 it was down to 66. Jules S-C Dumont D'Urville, *An Account of Two Voyages to the South Seas*, vol. I, trans. Helen Rosenman (Melbourne, 1987), 236; WPHC 1290/30: AR Santa Cruz, 1929.
130. WPHC 52/23: Greene to Garrick, Caldwell and Ellis, 8 Feb. 1923; WPHC 1023/22: Middenway to RC, 24 Feb. 1922.
131. Heath, 'Land policy', 311–12.
132. Navanora seems to have been the woman called Namunora in MacQuarrie, *Vouza*, 48–65 and Nora in DP: Deland to Everyone, 2 Feb., 3 May 1927.
133. WPHC 1359/34: DO's diary, Santa Cruz, 10 Jan. 1934.
134. Ben Ramoli of Buma, interview, 1992.
135. WPHC 1359/34: DO's diary, Santa Cruz, 7 May 1934.
136. Bennett, *Wealth*, 146
137. BSIP F 5: Ashley, Timber Concessions in the Solomon Islands, day and month obscured, 1938.
138. Bennett, *Wealth*, 219–40.
139. BSIP FO 5/45: Walker, Draft Report on Forest Resources Survey, 15 Feb. 1946.
140. WPHC F45/2: Marchant to HC, 22 Dec. 1941 and encls.

Chapter 5

1. D'Urville, *Account of Two Voyages*, 225–6, 230–1, 236.
2. VR: Corres. 1931: Sec. to Scullin, 12 Aug. 1931; WPHC 1023/22: Middenway to RC, 24 Feb. 1922.
3. WPHC 1023/22: Middenway to RC, 24 Feb. 1922. Several names used by the Company and the early government officials have been replaced by names given by the local people. The old name for Mt Popokia was Kapogo. Sapolombe is also called Ramboe.
4. WPHC 1422/29: ARSC 1928; WPHC 1290/30: ARSC 1929; VR: Corres. to and from 1932–1936: Wilson to Sec. 19 Dec. 1933, Reports: Beckett, Report on Logging operations at Vanikoro, Dec.

Notes to pp. 92–95

1954; Colin Allan, *Solomons Safari*, part 2 (Christchurch, 1990), 144. Rainfall has reached an annual extreme of 7,900 mm (D. R. Stoddart, 'Biogeography of the Tropical Pacific', *Pacific Science* 46, no. 2 (1992), 285–7).

5. WP: Hill to Woodford, 22 June 1925; Ashby, *Blackie*, 60–88. Ashby's memory fails at times on names and is a year too early regarding the establishment of the Saboe Bay camp.

6. WPHC 228/25: Kane to HC, 24 Dec. 1924.

7. VR: Court to Butler, 10 Oct. 1928.

8. Roche, *History of New Zealand Forestry*, 107–15.

9. VR: Corres. 1931: Sec. to Scullin, 12 Aug. 1931.

10. VR: Corres. to and from, 1925–1932: Faithfull, Report of Timber on Vanikoro Island, n.d., c. Aug. 1925; Corres. 1927– 1931: Butler, General Report, Aug. 1926; Reports General, 139(a); Wilson Report, 19 Dec. 1933; Beckett, Report on Logging Operations at Vanikoro, Dec. 1954; WPHC 245/33: Sykes, Report on Tour of Inspection of the Forests of BSIP, 1932.

11. VR: Corres. 1925–1932: Faithfull, Report of Timber on Vanikoro Island, n.d., ca. Aug. 1925. These estimates were subsequently revised upwards, for example in 1934, to 300,000,000 su.ft. of kauri remaining after nine years of extraction (BSIP 9/111/2: ARSC, 1934).

12. *Forest Giants from Vanikoro* (Melbourne, [1938?]).

13. VR: Sec. to Gullett, 4 Feb. 1932. It seems that the company downplayed the use of kauri in building in this letter as it wished to show that kauri did not compete with comparable Australian timbers, and certainly did not mention koila, a hardwood, as it was asking for exemption from the new import duties on timber coming into Australia.

14. VR: Corres. to and from, 1932–1936: Minutes of Board of the Kauri Timber Co., Melbourne, 20 Dec. 1926, 2, 16 May 1938, 18 May 1939; Sec. to Wilson, 20 June, 19 July 1933; Corres. to and from 1937–1939: Boye to Sec. 12 Oct. 1937.

15. *Forest Giants*.

16. BSIP 9/111/2: ARCS, 1930; BSIP 9/111/3: QRSC, Mar. 1929, Aug. 1932, June 1937, Mar 1938; VR: Reports, 137: Davidson, Report by Victorian Manager on logs from Vanikoro, 21 July 1954.

17. For example, from July 1926 to end of February 1927, the company imported goods valued at A£6,190, duty-free (WPHC 1525/27: Middenway to RC, 10 Mar. 1927).

18. WPHC 52/23: Greene to Carrick et al., 8 Feb. 1923.

19. DP: Annie Deland to Everyone, 1 Jan. 1928; BSIP 9/1/5: DO to Gov. Sec. 14 Oct. 1926; Diary SC, 9 Aug. 18 Dec. 1934, 25 June 1935; WPHC 1290/30: ARSC 1930; WPHC 1525/32: RC to HC, 10 May 1932; BSIP 19/16: QRSC, June 1935.

20. Bennett, *Wealth*, 150–66; DP: Deland to Everyone, 22 Jan., 3 May 1927, 12 June 1928; VR: Vanikoro Corres: Smith to Sec. 18 Aug. 1928.

21. VR: Gov. depts to and from, 1926: de Bondy to Sec., 17 Apr. 1926; Gov. depts to and from, 1932–1934: Sec. to Sec. of Gov., 7 Aug. 1929; Corres. 1927–1931 (sic): de Bondy to Sec., 31 Dec. 1925; Court to C'man, n.d. 1926; Court to C'man, 15 Nov. 1926; Report, Vanikoro, 9 Apr. 1931; Manager to Sec., 16 Feb. 1928, Smith to Sec., 16 Feb. 1928; Butler, Report re Vanikoro, Aug. 1926; Director's Minute Book, 27 Jan. 1931; Corres. 1932–1936: Dawe to Sec. 27 Dec. 1932; Employees' Corres. 1932–1953: Sec. to Woy Sang Tuen Co. 5 May 1933; Corres. to and from 1949–1960: McEwin to Sec., 31 July 1951, McEwin to Sec, 8 Jan 1953; Corres. 1949–1950: Kerr Bros to Sec., 2 Feb. 1950; Reports monthly and general, 1949–1963: Filewood to Sec., Dec. 1956, Report, Vanikoro, 23 Apr. 1959; Rogers, Report Vanikoro, Sept. 1961; BSIP 9/111/2: ARSC 1930–1944; BSIP 9/1/5: RC, Inspection Notes Vanikoro, 1929; Diary, SC, 1934; BSIP 9/111/3: QR 1929–1933; WPHC 2894/27: ARSC 1926; WPHC 1422/29: ARSC 1928; WPHC 1290/30: ARSC 1929; BSIP 9/1: ARSC Tour Report, Dec. 1950, ARED, 1954; WPHC 1121/24: AR Labour Department (hereinafter LD) 1923; WPHC 1197/25: ARLD 1924; WPHC 1170/26: ARLD 1925; WPHC 1510/27: ARLD 1926; WPHC 1835/28: ARLD 1927; WPHC 1426/29: ARLD 1928; WPHC 809/30: ARLD 1929; WPHC 755/31: ARLD 1930; WPHC 1228/32: ARLD 1931; WPHC 506/33: ARLD 1932; WPHC 920/34: ARLD 1933; WPHC 1612/35: ARLD 1934; WPHC 1598/36: ARLD 1935; WPHC 2744/37: ARLD 1936; WPHC 1638/38: ARLD 1937; WPHC 2469/40: ARLD 1939; WPHC 2399/41: ARLD 1940.

Notes to pp. 96–99

22. VR: Gov. depts to and from 1926, Cowan to de Bondy, 28 Dec. 1925; Curtis, Report [on Labour], June [?] 1928; Butler, Report, Vanikoro, Aug. 1926; Zinneck to Curtis, 26 Oct. 1926; Cowan to Curtis, n.d. [Oct.] 1926, and photographs in collection; Sec. to Pilling, 10 Dec. 1925.
23. WPHC 1359/34: Diary SC, 14 May 1934; Bennett *Wealth*, 172–3; BSIP 9/111/2: ARSC, 1934; VR: Employees' Corres. 1932–1953: Kelly, n.d. [ca. June 1935].
24. See note 22.
25. BSIP 9/V/5: Ashley, Inspection Notes, Sept. 1929.
26. Ashby, *Blackie*, 77; VR: Island Reports, 1928: Smith to Sec. 16 June 1928.
27. Bennett, *Wealth*, 150–66; Judith Bennett, '"We do not Come Here to be Beaten": Resistance and the Plantation System in the Solomon Islands to World War II', in *Plantation Workers: Resistance and Accommodation*, ed. Brij V. Lal, Doug Munro and Edward Beechert (Honolulu, 1993),158–9; WPHC 1587/35: Dawe to Sec. 24 Jan. 1945; VR: Corres. to and from 1932–1936: Dawe to Sec. 24 Nov. 1934.
28. BSIP 9/III/3: QRSC, Mar. 1930.
29. PMB 553: E. Sandars, Papers.
30. WPHC 1525/27: Middenway to RC, 10 Mar. 1927.
31. BSIP 9/V/5: Ashley, Inspection Notes, 1929.
32. BSIP 9/III/3: QRSC, Mar. 1939; VR: Gov. depts to and from: Heffernan to Administrator, 31 Dec. 1925; Manager to RC, 16 Sept. 1926 and encs; Director's minute Book, 1934; Corres. to and from, 1932–1936; Dawe to Sec, 24 Nov. 1934; Kelly to Director, 6, 28 Mar., 29 June, 27 Aug., 9, 23, 25, 26, Sept., 30 Oct., 19 Nov. 1935; Mrs Kelly to Brunell, 19 July 1935; Corres. Gen. Misc. 1928–1931: Extracts from letters, Manager, Vanikoro, 13 May–17 Oct. 1929; WPHC 1422/29: ARSC 1928; BSIP 9/III/2: ARSC, 1930; WPHC 319/31: Kidson to HC, 20 Oct 1931; WPHC 1442/31: RC to HC, 1 June 1931.
33. DP: Deland to Everyone, 26 Nov. 1926, 22 Jan., 2 Feb. 1927, 11 Jan. 1928; VR: Smith to Sec. 16 June 1928, Corres. 1927–1931; WPHC 1525/27: Middenway to RC, 10 Mar. 1927; WPHC 2122/36: Hetherington to RC, 25 Aug. 1936.
34. VR: Vanikoro Corres. 1927–1931: Sec. to Curtis, 25 May 1927; Smith to Sec. 18 Aug. 1928; WPHC 2122/35: Hetherington to ADC, 25 Aug. 1936; Crichlow to Hetherington, 12 Dec. 1936.
35. WPHC 1525/27: Middenway to DC, 10 Mar. 1927; WPHC 2122/36: Hetherington to ADC, 25 Aug. 1936; VR: Corres. 1927–1931: Smith to Sec. 14 June 1928; DP: Deland to Everyone, 1 Jan. 1928.
36. WPHC 1359/34: Diary SC, 14 May, Aug. 1934; BSIP 9/111/2: ARSC 1934; VR: Employees' Corres. 1932–1953: Kelly, n.d. [ca. June 1935].
37. WPHC 1587/35: Diary SC, May–Nov. 1936; BSIP 9/111/3: QRSC, Apr., June, 1936; WPHC 2122/36: Hetherington to ARC, 25 Aug. 1936 and encls; BSIP 9/111/2: ARSC 1941; VR: Corres. to and from 1932–1936: Wilson to Sec. 16 Aug. 1933, Courtney to Miller, 11 July 1936, Courtney to Sec., 14 July 1936.
38. VR: Gov. depts to and from: de Bondy to Sec., 17 Apr. 1926; Sec., to Sec. of Gov., 7 Aug. 1929; de Bondy to Sec., 31 Dec. 1925; Court to C'man, n.d. 1926; Court to C'man, 15 Nov. 1926; Corres. 1927–1931: Report, Vanikoro, 9 Apr. 1931, Manager to Sec., 16 Feb 1928; Smith to Sec., 16 Feb. 1928, Butler, Report re Vanikoro, Aug. 1926; Corres. 1932–1936: Director's Minute Book, 27 Jan. 1931; Dawe to Sec. 27 Dec. 1932; Employees' Corres. 1932–1953: Sec. to Woy Sang Tuen Co. 5 May 1933; Corres. to and from 1949–1960: McEwin to Sec., 31 July 1951, McEwin to Sec, 8 Jan 1953; Corres. 1949–1950: Kerr Bros to Sec., 2 Feb. 1950; Reports monthly and general, 1949–1963: Filewood to Sec., Dec. 1956, Report, Vanikoro, 23 Apr. 1959; Rogers, Report Vanikoro, Sept. 1961; WPHC BSIP 9/111/2: ARSC 1930–1944; WPHC BSIP 9/1/5: RC ?, Inspection Notes Vanikoro, 1929; Diary, SC, 1934; WPHC BSIP 9/111/3: Quarterly Reports 1929–1937; WPHC 2894/27: ARSC 1926; WPHC 1422/29: ARSC 1928; WPHC 1290/30: ARSC 1929; WPHC BSIP 9/1: ARSC Tour Report, Dec. 1950, AR Eastern District, 1954; WPHC 1121/24: ARLD 1923; WPHC 1197/25: ARLD 1924; WPHC 1170/26: ARLD 1925; WPHC 1510/27: ARLD 1926; WPHC 1835/28: ARLD 1927; WPHC 1426/29: ARLD 1928; WPHC 809/30, ARLD 1929; WPHC 755/31: ARLD 1930; WPHC 1228/32: ARLD 1931; WPHC 506/33: ARLD 1932; WPHC 920/34: ARLD 1933; WPHC 1612/35: ARLD 1934; WPHC 1598/36: ARLD 1935; WPHC 2744/37: ARLD 1936; WPHC 1638/38: ARLD 1937; WPHC 2469/40: ARLD 1939; WPHC 2399/41: ARLD 1940.

Notes to pp. 101–108

39. VR: Corres. 1925–1932: Faithfull, Report of Timber on Vanikoro Island, n.d., ca. Aug. 1925; Butler, General Report, Aug. 1926; Smith to Sec. 18 Aug. 1928; Corres. to and from, 1932–1936: Wilson, Report on Result of Survey of Ambi Valley, 1 June 1933, Report of Timber Survey, 21 Dec. 1933 and encl.; Gov. depts, to and from, 1926, 1932–34 (sic): Sec. to Sec. of Gov.(?), 7 Aug 1929; BSIP 9/VII/1(a): Inquiry into the death of Towhamai, 14 Nov. 1928, enc; WPHC 1422/29: ARSC, 1928; DP: Deland to Everyone, 14 Dec. 1926, 1, 11 Jan. 1928; S. Frank Kajewski, 'A Plant Collector's Notes on the New Hebrides and Santa Cruz Islands', *Journal of the Arnold Arboretum* 11 (1930), 178. The timber-jacks used in Vanikolo were a New Zealand adaptation, developed in the hey-day of kauri logging in that country (A. H. Reed, *The New Story of the Kauri* (Wellington, 1964), 101–3).

40. VR: McEwin, Reply to Questionnaire, 20 June 1951.

41. Re New Zealand, see Reed, *The New Story*, 114–127.

42. Chris Hadley, *A Forester in the Solomon Islands* (Sussex, 1991), 99; VR: Director's Minute Book, 14 Nov. 1927; Corres. to and from 1935: Sec. to Manager, 29 May 1935, Dawe to Sec. 26 Sept. 1935; BSIP agreements, Jan. 1935–Aug. 1940: Sanders to Manager, Melb, 21 Dec. 1938, Agreement, 4 Aug. 1939; Reply to Questionnaire, 20 June 1951; *PIM*, 20 Feb. 1934, 35, 21 Feb. 1935, 29, 35.

43. Roche, *History*, 114–15, 129–30, 180, 187.

44. VR: Corres. to and from, 1932–1936: Butler, General Report Leading to Recommendations, Aug. 1926, Corres. 1925–1932; Wilson, Report on Result of Survey of Ambi Valley, 1 June 1933, Report of Timber Survey, 21 Dec. 1933 and enc.; BSIP 9/111/2: ARSC, 1937.

45. Ashby, *Blackie*, 83.

46. VR: Corres. 1925–1932: Butler, General Report Leading to Recommendations, Aug. 1926.

47. BSIP 9/111/3: QRSC, Mar 1929, enc. See also, VR: Corres. to and from, 1932–1936: Wilson, Report of Timber Survey, 21 Dec. 1933 and enc.; DP: Deland to Everyone, 11 Jan. 1928.

48. VR: Corres. to and from, 1932–1936: Wilson, Report of Timber Survey, 21 Dec, 1933 and enc.

49. BSIP 9/111/2: ARSC, 1937.

50. VR: Corres. to and from, 1932–1936: Wilson, Report of Timber Survey, 21 Dec, 1933 and enc.

51. BSIP 9/111/2: ARSC, 1937.

52. VR: Reports General: McEwin, Report on Reconnaissance and Timber Cruise, 5 Oct. 1947.

53. WPHC 1525/27: Middenway to RC, 10 Mar. 1927.

54. VR: Corres. 1925–1932: Faithfull, Report of Timber on Vanikoro Island, n.d., ca. Aug. 1925. See also DP: Deland to Everyone, 14 Feb. 1927.

55. VR: Corres. 1925–1932: Butler, General Report, Aug. 1926, Smith to Sec. 18 Aug. 1928; Corres. 1931: Sec. Melb. to Scullin, 12 Aug, 1931; Corres. to and from, 1932–1936: Dawe to Sec. 27 Dec. 1932; WPHC 1422/29: ARSC, 1928.

56. VR: Corres., General Misc, 1928–1931: Robinson to Sec., 25 Feb., 5 Mar., 1 May, 10 June, 25 Aug. 1931; Scrymgour to Mala Dev. Co. 9 Mar. 1931; Gov. depts to and from, 1926-1934: Sec. to Gullett, 4 Feb. 1932. See also WPHC 1214/32: AR Malaita 1931. Some of this timber was intended to repair Carpenter's vessel, *Whyalong*. It was never repaired and ended as a hulk at Tulagi (*PIM*, 17 Apr. 1935, 71).

57. One superfoot of log equals a twelfth of a Hoppus cubic foot or 0.106 of a true cubic foot. 100 superfeet = 0.301 cubic metres.

58. VR: Gov. depts to and from, 1926–1934: Sec. to Gullett, 4 Feb. 1932; Hall to Sec., 23 Dec. 1932; Synan to Director, 27 Feb. 1933.

59. VR: Corres. to and from: Sec. to Wilson, 20 June, 1933.

60. BSIP 9/III/2: ARSC 1930; WPHC 319/31: Kidson to DO, 11 Aug. 1931, enc.

61. WPHC 882/25: Memorandum of Agreement, 24 June 1926; WPHC 319/31: Ashley to HC, 22 Dec 1930, enc.

62. VR: Butler, Report of Vanikoro, Aug 1926.

63. WPHC 319/31: Court to Sec. to the Gov. 28 Sept. 1931, Kidson to HC, 20 Oct. 1931, encls; VR: Butler, Report of Vanikoro, Aug. 1926; WPHC 1023/22: Middenway to RC, 24 Feb. 1922; BSIP 9/III/2: ARSC, 1930; BSIP 9/III/3: QRSC, Sept. 1937.

64. WPHC 319/31: Court to Sec. to the Gov. 28 Sept. 1931; VR: Corres. 1927–1931: Smith to Sec. 18 Aug. 1928. No botanical identification of *tovoleko* was found. It is not a Fijian word.
65. WPHC 319/31: Hubbard, Memo, 7 Mar. 1933.
66. WPHC 319/31: Barley to HC, 18 Apr. 1933 and encls; WPHC 1442/31: RC to HC, 1 June 1931. Filose was subsequently charged with procuring and committing assault on Solomon Islanders and fined in 1932, relating to events while he was district officer, Santa Isabel. He resigned soon after (WPHC 2382/31; WPHC 584/33).
67. CO 225/205/40401: Minutes on file, c. Aug. 1925. See also WPHC 882/25: Amery to HC, 17 Mar. 1925 and encls. The colonial authorities were much more concerned about the wording of the agreement which pertained to the government collecting royalties 'on behalf of the native owners'. This could be interpreted as implying all royalties were to go to 'a few score backward natives which would obviously be improper' (CO 225/205/40401: Minutes, c. Aug. 1925).
68. WPHC 1244/21, Kane to HC, 22 Aug. 1921. Another Protectorate officer, A. Middenway, who reported on the terms of the licence as applied to Vanikolo conditions in 1922, also recommended a royalty based on superficial measurement, not trees felled (WPHC 1022/22: Middenway to RC, 24 Feb. 1922).
69. Troup's major publications included *The Silviculture of Indian Trees* 3 vols (Oxford, 1921), *Silvicultural System* (Oxford, 1928) and *Forestry and State Control* (Oxford, 1938). See also, *Empire Forest Review* 18 (1939), 187–188.
70. WPHC 245/33: Sykes, Report of a tour, 1932; CO 225/259/93589: Troup to Under Sec. of State, 22 Feb. 1932; CO 225/271/18739: Sykes, Comment, 7 Jan. 1933; Troup, Forest Exploitation in the Solomon Islands, 24 Feb. 1933; Peason to Furse, 7 March 1933; CO 225/284/38829: Fletcher to Sec. of State, 16 May 1934; Troup, Comments, 27 Nov. 1934; CO 83/207/10: Troup, Note on a discussion of the question of Forestry in Fiji, 12 Jan. 1935. Sykes formulated a forest policy for Fiji. Under Governor Richards, Fiji got its first forestry officer, J. S. Smith, to set up the new department in 1937 (W. A. Gordon, *The Law of Forestry* (London, 1955), 454; CO 83/218/2: Richards to Sec. of State, 6 Jan. 1937, and encls.).
71. VR: Corres. 1927–1931: Manager's report, 9; WPHC 319/31: Court to Sec. to the Gov. 28 Sept. 1931; WPHC 319/31: Barley to HC, 18 Apr. 1933 and encls; BSIP 9/III/2: ARSC 1932; BSIP 9/III/3: QRSC, 1933–1937.
72. BSIP F 28/4: Ashley, Memo: Timber Concessions in the Solomon Islands, n.d. ca. 1937 and encl.
73. VR: General Corres. 1935–1941: Burnell to Chairman, 15 Oct. 1937; Minutes of the Board of the Kauri Timber Co. 18 Mar., 4 Nov. 1937.
74. BSIP 9/V/I: RC, Notes on Tour, May 1940.
75. Cf. Walker, *Forests*, 23.
76. CO 225/271/18739: Troup, Forest Exploitation in Solomon Islands, 24 Feb. 1933, and encls; WPHC F45/2: Marchant to HC, 22 Dec. 1941 and encls; BSIP F 28/4: Smith to HC, 17 May 1941, and encls.
77. VR: Memorandum of Agreement, 30 May 1940.
78. WPHC 1244/21: Kane to AHC, 22 Aug. 1931.
79. WPHC 572/29: Kajewski to RC, 4 Dec. 1928. Kajewski classified a variety of *Calophyllum* (*baula, buni soloso*) which bears his name.
80. WPHC 572/29: Garvey to Sec. to the Gov., 10 Dec. 1928, Fletcher to Sec. of State, 12 Jan. 1932, Ashley to HC, 31 Oct. 1930; *Forest Giants*.
81. WPHC 245/22: Sykes, Report of a tour, 1932.
82. BSIP 9/III/2: ARSC 1932.
83. BSIP 9/I; SC Tour, Apr. 1945; Walker, *Forests*, 52.
84. It is not a little ironic that recent forestry specialists cite Vanikolo as an example of a forest that has recovered from logging. Had the company logged using the methods commonly practised from 1985 to 1995, very little kauri or any other valuable timber would remain except any that was deliberately planted (*SINFRI*, 1: 48).

Notes to pp. 111–117

85. BSIP F 28/14 Part 1: Officer-in-Charge to RC, 21 June 1945; VR: Corres. Misc. general: Sec. to Ellis, 4 Dec. 1925.
86. AR BSIP, 1928, 6.
87. WPHC 184/23: Kane to HC, 28 Dec. 1942.
88. Ben Tua of Mbuma, interview, 1992.
89. WPHC 2531/29: Garvey, Log of Tulagi, 23 Mar. 1929.
90. WPHC 83/32: Garvey, The Depopulation of Vanikoro, Jan. 1932. Seven of the 83 were away seeking plantation work (WPHC 1023/22: Middenway to RC, 24 Feb. 1922).
91. VR: Agreements 1935–1944: Ashley to Sec., 17 Apr. 1935. WPHC 1594/35: Ashley to Sec., 17 Apr. 1935 and encls.
92. BSIP F 28/14 Part 1: Officer-in-Charge to RC, 21 June 1945; BSIP F 28/16: Walker, Interim Report on Forests, 15 Feb. 1945.
93. WPHC 1359/34: Diary SC, 7 May 1934; WPHC 1589/35: ARSC 1934.
94. BSIP 9/1: C of Lands to DC, 3 Feb. 1951.
95. BSIP 9/1: Ramoli to [DO] n.d. ca. 1947.
96. BSIP F 28/17: RC to HC, 3 Jan. 1952.
97. BSIP F 28/17: Memorandum of Agreement, July 1953, encl.
98. VR: Kerr Bros., Corres. 1961–1964: Miller to Sec. 25 Mar. 1960.
99. BSIP 9/III/3: QRSC, Mar. 1940.
100. BSIP 9/V/1: Notes on RC's tour, May 1940; BSIP F28/14: Archer to Parkinson, 9 July 1945.
101. VR: Director's Minute Book, 1929–1941; Minutes of the Board of the Kauri Timber Company, 1924–1939; Roger, Vanikoro Branch report, Sept. 1961.
102. Bennett, *Wealth*, 286–92.
103. VR: Corres. to and from, 1939–1949: Boye to members of staff, 9 Feb. 1942, Boye to Sec. 19 Feb. 1942, Boye to Gibson et al., 17 Mar. 1942; Boye to Sec. 17 Mar., 2, 26 May, 24 June, 18 Sept., 31 Dec. 1942, 25 Mar., 5 July, 21 Sept. 1943. Ruby Boye at age 51 was commissioned as the only honorary third officer level in the Women's Royal Australian Navy (WRAN) to give her combatant status so the Japanese would not execute her as a spy. She was later decorated. (*The Australian Women's Weekly*, Feb. 1988,193–195.)
104. VR: Vanikoro Branch, 1939–1943: Wilson to Boye, 20 July, Boye to Sec. 26 Sept. 1943, Boye to Sec. 27 Oct. 1944; WPHC F 45/1: Sweeney to BSIP Gov., 7 July 1944, Nicoll to Sec. of State, 29 May 1947.

Chapter 6

1. See generally, Karl C. Dod, *The Corps of Engineers: The War against Japan* (Washington, 1966); John Rentz, *Marines in the Central Solomons* (Washington, 1952).
2. Tom Elkington, interview,1976.
3. Dod, *Corps*, 161, 210–4, 266–7
4. National Archives of New Zealand (hereinafter NZ): F T/C 1/40: Timber Controller to Commissioner for State Forests, 5 Nov. 1945.
5. L. T. Carron, *A History of Forestry in Australia* (Canberra, 1985), 298–301; Nancy M. Taylor, *The New Zealand People at War: The Home Front*, vol. 2 (Wellington, 1986), 787–813. House construction in New Zealand had suffered already during the Depression.
6. Third Division Historical Committee, *Pacific Pioneers: The Story of the Engineers of the New Zealand Expeditionary Force in the Pacific* (Dunedin, 1945), 43–45; NZ: WWII 1 da: Diary, 37 Field Park Coy, 1 Oct. 1943, Z130/1/1-14; A. Murray, Diary 3 NZ Div. Engrs, 22 Dec. 1943, Z130/1/1-15.
7. See for example, Rentz, *Marines*, 57; Dod, *Corps*, 208, 249, 250.

Notes to pp. 119–129

8. Walker, *Forests*, 26–30; BSIP F28/12–1: Goodsir to RC 3 Sept. 1945; Don Stuart, pers. comm., 26 Nov. 1994.
9. NZ: Air 265, 1: Record Book, June 1944, No. 1 Islands Work Squadron.
10. NZ: Air 1273 ix 2101/w: SWE to AOC, 8 Jan. 1945.
11. NZ: Air 265 1: No. 1 Islands Works Squadron, Record Book, May 1945.
12. Don Stuart, pers. comm., 26 Nov. 1994; NZ: Air 127, 3, viii, Vol. 2: Senior Works Engineer to Senior Accountant, 13 June 1945.
13. NZ: Air 127, 3, viii, Vol. 2 : Senior Works Engineer to Senior Accountant, 13 June 1945.
14. NZ: Air 265 1: No. 1 Islands Works Squadron, Record Book, Apr. 1944.
15. NZ: Forestry TC 1/40: Sando, Report on Inspection of Islands, Sept.–Oct. 1943.
16. BSIP F28/12–1: Goodsir to RC, 31 Aug. 1945; BSIP F28/4 Part 1: McLaughlin to Sec. HC, 7 Aug. 1946; NZ: Air 265 1: No. 1 Islands Works Squadron, Record Book, 1 Dec. 1943–5 Sept 1945.
17. *Building the Navy's Bases in World War II: History of the Bureau of Yards and Docks and the Civil Engineer Corps, 1940–1946*, vol. 2 (Washington, 1947), 251, 252–6.
18. Charles H. Stoddart, 'The Forests of the Solomon Islands', *Journal of Forestry* 44, no. 12 (1946), 1041–3.
19. NZ: Forestry TC 1/40: Sando, Report on Inspection of Islands, Sept.–Oct. 1943.
20. WPHC F48/89, Hyne, notes, 18 Aug. 1944; Dod, *The Corps*, 208, 209, 269; Bennett, *Wealth*, 291; Billings Papers: Billings, Re sawmill and places where they cut timber, c. 1944; Air Commander to Director of Works, 3 Feb. 1945; BSIP F28/14/1: Order, requisition of Batuna Sawmill, 26 Oct. 1943 and enclos.
21. Dethridge had joined the Australian Infantry Forces, but like most able-bodied Australian males who had lived in the Solomons and had local knowledge, he was assigned to the SI Defence Force to oversee Solomon Islands Labour Corps men. BSIP: F 28/15 Part I: Dethridge to Sandars, 19 Jan. 1945 and encl.; Bennett, *Wealth*, 281–2, 293–4.
22. BSIP: F 28/15 Part I: Sandars, Notes attached to Dethridge to Sandars, 16 July 1945; Walker, *Forests*, 13.
23. BSIP F28/15 Part II: DC to Sec. to Gov. 23 Oct. 1946 and encl. Chan Wing had extensive engineering experience and managerial ability and was later to establish several businesses in Honiara (*Solomon Star*, 6 Oct. 1989).
24. BSIP F28/12–1: Walker to RC, 15 July 1945.
25. WPHC F45/2/1: Vaskess, notes 23 July 1945.
26. BSIP F28/12–1: Goodsir to RC, 31 Aug. 1945.
27. Bennett, *Wealth*, 303–304; BSIP F28/12–1: Vaskess to RC, 22 Jan. 1944; Allan, *Solomons Safari*, 113.
28. See for example, the claims made by the Laulasi people, Allan, *Solomons Safari*, 152; *Solomons Star* 30 Aug. 1995; 1 Sept. 1995.
29. BSIP F28/16: Walker, Interim Report on Forests, 15 Feb. 1945, 9, f/n.
30. Hansell and Wall, *Land Resources*, 44 and Map 4d. See also, Schenk, *Forest Ecology*, Table Two, Appendix One.
31. Laracy, *Pacific Protest*; Bennett, *Wealth*, 292–9.
32. Clement cited in Ben Burt, *Tradition and Christianity: The Colonial Transformation of a Solomon Islands Society* (Camberwell, Victoria, 1994), 176.
33. Clement O'ogau cited in Ben Burt, Ibid.
34. BSIP 1/111, F28/12–1: Davies to DC, Malaita, 24 Apr. 1946.
35. Laracy, *Pacific Protest*; Bennett, *Wealth*; Burt, *Tradition and Christianity*, 175–200; Lasaqa, *Melanesians' Choice*, 48–9; WPHC F33/41, Vol. 1: Donald, Development of Guadalcanal Plains, 1947. People followed the new agricultural models provided by invading armies elsewhere in Melanesia to earn cash and provide dietary variety. In Bougainville and other places in Papua New Guinea the Melanesians emulated the rice farms of the Japanese, usually on a communal basis (John Connell, *Taim bilong mani: The Evolution of Agriculture in a Solomon Island Society* (Canberra, 1978), 77–110).

36. *PIM*, Oct. 1947, 71; Laracy, *Pacific Protest*: Bennett, *Wealth*; Burt, *Tradition and Christianity*, 175–200.
37. NZ: Forestry TC 1/40: Sando, Report on Inspection of Islands, Sept.–Oct. 1943. The Americans had similar concerns and surveyed several island resources with the co-operation of Douglas Oliver, then of the Board of Economic Warfare, but formerly and subsequently peacetime anthropologist. RC Noel was most suspicious of 'our old friend Oliver and his gang', believing they had 'elaborate plans for exploiting all sorts of articles of strategic importance in the Pacific' (WPHC CF 33/22: Noel to Mitchell, 12 July 1943 and encls).
38. NZ: Forestry TC 1/40: Sando, Report on Inspection of Islands, Aug.–Sept., Sept.–Oct. 1943 and encl.
39. Walker, *The Forests*, 4; BSIP F 28/15 Part 1: Walker to Sec to Gov., 25 Nov. 1945, 6 Apr. 1946.
40. WPHC 1427/1935: Hall to HC, 23 Oct. 1945.
41. BSIP F 28/12 Part 1: Sec of State for Colonies, Memo, 28 Oct. 1944, cited in Walker to RC, 2 Sept. 1945.
42. Bennett, *Wealth*, 301–2, 312.
43. Walker, *Forests*, 4. For examples of publications, see J. H. Kraemer, *Native Woods for Construction Purposes in the Western Pacific Region* (Washington, 1944); W. N. Sparkhawk, *Notes on Forest and Trees of the Central and Southwest Pacific area* (New York, 1945); Stoddart, 'The Forests of the Solomon Islands'.
44. BSIP F28/14 Part 1: Dethridge to Noel, 29 Apr.; Curran to RC 19 May; Officer-in-charge to RC, 21 June; Archer to Parkinson, 9 July; Horpel to RC, 21 Aug. 1945; HC to RC, 20 Mar.; Walker to Sec. to Gov., 25 Oct.; Walker to RC 2 Sept. 1945; Kleran (?) to Sec. to Gov., 17 Dec. 1945; Walker to Sec of Gov., 6 Apr. 1946.
45. *Empire Forestry Review* 29 (1948), 174.
46. BSIP F28/12–1: Walker, Memo on forest resources, 3 Sept. 1945; Walker to RC, 2 Sept. 1945 and notes; Goodsir to RC, 3 Sept. 1945; RC to HC, 27 Sept. 1945.
47. BSIP F28/12–1: Walker, Memo on forest resources, 3 Sept. 1945.
48. BSIP F28/12–1: Walker, Notes of an interview with the HC, 16 Apr. 1946; BSIP F28/16: Walker, Interim Report on Forests, 15 Feb. 1945, endnote; Laracy, *Pacific Protest*, 20–1.
49. BSIP F28/12–1: Walker, Notes of an interview with the HC, 16 Apr. 1946.
50. WPHC 1616/38: Ashley to HC, 22 Mar. 1938; BSIP 9/III/ 3: QRSC, Mar. 1940: BSIP 9/V/1: Notes on RC's tour, May 1940; BSIP F 28/14 Part 1: Jones to RC, 20 Apr. 1946; DC to RC, 21 June 1946; Walker to ARC, 15 July 1946; Walker, *Forests*, 50; BSIP F28/16: Walker, Interim Report on Forests, 15 Feb. 1945. In 1938, Jones was more interested in 'rosewood' (*Xylocarpus granatam*) than kauri probably because it was in the swamps and more accessible (WPHC 1616/38: Ashley to HC, 22 Mar. 1938).
51. Advisers in the Colonial Office shared this view, as did Forester Smith in Fiji. CO 225/338/86555/1A: Minutes, Robertson, 3 Aug. 1944. See also J. N. Oliphant, 'The Development of more Intensive Use of Mixed Tropical Forest', *Empire Forestry Review* 16 (1937), 29–37.
52. BSIP F28/14 Part 1: Dethridge to Noel, 29 Apr.; Curran to RC, 19 May; Officer-in-charge to RC, 21 June; Archer to Parkinson, 9 July; Horpel to RC, 21 Aug. 1945; HC to RC, 20 Mar.; Walker to Sec. to Gov., 25 Oct.; Walker to RC, 2 Sept., 1945; Walker to Sec. to Gov., 6 Apr. 1946.
53. BSIP F28/14 Part 1: Sec. to Gov. to Curran, 16 July 1945.
54. BSIP F28/14 Part 1: ARC to HC, 9 Apr. 1946; BSIP F28/16: Grantham to Sec. of State for Colonies, 13 May 1946.
55. Bennett, *Wealth*, 292–301; Walker,*Forests*, 15.
56. Troup, *Colonial Forest Administration*, 382.
57. Viscount Novar, 'Inaugural Meeting of the Empire Forestry Association', *Empire Forestry* 1 (1922), 5; N. D. G. James, *A History of English Forestry* (Oxford, 1981), 139, 194; Madhav Gadgil and Ramachandra Guha, 'State Forestry and Social Conflict in British India', *Past and Present* 123 (May 1989), 145; R. G. Albion, *Forests and Sea Power. The Timber Problem of the Royal Navy* (Cambridge,

Mass., 1926), 35–6, 363–8; Michael Adas, 'Colonization, Commercial Agriculture and the Destruction of the Deltaic Rainforest of British Burma in the late Nineteenth Century', in *Global Deforestation in the Nineteenth Century*, ed. R. P. Tucker and J. F. Richards (Durham, 1983), 96–7; Madhav Gadgil and Ramachandra Guha, *This Fissured Land: An Ecological History of India* (Berkeley, 1992), 118–19.

58. Richard Grove, 'The Origins of Environmentalism', *Nature* 345, no. 3 (May 1990), 12; Grove, *Green Imperialism*, 1–72.

59. Kurt Mantel, 'History of the International Science of Forestry', *International Review of Forestry Research* 1 (1964), 2–27; James, *History*, 194–196, 199; Nancy Lee Peluso, *Rich Forests, Poor People: Resource Control and Resistance in Java* (Berkeley, 1992), 44–78.

60. James, *History*, 194–6, 199; Gadgil and Guha, *Fissured Land*, 122–3.

61. Brian Harrison, ed., *The History of Oxford, Vol 8*, (Oxford, 1995), 147–148; Roderick P. Neumann, 'Forest Rights, Privileges and Prohibitions: Contextualising State Forestry Policy in Colonial Tanganyika', *Environment and History* 3 (1997): 45–68; R. A. Cline-Cole, 'Political Economy, Fuelwood Relations, and Vegetation Conservation: Kasar Kano, Northern Nigeria, 1850–1975', *Forest and Conservation History* 38, no. 2 (1994), 67–78; Peluso, *Rich Forests, Poor People*.

62. The journal later became the *Empire Forestry Review* and more recently the *Commonwealth Forestry Review*. Colonial foresters were in constant communication with each other through such journals and by interchange of personnel. When the Americans took over the Philippines from Spain in 1898 they modelled their Forestry Department and procedures on the British in India. Gifford Pinchot, chief of the US forest service in 1898, was deeply influenced by Brandis. A leading forester in the Philippines, Foxworthy, was invited to Malaya after World War One to set up research facilities and he brought data to the British on the forest types typical of Malesia, so very different from those of India, as well as information relevant to the marketing of timbers of the region. Richard P. Tucker, 'Managing Subsistence use of the Forest: The Philippine Bureau of Forestry, 1904–60', in *Changing Pacific Forests: Historical Perspectives on the Forest Economy of the Pacific Region*, ed. John Dargavel and Richard P. Tucker (Durham, North Carolina, 1992), 109; H. Steen, *The U. S. Forestry Service* (Seattle, 1976), 48.

63. Troup, *Colonial Forest Administration*, 8.

64. See Chapter 5; *Empire Forest Review* 18 (1939), 187–8. As well as advising on forestry policy and practice, the Colonial Office consulted Troup regarding suitable candidates for forestry officers in the colonies.

65. Troup, *Colonial Forest Administration*, 9.

66. CO 225/338/86555/1A: Stanley to HC, 28 Oct. 1944, Minutes, Robertson, 3 Aug. 1944; BSIP F28/12–1: Walker to Sec. to Gov. 8 Jan. 1946; Bentley to Sec. to Gov., 14 Jan. 1946. Heath ('Land Policy') does not deal with Walker's views and he does not seem to have consulted the study. This is rather surprising as Walker's report was the first published government-sponsored report where land tenure is discussed, albeit in a limited way.

67. Bennett, *Wealth*, 262; WPHC 2839/39: Map and enclosures. The mining regulation was similar to Fiji's of 1931. Irrespective of local systems of land tenure, this and others throughout the British empire took their origin in the decision of the 1930 Labour government in Britain (D. T. Lloyd, *Land Policy in Fiji* (Cambridge, 1982), 241–5).

68. WPHC 2839/39: Taburia to HC, 29 Apr. 1939.

69. BSIP F28/16: Walker, Interim Report on Forests, 15 Feb. 1945.

70. BSIP F28/12–1: Kleran, (?) to Sec. to Gov. 17 Dec. 1945.

71. BSIP F28/12–1: Davies to DC, 24 Apr. 1946.

72. BSIP F28/16: ARC to HC, 26 Feb. 1946.

73. BSIP F28/16: Grantham to Sec. of State for Colonies, 13 May 1946.

74. BSIP F28/12–1: Walker, Notes of an interview with the HC, 16 Apr. 1946.

75. Walker, *Forests*, 58

76. CO 225/338/86555/1A: Stanley to HC., 28 Oct. 1944; Minutes, Orde Brown, c. Aug. 1944.

Notes to pp. 139–144

77. Heath, 'Land Policy', 282. Noel and Walker forgot the 35 years it took to achieve this in Fiji and the considerable indigenous resistance to it even in a chiefly society with far bigger, more hierarchal, and more culturally and linguistically homogenous units than the several and various small, semi-acephalous ones of the Protectorate (Lloyd, *Land Policy in Fiji*, 264; Heath, 'Land Policy', 285, f/n.).
78. Walker cited Lord Hailey to give more weight to his ideas against the development of individual ownership. Hailey, however, was not so much against the drift towards individual rights in land, as cautioning against too rapid a change to a new system and concomitant social upheaval (Walker, *Forests*, 57–8).
79. Ibid.
80. Ibid., 62, 71.
81. Ibid., 18
82. Ibid., 58.
83. Heath, 'Land Policy', 282–5.
84. BSIP F28/16: Noel, Memo on Forest Report, 30 Dec. 1946.
85. Heath, 'Land Policy', 282–5.
86. ARFD 1953–1954.
87. Walker, *Forests*, 85.

Chapter 7

1. K. W. Trenaman, 'Forestry in the Solomon Islands since 1944' (Munda, Solomon Islands, Jan. 1959, cyclostyled).
2. BSIP F 28/14, Part I: HC to ARC, 28 June 1949.
3. BSIP F 28/14, Part II: Seton to Sec. to Gov., 15 Aug. 1952.
4. Trenaman, 'Forestry'.
5. Heath, 'Land Policy', 328.
6. WPHC F 10/14 Vol. 1: Director of Public Works to Sec., 26 May 1944 and encls.
7. WPHC F 10/14/8: ARC to HC, 8 July 1945 and encls.
8. BSIP F28/13: Sec. to Gov. to Sec. SDA, 6 Apr. 1950, Sec. to Sec. to Gov., 2 May 1950; BSIP F 28/18: Note to Rushbrook, Levers, n.d., ca. Feb. 1951; WPHC F 10/14/8: Act. RC to HC, 4 July 1946 and encls; Walker, *The Forests*, 34-5; WPHC F 45/23: Logie, Activities of the Forestry Dept (hereinafter FD) 1953; WPHC F 45/5/11: Logie to DC, 19 May 1953.
9. BSIP F 28/15 Part II: HC to RC, 17 Apr. 1948 and encls.
10. BSIP F 28/18: Gregory-Smith to Thomas, 22 Mar. 1951.
11. BSIP F 28/18: Hughes? to Sec. for Development and Native Affairs (hereafter DNA), 23 Aug. 1950; DC, Notes on discussion with N. Wallis, 2 Oct. 1950, Wallis to RC, 14 May 1951; RC to HC, 6 June 1951; Brett Hilder, 'Timber from Guadalcanal', *Walkabout* (1 May 1952): 17–20.
12. BSIP F 28/18: Hughes? to Sec. for DNA, 23 Aug. 1950; HC to RC, 20 July 1951 and Sec. of State, n.d., Comments; Hay et al. to DC, 15 Aug. 1950. Sustainable logging here means a timber yield that is produced continuously under a specified management programme. It does not imply ecological sustainability.
13. BSIP F 28/18: Sec for DNA to Levers, 4 Sept. 1950.
14. BSIP F 28/18: RC, Notes on discussion with N. Wallis, 2 Oct. 1950.
15. WPHC F45/5/7: Brief for Visiting Minister of State, Forestry: BSIP, 30 July 1951.
16. BSIP F 28/18: RC to HC, 18 Aug. 1951.
17. Walker, *Forests*, 92, 167.
18. BSIP F 28/18: Hughes? to Sec. for Dev. and Native Affairs, 23 Aug. 1950; HC to RC, 20 July 1951 and Sec. of State, n.d., Comments.

19. Some of this was bought by one of the company's principals, N. Wallis, a sawmiller in Sydney. Hilder, 'Timber', 17–20; *PIM*, June 1952, 68–9; BSIP F 28/18: Wallis to RC, 14 May 1951; RC to HC, 6 June 1951; LTO: 192-001-8, No. 91/96, Lot 1 of LR 395/5 and No. 113/66.
20. BSIP F 28/18: RC to HC, 24 May 1951, HC to RC, 5, 20 July 1951, and Sec. of State, n.d., Comments.
21. BSIP F 28/18: HC to RC, 20 July 1951; WPHC 61/34: Sec. of State to HC, 17 July 1951.
22. BSIP F 28/18: Sec. of State to HC, 13 Aug. 1951, Treasurer's Notes, 6 Nov. 1951; BSIP F 28/19 Part II: Sec. of State to HC, 26 July 1952, HC to ARC, 22 Sept. 1952; WPHC F 61/34: Sec. to Financial Sec. 7 Mar. 1953, encl.; VR: Corres. to and from, 1949–1953: Mc Ewin to Sec. Melb., 14 Dec. 1952.
23. WPHC F 61/34: Sec. to Financial Sec., 7 Mar. 1953; A. M. S. to Rednell, 22 Aug. 1954.
24. WPHC F 61/34: A. M. S. to Rednell, 22 Aug. 1954.
25. WPHC F 61/34: Sec. to Financial Sec., 7 Mar. 1953.
26. BSIP F 28/18: RC to HC, 25 May 1951 and encls.
27. Hilder, 'Timber', 17–20.
28. BSIP F 28/14 part II: Davies to Senior Ag. Officer, 27 June 1952, Note, 16 July, 1952; WPHC F 61/34: Note on Sec. to Sec. to Gov., 15 Dec. 1952; Bennett, *Wealth*, 135-8; Heath, 'Land Policy', 327–33.
29. BSIP F 28/19: Pullen to DC, 12 Aug. 1952. See also WPHC F 61/34: RC to HC, 15 Feb. 1952.
30. Bennett, *Wealth*, 304; BSIP F 28/19: DC to Pullen, 20 Aug. 1952.
31. For example, see BSIP F 1/111 F 15/6, Part 2: DO, 'How you can earn money', 20 Dec. 1947, encl, Development Plan, San Cristoval, Aug. 1948.
32. BSIP F 28/19: Pullen to DC, 12 Aug. 1952, DC to Pullen, 20 Aug. 1952.
33. BSIP F 28/19: DC to Tenaru Timbers, 4 Sept. 1952.
34. BSIP F 28/19: Pullen to DC, 12 Aug. 1952.
35. BSIP F 28/19: DC to Sec. to Gov., 20 Aug. 1952 and encl.; WPHC F45/23: Logie, Activities of the FD, 1953.
36. Hilder, 'Timber', 17–20; *PIM*, June 1952, 68–9; Dec. 1954, 125–6; ARFD, 1952, 1953–1954, 1956–1960; WPHC F 61/34: Sec. to Financial Sec., 7 Mar. 1953.
37. ARFD, 1953–1954, 1956–1960; BSIP Newsletter, 30/57, Oct. 1957; WPHC CF 164/3/13 vol. 1: Trenaman, Investigations in New Guinea and Australia on Timber Trade prospects, 9 Feb. 1959.
38. FDAR 1961–1964.
39. WPHC F45/5/7: Establishment of a FD. 17 May 1949; Progress report, 1950: Freeston to Sec. of State, 8 Mar. 1949 and encls; Trenaman, 'Forestry'.
40. See for example, Hadley, *Forester*, 29.
41. BSIP F 28/21: Logie to Sec. to Gov., 13 Nov. 1952.
42. BSIP F 28/21: Logie to Sec. to Gov., 13 Nov. 1952; FDAR 1956.
43. BSIP F 28/21: Logie to Sec. to Gov., 13 Nov. 1952; WPHC F45/23: Draft Queen's Regulation relating to forests, and encls, 1953.
44. WPHC F45/23: A further note on forestry policy, 10 Feb. 1954.
45. Billings Papers: Billings, Survey of West Cape, 19 Nov. 1944.
46. WPHC F45/23: Activities of the FD, 1953; BSIP F 28/21: RC to Sec. of State, 21 Nov. 1952.
47. The Council consisted of appointees, having its first Solomon Islander members nominated in 1950. The government was concentrating on the development and strengthening of the local councils, rather than central government. In the early 1950s, five official members, including the president (High Commissioner), five non-official expatriate members, and five Solomon Islanders were on the Council. All were nominated. Bennett, *Wealth*, 305–6, 318–20; *Minutes of Meeting of Advisory Council, 1951* (Suva, 1952).
48. WPHC F45/23: Draft Queen's Regulation, 1953.
49. FDAR 1953–1954.

50. BSIP F 45/23: Logie, Views and suggested future programme, c. 1953, encl.
51. BSIP F 45/23: Draft Queen's Regulation, 1953.
52. FDAR, 1953–1954, 1955, 1956; WPHC F45/5/9: Logie to Sec. to Gov., 31 Dec. 1952; WPHC F 45/5/11: Logie to DC 19 May, 2 Nov. 1953; WPHC F45/5/12: Logie to Chief Sec. 24 Dec. 1953; WPHC F45/23: Logie, Activities of the FD, 1953; WPHC F 45/5/1: Logie to Sec. to Gov., 31 Dec. 1952. The Protectorate paid for the education of the Solomon Islanders in Fiji.
53. WPHC F45/2/5: Davidson to Sec. to Gov., 9 Jan. 1953 and notes.
54. SDA: SI Minutes, 83rd Committee meeting, 3 Aug. 1931, 92nd Committee meeting, Feb. 1932.
55. Hugh Laracy and Geoffrey White (eds), *O'o: A Journal of Solomon Islands Studies* 4 (1988), 110.
56. Walker, *Forests*, 25; WPHC F45/5/11: Logie to DC, 19 May 1953.
57. Amherst and Thomson, *Discovery*, vol. 1, 128.
58. WPHC F45/2/5: Davidson to Sec. to Gov., 9 Jan. 1953, and notes.
59. WPHC F45/23: Logie, Views and future programme, 1953.
60. WPHC F 45/23: A further note of forestry policy for the informal meeting of the Advisory Council, 10 Feb. 1954.
61. WPHC F45/5/9: Logie to Chief Sec., 7 Jan. 1953 and encls.
62. Trenaman, 'Forestry'.
63. BSIP F28/16: Walker, Interim report on Forests, 15 Feb. 1945.
64. Alexander, *From the Middle Temple*, 269.
65. VR: Smith to Sec. 18 Aug. 1928, Island reports.
66. DP: Deland to Everyone, 11 Jan. 1928.
67. WPHC 1522/33: AR Nggela and Savo, 1932,
68. WPHC 3726/31: RC, Tour, 30 Dec. 1929–13 Jan. 1930; Bennett, *Wealth*, 219–40.
69. BSIP F 28/15, 1/111: Super. of Police to C. of Works, 5 July 1951.
70. WPHC F45/23: Logie, Activities of the FD, 1953; WPHC F 45/29: Wallis to Fin. Sec., 5 Oct. 1953 and encl.
71. BSIP 31/11: Keevil to CFO, 23 Oct. 1958; ARFB 1956.
72. VR: Rogers, Report on Vanikoro Branch, Sept. 1961.
73. VR: Corres. to and from 1949–1960: Guest to Bishop, 22 July 1955, Notes on Filewood's Report of June 1955; Haling to Sec., 23 Oct. 1955; Filewood to Sec. 30 Sept. 1955.
74. Bennett, *Wealth*, 308; VR: Monthly reports: Apr. 1954.
75. VR: Corres. to and from, 1949–1960: McEwin to Sec., 4, 12 Apr., 23 May 1951; Corres., 1953–1956: McEwin to Sec., 18 Jan., 30 Mar. 12 Apr. 1954; Monthly reports: July, Sept. 1951, Mar., June 1952; Reports general: Beckett, Report on logging operations, Dec. 1954.
76. VR: Corres., 1949–1960: McEwin to Sec., 331 July, 1951, 8 Jan., 10, 24 Feb. 1953; Moody to Sec. 3 Feb. 1952; Reports general: Beckett, Report on logging operations, Dec. 1954, WPHC F 15/52: DC to Sec. of Gov., 11 Jan. 1953: BSIP 9/111/1: AR Eastern District; Allan, *Solomons Safari*, 144–5; *PIM*, Feb. 1953, 16; Mar. 1953, 133; Apr. 1953, 45; Jan. 1955, 80.
77. Hadley, *Forester*, 92.
78. *PIM*, Mar. 1953; Allan, *Solomons Safari*, 140–1.
79. VR: Filewood, Report on Logging Operations, 30 June 1955,
80. VR: Reports general: McEwin, Proposed logging policy for Vanikoro and plan for general administration, 24 June 1953. A draw is a shattered log end or part thereof and is similar to a drag, which is a section of the outer tree torn off the side of the log and left on the stump. Shakes are circular cracks, sometimes existing in the living tree, sometimes caused by shocks received in felling. These are also called ring shakes.
81. VR: Corres. to and from, 1949–1960: McEwin to Sec., 20 Mar. 1950; McEwin, Proposed logging policy for Vanikoro and plan for general administration, 24 June 1953; WPHC F 45/5/9: Logie to Sec. to Gov., 31 Dec. 1952.
82. VR: Davidson, Report by Victorian manager on logs from Vanikoro, 21 July 1954.

Notes to pp. 156–158

83. VR: Gen. Corres., A to L, 1948–1949: Ass. Sec. to Haling, 24 Sept. 1948 and encls; Reports general: McEwin, Proposed logging policy, 24 June 1953; Beckett, Report on Logging operations at Vanikoro, Dec. 1954; Saxton (?), Report on Vanikoro and Ndeni, Sept. 1963.

84. VR: Reports general: Filewood, Report on logging operations, 30 June 1955; Jeffreys, Report on visit to Vanikoro, Jan.–Aug. 1957; Monthly reports: Bennett, 13 Oct. 1960.

85. WPHC F 45/5/9: Logie to Sec. to Gov., 31 Dec. 1952.

86. WPHC F 45/5/9: Logie to Sec. to Gov., 31 Dec. 1952; VR: Reports General: Beckett, Report on logging operations, Dec. 1954.

87. VR: Davidson, Report by Victorian manager on logs from Vanikoro, 21 July 1954; Notes of a visit to Honiara and Vanikoro, Aug. 1954.

88. VR: Rogers, Report on Vanikoro Branch, Sept. 1961; Henderson, Notes, n.d. Real profits are difficult to estimate, but cost of 100 superfeet of kauri landed in Australia in the mid-1950s was about 116s to 125s and kauri, top quality, retailed at 250s in 1957, but the landed cost does not allow for losses on defective logs, nor the overheads in running the Australian business which would have reduced the profit in Australia to less than 30%. VR: Reports general: Hanson, Royalty appraisal, Vanikoro island, Oct. 1958; WPHC Corres., 1957–1964: Chief Sec. to Deputy C'man, 7 Nov. 1957; Act. Chief Sec. to Sec. 19 May 1958 and encls; Corres. 1956–1960: Newmarket Plywood, Costs April 1944, Newmarket Plywood Co.; Greentree to Sec., 1 Oct. 1957; Corres. 1952–1954 : Greentree to Ploog, 15 Mar. 1954; Greentree to Sec., 14 May 1954, 8 June 1954.

89. VR: Monthly reports: Jan., May 1956; Corres. to and from 1949–1960: Filewood to Sec. 29 May, 9 July 1956; Bennett (?) to Sec., 3, 15 Oct., and Dec. 1960; Letters from manager, 1960–1964: Powell to Sec. 3 June, 6 Nov. 1963; Rogers to Taylor, 21 Nov. 1963; ARFD, 1956.

90. VR: WPHC Corres., 1957–1964: Adner to Sec., 14 May, 25 Mar. 1957, Chief Sec. to Deputy C'man, 9 Nov. 1957; BSIP Gov.: Rogers, Broad Summary of present Honiara Negotiations, 11 Nov. 1958.

91. FDAR, 1956–1960; WPHC CF 164/2/1 vol. 2: Note, Trenaman, 12 July 1957; Hall to Gutch, 8 July 1958. The assessor, Hanson, recommended a duty of 12s 1d per 100 superfeet (on 3 million superfeet) or 21s (on 5 million superfeet). He costed kauri at 105s 5d to 101s 4d landed in Melbourne and including profit to the producing Vanikolo branch. The internal correspondence of the company, upon which subsidiary firms made decisions to buy or not buy, reveals the cost as about 116s to 125s. Hanson calculated f.o.b. Vanikolo cost at 54s 4d to 60s 5d, while the company maintained that 78s f.o.b. cost was more accurate. Shipping cost was about 45s to Melbourne which both parties agreed on. This would explain the government's reduction to a duty of 9s, still not an unreasonable rate. VR: Reports general: Hanson, Royalty appraisal, Vanikoro island, Oct. 1958; WPHC Corres., 1957–1964: Chief Sec. to Deputy C'man, 7 Nov. 1957, Act. Chief Sec. to Sec. 19 May 1958 and encls.

92. VR: WPHC Corres, 1957–1964: Chapman to Manager, 29 Dec. 1961, Trenaman to Sec., 14 May 1963.

93. FDAR, 1956, 1957; WPHC CF 164/2/1/ vol. 3: Trenaman, The Kauri forest of Vanikoro, Tevai and Santa Cruz, Aug. 1958.

94. VR: Saxton (?), Report on Vanikoro and Ndeni, Sept. 1963; WPHC Corres., 1957–1964: Beckett to Rogers, 2 June 1961; See also Filewood, Report, June 1955 and attached notes, 22 July 1955.

95. Ben Tua, interview, 1992.

96. VR: Beckett, Report on logging operations at Vanikoro, Dec. 1954; Saxton (?), Report on Vanikoro and Ndeni, Sept. 1963.

97. VR: WPHC Corres., 1957–1964: Chapman to VR, 12 Feb., 12 June 1962, Trenaman to Sec., 23 Dec., Trenaman to Rogers, 16 Mar. 1964; Gass to Sec. 27 June 1960; Beckett to Rogers, 2 June 1961. Keith Trenaman, head of the Forestry Department, seems to have been a stickler for correct procedure, so much so that the company hid the fact that they were leaving unlogged a good patch of kauri in the Middle Ridge area. When Trenaman was on leave (Nov. 1961–July 1962), acting Chief Forestry Officer John Chapman was more willing to turn a blind eye to unworked timber in old sites and settled for thorough working of existing logging sites only. VR: Chapman to VR, 12 Feb. 1962, Chapman to Rogers, 4 June 1962, Chapman to VTC, 12 June 1962; Letters from manager, 1960–1964: Powell to Sec. 27 Apr., 26 Nov. 1962.

98. VR: Letters from managers, 1960–1964: Greentree to Rogers, 9 Mar. 1962; Saxton (?), Report on Vanikoro and Ndeni, Sept. 1963.
99. VR: WPHC Corres., 1957–1964: Rogers to Trenaman, 16 June, 11, 18 Aug., 13 Sept. 1961.
100. Michael Roche, J. Dargavel and J. Mills, 'Tracking the Kauri timber Company from Kauri to Karri to Chatlee', in *Australia's Ever-changing Forests 2*, ed. J. Dargavel and S. Feary (Canberra, 1993), 187–204; VR: WPHC Corres., 1957–1964: Rogers to Trenaman, 11 Aug. 1961.
101. VR: General manager's Reports: Newmarket Ply. Co., Monthly Report, Sept. 1960.
102. VR: Corres. in and out, 1952–1954: Greentree to Sec., 15 Mar. 1954.
103. VR: WPHC Corres., 1957–1964: Chief Accountant, Costs and production totals, 30 June 1958, enc; Acting Chief Sec. BSIP to VR, 19 May 1958.
104. VR: Saxton (?), Report on Vanikoro and Ndeni, Sept. 1963.
105. Allan, *Solomons Safari*, 145.
106. Bennett, *Wealth*, 301–18.
107. Hadley, *Forester*, 282; WPHC CF 164/2/1 vol. 1: Appendix to No. 30 of 15/3/57.
108. VR: WPHC Corres., 1957–1964: Rogers to Trenaman, 1 Sept. 1960.
109. FDAR 1964.

Chapter 8

1. WPHC SF 178/5/6: Anderson, Notes and impressions, July–Oct. 1957; General Assembly Resolution 1514 (XV) of 1960, United Nations Organisation; Old Files, Ministry of Natural Resources, (hereafter OF), M 21/1/1: Trenaman to Chief Sec. 18 Oct. 1964; OF, F32: Trenaman, Land for forestry, Minutes of Government House (hereafter GH) meeting and appendix, 27 Oct. 1966.
2. Jan G. Laarman, 'Export of Tropical Hardwoods in the Twentieth Century', in *World Deforestation in the Twentieth Century*, ed. John F. Richards and Richard P. Tucker (Durham, 1988), 151–6; Francois Nectoux and Yoichi Kuroda, *Timber from the South Seas: An Analysis of Japan's Tropical Timber Trade and its Environmental Impact* (Gland, Switzerland, 1989), 32.
3. FDAR 1960.
4. R. and V. Routley, 'Destructive Forestry in Australia and Melanesia', in *The Melanesian Environment*, ed. John Winslow (Canberra, 1977), 392, 394.
5. The now classic statement of this is Jack Westoby, 'The Role of Forest Industries in the Attack on Economic Underdevelopment', in Food and Agriculture Organisation (hereinafter FAO), *The State of Food and Agriculture 1962* (Rome, 1962), 88–128.
6. Heath, 'Land policy', 311–12.
7. It is not known how Allan arrived at the figure of 25 years. Perhaps it reflected a generation span. It certainly came close to the 30 years estimated much later in 1995 for garden land to return to 'primary forest' (*SINFRI* 1, 8).
8. Heath, 'Land policy', 311–18.
9. FDAR 1958-1960; OF, M 21/1/1: Minutes of GH meeting, 30 Mar. 1960, cited in Trenaman, *The Forest Estate*, Technical Note, 1/62, Aug. 1962.
10. FDAR 1956, cyclostyled, author's copy; Forestry Adviser to Sec. of State, cited in Trenaman, 'Forestry'; OF, M21/1/1: Trenaman, *Land for Forestry*, FD Technical Note No. 7/63, November 1963, encl.
11. OF, F32: Trenaman (?), Draft Memo for a GH meeting, Forestry Policy, encl. in Minutes of a GH meeting, 27 Oct. 1966; FDAR 1965, 1968.
12. FDAR 1956, cyclostyle, author's copy. Trenaman's Estate concept confirmed Logie's draft.
13. FDAR 1956–1965.

Notes to pp. 165–169

14. AR BSIP 1975, 17, 127. To make the 'vacant' land concept 'more acceptable politically' to groups with former interests in the land (before 1933), the government planned to make ex-gratia payments to local councils (OF, F1/8 25 Vol. 1: Trenaman to McAdam, 16 Sept. 1957).

15. OF, F1/8, vol. 1: Trenaman, Explanatory note on draft legislation, encl., 26 Oct. 1959.

16. OF, F/8, vol. 1: Trenaman to Dun, 26 Oct. 1959, Forests Regulation 1960, and encls. There were elements within the colonial government who were convinced that the vacant land principle would never be viable, yet they remained silent at this time (K. W. Trenaman, pers. comm., 10 Sept. 1994).

17. OF, M 21/1/1: Trenaman, *Land for Forestry*, FD Technical Note No. 7/63, Nov. 1963, encl.

18. Ian Heath points out that it might have been politically more acceptable if structures recommended by Allan had been set up. The Land Trust Board was an emasculated version of his original Board and the local land committees, apart from the local government councils, were not set up. Allan also emphasised the need for great tact, a quality often missing in otherwise conscientious colonial civil servants (Heath, 'Land Policy', 312–20).

19. Cited in Colin Allan, *Customary Land Tenure in the British Solomon Islands Protectorate* (Honiara, 1957), 287.

20. WPHC CF 164/2/9: Dennis to Chief Sec. 6 June 1959; Trenaman to Chief Sec., 28 Dec. 1959.

21. WPHC CF165/8/2: Davidson to Val, 17 Aug. 1955.

22. Heath, 'Land Policy', 327–39. Re centrality of Kolombangara to people's attitude well beyond the island itself, see, OF, F139: Trenaman to Chief Sec. 21 Dec. 1963.

23. Allan, *Solomons Safari*, 172–4.

24. Lands and Titles (Amendment) Ordinance, 1964; Heath, 'Land Policy', 391–20; Anthony Hughes, pers. comm., 26 June 1999.

25. OF, M 21/1/1 and BSIP F28/12–1: Trenaman, *The Forest Estate*, Technical Note, 1/62, and encl., Aug. 1962.

26. Bennett, *Wealth*, 148.

27. OF, F128: Chief Forestry Officer (hereinafter CFO) to Chief Sec. 5 July 1960.

28. The government land (28,816 acres) on Tetepare had some patches of commercial forest, mainly *Pometia*, but the steepness of the land and its shallow soil were found to be such a drawback that the government did not offer it to logging companies. OF, F128: C. of Lands to Chief Sec., 24 June 1960; Trenaman to Chief Sec. 5 July 1960; Trenaman to C. of Lands, 23 May 1961; F107: Note of a meeting at GH, 5 Mar. 1963.

29. This harked backed to High Commissioner Grantham's suggestion of 1946 which F. S. Walker had reluctantly accepted. OF, M 21/1/1 and BSIP F28/12–1: Trenaman, *The Forest Estate*, Technical Note, 1/62, and encl., Aug. 1962; Walker, Notes of an interview with the HC, 16 Apr. 1946. Allan supported this idea of payment for land that was not vacant for public purposes like forestry (Allan, *Customary Land Tenure*, 293–4); FDAR 1963.

30. Edvard Hviding and Graham B. K. Baines, 'Community-based Fisheries Management, Tradition and the Challenges of Development in Marovo, Solomon Islands', *Development and Change* 25, (1994), 34.

31. OF, F212: Kera to Chief Sec. 13 Sept. 1965, and encls; see also, Russell Parker, *Maekera: The Life Story of Hereditary Chief Nathan Kera and the Saikile Community of Solomon Islands* (Tenterfield, New South Wales, 1994), 34–6.

32. OF, M21/1/1: Trenaman to C. of Lands and Surveys, et al., 18 Oct. 1965; OF F32: Trenaman (?), Draft Memo for a GH meeting, Forestry Policy, encl. in Minutes of a GH meeting, 27 Oct. 1966.

33. OF, F212: Kimi, letter handed to visiting Members, 5 Aug. 1965.

34. FDAR 1962, 1963, 1964; OF, F216: Curtin to Trenaman, 30 Sept. 1966.

35. OF, F32: Kera to CFO, 9 Apr. 1963. R. T. Kera was to become the first Conservator of Forests in 1976 (FDAR, 1976, 1980); OF, F128: Nano to Trenaman, 22 Jan. 1964; OF, F107: Alisae to CFO, 4 June 1963, and encls. See for a similar request for Vuna, OF, F 216: Bengough, A brief summary of BSP Co.'s Survey of Vangunu Island, May 1965.

36. OF, F212: Hunter to C. of Lands, 13 Sept. 1965, and encls.

37. OF, F107: Alisae to CFO, 4 June 1963, and encls.

38. OF, F128: C. of Lands to DC Central, 12 Nov. 1963. Bryan was hardly disinterested. Representing the Pacific Timber Company, he was promised first option on the timber rights the government hoped to obtain near Rere-Kaoka in 1959, providing the company did a timber survey of Vangunu. Trenaman was on leave and others made this deal. The company failed to do the survey. When there was talk of the company transferring the option to the Australian Colonial Sugar Refinery, Trenaman became suspicious of the undercapitalised company, though the project seems to have collapsed while waiting the declaration of Forest Areas (WPHC CF 164/2/9: Dennis to Chief Sec., 6 June 1959 and encls.).

39. OF, F216: Nano to Trenaman, 15 Apr. 1965; Trenaman to Managing. Director, Lever's, 21 Nov. 1960 and encls; OF, F139 Vol. I: Notes, GH Discussion, 24 Nov. 1962.

40. N. J. Fox, 'His Schools will also Teach Skills', *Marist Messenger*, April 1967, 2–3; M. Cruickshank, 'Work in Timber is now a Going Concern', *Marist Messenger*, Aug. 1968, 23–6.

41. FDAR 1965; OF: Summary for the first quarter 1966, F 112/1/3.

42. OF, F129: Minutes of meeting, Financial Sec.'s Office, Honiara, 1 Nov. 1965. See also AR BSIP, 1953/4, photograph of boat-building at Gizo, opposite p. 10.

43. This was to the Vangunu people. OF, M21/1/1: Trenaman, *Land for Forestry* and postscript, 18 Nov. 1963. See also FDAR 1963,14.

44. WPHC BSIP CF 164/2/9: Minutes of a Meeting on Forestry Development, 2 Dec. 1959.

45. OF, F72: Trenaman to Swabey, 12 Mar. 1963. The Kalena Timber Co. soon replaced Mears and Ireton.

46. FDAR 1969. One of the largest was 1.2 million hectares in East Kalimantan – almost half the total land area of the Solomons (*Australian Timber Journal*, Dec. 1969, 129–43, 145; July 1970, 94).

47. OF, F1/8 25 Vol. 1: Trenaman to Walker, 16 Nov. 1967; FDAR 1963–1966.

48. FDAR 1963–1966; OF: Bengough, A brief summary of BSP Co.'s Survey of Vangunu Island, May 1965; Kuribashi to Trenaman, 26 May, 24 Nov. 1965; Trenaman to Pres. NRK, 9 Dec. 1965, F 216; G. D. Gilbert, 'Report on Visit to the Solomon Islands, June 1966' (Honiara, 1996, cyclostyled); G. J. Pleydell, 'Solomon Islands Timber: Developing a New Export', *Progress* 83 (1969), 81–2.

49. OF, F32: Trenaman (?), Draft Memo for a GH meeting, Forestry Policy, encl. in Minutes of a GH meeting, 27 Oct. 1966; Pleydell, 'Solomon Islands Timber', 82–5; FDAR 1967.

50. Twomey cited in Ian Heath, 'Land Policy', 363. The pros and cons of this scheme are discussed in Peter Larmour (ed.), *Land in Solomon Islands* (Suva, 1979) and by Ian Heath, 'Land Policy', 352–72.

51. FDAR 1958-1969; OF, F72: Trenaman to Whyatt-Smith, 1 July 1970; OF, F26/11/20: Trenaman, Forestry Development and Development Programme, Apr. 1970; OF, F306: Trenaman to Gane, 27 June 1968; F. C. Lembke, 'Forestry in the Solomon Islands', *The Australian Timber Journal and Building Products Merchandiser* 36, no. 8 (Sept. 1970), 46–8.

52. OF, F32: Trenaman to Chief Sec., 17 May 1966; OF, F128: Trenaman to C of Lands, 2 Aug. 1972.

53. OF, F32: Trenaman to Chief Sec., 17 May 1966. See also, OF, F32: Draft Memo for a GH meeting, Forestry Policy, encl. in Minutes of a GH meeting, 27 Oct. 1966.

54. BSIP, F112/1/6: Tour with CFO, north Malaita, 24–26 May 1961; OF, F32: Minutes of GH meeting, 26 Mar. 1964; OF, 4 May, M21/1/1: Trenaman to Chief Sec., 15 Apr. 1965, F32; Trenaman, *Land for Forestry*. See, for example, Trenaman's comprehensive analysis and statement, OF, F32: Draft Memo for GH meeting and attachment, 26 Oct. 1966.

55. R. Eresi in Solomon Islands National Parliament, *Debates*, meeting, Mar. 1980, 189–90; FDAR 1967.

56. OF, M 22/1/1: Public Notice, 21 June 1968.

57. OF, F 306: J. Bryan on behalf of I. Tagoa, Timon Taria, P. M Kaomaue, J. Mali, J. Bosa, Reuben Bula, Charles Mau, Andsen Cook N, John Golagola, Paul Pirua, Tago, Ronouania, to T. Russell, 16 July 1968. See footnote 38, for J. Bryan's background.

58. OF, F306: Tagoa to Select Committee, 6 Aug. 1968.

Notes to pp. 175–178

59. OF, F212: Balesi to [Trenaman], 13 Nov. 1968.

60. OF, F 209: Collis to Landholders, 20 Oct. 1968; OF 127: Jack to Munro, 28 Dec. 1968; Report of a Special Select Committee appointed by the Leg. Council, Legislative Council Paper, No. 86 of 1968; BSIP Legislative Council *Debates*, 6 June 1968; FDAR 1968; OF, F 306: Select Committee on Forestry, and encls.

61. See his statement in FDAR, 1968, 1.

62. OF, F128: Trenaman to C. of Lands, 2 Aug. 1972. Trenaman had been on overseas leave from June to October 1969.

63. Under the Lands and Titles Ordinance, a 'profit' could only be granted in a registered estate, that is, not customary land. So, the customary land was actually purchased by the government, then literally in the same instant, this registered perpetual estate was transferred back to the vendors who would then grant the government a 'profit' to cut the timber. Peter Larmour, 'The North New Georgia Corporation', in *Land, People and Government*, ed. Peter Larmour, Ron Crocombe, and Anna Taugenga (Suva, 1981), 137.

64. OF, F32, F2/2: Trenaman, Forestry – Information Note, July 1970.

65. Planning Unit, *Sixth Development Plan 1971–1973* (Honiara, 197), 70–1; FDAR 1971.

66. BSIP, Forests and Timber Ordinance, Oct. 1972, No. 11 of 1972.

67. Although not specifically confined to forestry, the colonial government, in 1972, tried to get the Land Development (Control) Bill through the Governing Council, which would have given the government control over the kind of development to be carried out on land. It was defeated. Heath, 'Land Policy', 405.

68. OF, F128: Trenaman to C. of Lands, 2 Aug. 1972.

69. OF, F128: Twomey, Public Notice, Land and Titles Amendment Bill, 26 Oct. 1972; Heath, 'Land Policy', 388–99, 401–6.

70. For similar patterns in recent times among the Marovo Lagoon *butubutu*, see Hviding, *Guardians*, 311.

71. See for example, Legislative Assembly, *Debates*, May 1975, 352–69, May 1979, 800–2; Larmour, 'North New Georgia Corporation', 134.

72. Cited in OF, F 206, F2/3: Trenaman, Forestry Policy Review Committee, background paper, Jan. 1975.

73. OF, F32, F2/2: Trenaman, Forestry-Information note, July 1970.

74. OF, F192: Copy of Licence No. T/3/70, 1O Oct. 1970, encl.

75. BSIP F26/11/20: Trenaman, Forestry policy and development Programme, Apr. 1970.

76. FDAR 1969.

77. FDAR 1960-1980; *BSIP Newsletter*, 6/62, June 1961.

78. BSIP F26/11/20: Pepys-Cockerell, Business Enterprise among the Malaitans, Dec. 1962, encl.; FDAR 1957, 1958. Arson is a common form of resistance or protest on Malaita, especially in conflict of property rights.

79. BSIP F112/1/4: Pepys-Cockerell to Trenaman, 5 Sept. 1961, Trenaman to Pepys-Cockerell, 11 Sept. 1962. See also, BSIP F112/1/4: DC to Idufuasi, 21 July 1965; Filor to Bengough, 3 Aug. 1966.

80. FDAR 1967.

81. *Wood World*, Jan. 1970, 4.

82. FDAR 1967. The administration had a boat-building school at Auki, Malaita, ran short ad hoc training courses, and sent a handful of boys to Australian-financed Idubada Technical Institute in Papua. The technical institute in Honiara did not open until 1969 in the wake of scathing reviews of technical education – or the lack of it – in 1966 and 1967 and expanding man-power needs. Pleydell, 'Solomon Islands Timber', 84; *PIM*, Aug. 1966, 45–6, Mar. 1969, 28; B. S. Palmer, 'The Interaction of Churches and State in the British Solomon Islands over the Designated Schools Scheme, 1964–1967' n.d., n.p., cyclostyled.

83. Lembke, 'Forestry', 31, 35. See also AR BSIP 1964/5, 34, 1966, 32; OF, F129: Acting Chief Sec. to Trenaman, 1 Feb. 1963 re importing Bougainvillean workers.

Notes to pp. 178–181

84. OF, M21/1/1: Information Brief for the Minister of Trade, Industry and Labour, n.d., [1974].
85. OF, F72: Trenaman to Wyatt-Smith, 11 July 1973.
86. BSIP F112/1/1: Finnemore to Conservator of Forests (hereafter COF), 14 Feb. 1971; BSIP F112/1/7: Funifaka to Gauwane, 23 Jan. 1974; Martin to Laqueta, 5 Dec. 1974; Trenaman, Forestry – A Development Projection, May 1973.
87. OF, M21/1/1: Trenaman, Forest Policy, n.d., c. 11 Nov. 1974.
88. OF, F99: Trenaman to Establishment Sec. 13 July 1970.
89. BSIP F112/1/7: Trenaman, Forestry – A Development Projection, May 1973; Trenaman to Wyatt-Smith, 11 July 1973, F72; CFO to Damirara, 15 Feb. 1960; BSIP F 112/1/6: ?, Tour with CFO, North Malaita, 24–26 May 1961 ; BSIP F112/1/4: Filor to Bengough, 3 Aug. 1966; BSIP F112/1/7: Butler to Ag. and Industrial Loans Board, May 1973. Elsewhere in Melanesia in the late 1960s, United Nations' advice supported the concepts of greater indigenous participation in the timber industry through adding value locally and of small-scale logging as opposed to highly mechanised large scale extraction, or at least as a viable adjunct to it. See S. D. Richardson, *The Role of Forest-based Industries in the Economic and Social Development of West Irian* (New York, 1968).
90. For example, see OF, M21/1/1: Trenaman, *The Forest Estate*, Aug. 1962; Trenaman, *Land for Forestry*, Nov. 1963; Notes, GH, 24 Nov. 1962; OF, F32: Memo for GH meeting, and encls, Minutes of meeting, 26 Mar. 1964.
91. See E. F. Schumacher, *Small is Beautiful: A Study of Economics as if People Mattered* (London, 1973), for arguments in favour of intermediate or appropriate technology. As an aid and development concept, it arrived in the Solomons in the mid-1970s, exemplified in B. Stephen Oxenham (ed.), *The Road Out: Rural Development in Solomon Islands* (Suva, 1981), especially 57–68, 91–7, and, to some extent, in the Office of the Prime Minister, *National Development Plan for the Solomons, 1975–1979* (Honiara, 1975).
92. OF, F32: Gina to Chairman, Forestry Policy Review Committee, 3 Feb. 1975.
93. BSIP F112/1/7: Butler to Ag. and Industrial Loans Board, 31 July 1973; see also BSIP AR, 1973, photograph opposite p. 37.
94. *Annual Report of the Agricultural and Industrial Loans Board, 1973* (Honiara, 1974), 14.
95. BSIP F112/1/7: Trenaman to DC, Malaita, 19 Apr. 1973.
96. BSIP F112/1/7: Trenaman, Forestry – A Development Projection, May 1973, encl.
97. BSIP F112/1/7: Trenaman to Gen. Manager, Allardyce, 27 Mar. 1973 and encl., F192; Trenaman to DC, Malaita, 19 Apr. 1973; Trenaman, Forestry – A Development Projection, May 1973; See also, FDAR, 1976, 3; Legislative Assembly, *Debates*, May–June 1979, Kenilorea, 846; Tovua, 944. This was also the attitude of the Ag. and Industrial Loans Board (*Annual Report of the Agricultural and Industrial Loans Board*, 1973, 140).
98. BSIP F112/1/7: Trenaman, Forestry – A Development Projection, May 1973, encl.
99. OF, F192: Trenaman to Gen. Manager, Allardyce, 27 Mar. 1973 and encl.; BSIP F112/1/7: Trenaman to DC, Malaita, 19 Apr. 1973; Trenaman, Forestry – A Development Projection, May 1973; OF, M21/1/1: Information Brief for Min. of Trade, Industry and Labour, Forestry Policy, 10 Nov. 1974; FDAR 1973–1975; J. G. Groome, *Technical Appendices for the Malaita Forestry Development Study* (Suva, 1982), 106; Martin Bennett et al., *Forestry Project Identification Mission, May–June 1991*, (Herts., Great Britain, 1991), 108–9; A. J. Leslie, *Forest Economics, British Solomons Islands Protectorate: Project Findings and Recommendations* (Rome, 1975), 15–19.
100. *Forestry Policy*, White Paper: BSIP 12, (Honiara, 1968).
101. OF, F2/3, F306: Forest Policy Review Committee, Terms of Reference, n.d., c. Dec. 1974.
102. OF, F2/2, F32: Gauwane, Report of the Forest Policy Review Committee, 14 July 1975; Statement by Min. in moving a motion to adopt the Report, July 1975 and encls; OF, F2/3, F306: Memo. by the Min. of Natural Resources, 23 July 1975. Trenaman felt betrayed when the committee members later reversed their stance on government purchase of timber rights.
103. OF, F 306: Osifelo to Sec., 31 Jan. 1975. A controlled forest has yet to be declared.

Notes to pp. 182–187

441

104. In the amendment, the phrase 'conservator of forests' was used despite the official head being known as the chief forestry officer since 1974. FDAR 1977; Peter Larmour, 'North New Georgia Corporation', 142. The essence of this amendment was drafted by Ken Marten, FD (Ken Marten, pers. comm, Nov. 1995). I could find no published copy of the 1977 parliamentary debates relating to the amendment.

105. *Solomon Islands New Drum,* Honiara, 25 Nov. 1977; Forests and Timber (Amendment) Ordinance 1977; FDAR 1976. Ken Marten was acting CFO for twelve months after Trenaman's departure. R. T. Kera then became head of department (Ken Marten, pers. comm., Oct. 1996).

106. For a similar view, see M. Rangarajan, 'Imperial Agendas and India's Forests: The Early History of Indian forestry, 1800–1878', *The Indian Economic and Social History Review* 31, no. 2 (1994), 153.

107. WPHC SF 178/5/6: Anderson, Notes and impressions, July–Oct. 1957; WPHC CF165/8/2: Davidson to Val, 17 Aug. 1955; OF, M 21/1/1: Trenaman, *The Forest Estate*, Technical Note, 1/62, Aug. 1962.

108. Ruhi Grover, 'Rhythms of the Timber Trade: Forests in the Himalayan Punjab, 1850–1925' (Ph. D. diss. University of Virginia, 1997).

109. The process of enclosure the Britain in the eighteenth and early nineteenth centuries involved the fencing in of common pasture and grazing lands, as well as scattered agricultural plots. This benefited the landowners who had clear title to formerly common lands; it certainly increased productivity and thus enable the landowners to demand higher rents of their tenants. Coupled with a population increase, however, it saw very small farmers, cottagers, and squatters driven off the lands and into the slums of the new industrial centres or to Britain's colonies.

110. FDAR 1968-1975. See chap. 9, n. 25.

Chapter 9

1. FDAR 1957.
2. FDAR 1958, 1960.
3. FDAR 1958.
4. FDAR 1958, 1968.
5. In surveying likely land for the Forest Estate from 1965, the Department notionally demarcated forests within 1.5 miles of the coast and within 1.5 mile radius from any village as being outside any logging concessions in regard to timber stocks, but Trenaman omitted this margin in the Forest Areas description of 1968 (K. Marten, pers comm., 8 Nov. 1997).
6. BSIP F28/12 Part I: Progress Report on CDW Scheme for the year 1947; Depts of Ag., Lands and Forestry, *Report of the Interdepartmental Committee of Shifting Cultivation and Soil Exhaustion in West Kwara'ae* (Honiara, June 1969); D. Kausimae, Legislative Assembly of BSIP, *Debates*, May 1975, 575–6; FAAR 1969; K. W. Trenaman, pers. comm., 10 Sept. 1994.
7. FDAR 1975.
8. FDAR 1976.
9. K. W. Trenaman, pers. comm., 10 Sept. 1994. Tractor/bulldozer logging caused more damage, 15 to 25% of the area in comparison to yarder/hauler logging damage at about 7.5% (K. Marten, pers. comm. 14 Nov. 1997).
10. I. S. Webb, *The Influence of Logging Operations on the Soils of Kolombangara* (Honiara, 1973); *The Effects of Tractor Logging on some Soil Properties and on the Growth of Tree Crops* (Honiara, 1974).
11. OF, F 139: Trenaman to Chief Sec., 7 Mar. 1964.
12. Ecological sustainability means that the structure and biotic diversity of the forest are maintained, logging disturbing the forest for only a short time, in much the same way as natural changes and processes do. Many would argue that this will result in the greatest long-term economic return from the diversity of forest 'products' (including such things as clean water, healthy reefs, etc.) to all the population. Economic sustainability means keeping timber plantations going as improve-

ments upon or compensation for the natural forest, the aim being relatively short-term and visible monetary return solely from logging, for mainly the export market.

13. FDAR 1970; K. W. Trenaman, pers. comm., 10 Sept. 1994; Tim Thorpe, 'The Effectiveness of Aid Delivery: A Comparative Study' (M. Phil., draft, Massey University, 1994), 52.

14. FDAR 1956–1969.

15. OF, F139: Trenaman to Evans, 8 Jan. 1963. Evans was a logger from the United Africa (Timber) Company, a subsidiary of Unilever, the parent company which developed from the original William Lever holdings in England and overseas. Pleydell, 'Solomon Islands Timber', 80–1; 'Solomons Operation Works for Pacific', *World Wood* 11, no. 1 (Jan. 1970), 3.

16. OF, F9/1: Dennis to Chief Sec., July(?) 1962, 2 Sept. 1964; Trenaman to Chief Sec., 30 Dec. 1970; Swabey to Trenaman, 3 Mar. 1961; Trenaman to Whitmore, 28 Mar. 1974; FDAR 1965; Chaplin and Neumann, *Afforestation*; Allan, *Solomons Safari*, 55–6. See also *Sun* (Honiara), 28 Oct. 1982, 9. *Brousonnettia papirifera* escaped from the Forestry plots, following the March 1967 cyclone (K. Marten, pers. comm., 14 Nov. 1997).

17. Whitmore, *Change with Time*, 66.

18. Whitmore, Ibid. Whitmore's assessments after 1965 were largely done by the Department (K. Marten, 14 Nov. 1997).

19. OF, F32, F2/2: Gauwane, Report of the Forestry Policy Review Committee, 14 July 1975.

20. BSIP F28/12–1: Trenaman, The Forest Estate, Technical Note, 1/62, and encl., Aug. 1962.

21. FDAR 1962, 11.

22. OF, F 139, Vol. 1: Notes of discussion at GH, 24 Nov. 1962; Trenaman to Evans, 18 May 1963; Trenaman to Chief Sec., 21 Dec. 1963, 7 Mar. 1964; Act. Chief Sec. to Sec. Ag., 8 May 1964; Proposal for settlement of land matters on Kolombangara, 11 May 1965; OF, F32: Memo. for GH meeting, and encls, Minutes of meeting, 26 Mar. 1964.

23. OF, F32: [Trenaman], Aspects of Forest Policy, enc., 26 Mar. 1964. See also FDAR 1963, p. 2.

24. OF, F32: Note of a discussion at GH, 9 Apr. 1964.

25. OF, F306: Trenaman to Financial Sec., 13 July 1970. The 1200–1500 square miles estimate of the original high-value commercial forest was reduced downwards to 800 square miles as foresters discovered more about timber stocks throughout the 1960s. In 1970, Trenaman calculated tentatively 350 square miles were needed to produce (or reproduce the original volume from 800 square miles) under line-planted new tree crops. The last time he publicly gave a target figure, in the Departmental Report of 1970, it was 500 square miles. The actual area of the Forest Reserve at the time had been over-estimated and was closer to 250 than 325 square miles. Thus for Trenaman in 1970, another 100 to 250 square miles of public land were required to achieve a sustainable commercial public forest, all else being equal. OF, F306: Trenaman to Financial Sec. 13 July 1970; FDAR 1970; FDAR 1965–1970.

26. Lembke, 'Forestry', 46, 48; 'Levers to Increase Log Exports to Australia', *The Australian Timber Journal and Building Products Merchandiser*, Sept. 1970, 29–33; FDAR 1964. See also FDAR 1967.

27. *Forestry Policy*: White Paper, BSIP, 12, 32.

28. By 1970 William Piaito, of Santa Cruz, had achieved the rank of Forester and was awarded the British Empire Medal for his services to forestry (FDAR 1954–1970). In the 1970s, Nelson was better known as David Kausimae.

29. Hadley, *Forester*; FDAR 1954–1980. On Vanikolo, Hadley set up a two-acre experimental plot planted with kauri wildings in mid-1954. It seems that this plot was damaged by fire in a later clearing operation. Trials of other plantation species, including *Swietenia macrophylla* were made in 1958 at Vanikolo (Hadley, 85–7; FDAR 1958; Graham Chaplin, *Silvicultural Manual for the Solomon Islands* (London, 1993), 165).

30. FDAR 1954–1962.

31. In FD practice, *Campnosperma brevipetiolatum* is known as *brevipetiolata*, though the former is more correct.

32. FDAR 1966–1977.

33. FDAR 1967–1970; K. D. Marten, 'Silvics of Species: Agathis macrophylla' (Honiara, 1970, cyclostyled).
34. FDAR 1966, 1968.
35. FDAR 1969, 1970; Hansell and Wall, *Land Resources*.
36. FDAR 1967; Chaplin, *Silvicultural Manual*, 11–14, 281.
37. G. R. Watt and K. W. Trenaman, *Timber Planting Programme Projection 1974 to 1985 and Log Production Cost Estimates* (Honiara, 1974), 3; Lembke, 'Forestry', 48; P. J. Wood and G. R. Watt, *An Evaluation of British Aid to Assist Forest Development in the Solomon Islands, 1965–1980* (London, 1982), 17.
38. H. C. Dawkins, *The Management of Natural Tropical High Forest with Special Reference to Uganda* (Oxford, 1958).
39. Wood and Watt, *Timber Planting*, 16.
40. Visiting experts came to the same conclusions (Gilbert, 'Report on Visit', 20).
41. K. W. Trenaman, pers. comm. 10 Oct. 1994. *Taungaya* working involves tending of crops planted between trees which are thus maintained so that weeds and dense growth do not choke the tree seedlings.
42. OF, F2/2, F32: Watt to Spears, 22 Mar. 1974; Lembke, 'Forestry', 48.
43. OF, F 139: Note of a discussion in the Act. HC's Office, 14 Apr. 1964.
44. Pleydell, 'Solomon Islands Timber', 88; BSIP F26/11/20: Trenaman, Forestry Policy and Development Programme, Apr. 1970; OF, F7/4: Visit of the Forestry Advisor, J. Whyatt-Smith, 24 Feb.–3 Mar. 1975; FDAR 1968–1976; CS.
45. FDAR 1967.
46. *Forestry Policy*.
47. Extraction rate is usually based on the annual allowable timber cut, calculated by dividing the available merchantable primary timber volume by the number of years the resource is required to last. The number of years the resource is required to last is a government policy decision. O. Wahl, *Local Timber Processing Study, Phase II, Solomon Islands*, (Rome, 1979), 42.
48. FDAR 1968; K. W. Trenaman, *Timber Development Prospects in the Solomons – Based Mainly on the Findings of a Pilot Study by FAO/ESCAFE on Timber Trends and Prospects in the Asia-Pacific Regions*, (Honiara, 1960); *Timber Development Prospects in the Solomons with Appendix of Summary of Asia and Oceania Study by FAO*, (Honiara, 1963); Lembke, 'Forestry', 46, 48.
49. Both Solomons and Australia's Papua New Guinea in this era took on board the recommendations of the World Bank or its agency, the International Bank for Reconstruction and Development for much greater utilisation of the forest resource. In the Protectorate, this became a major focus of the Sixth Development Plan. D. Lamb, *Exploiting the Tropical Rainforest: An Account of Pulpwood Logging in Papua New Guinea* (Paris, 1990), 29–38; International Bank for Reconstruction and Development, *The Economy of the Solomon Islands* (Washington, 1969), ii, 11–12; BSIP, *Sixth Development Plan, 1971–1973*,. 5,11, 13, 65–78.
50. FDAR 1968–1976; OF, F 2/3, F 306: Trenaman, Forest Policy Review Committee, General Background Paper, Jan. 1975.
51. OF, F32, F 2/2: Gauwane, Report of the Forestry Policy Review Committee, 14 July 1975, and encls; OF, F 2/3, F 306: Trenaman, Forest Policy Review Committee, General Background Paper, Jan. 1975, and encls; FDAR 1970–1976.
52. Leslie, *Forestry Economics*, 24–5. Usually, softwoods are used for pulp. However, some hardwoods, especially eucalypts can be used for certain grades of pulp.
53. K. W. Trenaman, pers. comm., 10 Oct. 1994; Leslie, *Forestry Economics*:; Watt and Trenaman, *Timber Planting*. See bibliography for Watt's research papers.
54. G. R. Watt, *Social Cost Benefit Analysis of Wood Processing Industries that could be Set up to Utilize Logs from the Replanting Programme* (Honiara, June 1974); *A Note on the Benefits which could be Derived from Harvesting and Processing of Timber from the Existing Resource, from Different Areas of Plantation, and from 130,000 Acres of Plantation on Public Land* (Honiara, 1974); Leslie, *Forestry Economics*; OF, F 7/4: Visit of Forestry Advisor, 24 Feb.–3 Mar. 1975; Wahl, *Local Timber Processing*, 6–10.

Notes to pp. 199–203

55. Wood and Watt, *Evaluation*; K. W. Trenaman, pers. comm.,10 Oct. 1994; Whitmore, *Change with Time*, 62–3; Watt and Trenaman, *Timber Planting Programme*, 3; *A review of Research Progress: 1981–1986* (Honiara, 1986), 23; Chaplin, *Silvicultural Manual*, 8–9, 84–9.

56. Cf. New Zealand forestry's marriage to *Pinus radiata*.

57. Wood and Watt, *Evaluation*.

58. For example, see *PIM*, Feb. 1969, 121–2; BSIP F26/11/20: Trenaman, Forestry Policy and Development Programme, Apr. 1970.

59. Chaplin, *Silvicultural Manual*, 7–8.

60. As early as 1946, the New Zealand military foresters had sent logs of five indigenous species to New Zealand for testing for veneer and plywood use. The species tested at the Ellis Veneer Co.'s Mananui factory were *Pometia pinnata, Vitex cofassus, Terminalia Brassii, Calophyllum* spp., *Gmelina* spp. (BSIP F28/14, Part 1: Walker to Sec. of Gov. 6 Apr. 1946).

61. FDAR, 1967, p. 10.

62. Lembke, 'Levers', 29–35.

63. FDAR 1960–1979.

64. Whitmore, *Guide*, v, vii.

65. FDAR 1963–1970; Schenk, *Forest Ecology*, 24.

66. FDAR 1969–1976; Commonwealth Forestry Association, 'Forest Policy in the British Solomon Islands Protectorate', *Commonwealth Forestry Review* 48 (1969), 190–2; K. W. Trenaman, pers. comm. 10 Oct. 1994.

67. OF, F32: Trenaman to Whyatt-Smith, 1 July 1970.

68. OF, F F32: Memo for GH meeting, and encls, Minutes of meeting, 26 Mar. 1964

69. FDAR, Summary, 1968.

70. FDAR 1965; OF, F32: Memo for GH meeting, and encls, Minutes of meeting, 26 Mar. 1964; OF, F139: Notes, GH, 24 Nov. 1962, Notes, Act. HC's Office, 14 Apr. 1964. The one exception prior and subsequent to this has been the Vanikolo regeneration work where government funds were spent on customary land.

71. OF, F1/8, 25 Vol. III: Trenaman to Spence, 26 July 1966.

72. FDAR 1967; OF, F32: Memo for GH meeting, and encls, Minutes of meeting, 26 Mar. 1964; Wood and Watt, *Evaluation*, 9.

73. As an outcome of the 1969 Ordinance, royalty was 0.75 cents per cubic foot. In March 1974, the levy was raised to 4.5 cents as the market had improved. OF, M21/1/1: Information Brief for the Min. of Trade, Industry and Labour, Forest Policy, 1974, encl.; BSIP F112/1/3: Trenaman, Forestry Information note, July 1970.

74. Wood and Watt, *Evaluation*, 10; FDAR 1976–1978. Duty was 3.5% on super-small logs.

75. OF, F32: Manedao and Hebala to Financial Sec., 5 Apr. 1972.

76. BSIP F112/1/7: Trenaman, Forestry– A Development Projection, May 1973; FDAR 1976. See also Lembke, 'Forestry', 48.

77. FDAR 1970–1977; OF, F 2/2, F32: Trenaman to Whyat-Smith, 11 July 1973; Trenaman to Financial Sec., 25 Mar. 1974; Watt to Spears, 22 Mar. 1974; Watt and Trenaman, *Timber Planting*.

78. FDAR 1978. The committee members were C. Gauwane, J. Augusta, M. Kelesi, M. E. Martin, G. Ngumi, G. Siama, and L. Wittington.

79. OF, F 2/3, F306: Report of the FPRC, 14 July 1975; FDAR 1975–1980.

80. OF, F 2/3, F306: Memo. by the Min. of Natural Resources, 23 July 1975.

81. Revised figures based on updated compartment records published in 1986 fixed the area replanted at 15,817.4 hectares (FDAR 1986). The latter figure has been accepted by Chaplin (Chaplin, *Silvicultural Manual*, 7).

82. OF, F32: Trenaman to Whyatt-Smith, 1 July 1970; Wood and Watt, *Evaluation*, 35, 54–6; H. M. Shelton, J. H. Schottler, and G. Chaplin, *Cattle Under the Trees in the Solomon Islands* (Brisbane, 1987),1–2. CDW aid to reforestation was exclusive of the general funding of other FD projects, including the

total funding of the Department itself from 1954 to 1956, to the extent of 33% to 75% of its expenditure in the years 1957 to 1965 (FDAR 1954–1965). At Independence and for a few years after, the Solomon Islands dollar was equivalent to the Australian dollar. In 1997, a Solomons dollar bought about 33 cents Australian.

83. FDAR 1975, 1976.

84. M. Bigger, *Forest Entomology in the Solomon Islands: A Report on Research carried out between 1979 and 1983*, (London, 1984), 1.

85. FDAR 1976–1978; R. Macfarlane, G. V. H. Jackson and K. D. Marten, 'Die-back of Eucalypts in the Solomon Islands', *Commonwealth Forestry Review* 55 (1975), 133–8.

86. FDAR 1978; Bigger, *Forest Entomology*, 1, 14–15; Chaplin, *Silvicultural Manual*, 109–18.

87. Bigger, *Forest Entomology*, 1–16.

88. Whitmore, *Change with Time*, 29–30, 68–9; Whitmore, *Tropical Rain Forests of the Far East* (Oxford, 1984), 233–5.

89. Lembke, 'Levers', 31, 37, 41. Super-small tree mid-diameter was no less than 28 centimetres or about 11 inches, a girth of about 2 feet 10 inches. OF: F17/1 Report on a visit by R. H. Kemp, 7–14 Oct. 1970.

90. Wood and Watt, *An Evaluation*, 6–7. Specific goals were 1976: 1,650 ha; 1977: 3,644 ha; 1978: 4,858 ha; 1979: 5,668 ha, a total of 15,790 ha. Wood and Watt, *Evaluation*, 15.

91. Watt and Trenaman, *Timber Planting*; Wood and Watt, *Evaluation*, 23. Trials showed that, since only one out of five trees would reach saw log size, the optimal spacing should be 10 metres between lines with 5 metres between seedlings, resulting in 500 stems per hectare. ODA rejected this because they did not believe the Department could maintain the supply of seedlings (K. D. Marten, pers. comm. Sept. 1994).

92. K. W. Trenaman, pers. comm., 10 Sept. 1994. Whitmore makes the nice distinction between clear-felling in the Tropics and in temperate zones. Much more standing timber is left in the normal tropical operation as it is usual that only certain valued species are wanted. However, where small sizes are acceptable, most commonly in the pulp industry, logging operations leave fewer trees and seedlings standing, although non-acceptable species are left (Whitmore, *An Introduction*, 174).

93. Wood and G. R. Watt, *Evaluation*, 36–42; Ken Marten, pers. comm. Oct. 1996; P. E. Neil, 'Climber Problems in Solomon Islands Forestry', *Commonwealth Forestry Review* 63, no. 1 (1984): 29–33.

94. Wood and Watt, *Evaluation*, 42, 55–61.

95. FDAR 1974, 1975; K. W. Trenaman, pers. comm., 10 Sept. 1994.

96. Leslie, *Forest Economics*, 22.

97. Wood and Watt, *Evaluation*, 21–3, 39–46, 61–2, 81–2. See also FDAR 1988.

98. FDAR 1980; T. Fraser and A. Larsen, *Consultants' Final Report: Evaluation of Reforestation Project, North New Georgia* (Rotorua, New Zealand, 1981),13–15; Chaplin, *Silvicultural Manual*, 13; Wood and Watt, *Evaluation*, 64–85; K. D. Marten, pers. comm., Sept. 1994. There were also some discrepancies regarding area under trees, reflected in slight variations in the replanting figures in the Annual Reports in the 1970s and 1980s.

99. Foreign service official, cited in Malcolm McKinnon, *Independence and Foreign Policy: New Zealand in the World Since 1935* (Auckland, 1993), 256.

Chapter 10

1. OF, F72: Trenaman to Swabey, 12 Mar. 1963; Devon Minchin, pers. comm., Sept. 1996.

2. Mears' company, Nasipit, was based in Manila. Ireton's holdings were Kennedy Bay, North Borneo and Associated Timber Industries, Hong Kong. OF, F72: Trenaman to Swabey, 12 Mar. 1963; OF, F139: Bim [Davies] to Tench, 30 Apr. 1963; Register of Companies, Honiara (hereinafter RCo): Kalena Timber Company, CRO No. 5 of 1964.

3. FDAR 1962–1977; Devon Minchin, pers. comm; OF, F139: Trenaman to DC Western, 4 Dec. 1962; Trenaman to Evans, 8 Jan. 1963; F72: Trenaman to Swabey, 12 Mar. 1963; OF, F192: Minchin to CFO, 22 Mar. 1963; RCo: Allardyce Lumber Co., No. 2 of 1964; Raj Kumar, *The Forest Resources of Malaysia: Their Economics and Development* (Singapore, 1986), 70–1.

4. Devon Minchin, pers. comm., Sept.–Oct. 1996; FDAR 1961; Brookfield with Hart, *Geographical Interpretation*, 196. Richardson in 1968 calculated that to conduct an economic operation in Melanesia, the minimal requirement was a resource of 60 cubic metres per hectare (86 cubic feet per acre), logged at 150,000 cubic metres (5.3 million cubic feet) per year for twenty years at a f.o.b. cost no greater than about US$9 per cubic metre (Richardson, *The Role of Forest-based Industries*, 73). With bark removed about 10 to 20% is lost, so the Isabel sample was probably on the margin of profitability until the roading problems were solved.

5. Devon Minchin, pers. comm.

6. OF, F32: Trenaman, The future of the Allardyce Lumber Company, July 1972; OF, F128: Trenaman to C. of Lands, 2 Aug. 1972; OF, F7/4: Visit of Forest Adviser, Feb.–Mar. 1975; OF, F192: Allardyce's license, 1 Oct. 1970, Proposal for a joint venture, n.d., [c.1970], Trenaman to Gen. Manager, 6 Mar. 1973; FDAR 1964–1966.

7. For this, the company charged a 5% management fee on the export value of their logs which, by agreement, was deducted from the 7.5% reforestation levy (Devon Minchin, pers. comm., 25 Nov. 1996).

8. *Solomon Star* (hereinafter *SS*), 8. Sept. 1995; Confidential sources (CS).

9. James S. Fingleton, *Assistance in the Review of Forestry Policy and Legislation* (Rome, 1989), 4–5.

10. For his views on devolution, see Solomon Mamaloni, 'Politics of Rural Development: A Political Structure for Rural development', in *The Road Out: Rural Development in the Solomon Islands*, ed. Stephen Oxenham (Suva, 1981), 79–85.

11. FDAR 1973–1978; *Debates*, Feb. 1980, 706–8; CS. Walter's father was Fred Jones, the pre-war trader; his mother was from the Reef Islands

12. FDAR Summary, 1972; FDAR 1977–1979; Devon Minchin, pers. comm.

13. *News Drum*, 2 Apr. 1982; CS.

14. CS; Kikolo, *Debates*, May 1975. See also Brookfield with Hart, *Geographical Interpretation*, 196.

15. Hviding, *Guardians*, 317–19.

16. FDAR 1967–1980; Parker, *Maekera*, 39–41; CS.

17. Bennett, *Wealth*, 125–43, 151–66, 303–9.

18. *Wood World*, Jan. 1970, 3; Lembke, 'Levers', 26–9; OF, F139: Trenaman to DC, Western, 4 Dec. 1962; Evans to FD, 13 Jan. 1963.

19. OF, F139: Evans to Morris, 12 Mar. 1963.

20. OF, F139: Trenaman to Chief Sec., 7 Mar. 1964; Brief for Thorpe, Collins and Walton, 28 Feb. 1967; OF, F209: Thorpe to Forster, 15 June 1967; Lembke, 'Levers', 25; FDAR 1970.

21. Larmour, 'North New Georgia Corporation', 135; FDAR 1981; Wahl, *Local Timber Processing*, 77–9.

22. Tony Marjoram and Sue Fleming, 'Levers in the Solomon Islands', *Raw Materials Report* 2, no. 1 (1982), 27.

23. For this and all related Acts, see Ranjit Hewagama, *The Revised Laws of Solomon Islands: Prepared under the Authority of the Revisesd Edition of the Laws Act 1995*. (Hertford, 1996, 9 vols.)

24. Larmour, 'North New Georgia Corporation', 133–44; Fraser and Larsen, *Consultants' Final Report*, 4, 7–8.

25. *Debates*, May–June 1979.

26. *Seijama v Luna*, 24 Feb. 1981, *Solomon Islands Law Reports*, 1982; *Solomons Tok Tok*, 12 Mar. 1980; *News Drum*, 20 June 1980; CS.

27. *Solomons Tok Tok*, 20 May 1981; 20 Jan. 1983.

28. *News Drum*, 2 Apr. 1982.

Notes to pp. 220–228

29. *Solomons Tok Tok*, 20 Jan. 1983; Gordon Rence, 'Land, Trustees and Timber', *Pacific Perspective* 8, no. 2 (1979), 18–19.
30. FDAR 1980, 1981; *SS*, 8 June 1993; CS.
31. *Sun*, 28 Oct. 1982; *SS*, 10, 24 Sept. 1982. Tausinga returned from studies at the University of Papua New Guinea to assist in negotiations.
32. Hviding, *Guardians*, 318.
33. *Sun*, 28 Oct. 1982; *SS*, 10, 24 Sept. 1982.
34. CS. See also, *News Drum*, 10 Apr. 1982.
35. *Solomons Tok Tok*, 17 Sept. 1984; G. B. K. Baines, 'Traditional Resource Management in the Melanesian South Pacific: A Development Dilemma', in *Common Property Resources: Ecology and Community-based Sustainable Development*, ed. Firket Berkes (London, 1989), 284; Education Unit, *The Use of Forests* (Honiara, 1984), 55, 62–9.
36. *Solomons Tok Tok*, 17 Sept. 1984; *SS*, 2 Sept. 1983,15 Mar., 12 July 1985; CS; *News Drum*, 2 Apr. 1982.
37. Paul Scobie, 'Angry Solomon Islanders want to Regain Control of their Forests', *Habitat* 10 (1982), 5–6; Paul Scobie, 'Solomon Islands Rainforest – A Chance for Conservation', *National Parks Journal* 31 (1987), 15–16; Paul Scobie, pers. comm., Mar. 1997.
38. *News Drum*, 2 Apr. 1982; *SS*, 22 Aug. 1986; CS.
39. See, for example, *Sun*, 28 Oct. 1982.
40. *SS*, 22 Aug. 1986.
41. Local processing had been a policy desideratum since 1975, but from 1980 the FD had made this a condition of both Kalena's and Allardyce's licenses. The legal status of this requirement for Lever's under the North New Georgia Act is not clear, though the Division throughout 1981–1983 urged Lever's to mill 20% of its cut and/or reforest, a move that they increasingly resisted while the market in sawn timber was depressed and, more important, the security of investment on north New Georgia customary land became more and more remote. The Division did not press the matter at the time because to have done so would probably have caused the company to leave the country. CS; Wahl, *Local Timber Processing*, 10; FDAR 1981, 1982; Fraser and Larsen, *Consultants' Final Report*, 116–127).
42. Nectoux and Kuroda, *Timber*, 34–6; FDAR, 1970–1983; Kumar, *Forest Resources of Malaysia*, 42–5; George J. Aditjondro, 'Problems of Forestry and Land Use in the Asia-Pacific Region: The Irian Jaya Experience', in *The Ethics of Development: Choices in Development Planning*, ed. Philip J. Hughes and Charmian Thirlwall (Port Moresby, 1988), 111.
43. Pleydell, 'Solomons Islands timber', 87.
44. Cited in Ombudsman's Office, *Report of the Ombudsman* (hereafter RO) *1989* (Honiara, 1989), 23.
45. Such advice was not given subsequently by provincial governments. It was not the statutory duty of the FD to offer it, unless specifically asked by landholders or loggers.
46. *RO, 1989*, 25–6.
47. The Minister also had power to regulate tree felling under Section 33 of the Forest and Timber Act, 1969. This could have been interpreted to control logging methods.
48. *RO, 1989*, 18–42.
49. *SS*, 5 July 1988. About the same time, B. K. Maurice obtained a licence to log in west Guadalcanal with a quota of 30,000 cubic metres a year (ARFD 1984–1987; *Solomon Star*, 18 Nov. 1983).
50. Hilly, Alebua, in *Debates*, May–June 1984.
51. *RO*, 33.
52. Alebua, *Debates*, May–June 1984; CS.
53. *RO, 1991* (Honiara, 1991),14. Longgu–Valesi is ward 12, east Guadalcanal.
54. CS; *SS*, 22 Oct. 1982.
55. Landholders included Bara Buchanan and Rose Dettke.

56. *Fugui and another* v. *Solmac Construction Company Ltd. and Others*, *The Solomon Islands Law Reports*, High Court of Solomon Islands, 11 Oct. 1982. Fugui originally applied for a motion for a remedy under s. 8 of the Constitution of Solomon Islands relating to the protection of people from deprivation of their property. J. C. Corrin, 'Abrogation of the Rights of Customary Land Owners by the Forest Resources and Timber Utilisation Act', *Queensland University of Technology Law Journal* 8 (1993), 139. See also *Tokotok/Sun* Oct. 1982; *SS*, 8, 22, 29 Oct., 3 Dec. 1982. Soon after, Bele resigned because of 'family reasons'; Tegavota left the Attorney-General's office 'because there was no incentive in his job' (*SS*, 5 Nov. 1982).

57. *Tokotok/Sun*, Oct. 1982. Wards 18 and 19 are known as Burianiasi and Nafinua.

58. *SS*, 8, 22, 29 Oct., 3 Dec. 1982; FDAR 1982–1985.

59. BSIP F112/1/5: Bobai to Clerk, Malaita Council, 2 July 1976; *SS*, 4 June 1982; Groome and associates, *Technical Appendices*, 115, 137; Burt, *Tradition and Christianity*, 220–1.

60. *Solomons Tok tok*, 20 Jan. 1983; *SS*, 30 Sept. 1983; CS; Groome and associates, Ibid.; Kevin Clark, 'Popular Participation in Rural Development: The New Zealand/Solomon Islands Customary Land Reforestation Project on Malaita' (M. Phil., Massey University, 1991), 123–4.

61. The Buma Sawmilling Association took over the mill from the Catholic Church in 1978, but by 1982 was fairly rundown (Groome and associates, *Technical Appendices*, 112).

62. Peter Salaka who had resigned as a director of Cape Esperance in December 1981, reappeared in late 1983, by which time he was no longer Minister of Natural Resources.

63. *RO, 1988* (Honiara, 1988), 25–6, 31; FDAR 1981–85; *SS*, 5 July 1990; CS.

64. BSIP, F441/1/15: Public meeting of Governing Council, Nov. 1972.

65. OF, F128: Trenaman to C. of Lands, 18 May 1961; *SINFRI*, 1: 42–43; RCo: Integrated Forest Products, CRO 4/1982. IFI was registered as Integrated Forest Products.

66. *Sun*, 10 Sept. 1982.

67. Kinika, *Debates*, May–June 1984.

68. CS; Kinika, *Debates*, May–June 1984; RCo: Rural Industries Ltd. CRO 31/1982; *PIM*, Aug. 1989, 61.

69. English translation: 'The form two was given out, put up, but you know for how long – not even one month! Then the provincial government wrote to the Ministry [of Natural Resources], saying everything was correct, let them come, give them the logging permit. But the agreement the people signed was not in order. Discussion started in the morning and finished in the morning, for about one or two hours, then the signing. People signed the agreement too quickly. No break, only talk, then the paper was pushed under the noses of the big-men and they were told to sign it. It went off to the province then to the Ministry which issued the licence. So many people on Makira, including me who comes from that area, knew nothing about this and were alarmed to see that the Americans had come, 65 of them. But by the time I heard of it, it was too late, they were there.'

70. RCo: CRO 4/1982; CS.

71. *SS*, 29 Oct., 5 Nov. 1982; CS.

72. FDAR 1981–1985; *SS*, 2 Dec. 1988, 3, 8; CS. Poznanski was the first woman in parliament.

73. *The Use of Forests*, 62–9.

74. CS.

75. *SS*, 2 Sept. 1983, 15 Mar., 12 July 1985; CS.

76. FDAR 1986; Burt, *Tradition and Christianity*, 221.

Chapter 11

1. FDAR 1981.

2. R. Grynberg, S. Vemuri, and D. Wigson, 'Reconciling Ethical Behavour: Transfer Pricing in the Forest Industry of Papua New Guinea', in *The Ethics of Development: Choices in Development*

Notes to pp. 236–243

Planning, ed. Philip J. Hughes and Charmian Thirlwall (Port Moresby, 1988), 178. The authors outline the various forms of transfer-pricing: under-invoicing of exports; double-invoicing (part payment received in the home country and part in a tax haven); over-invoicing of imports; debt loading (a company can load itself with so much debt from related companies that it can never make a profit in the host country); and commissions and royalties paid to parent companies, particularly by manipulating the terms of credit between companies ('thin capitalisation'). See also Price Waterhouse, *Forestry Taxation and Domestic Processing Study* ([Honiara], 1995), 38–44.

3. Until its departure from Solomons, Lever's Pacific Timbers was one of the few companies that has ever declared a significant profit and paid corporate tax. Up to 1992, Allardyce paid tax for only one or two years of its stay in Solomons (CS).
4. Wahl, *Local Timber Processing*, 10; FDAR 1981, 1982.
5. A. I. Fraser, *Fiscal Aspects of the Timber Industry: Solomon Islands* (Rotorua, New Zealand, 1980).
6. Parker, *Maekera*, 39–41.
7. CS; FDAR 1984.
8. FDAR 1983.
9. FDAR 1985; Price Waterhouse, *Forestry Taxation*, 15.
10. FDAR 1983.
11. FDAR 1984.
12. FDAR 1980–1982; Alebua, *Debates*, May–June, 1984.
13. Office of the Prime Minister (hereinafter OPM), 'A National Forest and Timber Policy', in *Forestry and Foreign Investment: Government of the Solomon Islands* (Honiara, 1984), 24–6.
14. Tropa, *Debates*, May–June 1984.
15. For The Forest and Timber (Amendment) Act 1984, and all related Acts, see Hewagama, *The Revised Laws*.
16. Fingleton, *Assistance*, 17–18; Corrin, 'Abrogation of rights', 131–3.
17. The Forest and Timber (Amendment) Bill, *Debates*, May–June 1984. See also FDAR 1984.
18. Fingleton, *Assistance*, 43–4. The decision of the CLAC is final, but a challenge to the validity of the area council meeting or timber rights meeting is to the High Court. The CLAC may also hear appeals from the local courts which have exclusive jurisdiction over customary land claims within their defined geographic area. In 1985 a chiefs' committee was set up to hear customary land disputes in the first instance; if no agreement is reached the hearing can start afresh in the local court. The main outcome of this 1985 amendment to the Lands and Titles Act 1968 has been to add another tier to the procedure. Philip Ells, *Logging and Legal Access, Solomon Islands* (draft paper, Ely, 1996), 5–6.
19. FDAR 1987. The title Commissioner of Forests covered the Chief Forestry Officer who officially became Commissioner in 1988 (FDAR 1988).
20. Fingleton, *Assistance*, 40–1.
21. FDAR 1985.
22. Kinika and Alebua, *Debates*, May–June 1984.
23. FDAR 1984–88. In 1988, the establishment for the Division was 64, but there were only 45 salaried staff in the country, with eight overseas studying.
24. CS. For example, see *Sun*, Aug. 1982.
25. *RO, 1988*, 31.
26. FDAR 1985.
27. *RO,1988*, 25–52. Gaviro carried out the duties of the Commissioner, but did not have that title until 1988. (FDAR 1980–1988)
28. CS.
29. FDAR 1986.

30. Kenilorea had been forced out of the prime ministership in the wake of allegations that he had favoured his own home area, 'Are'are, in the allocation of overseas aid and relief material, as well as accepting French government funding to rebuild his own home, following Cyclone Namu in May 1986. His deputy, Alebua, became leader of the United Party and prime minister in December (Sam Alasia, 'Politics', in *Ples Blong Iumi: Solomon Islands, the Past Four Thousand Years*, ed. Hugh Laracy (Suva, 1989),150; *SS* 19 Dec. 1986, 13).

31. OF, M21/1/1: Information Brief for the Minister of Trade, Industry and Labour: Forest Policy, n.d. [1974], encl.

32. Eagon's directors in Dec. 1983 were Youngjoo Park, Mejeong-Haeng Choi, Kyesoo Juhn, all Koreans (RCo: Cro 50/83).

33. Choiseul was then part of Western Province.

34. Cited in *RO,1990*, 13.

35. *RO, 1990*, 11–19. FD officials subsequently denied that a licence had been issued for all of Choiseul (CS). It may be that the Ombudsman confused the Form One permission to negotiate with the licence. Certainly in 1997 Eagon continued to negotiate with landholders on Choiseul for logging rights.

36. *RO, 1990*, 18–19.

37. J. R. D. Wall, *Proposals for Land Development after Logging on North New Georgia, Solomon Islands*, Project Report 99 (London, 1980); Fraser and Larsen, *Consultants' Final Report*; *SS*, 2 Sept. 1983, 12 July 1985; CS.

38. J. D Tausinga, 'Our Land, Our Choice: Development of North New Georgia', in *Independence, Dependence, Interdependence: The First 10 Years of Solomon Islands Independence*, ed. Ron Crocombe and Esau Tuza (Honiara, 1992), 57.

39. *Allardyce Lumber Company and Ors v. Attorney-General and Ors.*, August 1989, Court records, Honiara; CS. Paia had worked for Lever's as a personnel and education officer, so knew more about the legalities than most others. (CS)

40. CS.

41. Hviding, *Guardians*, 319.

42. Tausinga, 'Our Land, Our Choice', 55–62.

43. *SS*, 11 Nov. 1994, 3; Hviding, *Guardians*, 319.

44. *SS*, 23 July 1993, 4. See Tausinga, 'Our Land, Our Choice'.

45. Kenilorea, *Debates*, July–August 1989.

46. FDAR 1986.

47. FAO, *Solomon Islands Forestry Development Project-Preparation Mission, Report of the FAO/World Bank Cooperative Programme Centre* (Rome, 1989).

48. FD, *Forest Policy Statement* (Honiara,1989); Neva Wendt, 'Environmental Problems in the South Pacific: The Regional Environment Programme Perspective', in *Resources, Development and Politics in the Pacific Islands*, ed. Stephen Henningham and R. J. May (Bathurst, New South Wales, 1992), 185–94.

49. *Debates*, July–Aug. 1989.

50. *Forest Policy Statement*, 5.

51. CS; Tim Thorpe, 'Effectiveness of Aid', 57.

52. Fingleton, *Assistance; Debates*, July–Aug. 1989.

53. CS.

54. FDAR 1989.

55. *Allardyce Lumber Company and Ors v. Attorney-General and Ors.*; *Debates*, 2–16 May 1990; Forest Resources and Timber Utilisation (Amendment) Act 1990; *SS*, 16 Feb. 1990. The May amendment (no. 7 of 1990) did not address the pivotal issue of 'the relationship between customary landowners' rights and the rights of the grantors of timber rights' (Corrin, 'Abrogation of rights', 137).

56. Ells, *Logging*, 11–12.

Notes to pp. 252–259

57. CS; Forest Resources and Timber Utilisation (Amendment) Act 1990.
58. *Debates*, 2–16 May 1990.
59. RO, 1988, 1989, 1990, 1991.
60. See for example, FDAR 1989, 19.
61. *SS*, 6 Jan., 1 Mar. 1995.
62. See, for example, *SS*, 6 Oct. 1989.

Chapter 12

1. FDAR 1981.
2. FDAR 1953–1981, and revised figures of replanted areas in FDAR 1986.
3. FDAR 1984. The Department between 1965 and 1970 studied natural regeneration after logging and, given the many variables, estimated that it would be 100–150 years for the volume of the original forest to be replaced (Marten, *Forest History in the Solomon Islands*).
4. FDAR 1984–1989; R. J. Cheatle, *Productivity and Status of Logged-over Soils at Barora, North New Georgia, Solomon Islands* (Honiara, 1988); R. J. Cheatle, 'Tree Growth on a Compacted Oxisol', *Soil Tillage Research* 19 (1991), 331–4; S. Iputu, *Overview of Forestry Research in Solomon Islands in Relation to Forestry Extension* (Munda, 1992), 5.
5. FAO, *Solomon Islands Forestry Development*.
6. E. D. Shield, *Plantation Opportunity Areas in the Solomon Islands* (Honiara, 1992), 1–2.
7. These figures are from the annual reports of the division, plus material supplied from the Timber Control Unit. This represents recovered timber, logs exported, and timber sawn – it does not include wastage which could average at about 20 percent.
8. FDAR 1981–1989; FAO, *Solomon Islands Forestry Development*.
9. World Bank, *Pacific Island Economies: Towards Efficient and Sustainable Growth Volume 5: Solomon Islands: Country Economic Memorandum* (New York, 1993); Shield, *Plantation*, 46–8, Appendix 7; Ministry of Finance, *1995 Forestry Review* (Honiara, 1996), 2. The amount actually logged, as opposed to the amount declared or recorded, would add another 122,673 cubic metres to this total.
10. Shield, *Plantation*, 46–7; Chaplin, *Silvicultural Manual*, 7–23.
11. FDAR 1984–1989; Chaplin, *Silvicultural Manual*, 7–23, 41, 78, 109, 205, 207; Shield, *Plantation*, 46–8; Rosemary Kinne, pers. comm. Aug. 1996.
12. In the 1970s, the title was the Common Market or the European Economic Community. To-day, it is the European Union (EU).
13. *SS*, 23 Mar. 1990; *Solomons Voice* (hereinafter *SV*), 10 May 1996; World Bank, *Pacific Island Economies*, 38–9; Thorpe, 'The Effectiveness of Aid', 103–20.
14. Thorpe makes a similar point, see ibid., 27.
15. Bennett et al., *Forestry Project*, 25–6, 79, 95–6; Central Bank of Solomon Islands, *Annual Report* (hereinafter *CBAR*) *1990* (Honiara, 1991), 11.
16. The TFAP had been the 1980s' baby of the World Bank, FAO, the United Nations Development Programme and the World Resources Institute, the last an environmental think-tank that within a couple of years withdrew its support when many third-world NGOs argued that the plan was set to promote export-oriented timber investments, camouflaged by small projects for the environment. At least until 1991, wherever it was put in place, deforestation accelerated or was not significantly reduced from former unsustainable levels. Bruce Rich, *Mortgaging the Earth: The World Bank, Environmental Impoverishment and the Crisis of Development* (London, 1994), 160–6.
17. Bennett et al., *Forestry Project*, 19–21; Thorpe, 'Effectiveness of Aid', 26, 121–2.
18. Commonwealth Development Corporation, *CDC: The Committed Investor* (London, [1988?]), 2.

Notes to pp. 259–263

19. The Australian Development Aid Bureau did an estimate of costs and benefits, but CDC and the government considered Lever's costings more realistic.

20. FDAR 1986–1989; S. E. G. Brook, *Kolombangara Forest Products Limited Solomon Islands* (London, 1991), 1–2; Thorpe, 'Effectiveness of Aid', 87–102; *SS*, 22 Aug. 1986, 7–8; 9 Sept. 1988, 6; 4 Nov. 1988, 4; 25 Nov. 1988, 3; 13 Jan. 1989, 8; 20 Jan. 1989, 6; 28 Apr., 5 May 1989; 2 June 1989, 6; 22 Dec. 1989; 9 Mar., 1 June, 20 July, 31 Aug. 1990; 12 Dec. 1990, 9; *CBAR, 1989* (Honiara, 1990), 54; *Pacific Island Monthly*, Nov. 1990, 13; CS. Re attitude to public land, see Torben Monberg, *The Reactions of People of Bellona Island towards a Mining Project* (Copenhagen, 1976), 55–8. The government received assistance in negotiations in 1988 from the UN Centre on Transnational Corporations – another form of aid.

21. *Islands Business*, Oct. 1995, 41. A trial shipment was logged in 1993. (*SS*, 16 June 1993).

22. *SS*, 9, 23, 30 June 1989; Wayne Wooff, interview, Aug. 1992.

23. Archives, Oct. 1998, <http://pidp.ewc.hawaii.edu/PIReport/>, Internet.

24. Wayne Wooff, interview, Aug. 1992; *Islands Business*, Oct. 1995, 41; *Forestry Review 1995*, 2.

25. JOAA, led by Fumio Tanaka, was doing one of its periodic tours to try to obtain 30,000 hectares of land to plant trees for the Japanese paper pulp industry, to be financed through the Japanese aid agency JICA as had been done with exotics in the Philippines, for example. Japan offered to build a forestry training facility in 1991 as part of getting access to huge tree plantation areas, despite the fact SICHE already had one. A year later they offered a timber processing facility, but made the same proviso, so the Solomons government did not take this up (CS; Rosemary Kinne, pers. comm., Aug. 1996).

26. Wards 2, 3, 4, 36 and 37. CS; J. G. Groome and associates, *Malaita Forestry Development Study* (Suva, 1982).

27. *SS*, 7 Sept., 4 Nov. 1994; 6 Jan. 1995; 2 May 1997.

28. OF, F32: Gina to Chairman, Forestry Policy Review Committee, 3 Feb. 1975; FDAR 1979; CS.

29. CS; FDAR 1981–83; *SS*, 8 Sept. 1995.

30. Groome and associates, *Malaita Forestry*; 'A National Forest and Timber Policy', 24–26.

31. Clark, 'Popular Participation', 125–126.

32. In theory 'existing customary social values and structures' were to form the basis of implementation of the project (Clark, 'Popular Participation', 12). Land registration was hardly in this category since it invariably cuts out the rights of some community members to usufruct; all the more so, in this case where the land would be removed from the common pool for 20–30 years. The registration was carried out because the project's designers believed it could become some kind of joint venture with the Division or the government (Ibid., 205).

33. Clark, 'Popular Participation', 144–55.

34. See also I. H. I. Manner, 'Buma Village, West Kwara'ae. Malaita, the Solomon Islands', in *Agroforestry in the Pacific Islands: Systems for Sustainability*, ed. W. C. Clarke and R. R. Thaman (Tokyo, 1993), 45.

35. See also, Francis Deklin, 'Papua New Guinea Forest Policy and Local Interest Groups', in *Resources, Development and Politics in the Pacific Islands*, ed. Stephen Henningham and R. J. May (Bathurst, New South Wales, 1992), 119–28.

36. Clark, 'Popular Participation', 146; For a discussion of aspects of this, see Ben Burt, 'Land Rights and Development: Writing About Kwara'ae Tradition', *Cultural Survival* 15, no. 2 (1991), 62–4.

37. Clark, 'Popular Participation', 141, 151.

38. Ibid., 139, 141–5.

39. Ibid., 202; Tim Thorpe, *Customary Land Reforestation Project: Draft Agreement, 20 May 1992* (Honiara, 1992).

40. FD, *Forestry Policy Statement* (Honiara, 1989), 19.

41. Though project appraisal addresses women's concerns, if this is conducted by males and in the presence of other males, women's input and concerns are likely to be muted. Tim Thorpe and

Ian Frazer, *Project Appraisal System: A System for Appraising Reforestation Projects on Customary Land* (Honiara, 1992); Thorpe, *Customary Land*; Clark, 'Popular Participation', 150–2.

42. FDAR 1982–1989; Author's notes of proceedings, Forestry Extension First Annual Conference, 19–21 Aug. 1992; S. Iputu, *Seed Collection and Supply* (Munda, 1992); Iputu, *Overview*; Thorpe and Frazer, *Project Appraisal*; Thorpe, *Customary Land*; Clark, 'Popular Participation', 139; Bennett et al., *Forestry Project*, 39–41; Ministry of Finance (hereinafter MOF), *1995 Forestry Review*, 9.

43. MOF, *1995 Forestry Review*, 9.

44. Australia had assisted with testing of wood properties in the 1950s, sent an adviser to instruct operators of heavy equipment in 1967, and provided forester education at Bulolo College, Papua-New Guinea in the 1970s. *BSIP Newsletter* 5/59 (May 1959), 6/59 (June 1959); FDAR 1967, 1969, 1970.

45. Major pre-World War Two planting companies had pioneered this approach to keep *Merremia* and other undergrowth from choking the young coconut palms, which, once the fronds get beyond the animal's reach, are not damaged by the rubbing of an itchy cow. However, this applied only on 'young' plantations because 'masses of roots and the shade from the full grown coconut trees greatly retards all other vegetable growth', as W. H. Lucas had noted in 1917. Aid agencies would do well to study a little history. BPA: Lucas, Memo to Director, 2 Nov. 1917.

46. Rosemary Kinne, pers comm., Aug. 1996.

47. FDAR 1980; FD, Ministry of Natural Resources. *A Review of Research Progress: 1981–1986* (Honiara, 1986), 4–5; Shelton, Schottler, Chaplin, *Cattle Under the Trees*. Lever's in the pre-war period had studied the effect of cattle stocking on soil compaction – yet no account was taken of this.

48. The Asian Development Bank conducted a feasibility study for Isabel in 1988, but framed support as loan-based, rather than by grant (Thorpe, 'Effectiveness of Aid', 52).

49. The Australian International Development Assistance Bureau (AIDAB), formerly ADAB, currently AUSAID.

50. Thorpe, 'Effectiveness of Aid', 55. The major Australian political parties formulated an environmental agenda about this time. The Australian Labour Party (ALP) adopted a policy for conservation of the tropical forests of Pacific Islands and of the Association of Asian Nations (ASEAN) as part of its platform as early as 1984. The ALP remained in power from then until 1996. AIDAB, *Changing Aid for a Changing World: Key Issues for Australia's Overseas Aid Program on the 1990s* (Canberra, 1992), 1; Elim Papadakis, *Environmental Politics and Institutional Change* (Cambridge, 1996), 96–118; Australian Government, *Australia's Overseas Aid Program, 1995/96* (Canberra, 1995),15; Australian Government, *General Audit Report number 13: AIDAB Aid to Papua New Guinea 1990–91* (Canberra, 1990), 44.

51. AIDAB, 1992, cited in Thorpe, 'Effectiveness of Aid', 53.

52. Thorpe, 'Effectiveness of Aid', 46–57; CS.

53. The consultants were all Australia-based: ACIL Australia Ltd, International Forest Environment Research and Management, and Environment and Resources Spatial Information Systems.

54. CS; Thorpe, 'Effectiveness of Aid', 58–60.

55. *SINFRI*, vol. 1, 2.

56. Thorpe, 'Effectiveness of Aid', 67–9, 73. See for list of publications, *SINFRI*, vol. 1, appendix 1.

57. Thorpe, 'Effectiveness of Aid', 77.

58. Ibid., 75–6. SIG means Solomon Islands Government.

59. For example, see A. Constantini, *Timber Control Project Solomon Islands: Forest Inspectorate Ranger Training Course* (Honiara, 1992); Bill Kranenburg, *Grading Solomon Islands Timber: Proposed Grading Rules* (Honiara, 1992).

60. *SS*, 11 Mar., 1, 8, 27 July 1994.

61. *SV*, 1 Apr. 1993; *SS*, 11 Mar. 1994. The consultants Forestry Technical Services (FORTECH) ran the project from May 1991, under the control of Taylor.

62. Thorpe, 'Effectiveness of Aid', 80–1. The author attended some of these courses and found them interesting for a lay-person, but they seemed somewhat superficial for anyone in the industry. Seminar on saw-milling for Forest Rangers, Speaker Alan Cameron, 27, 28 August 1992; Timber

Grading Seminar, Speaker Bill Kranenburg, Honiara, 1, 2 Sept. 1992. See also, Kranenburg, *Grading Solomon Islands.*

63. Samson Gaviro, interview, 9 Nov. 1992.

64. See for example, *SS*, 3 Mar. 1989. Australia also has contributed to education though its funding of curriculum development. The form two social studies text, *Our Forests*, published in draft form in 1983, raises issues related to logging and the ecological significance of the forests.

65. Before 1993 the intakes of between 16 and 18 had mainly been for the Division with two to four from KFPL. The 1993 intake consisted of eight sponsored by Kolombangara chiefs, five by Kolombangara Landowners' Trust, four by Allardyce, three from KFPL, two from Pacific Conservation and Development Trust, two from Nature Conservancy, two from World Wildlife Fund. Of the 14 logging companies approached for sponsorship for 1993, only three responded. Another measure of the logging industry's commitment to the future of Solomons' forests! Figures for 1993 kindly supplied by Rosemary Kinne, pers. comm., Aug. 1996; S. Vasu et al. 'Selection and Performance of Preservice and Inservice Candidates in Forestry', *Pacific-Asian Education* 5, no. 2 (1993), 67–72

66. Industry representatives were from Lever's, H. Paia in 1985, then Allardyce's C. Delaney, 1985–1987, later KFPL's W. Wooff from 1990. The environment and the NGO representatives were G. Dennis, 1985–1988, A. Baenesia from 1990, and B. Tele from 1991, representing the Environment Division of the Ministry of Natural Resources (Rosemary Kinne, pers. comm. Aug. 1996).

67. Rosemary Kinne, interview, 1992.

68. Bennett et al., *Forestry Project,* 15–16; FDAR 1981–1984.

69. The first three women with tertiary qualifications from SICHE graduated in late 1997 (*SS*, 5 Dec. 1997).

70. Rosemary Kinne, interviews 1992; Solomon Islands College of Higher Education, *School of Natural Resources, Handbook, 1992* (Honiara, 1992). For an example of one graduate's work see, J. Anderson Ifui, *Tekolea Nature Trail on the Poitete Reserve Forest Kolombangara* (Poitete, 1990).

71. Myknee Qusa Sirikolo, *The Honiara Botanical Gardens* (Honiara, 1992); Myknee Qusa Sirikolo, interview, 12 May 1992; *SS*, 31 Mar. 1995; ARFD 1966, 1967, 1987, 1989.

72. [Tania Leary], *A National Environmental Management Strategy for Solomon Islands*, draft plan, ([Honiara], c. 1992); *SS*, 2 July 1993.

73. Bennett et al., *Forestry Project,* 51; *SS*, 9 Nov. 1990; Rosemary Kinne, pers. comm. Aug. 1996.

74. Sosimo Kuki and Tim Thorpe, *Chainsaw Sawmilling* (Honiara, 1992); Thorpe, 'Effectiveness of Aid', 126–35; *SS*, 6 July 1995.

75. John Roughan, Abraham Baeanesia, interviews, Honiara, 1992.

76. Cited in Bennett et al., *Forestry Project,* 56.

77. Solomon Islands Development Trust, *Summary Reports, 1990, 1991* (Honiara, 1990, 1991); *SS*, 8 June 1990; John Roughan, Abraham Baeanesia, interviews, 1992; *Link* 23, Nov./Dec. 1991; no. 25, May/June 1992; Thorpe, 'Effectiveness of Aid', 136–51; *SV,* 9 Feb. 1996.

78. John Roughan, 'Solomon Islands Nongovernmental Organizations: Major Environment Actors', *Contemporary Pacific* 9, no. 1(Spring 1997), 157–66; *SINFRI*, vol. 1, 13.

79. CS; *SS*, 9 Nov. 1990, 26 July 1995; Bennett et al., *Forestry Project,* 55–6.

80. *SS*, 26 Nov. 1993; See also *SS*, 25 Jan. 1995.

81. *Link*, no. 25, May/June, 1992; *SS*, 26 Nov. 1993; 21 Dec. 1994; 12 May 1995, 27, 29 Nov. 1996; *Sol-tree Nius*, Feb., Apr. 1996; ADFD 1989; Ezekiel Alebua, interview, Honiara, 22 Oct. 1992.

82. *SS*, 21 Oct., 7 Dec. 1994; Bennett et al., *Forestry Project,* 54.

83. *SS*, 12 Oct. 1990, 13 June 1995; G. B. K. Baines, *Timber Control Unit Project Solomon Islands: Training for Forestry Extension services for Customary Landholders in Solomon Islands* (Canberra, 1992), 11.

84. *SS*, 23 Aug. 1995.

85. Pratap Chatterjee and Matthias Finger, *The Earth Brokers: Power, Politics and World Development* (London and New York, 1994), 71.

86. Kumpulan Emas Berhad, Annual *Report, Kumpulan Emas Berhad, 1995* (Kuala Lumpur, 1995), 14.
87. Honiara Municipal Authority, *Baonsa Book* (Honiara, 1990); *SS*, 19 Nov. 1993; Sheila Macbride-Stewart, 'Catholic Women's programmes in Malaita: Breaking the culture of silence through empowerment' (M. Phil. thesis, Massey University), 56–8, 75.
88. *SS*, 3 Feb. 1995.
89. Bennett et al., *Forestry Project*, 43; *SS*, 13 Oct. 1989; 2 Aug. 1995; Rosemary Kinne, pers. comm., Aug. 1996. This kind of fertilising and soil stabilisation is well established overseas and was advocated by René Dumont for Melanesia in the early 1970s. René Dumont, 'Some Reflections on Priorities in Melanesian Development', in *Priorities in Melanesian Development*, ed. Ronald J. May (Canberra, 1974), 15.
90. Thorpe, 'Effectiveness of Aid', xi–xvii.
91. Diplomatic jargon makers now label aid as 'development assistance', a term as misleading as its predecessor.
92. Chatterjee and Finger, *The Earth Brokers*, 70–1, 148–89; Rich, *Mortgaging the Earth*; Paul J. Nelson, *The World Bank and Non-Governmental Organizations; The Limits of Apolitical Development* (New York, 1995).
93. Thorpe, 'Effectiveness of Aid', 123–4.
94. Grynberg maintained that Duncan's economic analysis of downstream process and bans on export of logs entirely missed the *Realpolitik* prevailing in Papua New Guinea and Solomon Islands (Roman Grynberg, *SS*, 23 June 1995). See R. C. Duncan, *Melanesian Forestry Sector Study* (Canberra, 1995).
95. Regina Scheyvens, '"Engendering" Environmental Projects: The Case of Eco-timber Production in the Solomon Islands', *Development and Practice* 8, no. 4 (Nov. 1998), 439–53.
96. See also A. V. Hughes, *Climbing the Down Escalator: The Economic Conditions and Prospects of Solomon Islands* (Canberra, 1988), 9.
97. Australian Government, *Report of the Committee to Review Australian Overseas Aid Program* (Canberra, 1984), 167; *PIM*, Jan. 1996, 28–9, May 1996, 37–8; *Islands Business*, Jan. 1996, 12–13; *SS*, 25 Aug., 8 Sept. 1993. See also *CBAR, 1995* (Honiara, 1996), 12–13.
98. Bruce Knapman, 'Aid and the Dependent Development of Pacific Island States', *Journal of Pacific History* 21, no. 3 (1986), 147–8; Steve Hoadley, *The South Pacific Foreign Affairs Handbook* (Sydney, 1992), 29.
99. *SS*, 3, 8 Sept. 1993.

Chapter 13

1. *SS*, 18 Aug. 1989; *CBAR, 1990,* 31. The limitations on the provinces' powers are set in the Provincial Government Act 1981, Section 33. Re devolution, see Mamaloni, 'Politics of Rural Development', 79–85. After the establishment of the Provincial Development Unit (PDU) in the mid-1980s, the central government could co-ordinate various aid grants along with Stabex funds to small business projects, usually in the form of loans. These were allocated to people in remote areas, but the national unit administered them (*CBAR, 1991* (Honiara, 1992), 39).
2. *SS*, 5, 9 Apr. 1995, 29 Nov. 1996; *Katibas News*, July 1996.
3. Most of this has been facilitated by Development Bank loans. John Lauder, 'Communal Farm Development in the Solomon Islands, *Pacific Economic Bulletin* 2, no. 1 (1987), 11–16.
4. *Solomons TokTok*, 12–17 Feb., 1990; *SS*, 16 Feb. 1990; 17 Jan. 1997.
5. *SS*, 24 Mar. 1995.
6. CS; FDAR 1977–1989; Bennett et al., *Forestry Project*, 14, 25, 66–7.
7. *SS*, 15 Dec. 1993, 5; 31 Aug. 1994, 10–11.

Notes to pp. 280–287

8. *SS*, 19 Oct., 25 Nov. 1994, 27 Jan., 24 Mar. 1995; Price Waterhouse, *Forestry Taxation*, 14.

9. *SS*, 16 Aug. 1995. See also *SS*, 17 Mar. 1995 for Guadalcanal; Price Waterhouse, *Forestry Taxation*, 77.

10. Bennett et al., *Forestry Project*, 45–6.

11. *SS*, 24 Feb. 1995; Development of the Forestry Sector and Related Industries/Resources, Policy Paper no: 4/94, Policy Evaluation Unit, Office of the Prime Minister, 19 Dec. 1994, copy in author's possession.

12. *SS*, 6 Jan. 1995.

13. John Collee, 'Stripping the Pacific Island Rain Forests', *San Francisco Examiner*, 1 Jan. 1995, A14–A15; SWIFT – Netherlands foundation, *Solomon Western Islands Fair Trade Newsletter* 1 (June 1996), 1.

14. CS; John Beverley, interview, 1992: FDAR 1982–1986.(Deficits in any year could be carried over to the following year's quota, it would seem.)

15. CS; Bennett et al., *Forestry Project*, 73–4.

16. Edward Milner, 'The Chainsaw Massacre', *Post Courier, Weekend Magazine*, Port Moresby, 13 May 1994, 33.

17. Justice Palmer also made the point that the customary land claimants who did not want to dispose of their timber rights were not necessarily the true customary landholders (Ells, *Logging and Legal Access*, 11).

18. CS; *News Drum*, 13 Nov. 1981; *SS*, 23 June 1989; *Link*, May/June 1992; *New Zealand Herald*, 16 Oct. 1993; *Post-Courier*, Weekend Magazine, 13 May 1994; *Katibas News*, July 1996.

19. CS. The association benefited from the advice of an experienced accountant, Alex Painter, and his wife Patricia, who came from Australia under a volunteer scheme for six months (*SS*, 16 Feb. 1990; Parker, *Maekera*, 41).

20. Eric Havea, interview, 1 Aug. 1992; Baines, *Timber Control Unit*, 11.

21. *SS*, 7 Feb. 1986, 3 Mar. 1989; 19 Jan. 1994.

22. Bennett et al., *Forestry Project*, 53.

23. Leslie Boseto, interview, July 1992.

24. The Forest Certification Council has teeth, and declared a supposedly sustainable logging operation conducted by Isoroy/Leroy Gabon in Gabon to be a lie. <http://forests.org/recent/agagabon.txt>: Danielle Knight, Environment Gabon, InterPress Service, 18 Nov. 1997.

25. *SS*, 15 July 1994, 29 Nov. 1996; Martin Bennett et al., *Forestry Project*, 53–4; Leslie Boseto, interview, 1992; Alan Leadley, interview, Nov. 1994; Patagavara to Lewis, 4 Oct. 1994 and encls, supplied by A. Leadley; *Islands Business*, Aug. 1995, 6; Scheyvens, '"Engendering"'.

26. Rosemary Kinne, pers comm., Aug. 1996. Of course, despite KFPL's best efforts to retain as much biomass as possible after felling by not burning off 'waste' cuttings, the company's monoculture of an exotic cannot replace the varied forest species ecology on its plantations.

27. *SS*, 26 May 1995. These lands had been purchased from the estate of a planter in the 1960s, then held by the United Church, which returned them to the Pezoporo clans in the 1970s (LTO: 063–002–1, LR 197/1, 197/2, 197/3).

28. *Katibas News*, July 1996. See also *SV*, 25 July 1997. For an examination of the Vella Lavella people's attitude, see John M. McKinnon, *Bilua Report: Preliminary Results of Research into Rural Economic Development, Vella Lavella, BSIP* (Wellington, 1973).

29. *SS*, 11, 29 Mar., 6, 22, 27 Apr., 6, 11 May, 3 Aug. 1994, 7 Apr. 1995; Bennett et al., *Forestry Project*, 67. Mamaloni justified the creation of a para-military force in 1991, to patrol the volatile Bougainville-Solomons border.

30. Annette Lees, *A Representative Protected Forests System for the Solomon Islands: Report* (Honiara, 1991); *New Zealand Herald*, 30 May 1992; Edvard Hviding and Graham B. K. Baines, 'Community-based Fisheries Management, Tradition and Challenges of Development in Marovo, Solomon Islands', *Development and Change* 25 (1994), 27–9.

31. M. F. Olsen and M. H. Turnbull, *Assessment of Growth Rates of Logged and Unlogged Forests in Solomon Islands* (Honiara, 1993).
32. Cited in Forest Monitor Ltd, *Kumpulan Emas Berhad and its Involvement in the Solomon Islands* (Ely, UK, 1996), 7. See also, Deborah Snow and John Collee, 'Rape of an Island Paradise', *The Observer*, 29 Sept. 1996, 4–5
33. *SS*, 3, 5, 10, 12, 19 Aug. 1994, 28 Oct. 1994; *The Fiji Times*, 13 Aug. 1994.
34. Hviding, *Guardians*, 342.
35. Snow and Collee, 'Rape of an Island Paradise', 4. Cf. Hviding, *Guardians*, 343 ff.
36. *SS*, 9 Apr., 27 May 1994; Aldrick, *The Susceptibility of Lands*, 15; 'Tetepare under Threat', discussion paper for the Western environment summit, June 1994, author's copy; *Tol Blong Pasifik* 49, no. 2 (June 1995), 24. The site includes the government land the British refused to allow to be logged because of its steepness and shallow soils. (OF, F128: C. of Lands to Chief Sec., 24 June 1960; Trenaman to Chief Sec., 5 July 1960; Trenaman to C. of Lands, 23 May 1961; F107: Note of a meeting at GH, 5 Mar. 1963).
37. Wardlow Friesen, 'Melanesian Economy on the Periphery: Migration and Village Economy in Choiseul', *Pacific Viewpoint* 34, 2 (1993), 193–214.
38. *SS*, 23 Feb. 1990. Chirovanga, Mole and Subesube landholders subsequently (in October 1990) formed a Peoples Union to represent then on logging matters, with Kengava as adviser (*SS*, 19 Oct. 1990).
39. Michael Lomiri, interview, 8 Aug. 1992. It was during the Western Province environment week, 5–9 Dec. 1990 that Kengava learned that Eagon contracts paid less per log than any other logging company (Rosemary Kinne, pers. comm. Aug. 1996).
40. *SS*, 7 Sept. 1994. See also *SS*, 1 Feb. 1995.
41. Michael Lomiri.
42. *SS*, 7 Sept., 4 Nov. 1994; 6 Jan. 1995; *SINFRI* (Honiara, 1995), vol. 4, 42.
43. *SS*, 1 Mar. 1995.
44. *SV*, 3 Nov. 1995; *SS*, 4 Feb., 11 Mar. 1994.
45. *SS*, 19 Apr. 1995.
46. *SS*, 17 Nov. 1995, Apr. 1996; See also 19 July 1996
47. Alex Vickers, interview, 1992. Fr Vickers was based on Nila in Shortland (Alu) in 1997.
48. To 1997, no indigenous Catholic priest has presented himself as a candidate for bishop, though there are some experienced men working in parishes.
49. CS; Macbride-Stewart, 'Catholic Women's Programmes'. The United Church has been criticised for its lack of involvement of women in its SWIFT enterprise, except as unpaid labour in the family unit (Scheyvens, '"Engendering"').
50. The company lacked the finance to pursue its goals (WPHC CF 164/2/9: Dennis to Chief Sec., 6 June 1959).
51. Jejevo Communique 1985, copy in author's possession.
52. Shell Film Unit, *Fate of the Forests* (United Kingdom, 1982).
53. Provincial secretaries were public servants, paid by the central government (*SS*, 12 Jan. 1990).
54. *SS*, 2 Dec. 1988, 3, 8; CS. Boyers was one of the principals of Earthmovers Solomons Ltd, a civil engineering and logging company which had first started logging on Kohinggo in 1983. He appears to have been seeking Isabel logs to supply the mill at Pacific Timbers which concentrated on sawn timber from 1987/8. The Symes properties had proved their worth, producing over 9,000 cubic metres in log exports in 1985/6.
55. CS; *Solomons Tok Tok*, 12–17 Feb. 1990; *SS*, 31 Mar. 1989.
56. Rural Industries had in 1987 been used by Monarch Leasing Pty Ltd of Singapore as a front to get logging rights on Makira, to avoid the conditions of the new standard logging agreement. *RO, 1991* (Honiara, 1991), 14.

57. CS. Dodges such as transfer-pricing could have meant initial investment would never be recouped by the company.
58. For example, see *SS*, 30 Mar. 1990.
59. CS; *Link*, Nov./Dec. 1991. See also Forest Monitor Ltd, *Environmental and Social Impact Assessment: ITC Operations on Isabel Island, Solomon Islands* (Ely, 1997).
60. *SS*, 25 May, 24 June, 16 Sept. 1994; Ziza, 'Logging in Kokota' See also Forest Monitor Ltd, *Environmental and Social Impact*.
61. Kumpulan Emas Berhad, *Annual Report*, 13.
62. *The Borneo Post*, 21 Feb. 1994, Ecological Enterprises, 23 Feb. 1994, Gaia Forest Conservation Archives, Internet; *Sol-Tree Nius*, Feb. 1996.
63. *SINFRI* (Honiara, 1995) 5: 63. Axiom Forest Products, incorporated in the Virgin Islands at the time, controlled Silvania Products, Integrated Forest Industries, Rural Industries, and Isabel Timber Company in 1991. Axiom Forest Products was a wholly-owned subsidiary of Axiom Forest Resources Ltd, incorporated in Hong Kong. In September 1992, China Industrial International Corporation, incorporated in Bermuda, replaced Axiom Forest Resources. The former became China International Forest Products and was the major holding company of Axiom Forest Products. The holding company was sold to Lee Kuan Mang, a major shareholder in the Malaysian company, Mint Horizon Sdn Bhd. In October 1993 Kumpulan Emas, a Malaysian company with wide business interests bought China International Forest Products from Mint Horizon and became Emas Pacific, controlling the four logging companies in Solomons. (Forest Monitor Ltd, *Kumpulan Emas Berhad*).
64. CS; *SINFRI*, 5: 2, 15–17; RCo: Eastern Development Enterprises Ltd. CRO 10/1987; *SS*, 18 Feb., 2 Mar., 25 May, 24 June, 16, 23 Sept. 1994; 5 May, 31 Mar. 1995.

Chapter 14

1. *SS*, 18 Nov. 1983, 3; FDAR 1985.
2. CS. The Taiwanese bought into another company, Pan Ocean Forest Development which applied for a licence to log East Are'are and other areas on Malaita, but seems to have lost momentum as a result of the moratorium and the damage inflicted on the forest by cyclone Namu in 1986.
3. *RO, 1990, 1991*.
4. CS; *RO, 1989*, 61. The Annual Reports of Forestry Division do not appear to state the quota for Kayuken's Malaita operation, only for west Guadalcanal (FDAR 1986–1989); Groome and associates, *Malaita Forestry*, 24.
5. Alan L. Cameron, *Primary Processing of Logs and Timber Handling in the Solomon Islands, Timber Control Unit*, vol. 1 (Canberra, May 1992), i.
6. CS; Groome and associates, *Malaita Forestry*, 13–16; *SS*, 19 Aug. 1994, 8; Bill Kranenberg, *Grading and Training Report, Timber Control Unit Project* (Canberra, 1992); Cameron, *Primary Processing*; Alan L. Cameron, *Training Needs in the Sawmilling Industry in the Solomon Islands, Timber Control Unit* (Canberra, 1992); Bennett et al., *Forestry Project*, 108–10; *SS*, 6 July 1995; Coopers and Lybrand, *A Report on Timber Processing in the Solomon Islands and Strategies to Achieve a Profitable Industry through Sustainable Production Levels* (Honiara, 1993); FDAR 1982–1989.
7. FDAR, 1989; *Solomons Tok Tok*, 12–17 Feb. 1990.
8. See for example, *SS*, 23, 29 Mar. 1994.
9. CS; *The Use of the Forests*, 74–5; Bennett et al., *Forestry Project*, 108–10.
10. CS; *SS*, 23 May 1997.
11. CS; Roughan, 'Solomon Islands'; *SS*, 31 Mar. 1995, 17 Jan. 1997; RCo: Kayuken Pacific Ltd., CRO 50/1982, Waibona sawmilling and logging Co., CRO 59/1987.
12. CS; *Sol-Trust Nuis*, Feb. 1996; *SS*, 4 Nov. 1996; 15, 17 Jan. 1997.

Notes to pp. 304–314

13. CS; *Link*. Nov./Dec. 1991; Tim Thorpe, *Experience with Portable Sawmills in the Solomon Islands* (Honiara, 1992). Portable saw-mills had become very popular, particularly on Malaita, in the wake of cyclone Namu in 1986 as several relief aid agencies offered them to assist rebuilding, using the wind-felled trees.
14. *SV*, 18 June 1993; *SS*, 18 June 1993.
15. *SS*, 19 Aug. 1994.
16. This preceded the High Court ruling of May 1994 re legality of such taxes.
17. CS; *SS*, 19 Jan., 2 Feb., 14 June 1990. By 1996, Freeman's notorious business scams were well known in the Pacific (*Islands Business*, Mar. 1996, 96).
18. *RO, 1989*, 51.
19. CS; Bill Abana, 'Logging is no Development', *O'o: a Journal of Solomon Islands Studies* 2, no. 1 (1989), 62; *RO, 1989, 1991*; *SS*, 9 Sept. 1992. Note that Bill Abana, an agricultural officer, was killed in the same crash described earlier. He was due to take up a position on the council of SICHE and was very aware of the environmental damage done by logging (CS).
20. CS; *RO, 1991*, 14.
21. CS; FDAR 1987. Both Integrated Forest Industries and Rural Industries passed into ownership of Axiom in 1991 (see n. 63 in chap. 13).
22. *PIM*, August, 1989.
23. CS.
24. 600,000 hectares is over 20% of the land area of Solomons.
25. *SS*, 23 Mar., 22, 27 Apr., 15, 24 June, 22, 27 July, 10 Aug., 23 Dec. 1994. In 1984, the Division had planned phasing out round log exports – to be completed by 1995! OPM, 'A National Forest and Timber Policy', *Forestry and Foreign Investment: Government of the Solomon Islands* (Honiara, 1984), 24–6.
26. *SS*, 7 June 1994, 4; 15 June 1994, 4. See also 2 Feb., 14 June, 1990; 29 July, 5 Aug. 1994.
27. *SS*, Samson Oloni, letter to editor, 24 June 1994, 4.
28. *SS*, 27 July 1994, 13; 3 Aug. 1994, 14; 12 Aug. 1994, 5; *New Straits Times*, 8 Nov. 1994. Alebua appears to have been out of favour because he did not fully support the speed with which Hilly was introducing new forest policy.
29. Victor Totu, interview, 16 Sept. 1992.
30. *Solomons Tok Tok*, 24 July, 28 Aug. 1986; Rosemary Kinne, pers. comm., Aug. 1996.
31. *SS*, 25 May 1994.
32. *Solomons Tok Tok*, 20 Jan. 1989; Tim Thorpe, 'Effectiveness of Aid', 128; RCo: Dalsol Ltd. Cro 48/1983; *PIM*, Aug. *1989*, 51; *SS*, 25 May 1994.
33. Dalsol bought out B. K. Maurice, on Guadalcanal, apparently escaping thereby the effect of the 1983–1987 moratorium. ARFD 1987–1989; *Link*, Nov./Dec. 1991, 12.
34. *SINFRI*, vol. 2 (Honiara, 1995); E. Alebua, interview, 27 April 1992; *SS*, 30 June, 18 Aug. 1995.
35. *SS*, 25 Oct., 8, 29 Nov. 1995; *Sol-tree Nius*, Feb. 1996. Berjaya took this in its stride, finding a 'front' company in Suriname in 1996 in order to log up to a million hectares. <http://forests.org/recent/direcons.txt>, 18 Oct. 1996; <http://forests.org/recent/sarsayno.txt:>, 5 Mar. 1998, Gaia Forest Conservation Archives, Internet.
36. *SS*, 21 May 1993, 24 Mar. 1995.
37. *SS*, 19 Nov. 1993; *Pacific Island Monthly*, Feb. 1994.
38. *SS*, 21 May, 6 Aug. 1993; 13 Jan. 1995.
39. *SS*, 8 Sept. 1993.
40. *SS*, 5 Nov. 1993, 13 Jan., 8, 15, 17 Feb., 15, 24, 31 Mar., 5 Apr., 26 May, 1995; <http://forests.org/recent/bbriefin.txt>: Unofficial transcript of Background briefing, Radio National, Australian Broadcasting Corporation, 14 Jan. 1996, from Ecological Enterprises, 27 Jan. 1996, Gaia Forest Conservation Archives, Internet.
41. *SS*, 24 May 1995. Mamaloni had been suspicious of Greenpeace for some years (*SS*, 4 Aug. 1989).

42. *SS*, 24, 26, 31 May, 7, 23 June, 6 July 1995; *Pacific News Bulletin*, June 1995, p. 7; <http://forests.org/recent95.html>: Mar.–Dec. 1995, Gaia Forest Conservation Archives, Internet.

43. *SS*, 2, 4 Aug. 1995; New Zealand National Radio. Asia and Pacific News, May, June, July 1995, Jan. 1996; Development Studies Centre, *Te Amokura, Na Lawedua* 7, no. 2 (May–Aug. 1995), 6. Apa's death was not reported in the Solomons media until five weeks after it happened. (<http://forests.org/recent/bbriefin.txt>: Unofficial transcript of Background briefing, Radio National, Australian Broadcasting Corporation, 14 Jan. 1996.)

44. *Sol-tree Nius*, Feb. 1996. The Mamaloni government's promises of development also came to nothing (*SS*, 11 June 1997).

45. *SINFRI, 9* (Honiara, 1995).

46. *CS*; *SS*, 5 May, 15 June, 6, 27 Oct. 1995. See also *SS*, 10 Nov. 1989; *SV*, 19 Apr. 1996.

47. *SS*, Mar. 1989.

48. T. E. Barnett, 'Legal and Administrative Problems of Forestry in Papua New Guinea', in *Resources, Development and Politics in the Pacific Islands*, ed. Stephen Henningham and R. J. May (Bathurst, Australia, 1992), 94.

49. Ibid., 90.

50. OPM, Development of the Forestry Sector and related industries, 19 Dec. 1994., PP No. 4/94, photocopy.

51. *SS*, 24 Feb. 1995.

Chapter 15

1. World Bank, *Pacific Island Economies*, iii.

2. Ian Frazer, 'Resource Extraction and the Post-colonial State in Solomon Islands', in *Asia Pacific: New Geographies of the Pacific Rim*, ed. R. F. Watters and T. G. McGee (Vancouver, 1997), 318–34; *CBAR, 1995*, 27; *CBAR, 1996*, 29.

3. *CBAR, 1989–1996*.

4. Domestic debt excludes government debt to local businesses. *CBAR, 1991*, 33; *CBAR*, 1996, 47; <http://www.boh.com/econ/pacific/solaer.asp>: Bank of Hawaii, 'An economic assessment of the Solomon Islands' (Honolulu, 1994), Internet.

5. J. Fallon and C. Karabalis, 'Current Economic Trends in the South Pacific', *Pacific Economic Bulletin* 7, no. 2 (1992),10; Bank of Hawaii, 'An economic assessment'; *CBAR, 1996*, 33, 47.

6. *CBAR,1989–1998*; Australian Government, *The Solomon Islands Economy: Prospects for Stabilisation and Sustainable Growth* (Canberra, 1991); Bank of Hawaii, 'An Economic Assessment'; World Bank, *Pacific Islands Economies* ; Bennett et al., *Forestry Project*, 60.

7. Lauder, 'Communal Farm', 11.

8. Frazer, 'Resource Extraction'.

9. The statistical data for 1996 are not as accurate as former years (*CBAR 1996*, 16).

10. Frazer, 'Resource Extraction'; *CBAR, 1995–1996*.

11. *CBAR, 1989–1996*; Australian Government, *The Solomon Islands* ; World Bank, *Pacific Islands Economies*; South Pacific Commission, *Pacific Islands Population, Report* (Noumea, 1994), 64; *SS*, 19 May 1993; Ministry of Finance, *1995 Forestry Review*; Price Waterhouse *Forestry Taxation*, 12, 38–44.

12. *CBAR, 1991*, 8.

13. *CBAR, 1989–1991*. Real GDP is a measure of output volume and is an indicator of economic activity. This adjusts current values for production and employment back to a base year to show changes in volume. Nominal GDP is at current prices, the money values obtaining in the particular year of account.

14. Bennett, '"We do not come here"', 129–87.

15. W. J. Badcock, 'Agriculture in the British Solomon Islands Protectorate', *Agricultural Journal of Fiji* 17, no. 3 (1946), 63–70.
16. Walker, *The Forests*, 18.
17. OF, F206, F2/3: Trenaman, Forestry Policy Review Committee: General background Paper, Jan. 1975. See also OF, M 21/1/1: Information Brief for the Minister of Trade, Industry and Labour: Forest Policy, n.d. [1974], encl.
18. OF: For 2/3, 306, Vol. III: Forest policy review committee, n.d. [c. May 1978].
19. Groome and associates, *Malaita Forestry*.
20. Their *Guide to the Useful Plants of the Solomon Islands*, by C. P. Henderson and I. R. Hancock, was indicative of an appreciation of indigenous knowledge within a farming-systems approach. The United Kingdom provided the finance for this book through its ODA.
21. CS.
22. FD *Forest Policy Statement*, 6–8.
23. Ross M. Cassells details his methods and possible limitations in 'Valuation'. His calculations err on the side of conservatism as he did not place a monetary equivalent on water and soil resources, *tambu* sites and recreation. Water supplies and fisheries are commonly damaged by logging in Solomon Islands (Penelope Schoeffel, Gerard Fitzgerald and Alison Loveridge, *Forest Utilization in the Solomon Islands: Social Aspects and Issues* (Honiara, 1994), 83, 88–9).
24. Cassells, 'Valuation', 108–70.
25. Economic rent is the difference between the market value of the round logs and the investment cost in harvesting, transporting and processing the logs. The harvesting company has made no investment at the growing stage, so a major proportion of the price is represented by the economic rent. This value should accrue to the landholder or the community, not the logger (Price Waterhouse, *Forestry Taxation*, 17–27).
26. 'Tetepare under Threat', discussion topic for Western Province Environment Summit, June 1994.
27. Edward B. Barbier, Joanne C. Burgess, and Anil Markandya, 'The Economics of Tropical Deforestation', *Ambio* 20, no. 2 (1991), 55.
28. The work of Luca Tacconi and Jeff Bennett at the University of New South Wales has focused on this, an area of interest to AUSAID in Vanuatu. See for example, Luca Tacconi and Jeff Bennett, 'Biodiversity Conservation: The Process of Economic Assessment of a Protected Area in Vanuatu', *Development and Change* 26 (1995), 89–110.
29. Barbier, Burgess, and Markandya, 'The economics', 55; Joanne C. Burgess, 'Timber Production, Timber Trade and Tropical Deforestation', *Ambio* 22, no. 2–3 (1993), 138–9. See, for example of omissions, Price Waterhouse, *Forestry Taxation*.
30. For household size, see SIG, *Statistical Bulletin: Solomon Islands 1986 Population Census Report 1: (supplement) Summary of Population by Socio-Economic Characteristics Final Results* (Honiara, 1988); Schoeffel, Fitzgerald and Loveridge, *Forest Utilization*, 30.
31. William Peterson, *Population* (New York, 1975), 8–15; *CBAR 1996*, 8.
32. Maurice King, cited in A. J. McMichael, *Planetary Overload: Global Environmental Change and the Health of the Human Species* (Cambridge, 1993), 116.
33. Schoeffel, Fitzgerald and Loveridge, *Forest Utilization*, 41–7; Whitmore, *Introduction*, 30, 134, 155.
34. This is an extremely generous estimate. The 1931 population census returned a count of 94,000. K. Groenewagen, *Report on the Census of Population, 1970* (Honiara, 1972), 1–10.
35. Bennett, *Wealth*, 113–15, 151; Bennett, '"We do not Come Here"', 131.
36. Bennett, *Wealth*, 9; Bank of Hawaii, 'An Economic Assessment'; Schoeffel, Fitzgerald and Loveridge, *Forest Utilization*, 41–7.

37. *SINFRI*, vol. 1, i, 6–11; Bennett, *Wealth*, 10; M. M. Townsend, 'Problems of Land Tenure on Malaita', *Atoll Research Bulletin* no. 85 (1961), 28; J. J. Hibberd and J. Schenk, *Country Report for the 1992 UNCED Conference: Solomon Islands* (Apia, 1991), 93.

38. Bellona is a microcosm of this process which was evident even in the 1970s, (see Monberg, *The Reactions*, 5–21).

39. *SS*, 1992–1996.

40. Tom Rudel and Jill Roper, 'Regional Patterns and Historical trends in Tropical Deforestation, 1976–1990; A Qualitative comparative Analysis', *Ambio* 25 (1996), 165.

41. For much of the South Pacific islands there are simple indicators of a middle class – the sight of beggars on the town streets, the practice of aerobics and sales of related exercise equipment, and a Macdonalds outlet. Honiara has seen the first two, the big M's golden arches have yet to come.

42. Lamb, *Exploiting*, 160.

43. The Mono people's conquest of the original Alu people in the mid-nineteenth century was because of desire for resources (School of Oriental Studies, London: G.C. Wheeler, Mono-Alu notes. MS 184245).

44. David W. Gegeo, 'Tribes in Agony: Land, Development, and Politics in Solomon Islands', *Cultural Survival Quarterly* 15, no. 2 (1991), 53.

45. Schoeffel, Fitzgerald and Loveridge, *Forest Utilization*, 25–6.

46. For example, see Maenu'u, *Bib-Kami*, 13–14.

47. Hviding and Baines, *Fisheries management*; [George Scott], Land: Proposal for a government policy, encl., Customary Land Records Bill, 1992, revised first draft, 29 August 1990, Survey and Cartographic division, Ministry of Ag. and Lands, Honiara.

48. Smallholder herds peaked in numbers in 1978. These were largely a waste of time, resources, and money. In the few places where economics of scale operate, there could be some benefit. Recently, to feed mainly the urban demand, there have been calls to revive the industry. A new project for an aid-donor perhaps? (Bennett, *Wealth*, 334–5; *SS*, 8 June 1990).

49. See for example, Maenu'u, *Bib-Kami*; Bennett, *Wealth*. This pattern applies to lagoon and marine resources too (Hviding, *Guardians*, 321–5).

50. Re land sales see Bennett, *Wealth*, 125–49.

51. Marion W. Ward, *Women and Employment in Solomon Islands* (Canberra, 1995), 18–21; Hviding and Baines, *Fisheries Management*, 11; Macbride-Stewart, 'Catholic Women's Programmes', 51; Forest Monitor Ltd, 'Environmental and Social Impact', 10.

52. Proceedings in the law of tort involving nuisance is an obvious avenue when rivers are polluted, streams dry up, and reefs are damaged as a result of poor logging methods or indeed unauthorised logging; yet this remedy has not been activated to any extent.

53. Ells, *Logging*.

54. The Ombudsman's *Reports* are redolent of these situations. Many painful and vitriolic examples regularly appear in the press. See for recent examples: Vaghena, (alienated land) *SS*, 20 Jan. 1995, 7; 15 Feb. 1995, 14; 22 Feb. 1995, 12; 10 Mar. 1995, 3; for west Are'are,*SS*, 17 Mar. 1995, 5; 29 Mar. 1995, 4; 5 Apr. 1995, 4; 19 Apr. 1995, 4; 28 June, 1995, 4; for east Fataleka, *SS*, 4 Feb. 1994; 10 May 1995; 19 May 1995, 4; for small Malaita, *SS*, 18 June 1997, 1; for Mono, *Sol-tree Nius*, Feb. 1996; for Rendova, *SS*, 30 July 1997, 4.

55. See for example, Paul Theroux, *The Happy Isles of Oceania: Paddling the Pacific* (London, 1992),199–201; *SV*, 12 June 1997.

56. Baines, *Timber Control Unit*, 5.

57. A very different situation to say, India. For example, see Grover, 'Rhythms of the Timber Trade'.

58. See also Maev Collins, 'Social and Political Impact of Logging Projects: West Coast Manus, Papua New Guinea' (seminar paper, Australian National University, July 1990), 4. See also, Lamb, *Exploiting*, 181–6.

59. Baines and Hviding noted this even among rural dwellers who had formal education. Graham B. K. Baines and Edvard Hviding, 'Traditional Environmental Knowledge for Resource Management

in Marovo, Solomon Islands', in *Traditional Ecological Knowledge: Wisdom for Sustainable Development*, ed. Nancy M. Williams and Graham Baines (Canberra, 1993), 60.

60. R. R. Thaman, 'Ethnobotany of Pacific Island Coastal Plants', in *Fauna, Flora, Food and Medicine: Science of Pacific Island Peoples*, vol. 3, ed. John Morrison, Paul Geraghty, and Linda Crowl (Suva, 1994), 178.

Chapter 16

1. For an analysis of the fisheries, see, Frazer, 'Resource extraction'.
2. Jeffrey S. Steeves, 'Unbounded Politics in Solomon Islands leadership and Party Alignments', *Pacific Studies* 19, no. 1 (1996), 125–33.
3. *SS*, 30 June 1993. Houenipwela is also known as Hou. The former bank chairman, Tony Hughes, had taken a similar stance.
4. John Roughan, 'Get a Handle on the Future by Knowing the Past', (paper presented to the United Party, Honiara, 1996), 1; *PIM*, July 1993, 25.
5. The World Bank in 1996 calculated the loss in 1993 at a staggering SI$94 million (*SS*, 3 Oct. 1997).
6. *SV*, 16 Apr. 1993.
7. *Pacific Islands Business*, July 1993, 38; *PIM*, Nov. 1990, 11, Sept. 1993, 39, Dec. 1993, 17.
8. *SS*, 30 June 1993.
9. *SS*, 21, 30 July 1993. See also *SS*, 29 Sept. 1993.
10. *SS*, 5, 28 Jan., 11 May 1994; *The Straits Times*, 19 May 1994; Coopers and Lybrand, *A Report on Timber Processing*.
11. See for example, OPM, 'A National Forest', 24–6.
12. *SS*, 24 Sept. 1993.
13. The World Bank hardly comes with clean hands on the conservation front, as its several ill-considered industrial development and re-settlement projects in the third world over the years have caused massive environmental and social damage, at least to 1987. (Chatterjee and Finger, *The Earth Brokers*, 144–56; John Cleave, 'Environmental Assessments', *Finance and Development* 25 (1988), 44–7; Rich, *Mortgaging*).
14. There were certain economic contradictions implicit in this plan because it was premised on increasing exports. While having a certain economic logic, structural adjustment will have social repercussions on the most vulnerable. In the longer term, its emphasis on increasing foreign exchange earnings and production of exports will lead to more demands on the sea, land and ultimately the forests. What has been labelled as 'export-led collapse' by Oxfam in the wake of structural adjustment programmes in Africa soon could have a Melanesian counterpart (Rich, *Mortgaging*, 106–99; *CBAR, 1994, 1995*). The Solomon Islands government apparently has been wary of borrowing heavily from the World Bank and the IMF. It now may have no other option (*SS*, 17 Jan. 1997).
15. World Bank, *Pacific Island Economies*, 9, 31-32; *PIM*, Mar. 1994, 12–13; *SS*, 30 June, 22 Sept., 22 Oct. 1993, 4, 9, 16, 18, 23, 25 Feb., 9, 16, 18, 23, 25, 31 Mar., 22 Apr., 3, 10 Aug. 1994; *The Straits Times*, 19 May 1994; *Islands Business*, Feb. 1995, 27–8.
16. *SS*, 2 May 1993, 9, 11 Mar., 1, 8, 27 July 1994.
17. *SS*, 6, 18, 20, Aug. 1993, 7, 21, 24 Oct. 1994, 27 Jan. 1995; *CBAR, 1994*, 16–17.
18. *SV*, 18 Feb., 5 Mar. 1993; *SS*, 3 Aug. 1994.
19. *SV*, 5 Aug. 1994.
20. Hii had been a director of Rural Industries Ltd from 1987–1991 and of Integrated Forest Products in 1988 (RCo: CRO 31/1982, CRO 4/1982). Mega, in 1991, was said to have control of Dalsol,

Waibona, as well as IFI and RI. Hii and Audie Murphy Mamaloni were secretaries to Somma Ltd. in 1993 (*Link*, Nov./Dec. 1991; RCo: Registrar of Companies, Somma Ltd, Honiara.).

21. Hviding, *Guardians*, 338, 339. Colonial foresters had regarded the Marovo barrier islands as sacrosanct, lest erosion damage the reefs and lagoon.

22. *SS*, 9 Mar. 1994; Coopers and Lybrand, *A Report*. SIFIA recommended an annual cut of 600,000 cubic metres, but was placing its hopes for the future on KFPL and government plantations. The latter have now been privatised, and could be logged and not replanted. Moreover, increased local milling, notwithstanding lack of expertise and gross wastage, means less timber for foreign loggers, something SIFIA did not want.

23. *SS*, 22 July 1994, 28 June, 31 Mar., 4 Aug. 1995; *New Straits Times*, 7 Nov. 1994.

24. Founded in 1972, the South Pacific Forum consists of the independent states of the South Pacific, including Australia and New Zealand. It concerns itself with matters of common interest and is a voice for the region in the international arena. It is heavily financed by Australia and New Zealand.

25. Kumpulan Emas Berhad, *Annual Report*, 13. These actions were supported by the Western provincial government (*SS*, 12 Aug. 1994).

26. Some years before, the Solomons government had taken out a loan at a commercial rate of around seven to eight per cent with Export Finance Investment Corporation of Australia to purchase two purse-seiner ships. The Mamaloni government sold these in 1990, but the debt stood. The Australian government was to repay the debt to the Corporation (*CBAR, 1989*, 54; *SS*, 9 Mar., 20 July, 31 Aug. 1990, 10 Aug. 1994).

27. *SS*, 3, 10 Aug. 1994; *New Straits Times*, 4 Aug. 1994.

28. *SS*, 5 Aug. 1994. This did not go down well with the Malaysian leader, Dr Mahathir, who claimed Australia believed it owned the Pacific. Moreover, Australia's own environmental record before and since its federation in 1901 was hardly a model for the Pacific to emulate. Australia has all the fervour of a new convert. 'Land opening' was permitted in Australia without knowledge of the environmental effects as late as 1960, and forests were logged beyond long-term sustainablilty into the 1970s. Environmental consciousness was not part of the new federal Australia, but emerged in the 1960s. Ecologically sustainable development for Australia was enshrined in major federal policy only as late as 1989. However, tropical forest conservation had been Australian Labor Party policy as early as 1984. Kevin Frawley, 'Evolving Visions: Environmental Management and Nature Conservation in Australia', in *Australian Environmental History: Essays and Cases*, ed. Stephen Dovers (Melbourne, 1994) 66, 74; Papadakis, *Environmental Politics*, 95–129; *Insight*, 7 Aug. 1992, 12–13.

29. *Australian Financial Review*, 3 Aug. 1994.

30. *SS*, 5 Aug. 1994.

31. *SS*, 5, 10 Aug. 1994; *SV*, 5 Aug. 1994; Seminar on Bougainville, Honiara, 1992. Australia was former colonial ruler of Papua New Guinea and the copper mine on Bougainville had been opened amid landholder protest during the colonial period. Conflict continued over mining royalties until 1989, when civil war broke out between Bougainville groups and the state of Papua New Guinea. Australia had trained PNG forces and supplied helicopters that were used by PNG to kill Bougainvilleans. There were even rumours in Honiara in 1994 that Australia was to invade Solomons. <http://forests.org/recent/bbriefin.txt>: Unofficial transcript of Background briefing, Radio National, Australian Broadcasting Corporation, 14 Jan. 1996, from Ecological Enterprises, 27 Jan. 1996, Gaia Forest Conservation Archives, Internet.

32. *SS*, 10 Aug. 1994.

33. Governor-General Moses Pitakaka had been head of a commission of inquiry into alleged corruption set up by Hilly. Many believe Pitakaka was voted in as governor-general by the parliament, following Mamaloni's nomination of him for the post, 'as a sweetener to drop his investigation' because it was leading to prominent politicians and senior government officials (*Islands Business*, Feb. 1995, 27).

34. *SS*, 9, 14, 23 Sept., 5, 7 Oct., 11 Nov. 1994. Re Qurusu, see *SS*, 10 Mar. 1989.

35. *The Straits Times*, 7, 8 Nov. 1994; *SS*, 11 Nov. 1994.

Notes to pp. 345–354

36. *SS*, 26 May 1995; Asia and Pacific News, Radio New Zealand International 16 Jan. 1996; *Islands Business*, Feb. 1995, 27; <http://forests.org/recent/bbriefin.txt> Unofficial transcript of Background briefing, Radio National, Australian Broadcasting Corporation, 14 Jan. 1996. Several of Mamaloni's supporters ended up in court. Seri, Orodani, Maetia, Andresen and Zapo faced High Court hearings on charges of accepting bribes from a Honiara businessman for switching from the NCP to the SINURP. In November 1997, Musoata, Andresen and Luialamo were found guilty of corruption involving accepting the use of hire-cars. Allan Paul's case on much more serious charges concerning bribery, during the period Nov.–Dec. 1993 was still being heard at the time (*SV*, 17 May 1996; *SS*, 22 Oct., 12 Nov. 1997).
37. *SS*, 10 Aug. 1994, 12, 26, May, 4, 9, Aug. 1995; *CBAR, 1989–1995*; *Islands Business*, Feb. 1996, 50. There was talk of changing the law to get the bank to co-operate, but the government backed away in the face of wide-ranging disapproval, particularly from the financial sector. A. V. Hughes, 'Reflections on central banking in Solomon Islands', *Pacific Economic Bulletin* 10, no. 2 (1995), 72–3.
38. *SS*, 27 Jan. 1995; *PIM*, Dec. 1994, 28. The timber levy never did get to landowners' projects (Price Waterhouse, *Forestry Taxation*, 51, 77).
39. *SS*, 25 Nov. 1995.
40. *CBAR, 1995*, 41.
41. MOF, *1995 Forest Review; Islands Business*, Feb. 1996, 49–50; Price Waterhouse, *Forestry Taxation*, 21–2.
42. *SS*, 24 Feb. 1995.
43. OPM, Development of the Forestry sector. See also *The Independent*, 9 Feb. 1996.
44. CS; *Islands Business*, Feb. 1996, 50.
45. MOF, *1995 Forest Review*; Price Waterhouse, *Forestry Taxation*.
46. *SS*, 30 Apr. 1997.
47. *PIM*, May 1993, 18; *SS*, 19 May 1993, 14 Sept. 1994, 6 Jan., 3, 24 Mar., 26 Apr. 1995; 30 Apr. 1997. The NCP government had replaced the discretionary find with a constituency development fund (CDF) and promised to rationalise allocation, but the Mamaloni government re-routed funds to parliamentarians' projects (*SS*, 18 Aug., 8 Sept, 1993, 24 Mar., 26 Apr. 1995, 27 Mar., 30 Apr. 1997). Steeves makes the same point about party allegiances in 'Unbounded politics', 115–38.
48. *SS*, 8 Feb. 1995.
49. *SS*, 8 Feb. 1995. See also, for similar view of Bill Gina, *Islands Business*, Aug. 1995, 6
50. *SS*, 8, 24 Feb., 13 Apr. 1995; Price Waterhouse, *Forestry Taxation*.
51. *SS*, 22 July 1994; *SINFRI* vol.1, 33.
52. *SS*, 27 Jan. 1995. See also *Islands Business*, July 1996, 22.
53. *SS*, 28 Nov. 1997.
54. A common pattern – see Westoby, *The Purpose of Forests*, 265.
55. Editorial, *Sol-Tree Nuis*, Feb. 1996
56. *The Australian*, 7 June 1995.
57. *SS*, 6 Sept. 1995.
58. Bilney quoted in *Islands Business*, Oct. 1995, 26.
59. Australian Embassy, Fiji, *Australia-South Pacific News Letter*, 1 Nov. 1995. There is an interesting debate about who told whom what. Did the Australians tell Mamaloni that there was to be a cut in funds or did the Solomon Islands government tell Australia to keep its money as early as July 1995? Political face-saving seemed more important than saving the forests for the future. (*Solomon Nuis*, Dec. 1995; *SV*, 9, 16 Feb. 1996).
60. *SS*, 13 Sept. 1995, *The Australian*, 21 Sept. 1995, 9; CS.
61. *SV*, 3 Nov. 1995; *SS*, 2 Feb., 5 Oct. 1994, 8, 13 Sept. 1995, 12 Mar. 1997; CS; Thorpe, 'Effectiveness of Aid', 26; MOF, *1995 Forestry Review*, 9; *Sol-Tree Nius*, June 1996; Hviding, *Guardians*, 317; Price Waterhouse, *Forestry Taxation*, 116–17, 148. The EU reduced its aid spending across the board because of similar problems with the government (*SS*, 22 Jan. 1997).

62. *SV*, 24 Nov. 6, 13 Oct. 1995; Kumpulan Emas Berhad, *Annual Report*, 57; *SS*, 30 Aug. 1995. This prison saga dragged on into 1997 as Chan Wing did not deliver, so a desperate Finance Ministry initiated court action (*SS*, 1 Mar. 1996, 28 May, 31 Oct. 1997).

63. *SV*, 24 Nov., 13 Oct. 1995; *Courier Mail*, 29 Dec. 1995, 1 Jan. 1996; *PIM*, Feb. 1996, 10–11. Remissions and exemptions were not simply the prerogative of local companies.

64. *SV*, 16 Feb. 1993

65. <http://forests.org/recent/abcsolo.txt>: Sam Alasia, Interview, Transcript of Indian Pacific Radio National, Australian Broadcasting Commission, 3 Feb. 1996, Gaia Forest Conservation Archives, Internet.

66. See Austeo document, cited in *PIM*, Aug. 1997, 17.

67. Asia and Pacific News, Radio New Zealand International, 25 Jan. 1996. The full text of Hanley's speech was reported in *SS*, 2 Feb. 1996.

68. *CBAR, 1996*, 8. Both primary funders, Australia and the EU in 1995 ceased assistance to the Provincial Development Unit 's several small-scale projects because of misuse of funds (*CBAR, 1995*, 50).

69. *SV*, 9, 16 Feb. 1996.

70. *Islands Business*, Apr. 1996, p. 54; MOF, *1995 Forestry Review*, 2; *Katibas News*, July 1996; *CBAR, 1995*, 18; *SV*, 18 July 1997.

71. Forest Monitor Ltd, 'Environmental and Social Impact'.

72. *SV*, 8 Mar. 1996; *PIM*, Apr. 1996, 48–9: Asia and Pacific News, Radio New Zealand International, 21 July 1996. The author asked permission of the editor of the *Solomons' Voice* to reproduce a cartoon satirising Bilney, but was refused.

73. *Islands Business*, May 1996, 12.

74. *SS*, Feb.–May 1997;<http://forests.org/recent/bbriefin.txt>: Unofficial transcript of Background briefing, Radio National, Australian Broadcasting Corporation, 14 Jan. 1996. Misappropriation of public monies by politicians during Mamaloni's time emerged in late 1997 (*SS*, 2 Dec. 1997).

75. *SV*, 23 Feb., 1 Mar., 19 Apr., 3, 10, 17, 23, 31 May, 12 July 1996; *SS*, 6 Dec. 1996; *Islands Business*, Oct. 1995, 36, Feb. 1996, 49–50, June 1996, 54–6, July 1996, 56; Asia Pacific News, Radio New Zealand International, 22 July 1996.

76. *SV*, 1, 9 Apr. 1996; *CBAR, 1996*, 45.

77. *Islands Business*, Dec. 1995, 39

78. *Islands Business*, May 1996, 12.

79. *Islands Business*, June 1996, 57. The hunt continued into August, with Tulagi residents complaining of sanitation problems caused by the Field Force men (Asia Pacific News 15 Aug. 1996, Radio New Zealand International).

80. *SV*, 6, 16, 23 Aug., 6, 13 Sept. 1996; Foreign Correspondent, *Solomon Islands Logging Practice*, Video documentary, Australian Broadcasting Commission, Australia, 13 Aug. 1996. In order to counter adverse publicity, the SIFIA, with the co-operation of the government, set about making a video presenting their side of the logging issue. This was completed by January 1997 (statement of E. Kes, of SIFIA, Asia Pacific News, 31 Jan. 1997, Radio New Zealand International; *SS*, 31 Jan. 1997). See also *SS*, 30 July 1997.

81. <http://forests.org/recent/bbriefin.txt>: Unofficial transcript of Background briefing, Radio National, Australian Broadcasting Corporation, 14 Jan. 1996.

82. *SS*, 15 Sept. 1995; Prime Minister's Christmas Message, *Solomon Nuis*, Dec. 1995; *SV*, 21 Dec. 1995, 26 Jan., 9 Feb., 22, 29 Mar. 1996; *The Independent*, 9 Feb. 1996; *CBAR*, 1994, 65; *CBAR*, 1996, 9, 46; Hughes, 'Reflections', 70–3.

83. *SS*, 28 Feb. 1997. The High Court found the Act to be unconstitutional. Ulufa'alu planned to rescind it (*SS*, 29 Aug. 1997).

84. *SS*, 29 Jan., 14, 21, 27 Mar., 4, 9, 30 Apr., 7, 14, 16 May, 4, 18 July, 29 Aug., 12 Nov. 1997; *SV*, 18 Apr., 30 May, 18 July, 22 Aug. 1997.

85. *SV*, 29 Aug. 1997.

86. Sogavare was Permanent Secretary of the Ministry of Finance until he was sacked by Mamaloni because he objected to corruption. He subsequently attended university in New Zealand in 1995 (CS).
87. *SS*, 24, 26, 30 Sept., 10, 11, 15, 22, 31 Oct., 7, 14, 19, 23 Nov. 1997. Salvaged by the Central Bank and the FD, the forest database set up by TCU survived through 1996, but faced dissolution until another temporary deal was struck with SIFIA, representing itself as a responsible organisation. The end of the resource on customary land is in sight and the future lies with commercial plantations along the lines of KFPL or with reforestation of customary land that the Solomons government is unlikely to finance. A cynical view would be that SIFIA, whose members have creamed the forests, seems to be gearing up for co-opting aid donors to subsidise their industry, but then, capitalism always needs state support to flourish (*CBAR, 1996*, 17–18).
88. *SS*, 22, 31 Oct. 1997.

Conclusion

1. See also, Schoeffel, Fitzgerald and Loveridge, *Forest Utilisation*, 82–104.
2. Re Viru, see Hviding, *Guardians*, 317.
3. For example, see *SS*, 13 Dec. 1996.
4. Statement of E. Kes, of SIFIA, Asia Pacific News, 31 Jan. 1997, Radio New Zealand International; *SINFRI*, 1: 44, 48.
5. Tom Tietenberg, *Environmental and Natural Resource Economics* (Glenview, Illinois, 1988), 252–3. The increased price could just as easily be creamed off by the industry. One solution would be compensation, paid by the governments of the world to loggers (whether private companies or local communities) implementing sustainable regimes in the tropical forests. Such regimes while technically feasible are more costly to carry out than the typical methods seen in much of Solomons' logging and thus less immediately profitable to the industry. Production support for concessionaires carrying out these methods would be in the form of area-dependent payments for limited time periods. In total, the revenue needed to fund this plan is of the order of US$2.25 billion dollars annually for 20–30 years. All timber products exported from all counties could be taxed. In addition to the expected contributions via existing export taxes and royalties by governments in tropical countries and diversion of the North's current aid budget to tropical forests, this global tax would be used to pay the concession holders who do the sustainable logging and to fund the machinery to support, implement and monitor this scheme. The new tax would be about 1.5% of export value. In this way, the North could subsidise the sustainability of tropical forests and their ecological and genetic diversity, and prevent total diversion of the industry's sources from the tropics to temperate zones (Christian Friis Bach and Soren Gram, 'The Tropical Timber Triangle', *Ambio* 25, no. 3 (May 1996), 166–70).
6. Of course Northern governments would need to see that polluters do not simply buy their way out of their responsibilities to reduce the emission of greenhouse gases at source. And in the South and the North, existing forests should not qualify as carbon offsets, because they were not established for that purpose. This could be a real opportunity for countries like Solomons. The history of any particular forest would need to be examined. Exact quantification of carbon sequestration is needed to arrive at costings and monitoring systems put in place. Experiments along these lines are being conducted throughout the world, but most particularly in Malaysia (Pedro Moura Costa, 'Tropical Forestry Practices for Carbon Sequestration: A Review and Case Study from Southeast Asia', *Ambio* 25, no. 4 (1996), 279–83).
7. A. Runte, '"Worthless lands" – Our National Parks: The Enigmatic Past and Uncertain Future of America's Scenic Wonderlands', *American West* 10 (1973), 4–11; C. M. Hall. 'The 'Worthless Lands Hypothesis' and Australia's National Parks and Reserves', in *Australia's Ever Changing Forest: Proceedings of the First national Conference of Australian Forest History*, ed. K. J. Frawley and N. Semple (Canberra, 1988), 441–56; Paul Star, 'From Acclimatisation to Preservation: Colonists and the Natural World in Southern New Zealand, 1860–1890' (Ph.D. thesis, (draft), University of

Otago, 1997), pagination incomplete; M. Colchester, 'Beyond Participation: Indigenous People, Biological Diversity Conservation and Protected Area Management', *Unasylva* 47 (1996), 33–9; See, for Australia, Libby Robin, *Defending the Little Desert: The Rise of Ecological Consciousness in Australia* (Melbourne, 1998).

8. In ecophilosophy, Arne Naess made the distinction between 'shallow' and 'deep' ecology in 1972. Shallow ecology and the environmentalism it supports see human beings as separate from their environment and the centre of all value; it accepts mechanistic materialism as the dominant value in relation to the use of the environment and it accepts the economic growth paradigm that typifies all industrial and most developing nations. Deep ecology holds that human beings are but one member of a diverse biotic community and all natural phenomena are inter-related. Deep ecology rejects the economic growth ideology and its political, social, and economic practices, valuing instead those which support sustainability of the environment. In sum, deep ecology stresses an ecocentric rather than an anthropocentric view. Arne Naess, *Ecology, Community, and Lifestyle: Outline of an Ecosophy*, trans. and rev. David Rothenberg (New York, 1989).

9. Colchester, 'Beyond Participation', 32–4. For the Marovo people, see Hviding, *Guardians*, 355–75.

10. *SINFRI*, vol. 1, 51. See also Forest Monitor Ltd, 'Environmental'.

11. Nunn, *Environmental Change*, 189–310.

12. Bahn and Flenley, *Easter Island*; Nunn, *Environmental Change*, 290–2.

13. Peter Bellwood, *The Polynesians: Prehistory of an Island People* (London, 1978), 130–40. South Island Maori were not only foragers. They replanted, for future harvesting, the cabbage tree (*Cordyline australis*) after felling a mother plant to process for making sugar.

14. M. I. Weisler, 'Henderson Island Prehistory: Colonization and Extinction on a Remote Polynesian island', *Biological Journal of the Linnean Society* 56 (1995), 398–402.

15. Patrick V. Kirch, 'Polynesian Agricultural Systems', in *Islands, Plants, and Polynesians: An Introduction to Polynesian Ethnobotany*, ed. Paul Alan Cox and Sandra Anne Banack (Oregon, 1991), 119–20.

16. Jacques Barrau, *Subsistence Agriculture in Polynesia and Micronesia* (Honolulu, 1961), 18.

17. Yen, 'Agricultural Systems', 66; Kirch, 'Polynesian Agricultural Systems', 120–3.

18. Yen, cited in Kirch, 'Polynesian Agricultural Systems', 120. In relation to the Solomons, these figures should be seen as notional, not definitive. There is considerable variation on output per unit with area and crop as well as agricultural skills throughout Melanesia. See Brookfield with Hart, *Melanesia*, 89–92.

19. J. C, Roux, 'Traditional Melanesian Agriculture in New Caledonia and Pre-contact Population Distribution', in *Pacific Production Systems: Approaches to Economic Prehistory*, eds D. E. Yen and J. M. J. Mummery (Canberra, 1990), 172. Production per unit input of work is considerably reduced after the irrigation system is in place and is usually higher than in most dry-land gardens. (Matthew Spriggs, 'Why Irrigation Matters in Pacific Prehistory', in *Pacific Production Systems: Approaches to Economic Prehistory*, eds D. E. Yen and J. M. J. Mummery (Canberra, 1990), 176–7.

20. The classic statement regarding population growth inducing intensification is Ester Boserup, *The Conditions of Agricultural Growth: the Economics of Agrarian Change under Population Pressure* (London, 1965).

21. Hviding describes this process for much of the Marovo area (Hviding, *Guardians*, 75–8). See also W. C. Clarke and R. R. Thaman, eds, *Agroforestry in the Pacific Islands: Systems for Sustainability* (Tokyo, 1993), 41–8.

22. F. Braudel, *On History* (Chicago, 1980).

23. See for example, Christopher Browne with Douglas A. Scott, *Economic Development in Seven Pacific Island Countries* (Washington, D.C., 1989), 8, 30.

Bibliography

Interviews and correspondence

Alebua, Ezekiel, 27 Apr. 1992.
Baeanisia, Abraham, 22 Oct. 1992.
Beverley, John, 3 Nov. 1992.
Boseto, Leslie, July 1992.
Corrin, Jennifer, Correspondence, Aug. 1996.
Dennis, Geoff, Correspondence, Feb. 1995.
Ebinuwi, Mikelo, 1976.
Elkington, Tom, 1976.
Gaviro, Sam, 9 Nov. 1992.
Gina, Lloyd Maepeza, 29 July 1992.
Gina, Olive, 27 July 1992.
Havea, Eric, 1 Aug. 1992.
Hughes, A. V., Correspondence, 1999.
Kinne, Rosemary, Apr. 1992 and Correspondence, 1996.
Leadley, Alan, Nov. 1994.
Leleti, William, 12 Aug. 1992.
Liligula, Ruth, 8 May 1992.
Lomiri, Michael, 7 Aug. 1992.
Malalifu, Justus, 1976.
Marten, K. D., Correspondence, 1994–1997.
Minchin, Devon, Aug. 1996.
Nikolas, Joseph, 1976.
Normani, Joseph, 1976
Odofia, Joseph, 1976.
Piopio, Jacob,1976.
Plowman, Peter, 1975.
Ramoli, Chris, June 1992.
Roughan, John, 1992, 1998.
Sirikolo, Myknee Qusa, 12 May 1992.
Stuart, Don, Correpondence, Nov. 1994.
Trenaman, K. W., Correspondence, 1994.
Tua, Ben, June 1992.
Tutu, Victor, Sept. 1992.
Vickers, Alex, 1992.
Woof, Wayne, 13 Aug. 1992.

Bibliography

Manuscript sources

1. Private papers

Admiralty Library, London.
    MacGillivray, John. Private Journal aboard HMS Herald to the South Pacific. (On microfilm, National Library of Australia, Canberra).
Archives, University of Melbourne.
    Fairley, Rigby and Co. Ltd, Records, 1912–1920.
    Vanikoro Kauri Timber Company, Records, 1922–1964.
Division of Pacific and Asian History, Australian National University, Canberra.
    C. M. Woodford, Diaries and Papers on microfilm.
D. Franks, Carshalton, Surrey.
    Monckton, Eric, Papers. (On microfilm at PMB 1021).
Joan Laws, Leura, New South Wales.
    Sandars, E., Papers (PMB 553).
Ellen G. White Research Centre, Seventh Day Adventist Church, Cooranbong, New South Wales.
    Solomon Islands Mission, minutes of meetings, 1931–1948.
    Correspondence, 1941–1945.
    Photographs, 1920–1960.
National Library of New Zealand, Wellington.
    Hocart, Arthur M., Papers. (Also on microfilm Division of Pacific and Asian History, Australian National University, Canberra).
    Stanmore Correspondence, microfilm 175.
Nicholson Whaling Collection, Providence public library, Providence, Rhode Island, USA.
    Log of *Alfred Gibbs* (PMB 572).
Macmillan Brown Centre of Pacific Studies, Unversity of Canterbury, Christchurch, New Zealand.
    F. J. W. Billings, Papers.
Methodist Archives, Christchurch.
    Leadley, E.C., Diary, 1936.
Methodist Church of New Zealand, Penrose, Auckland.
    On microfilm of Pacific Manuscript Bureau (PMB).
    Jones, Lina, Correspondence, 1924 (PMB 946).
    Harkness, Effie, Correspondence, 1938 (PMB 946).
    Metcalfe, John R., General Letters, 1938 (PMB 68).
    Goldie, J. F. Correspondence, 1922–1951 (PMB 925).
Mitchell Library, Sydney.
    Methodist Overseas Mission Records (MOM).
    Synod Minutes, Solomon Islands District, 1914, MOM Box 174–190.
    Solomon George Brown, Letterbook, 24 May–2 June 1902.
    Goldie, J. F., Roviana Circuit Report, 1915.
    Sinclair, W. A., and J. W. Court, Report of the Representatives to the Solomon Islands Mission District, 1920.
National Archives of Solomon Islands, Honiara.
    Melanesian Mission and Church Diocesan records.
    Hopkins, A. I., Autobiography (PMB 557).
National Library of Australia, Canberra.
    Deland, Ann, Papers, MS 4725.
Noel Butlin Archives, Australian National Univeristy, Canberra.
    Burns Philp Records.
    Correspondence, Solomon Islands.
    Lucas, W. H., Reports on Solomon Island Plantation Properties, 1910.

Bibliography

Lucas, W. H., Notes on the Western Pacific and Australian Interests therein: Actual and Potential, 1917, South Seas Misc.
Old Dartmouth Whaling Museum, New Bedford, Mass., USA.
Log of *Two Brothers* (PMB 284).
Rhodes House, Oxford, UK.
Barrow, Lennox G., Outlying Interlude: an Account of BSIP, 1942–1947, (PMB 517).
School of Oriental and African Studies, London.
Wheeler, G. C., Mono Alu Notes, MS 184245.
Suffolk County Whaling Museum, Sag Harbour, Long Island, New York, USA.
Log of *Belle*, (PMB 680).
Unilever Historical Archives, London
Selected papers relating to the Pacific Islands Company and Lever's Pacific Plantations. (Microfilm, Division of Pacific and Asian History, ANU, Canberra).
Whaling Museum, Nantucket, Mass., USA.
Log of *Mohawk* (PMB 390).
W. R. Carpenters, miscellaneous correspondence, Solomon Islands, 1925–1931, in writer's care (PMB 1112).

2. Official papers

Great Britain
Records Section, Foreign and Commonwealth Office, Milton Keynes.
Western Pacific High Commission, Inwards Correspondence, 1875–1941, General Correspondence, F series, 1942–1954. (WPHC microfilm, 1875–1920 at National Library of New Zealand, Wellington).
Public Record Office, Kew.
Colonial Office Correspondence.
CO 225, Original Correspondence, Western Pacific, 1877–1913. (Microfilm at National Archives of New Zealand).
CO 83, Original Correspondence, Fiji, 1874–1946. (Microfilm at National Archives of New Zealand).
New Zealand
National Archives of New Zealand, Wellington.
World War II, series 1; Air Dept, series 118, 127, 265; Forestry, series F T/C.
Solomon Islands
National Archives of Solomon Islands, Honiara.
Western Pacific High Commission, BSIP Secretariat series, 1927–1952.
Forestry Division, Ministry of Natural Resources, Honiara.
Old files relating to forestry, 1960–1978.
Register of Companies Office, Honiara.
Registered logging companies.
Lands Titles Office, Honiara.
Land titles records.

Printed and other media sources

Abana, Bill. 'Logging is no development'. *O'o: a journal of Solomon Islands Studies* (University of the South Pacific) 2, no.1 (1989): 57–62.
Adas, Michael. 'Colonization, Commercial Agriculture and the Destruction of the Deltaic Rainforest of British Burma in the late Nineteenth Century'. In *Global Deforestation in the Nineteenth-Cen-*

tury World Economy, edited by R. P. Tucker and J. F. Richards, 95–110. Durham, NC: Duke University Press, 1983.

Aditjondro, George J. 'Problems of Forestry and Land Use in the Asia-Pacific Region: The Irian Jaya Experience'. In *The Ethics of Development: Choices in Development Planning*, edited by Philip J. Hughes and Charmian Thirlwall, 104–16. Port Moresby: University of Papua New Guinea, 1988.

Alasia, Sam. 'Politics'. In *Ples Blong Iumi: Solomon Islands, the Past Four Thousand Years*, edited by Hugh Laracy, 137–51. Suva: Institute of Pacific Studies of the University of the South Pacific, 1989.

Albion, R. G. *Forests and Sea Power. The Timber Problem of the Royal Navy, 1652–1862*. Cambridge, Mass.: Harvard University Press, 1926.

Aldrick, John M. *The Susceptibility of Lands to Deterioration in the Solomon Islands*. National Forest Resources Inventory Project, Working Paper no. 12. Honiara: AIDAB and Ministry of Natural Resources, May 1993.

Alexander, Gilchrist. *From the Middle Temple to the South Seas*. London: John Murray, 1927.

Allan, Colin Hamilton. *Customary Land Tenure in the British Solomon Islands Protectorate*. Report of the Special Lands Commission. Honiara: Western Pacific High Commission, 1957.

———. *Solomons Safari, 1953–58: Part II*. Christchurch: Nag's head, 1990.

Allardyce Lumber Company and Ors v. Attorney-General and Ors. High Court of Solomon Islands (August 1989).

Allen, J. et al. 'Pleistocene dates for the human occupation of New Ireland, northern Melanesia'. *Nature* 331 (1988): 707–9.

Amherst, Lord, of Hackney, and Basil Thomson, eds. *The Discovery of the Solomon Islands by Alvaro de Mendaña in 1568*. 2 vols. London: Hakluyt Society, 1901.

Anderson, Edgar. *Plants, Man, and Life*. Berkeley and Los Angeles: University of California Press, 1952.

Ashby, Ted. *Blackie: A Story of the Old-time Bushmen*. Wellington: A. H. and A. W. Reed, 1978.

Asia and Pacific News. New Zealand National Radio, June 1995–Jan 1997.

Australasian Record (Sydney), 1914–1934.

Australian Government. Parliamentary papers, 1917–1919. *British and Australian Trade in the South Pacific*, Vol. 5, 1916. Report of Interstate Commission, 1916.

———. Parliamentary papers, 1920–1921. Documents cited in Joint Committee of Public Accounts, *Interim Report on the Transactions of the War Service Homes Commissioner with Mr J. T. Caldwell*, 5 April 1921.

———. *Report of the Committee to Review Australian Overseas Aid Program 1984*. 1984.

———. *Auditor General Audit Report Number 13: AIDAB Aid to Papua New Guinea 1990–91*. 1990.

———. *Supplementary Evidence to AIDAB Submission to JCFAD Inquiry into Australia's Relations with Papua New Guinea*. Canberra, 1991.

———. *The Solomon Islands Economy: Prospects for Stabilisation and Sustainable Growth*. Canberra, 1991.

———. Australian International Development Assistance Bureau (AIDAB). *Changing Aid for a Changing World: Key Issues for Australia's Overseas Aid Program in the 1990s*. AIDAB, 1992.

———. *Australia's Overseas Aid Program, 1995/96*. Budget related paper no. 2. Commonwealth of Australia, 1995.

Australian Government. AIDAB and Ministry of Natural Resources, Solomon Islands. *Solomon Islands National Forest Resources Inventory: Forests of the Solomon Islands. National overview and Methods*. Vol. 1. Honiara: AIDAB and Ministry of Natural Resources, 1995.

———. *Solomon Islands National Forest Resources Inventory: Forests of the Solomon Islands: Guadalcanal Province*. Vol. 2. Honiara: AIDAB and Ministry of Natural Resources, 1995.

———. *Solomon Islands National Forest Resources Inventory: Forests of the Solomon Islands. Choiseul Province*. Vol. 4. Honiara: AIDAB and Ministry of Natural Resources, 1995.

———. *Solomon Islands National Forest Resources Inventory: Forests of the Solomon Islands. Isabel Province*. Vol. 5. Honiara: AIDAB and Ministry of Natural Resources, 1995.

———. *Solomon Islands National Forest Resources Inventory: Forests of the Solomon Islands: Rennell and Bellona Province*. Vol. 9. Honiara: AIDAB and Ministry of Natural Resources, 1995.

Australian Timber Journal and Building Products Merchandiser (Sydney) 1969–1970.

Bibliography

Bach, Christian Friis and Soren Gram. 'The Tropical Timber Triangle'. *Ambio* 25, no. 3 (May 1996): 166–70.
Badcock, W. J. 'Agriculture in the British Solomon Islands Protectorate'. *Agricultural Journal of Fiji* 17, no. 3 (1946): 63–70.
Bahn, Paul, and John Flenley. *Easter Island, Earth Island: A Message from our Planet*. London: Thames and Hudson, 1992.
Baines, Graham B. K. 'Traditional Resource management in the Melanesian South Pacific: A Development Dilemma'. In *Common Property Resources: Ecology and Community-based Sustainable Development*, edited by Fikret Berkes, 273–95. London: Belhaven Press, 1989.
―――. *Timber Control Unit Project, Solomon Islands: Training for Forestry Extension Services for Customary Landholders in the Solomon Islands*. Canberra: AIDAB and Forestry Technical Services Pty Ltd, June 1992.
Baines, Graham B. K., and Edvard Hviding. 'Traditional Environmental Knowledge for Resource Management in Marovo, Solomon Islands'. In *Traditional Ecological Knowledge: Wisdom for Sustainable Development*, edited by Nancy M. Williams and Graham B. K. Baines, 56–65. Canberra: Centre for Resource and Environmental Studies, Australian National University, 1993.
Balee, William. 'Indigenous History and Amazonian Biodiversity'. In *Changing Tropical Forests: Historical Perspectives on Today's Challenges in Central and South America*, edited by Harold K. Steen and Richard P. Tucker, 185–97. Durham, NC: Forest History Society, 1992.
Barbier, Edward B., Joanne Burgess, and Anil Markandya. 'The economics of Tropical Deforestation'. *Ambio* 20, no. 2 (Apr. 1991): 55–8.
Barnett, T. E. 'Legal and Administrative Problems of Forestry in Papua New Guinea'. In *Resources, Development and Politics in the Pacific Islands*, edited by Stephen Henningham and R. J. May, 90–118. Bathurst, Australia: Crawford House, 1992.
Barrau, Jaques. *Subsistence Agriculture in Melanesia*. Honolulu: Bernice P. Bishop Museum, 1958.
―――. *Subsistence Agriculture in Polynesia and Micronesia*. Honolulu: Bernice P. Bishop Museum, 1961.
Bayliss-Smith, Tim. 'Melanesian Interaction at the Regional Scale: Spatial Relationships in a Fluid Landscape'. In *Migrations and Transformations: Regional Perspectives on New Guinea*, edited by Andrew J. Strathern and Gabriele Stuzenhofecker, 295–331. Pittsburgh and London: University of Pittsburgh Press, 1994.
Bellwood, Peter. *The Polynesians: Prehistory of an Island People*. London: Thames and Hudson, 1978.
Bennett, Judith A. 'Cross-cultural Influences on Village Relocation on the Weather Coast of Guadalcanal, c. 1850–1950'. Master's thesis, University of Hawai'i, 1974.
Bennett, Judith A. ' Wealth of the Solomons: A History of Trade, Plantations and Society, Solomon Islands, c. 1800–1942'. Ph.D. diss., Australian National University, Canberra, 1979.
Bennett, Judith A. 'Oscar Svensen: a Solomons Trader among "the few"'. *Journal of Pacific History* 16, no 4 (1981): 170–89.
Bennett, Judith A. *Wealth of the Solomons: A History of a Pacific Archipelago, 1800–1978*. Honolulu: University of Hawai'i, 1987.
Bennett, Judith A. '"We do not come here to be beaten": Resistance and the Plantation System in the Solomon Islands to World War II'. In *Plantation Workers: Resistance and Accommodation*, edited by Brij V. Lal, Doug Munro and Edward D. Beechert, 129–87. Honolulu: University of Hawaii, 1993.
Bennett, Martin, et al. *Forestry Project Identification Mission, May–June 1991*. Draft Report. Herts., Great Britain: Overseas Development Administration, 1991.
Bigger, M. *Forest Entomology in the Solomon Islands: A Report on Research Carried Out between 1979 and 1983*. London: College House, Tropical Development and Research Institute, 1984.
Birket-Smith, Kaj. *An Ethnological Sketch of Rennell Island: A Polynesian Outlier in Melanesia*. 2nd edn. Copenhagen: The Royal Danish Academy of Sciences and Letters, 1956.
Black, Robert H. 'The Russell Islanders of the British Solomon Islands Protectorate'. Dipl. Anthrop., University of Sydney, 1963.
Bogesi, George. 'Santa Isabel'. *Oceania* 18, no. 3 (Mar. 1948): 208–32.

Bibliography

Boserup, Ester. *The Conditions of Agricultural Growth: The Economics of Agrarian Change under Population Pressure*. London: Allen and Unwin, 1965.
————. 'Environment, Population and Technology'. In *The Ends of the Earth: Perspectives on Modern Environmental History*, edited by Donald Worster, 23–38. Cambridge: Cambridge University Press, 1988.
Bowler, Peter J. *The Fontana History of the Environmental Sciences*. London: Fontana, 1992.
Braudel, F. *On History*. Chicago: University of Chicago Press, 1980.
Brook, James E. *Jim of the Seven Seas: A True Story of Personal Adventure*. London: Heath Cranton, 1940.
Brook, S. E. G. *Kolombangara Forest Products Limited Solomon Islands*. Forestry VA Report No. 2. London: Commonwealth Development Corporation, Oct. 1991.
Brookfield, H. C. with Doreen Hart. *Melanesia: A Geographical Interpretation of an Island World*. London: Methuen, 1971.
Browne, Christopher with Douglas A. Scott. *Economic Development in Seven Pacific Island Countries*. Washington, DC: International Monetary Fund, 1989.
Building the Navy's Bases in World War II: History of the Bureau of Yards and Docks and the Civil Engineer Corps, 1940–1946. Vol. 2. Washington: United States Government Printing Office, 1947.
Bulmer, R. N. H. 'Traditional Conservation Practices in Papua New Guinea'. In *Traditional Conservation in Papua New Guinea: Implications for Today*, edited by L. Morauta, John Pernetta, and William Heaney, 59–76. Boroko, Port Moresby, Papua New Guinea: Institute of Applied Social and Economic Research, 1982.
Burgess, Joanne C. 'Timber Production, Timber Trade and Tropical Deforestation'. *Ambio* 22, no. 2–3 (May 1993): 136–43.
Burman, Rickie. 'Time and Socioeconomic Change on Simbo, Solomon Islands'. *Man* 16, no. 2 (June 1981): 251–67.
Burt, Ben. 'Land Rights and Development: Writing About Kwara'ae Tradition'. *Cultural Survival* 15, no. 2, (1991): 62–4.
————. *Tradition and Christianity: The Colonial Transformation of a Solomon Islands Society*. Camberwell, Victoria: Harwood, 1994.
Burt, Ben, and Kwa'ioloa, *The Forest of Kwara'ae, 1998*. Photocopy of draft.
Burton, J. W. *The Fiji of To-day*. London: C. H. Kelly, 1910.
Cameron, Alan L. *Primary Processing of Logs and Timber Handling in the Solomon Islands, Timber Control Unit*. Project Report No. 3, Vol. 1. [Canberra]: AIDAB, May 1992.
————. *Training Needs in the Sawmilling Industry in the Solomon Islands, Timber Control Unit*. Project Report No. 3, Vol. 2. [Canberra]: AIDAB, May 1992.
Carron, L. T. *A History of Forestry in Australia*. Canberra: Australian National University Press, 1985.
Carter, George. *A Family Affair: A Brief Survey of New Zealand Methodism's Involvement in Mission Overseas, 1922–1972*. Auckland: Wesley Historical Society, 1972.
Cassells, Ross M. 'The Valuation of Subsistence Use of Tropical Rainforest on the Island of Choiseul, Solomon Islands: A Comparison between Subsistence Values and Logging Royalties'. M. Phil. thesis, Massey University, Palmerston North, 1992.
Central Bank of Solomon Islands, *Annual Reports 1989–1997*. Honiara: Central Bank, 1990–1998.
Chaplin, G. E. and A. J. Neumann. *Afforestation on the Guadalcanal Grasslands*. Forestry Note No. 18–2/87. Honiara: Forestry Dept., July 1987.
Chaplin, Graham. *Silvicultural Manual for the Solomon Islands*. London: Overseas Development Administration, Foreign and Commonwealth Office, 1993.
Chapman, M. D. 'Environmental Influences on the Development of Traditional Conservation in the South Pacific Region'. *Environmental Conservation* 12 (1985): 217–30.
Chatterjee, Pratap, and Matthias Finger. *The Earth Brokers: Power, Politics and World Development*. London and New York: Routledge, 1994.
Cheatle, R. J. *Productivity and Status of Logged-over Soils at Barora, North New Georgia, Solomon Islands*. Forestry Note No. 24–1/88. Honiara: Ministry of Natural Resources, Forestry Division, 1988.
————. 'Tree Growth on a Compacted Oxisol.' *Soil Tillage Research* 19 (1991): 331–4.

Bibliography

Cheyne, Andrew. *The Trading Voyages of Andrew Cheyne, 1841–1844*. Edited by Dorothy Shineberg. Canberra: Australian National University Press, 1971.
Clarence, Margaret. *Yield not to the Wind*. Sydney: Management Development Publications, 1982.
Clark, Kevin. 'Popular Participation in Rural Development: The New Zealand / Solomon Islands Customary Land Reforestation Project on Malaita'. M. Phil. thesis, Massey University, Palmerston North, 1991.
Clarke, W. C. 'Agro-deforestation in Melanesia'. *Cultural Survival Quarterly* 15, no. 2 (1991): 45–8.
Clarke, W. C., and R. R. Thaman, eds. *Agroforestry in the Pacific Islands: Systems for Sustainability*. Tokyo: United Nations University, 1993.
Cleave, John. 'Environmental Assessments'. *Finance and Development* 25 (Mar. 1988): 44–7.
Cline-Cole, R. A. 'Political Economy, Fuelwood Relations, and Vegetation Conservation: Kasar Kano, Northern Nigeria, 1850–1975'. *Forest and Conservation History* 38, no. 2 (1994): 67–78.
Codrington, Robert Henry. *The Melanesians: Studies in their Anthropology and Folklore*. Oxford: Clarendon Press, 1891.
Colchester, M. 'Beyond Participation: Indigenous Peoples, Biological Diversity Conservation and Protected Area Management'. *Unasylva* 47 (1996): 33–9.
Collee, John. 'Stripping the Pacific Island Rain Forests.' *San Francisco Examiner*, 1 Jan. 1995, A14–A15.
Collins, Maev. 'Social and Political Impact of Logging Projects: West Coast Manus, Papua New Guinea'. Seminar paper. Canberra: Department of Political and Social Change, Australian National University, July 1990. Photocopy.
Collinson, Clifford W. *Life and Laughter 'midst the Cannibals*. London: Hurst and Blackett, 1926.
Commonwealth Development Corporation. *CDC: The Committed Investor*. London: Commonwealth Development Corporation, [1988].
Commonwealth Forestry Association. 'Forest Policy in the British Solomon Islands Protectorate'. *Commonwealth Forestry Review* 48 (1969): 190–2.
Conklin, Harold, C. *The Study of Shifting Cultivation*. Washington, DC: Union Panamericana, 1963.
———. 'An Ethnographic Approach to Shifting Agriculture'. In *Environment and Cultural Behavior: Ecological Studies in Cultural Anthroplogy*, edited by Andrew P. Vayda, 221–33. New York: The Natural History Press, 1969.
Connell, John. *Taim bilong mani: The Evolution of Agriculture in a Solomon Island Society*. Development Studies Centre monograph, No. 12. Canberra: Australian National University, 1978.
Constantini, A. *Timber Control Project Solomon Islands: Forest Inspectorate Ranger Training Course*. Honiara: AIDAB and Forestry Technical Services, 1992.
Cooper, Matthew Owen. 'Langalanga Religion'. *Oceania* 43, no. 2 (Dec. 1972): 113–22.
Coopers and Lybrand. *A Report on Timber Processing in the Solomon Islands and Strategies to Achieve a Profitable Industry through Sustainable Production Levels*. Honiara: Solomon Islands Forest Industries Association, 1993.
Coote, Walter. *The Western Pacific*. 1883. Reprint, New York: Praeger, 1969.
Corrin, J. C. 'Abrogation of the Rights of Customary Land Owners by the Forest Resources and Timber Utilisation Act'. *Queensland University of Technology Law Journal* 8 (1993): 131–40.
Corris, Peter. *Passage, Port and Plantation: A History of Solomon Islands Labour Migration, 1870–1914*. Melbourne: Melbourne University Press, 1973.
Costa, Perdo Moura. 'Tropical Forestry Practices for Carbon Sequestration: A Review and Case Study from Southeast Asia'. *Ambio* 25, no. 4 (June 1996): 279–83.
Courier Mail (Brisbane, Australia) Dec. 1995– Jan. 1996.
Cruickshank, M. 'Work in Timber is now a Going Concern'. *Marist Messenger* 38, no. 8 (1968): 23–6.
Curriculum Development Centre, Department of Education. *The Use of the Forests*. Honiara: Curriculum Development Centre, c. 1984.
Dahl, L. *Regional Ecosystem Survey of the South Pacific Area*. Noumea: South Pacific Commission, 1980.
Dargavel, John. *Fashioning Australia's Forests*. Melbourne: Oxford University Press, 1995
Dauverge, Peter. *Shadows in the Forest: Japan and the Politics of Timber in Southeast Asia*. Cambridge, Mass: MIT Press, 1997.
———. 'Corporate Power in the Forests of Solomon Islands.' *Pacific Affairs* 71, no. 4 (Winter 1998/9): 524–46.

Bibliography

Davidson, J. W. *Peter Dillon of Vanikoro: Chevalier of the South Seas*. Edited by O. H. K. Spate. Melbourne: Oxford University Press, 1975.

Dawkins, H. C. *The Management of Natural Tropical High Forest with Special Reference to Uganda*. Imperial Forestry Institute Paper No. 34. Oxford: Imperial Forestry Institute, 1958.

Deklin, Francis. 'Papua New Guinea Forest Policy and Local Interest Groups'. In *Resources, Development and Politics in the Pacific Islands*, edited by Stephen Henningham and R. J. May, 119–28. Bathurst, New South Wales: Crawford House, 1992.

Dennis, G. F. 'Nature Notes on the Solomons'. *The Journal of the Solomon Islands Museum Association* 1 (1972): 31–42

Development Studies Centre. *Te Amokura, Na Lawedua* 7, no. 2 (May–Aug. 1995). Palmerston North: Development Studies Centre, Massey University, New Zealand.

Dillon, Peter. *Narrative of La Perouse's Expedition*. Vol. 2. London: Hurst, Chance, 1829.

Dod, Karl C. *The Corps of Engineers: The War against Japan*. Washington: Office of the Chief of Military History, United States Army, 1966.

Dumont D'Urville, Jules S-C. *An Account of Two Voyages to the South Seas*. Vol. I. Translated by Helen Rosenman. Melbourne: Melbourne University Press, 1987.

Dumont, Rene. 'Some Reflections on Priorities in Melanesian Development'. In *Priorities in Melanesian Development*, edited by Ronald J. May, 23–6. Sixth Waigani Seminar. Port Moresby: University of Papua New Guinea, Port Moresby; and Canberra: Australian National University, 1974.

Duncan, Ronald. C. *Melanesian Forestry Sector Study*. Canberra: Australian International Development Assistance Bureau, 1995.

Dureau, Christine May. 'Mixed Blessings: Christianity and History in Women's Lives on Simbo, Western Solomons'. Ph. D. diss., Macquarie University, 1994.

Ells, Philip. *Logging and Legal Access, Solomon Islands*. Draft paper. Ely, Cambridgeshire: Forests Monitor Ltd, July 1996.

Empire Forest Review, 1939–1948.

Enright, H. J., and C. Cosden. 'Unstable Archipelagos: South-west Pacific Environment and Prehistory since 30,000 B.P.'. In *The Naive Lands: Prehistory and Environmental Change in Australia and the Southwest Pacific*, edited by John Dobson, 160–98. Melbourne: Longman Cheshire, 1992.

Fallon, J. and C. Karabalis. 'Current Economic Trends in the South Pacific'. *Pacific Economic Bulletin* 7, no. 2 (1992): 1–18.

Falvey, D. A. et al. 'Petroleum Prospectivity of Pacific Islands Arcs: Solomon Islands and Vanuatu'. *The Apea Journal* 31, no. 1 (1991): 191–212.

Fate of the Forests. Shell International Petroleum Co., UK: Shell Film Unit, 1982.

Filer, Colin, with Nikhil Sekhran. *Loggers, Donors and Resource Owners*. Port Moresby: Papua New Guinea National Research Institute; London: International Institute for Environment and Development, 1998.

Fingleton, James S. *Assistance in the Review of Forestry Policy and Legislation*. Rome: FAO, 1989.

Finsch, Otto. 'Uber Naturprodukte der westichen Sudsee, besonders der deutschen Schutzgebiete'. *Deutsche Kolonialzeitung. Organ des deutschen Kolonialvereins* 4 (1887): 17–19.

Firth, Raymond. 'Anuta and Tikopia: Symbiotic Elements in Social Organization'. *Journal of the Polynesian Society* 63 (1954): 88–131.

_____. *Social Change in Tikopia*. London: George Allen and Unwin, 1959.

Fisk, E. K. 'The Economic Structure,' In *New Guinea on the Threshold: Aspects of Social, Political and Economic Development*, edited by E. K. Fisk, 23–43. London: Longmans, 1966.

Flannery, Timothy Fridtjof. *The Future Eaters: An Ecological History of the Australasian Lands and Peoples*. Chatswood, New South Wales: Reed Books, 1994.

Foanaota, Lawrence, 'Burial Sites on Vella Lavella Island'. *Journal of the Solomon Islands Museum Association* 2 (1974): 22–33.

Food and Agricultural Organisation. *Solomon Islands Forestry Development Project–Preparation Mission*. Report of the FAO/World Bank Cooperative Programme investment Centre. Rome: FAO, 1989.

Forest Giants from Vanikoro. Melbourne: Commercial Boxes Pty Ltd, Peacock Bros, [1938].

Forest Monitor Ltd. *Kumpulan Emas Berhad and its Involvement in the Solomon Islands*. Ely, Cambridgeshire: Forest Monitor Ltd, 1996.

Bibliography

Forest Monitor Ltd. *Environmental and Social Impact Assessment: ITC Operations on Isabel Island, Solomon Islands*. Ely, Cambridgeshire: Forest Monitor Ltd, April 1997.
Fox, Charles E. *The Threshold of the Pacific: An Account of the Social Organization, Magic and Religion of the People of San Cristoval in the Solomon Islands*. London: Kegan Paul, 1924.
Fox, N. J. 'His Schools will also Teach Skills'. *Marist Messenger* 37, no. 4 (Apr. 1967): 2–3.
Fraser, A. I. *Fiscal Aspects of the Timber Industry: Solomon Islands*. Rotorua, New Zealand: Institute of Forest Science, 1980.
Fraser, T., and A. Larsen. *Consultants' Final report: Evaluation of Reforestation Project, North New Georgia*. Rotorua, New Zealand: Chandler, Fraser and Larsen, 1981.
Frawley, Kevin. 'Evolving Visions: Environmental Management and Nature Conservation in Australia'. In *Australian Environmental History: Essays and Cases*, edited by Stephen Dovers, 55–78. Melbourne: Oxford University Press, 1994.
Frazer, Ian. 'Resource Extraction and the Post-colonial State in Solomon Islands'. In *Asia Pacific: New Geographies of the Pacific Rim*, edited by R. F. Watters and T. G. McGee, 318–34. Vancouver: University of British Columbia Press, 1997.
Friesen, Wardlow. 'Melanesian Economy on the Periphery: Migration and Village Economy in Choiseul'. *Pacific Viewpoint* 34, no. 2 (1993): 193–214.
Fugui and another v. Solmac Construction Company Ltd. and Others. High Court of Solomon Islands, *The Solomon Islands Law Reports* (11 Oct. 1982).
Gadgil, Madhav, and Ramachandra Guha. 'State Forestry and Social Conflict in British India'. *Past and Present* 123 (May 1989): 141–77.
Gadgil, Madhav, and Ramachandra Guha. *This Fissured Land: An Ecological History of India*. Berkeley and Los Angeles: University of California Press, 1993.
Gegeo, David W. 'Tribes in Agony: Land, Development, and Politics in Solomon Islands'. *Cultural Survival Quarterly* 15, no. 2 (1991): 53–6.
Gilbert, G. D. *Report on Visit to the Solomon Islands, June 1966*. Honiara: Forestry Department, 1966. Cyclostyled.
Golden, Graeme A. *The Early European Settlers of the Solomon Islands*. Melbourne: Published by the author, 1993.
Goldie, J. F. 'The People of New Georgia, their Manners and Customs and Religious Beliefs'. *Proceedings, Royal Society of Queensland* 22 (1909): 23–30.
Gordon, W. A. *The Law of Forestry*. London: Her Majesty's Stationery Office, 1955.
Grimshaw, Beatrice. *In the Strange South Seas*. London: Hutchinson and Co., [1908].
Groenewegen, K. *Report on the Census of Population, 1970*. Honiara: Western Pacific High Commission, British Solomon Islands Protectorate, [1972].
Groome, J. G. and associates. *Malaita Forestry Development Study*. Suva: United Nations Development Advisory Team, Economic and Social Commission for Asia and the Pacific, 1982.
———. *Technical Appendices for the Malaita Forestry Development Study*. Suva: United Nations Development Advisory Team, Economic and Social Commission for Asia and the Pacific, 1982.
Groube, Les M. 'The Taming of the Rainforests: A Model for Late Pleistocene Forest Exploitation in New Guinea'. In *Foraging and Farming*, edited by David R. Harris and Gordon C. Hillman, 292–303. London: Unwin Hyman, 1989.
———. 'Contradictions and Malaria in Melanesian and Australian Prehistory'. In *A Community of Culture: The People and Prehistory of the Pacific*, edited by Matthew Spriggs, et al., 164–86. Canberra: Australian National University, 1993.
Grove, Richard. 'The Origins of Environmentalism'. *Nature* 345, no. 3 (1990): 11–15.
———. *Green Imperialism: Colonial Expansion, Tropical Island Edens and the Origins of Environmentalism, 1600–1860*. Cambridge: Cambridge University Press, 1995.
Grover, Ruhi. 'Rhythms of the Timber Trade: Forests in the Himalayan Punjab, 1850–1925'. Ph. D. diss., University of Virginia, 1997.
Grynberg, R., S. Vemuri, and D. Wigson. 'Reconciling Ethical Behavour: Transfer Pricing in the Forest Industry of Papua New Guinea'. In *The Ethics of Development: Choices in Development Planning*, edited by Philip J. Hughes and Charmian Thirlwall, 175–203. Port Moresby: University of Papua New Guinea, 1988.
Guppy, H. B. *The Solomon Islands and their Natives*. London: Sonnenschein, 1887.
Hadley, Chris. *A Forester in the Solomon Islands*. Lewes, Sussex: The Book Guild, 1991.

Bibliography

Hall, C. M. 'The "Worthless Lands Hypothesis" and Australia's National parks and reserves'. In *Australia's Ever Changing Forest: Proceedings of the First National Conference of Australian Forest History*, edited by K. J. Frawley and N. Semple, 441–56. Canberra: Australian National University, 1988.

Hansell, John R. F., and John R. D. Wall, *Land Resources of the Solomon Islands. Vol. 4: New Georgia group and the Russell Islands*. Surbiton, Surrey: Land Resources Division, Ministry of Overseas Development, 1976.

Harrison, Brian, ed. *The History of Oxford*. Vol. 8, Oxford: Clarendon, 1995.

Havinden, Michael, and David Meredith. *Colonialism and Development: Britain and its Tropical Colonies, 1850–1960*. London: Routledge, 1993.

Heath, Ian. 'Land Policy in Solomon Islands'. Ph. D. diss., La Trobe University, 1979.

Henderson, C. P., and I. R. Hancock. *A Guide to the Useful Plants of Solomon Islands*. Honiara: Ministry of Agriculture and Lands, 1988.

Hermant, P., and R. W. Cilento. *Report of the Mission Entrusted with a Survey on Health Conditions in the Pacific Islands*. Geneva: League of Nations Organization, 1929.

Hewagama, Ranjit. *The Revised Laws of Solomon Islands: Prepared under the Authority of the Revised Edition of the Laws Act 1995*. 9 vols. Hertford: Caxton Hill, 1996.

Hibberd, J. J. and J. Schenk. *Country Report for the 1992 UNCED Conference: Solomon Islands*. Apia, Samoa: South Pacific Regional Environment Programme, 1991.

Hickie, F. J. 'Coconut planting in the British Solomons'. *The Planters' Gazette* (Dec. 1922): 11–12.

Hilder, Brett. 'Timber from Guadalcanal'. *Walkabout*, 1 May 1952, 17–20.

Hilliard, David. *God's Gentlemen: A History of the Melanesian Mission, 1949–1942*. Brisbane: University of Queensland Press, 1978.

Hoadley, Steve. *The South Pacific Foreign Affairs Handbook*. Sydney: Allen and Unwin, 1992.

Hogbin, H. Ian. *Experiments in Civilization: The Effects of European Culture on a Native Community of the Solomon Islands*. London: George Routledge and Sons, 1939; Reprint, Routledge and Kegan Paul, 1969.

———. *A Guadalcanal Society: The Kaoka Speakers*. New York: Holt Rinehart and Winston, 1964.

Honiara Municipal Authority. *Baonsa Book*. Honiara: Honiara Sup Sup Gaden, Honiara Municipal Authority, 1990.

Hopkins, A. I. *In the Isles of King Solomon: An Account of Twenty-five Years Spent amongst the Primitive Solomon Islanders*. London: Seeley, Service, 1928.

Hughes, A. V. *Climbing the Down Escalator: The Economic Conditions and Prospects of Solomon Islands*. Islands/Australia working Paper No. 88/2. Canberra: National Centre for Development Studies, Australian National University, 1988.

———. 'Reflections on Central Banking in Solomon Islands'. *Pacific Economic Bulletin* 10, no. 2 (1995): 70–3.

Hviding, Edvard. *Guardians of Marovo Lagoon: Practice, Place and Politics in Maritime Melanesia*. Honolulu: University of Hawai'i, 1996.

Hviding, Edvard, and Graham B. K. Baines. *Fisheries Management in the Pacific: Tradition and the Challenges of Development in Marovo, Solomon Islands*. Geneva: United Nations Research Institute for Social Development, 1992.

———. 'Community-based Fisheries Management, Tradition and Challenges of Development in Marovo, Solomon Islands'. *Development and Change* 25 (1994): 13–39.

Ifui, J. Anderson. *Tekolea Nature Trail on the Poitete Reserve Forest Kolombangara*. Poitete, Solomon Islands: Ministry of Natural Resources, Forestry Division, 1990.

Insight: Australian Foreign Affairs and Trade Issues. Canberra: Overseas Information Branch, Department of Foreign Affairs and Trade, 7 Aug. 1992.

Integrated Forest Industries. *Financial Statement of Integrated Forest Industries (SI) Ltd*. n. p., 30 June 1983.

International Bank for Reconstruction and Development. *The Economy of the Solomon Islands*. Washington: International Bank for Reconstruction and Development, East Asia and Pacific Dept, Feb. 1969.

Iputu, S. *Overview of Forestry Research in Solomon Islands in Relation to Forestry Extension*. Munda, Solomon Islands: Ministry Of Natural Resources, Forestry Division, 1992.

Bibliography

———. *Seed Collection and Supply*. Munda, Solomon Islands: Ministry Of Natural Resources, Forestry Division, 1992.
Irwin, Geoffrey. *The Prehistoric Exploration and Colonisation of the Pacific*. Cambridge: Cambridge University Press, 1992.
Islands Business (Suva, Fiji), 1990–1996
Ivens, Walter G. *Melanesians of the South-east Solomon Islands*. London: Kegan Paul, Tench, Trubner, 1927.
———. *The Island Builders of the Pacific: How and why the People of Mala Construct their Artificial Islands: the Antiquity and Doubtful Origin of the Practice, with a Description of the Social Organisation, Magic and Religion of their Inhabitants*. London: Seeley, Service, 1930.
———. 'Flints in the South-East Solomon Islands'. *Journal of the Royal Anthropological institute of Great Britain and Ireland* 61 (1931): 421–4.
Jackson, K. B. 'Tie Hokara, Tie Vaka, Black Man White Man: A Study of the New Georgia Group to 1925'. Ph. D. diss., Australian National University, 1978.
James, N. D. G. *A History of English Forestry*. Oxford: Blackwell, 1981.
Johnson, Martin. *Through the South Seas with Jack London*. London: T. Werner Laurie, 1913.
Kajewski, S. Frank. 'A Plant Collector's Notes on the New Hebrides and Santa Cruz Islands'. *Arnold Arboretum Journal* 11 (1930): 172–180.
Katibas News (Honiara), 1996
Keesing, R. M. *Kwaio Religion: The Living and the Dead in a Solomon Islands Society*. New York: Columbia University Press, 1982.
Keesing, R. M., and Peter Corris. *Lightning Meets the West Wind: The Malaita Massacre*. Oxford: Oxford University Press, 1980.
King, P. *Solomon Islands National Forest Resources Inventory: Project Implementation Document*. National Forest Resources Inventory Project, Working Paper No. 10. Honiara: AIDAB and Ministry of Natural Resources, 1991.
Kirch, Patrick V. *The Evolution of the Polynesian Chiefdoms*. Cambridge: Cambridge University Press, 1984.
———. 'Polynesian Agricultural Systems'. In *Islands, Plants, and Polynesians: An Introduction to Polynesian Ethnobotany*, edited by Paul Alan Cox and Sandra Anne Banack, 113–33. Portland, Oregon: Dioscorides, 1991.
Kirch, Patrick Vinton, and D. E. Yen. *Tikopia: The Prehistory and Ecology of a Polynesian Outlier*. Honolulu: Bishop Museum Press, 1982.
Knapman, Bruce. 'Aid and the Dependent Development of Pacific Island States'. *Journal of Pacific History* 21, no. 3 (1986): 139–53.
Knibbs, S. G. C. *The Savage Solomons: as They Were and Are. A Record of a Head-hunting people Gradually Emerging From a Life of Savage Cruelty and Bloody Customs, with a Description of their Manners and Ways and the Beauties and Potentialities of their Islands*. London: Seeley, Service, 1929.
Kraemer, J. H. *Native Woods for Construction Purposes in the Western Pacific Region*. Rev. edn, No. P-101. Washington: United States Navy Department, Bureau Yards and Docks, 1944.
Kranenberg, Bill. *Grading and Training Report, Timber Control Unit Project*, Project Report No. 2. Canberra: AIDAB, 1992.
———. *Grading Solomon Islands Timber: proposed grading Rules*. Honiara: Forestry Division, Timber Control Unit and Fortech, Canberra, 1992.
Kuki, Sosimo, and Tim Thorpe. *Chainsaw Sawmilling*. Honiara: Ministry of Natural Resources, 1992.
Kumar, Raj. *The Forest Resources of Malaysia: Their Economics and Development*. Singapore: Oxford University Press, 1986.
Kumpulan Emas Berhad. *Annual Report, Kumpulan Emas Berhad, 1995*. Kuala Lumpur, Malaysia, 1995.
La Caledonie (Noumea, New Caledonia), 11 May 1898.
Laarman, Jan G. 'Export of Tropical Hardwoods in the Twentieth Century'. In *World Deforestation in the Twentieth Century*, edited by John F. Richards and Richard P. Tucker, 151–6. Durham: Duke University Press, 1988.
Lamb, D. *Exploiting the Tropical Rainforest: An Account of Pulpwood Logging in Papua New Guinea*. Man and the Biosphere series, vol. 3. Paris: UNESCO, 1990.

Bibliography

Laracy, Eugenie, and Hugh Laracy. 'Beatrice Grimshaw: Pride and Prejudice in Papua'. *Journal of Pacific History* 12, no. 3 (1977): 154–76.
Laracy, Hugh. *Marists and Melanesians: A History of Catholic Missions in the Solomon Islands*. Canberra: Australian National University, 1976.
Laracy, Hugh, ed. *Pacific Protest: The Maasina Rule Movement, Solomon Islands, 1944–1952*. Suva: Institute of Pacific Studies of University of the South Pacific, 1983.
Laracy, Hugh, and Geoffrey White. *O'o: A Journal of Solomon Islands Studies*. Vol. 4 (1988). Honiara: University of the South Pacific.
Larmour, Peter, ed. *Land in Solomon Islands*. Suva: Institute of Pacific Studies of University of the South Pacific and Ministry of Agriculture and Lands, Solomon Islands, 1979.
Larmour, Peter. 'The North New Georgia Corporation'. In *Land, People and Government*, edited by Peter Larmour, Ron Crocombe and Anna Taugenga, 133–46. Suva: Institute of Pacific Studies of University of the South Pacific, 1981.
Lasaqa, I. Q. *Melanesians' Choice: Tadhimboko Participation in the Solomon Islands Cash Economy*. New Guinea Research Unit Bulletin No. 46. Canberra: Australian National University, 1972.
Latukefu, Sione. 'Pacific Islander Missionaries'. In *The Covenant Makers: Islander Missionaries in the Pacific*, edited by Doug Munro and Andrew Thornley, 17–40. Suva: Pacific Theological College; Institute of Pacific Studies of University of the South Pacific, 1996.
Lauder, John. 'Communal Farm Development in the Solomon Islands'. *Pacific Economic Bulletin* 2, no. 1 (1987): 11–16.
[Leary, Tania]. *A National Environmental Management Strategy for Solomon Islands*. Draft plan, [Honiara]: South Pacific Regional Environment Programme, c. 1992.
Lee, K. E. 'Some Soils of the British Solomons Protectorate'. *Philosophical Transactions of the Royal Society Bulletin* 255 (1969): 211–57.
Lees, Annette. *A Representative Protected Forests System for the Solomon Islands: Report*. Environment Division, Ministry of Natural Resources, Honiara, on behalf of Australian National Parks and Wildlife Service. Nelson, New Zealand: Maruia Society, 1991.
Lembke, F. C. 'Forestry in the Solomon Islands'. *The Australian Timber Journal and Building Products Merchandiser* 36, no. 8 (1970): 46–8.
———. 'Levers to Increase Log Exports to Australia'. The *Australian Timber Journal and Building Products Merchandiser* 36 (Sept. 1970): 16–43.
Leslie, A. J. *Forest Economics, British Solomons Islands Protectorate: Project Findings and Recommendations*. Rome: FAO, United Nations Development Programme, 1975.
Lesson, Pierre Adolphe. 'Vanikoro et ses Habitants'. *Revue d' Anthropologie* 4 (1876): 252–72.
Link (Honiara), no. 23, Nov.–Dec. 1991; no. 25, May–June 1992.
Lloyd, D. T. *Land Policy in Fiji*. Cambridge: Piggott Printers, 1982.
Luxton, Clarence Thomas James. *Isles of Solomon: A Tale of Missionary Adventure*. Auckland: Methodist Foreign Missionary Society of New Zealand, 1955.
McCall, G. 'Little Ice Age: Some Speculations for Rapanui'. *Rapa Nui Journal* 7 (1993): 65–70.
Macbride-Stewart, Sheila. 'Catholic Women's Programmes in Malaita, Solomon Islands: Breaking the Culture of Silence through Empowerment'. M. Phil. thesis, Massey University, Palmerston North, 1996.
Macfarlane, R., G., V. H. Jackson, and K. D. Marten. 'Die-back of Eucalypts in the Solomon Islands'. *Commonwealth Forestry Review* 55 (1975): 133–8.
McKinnon, Malcolm. *Independence and Foreign Policy: New Zealand in the World Since 1935*. Auckland: Auckland University Press, 1993.
McKinnon, John M. *Bilua Report: Preliminary Results of Research into Rural Economic Development, Vella Lavella, BSIP*. Wellington: Victoria University, 1973.
McLaren, Jack. *My Odyssey*. London: Jonathan Cape, 1923.
MacQuarrie, Hector. *Vouza and the Solomon Islands*. Sydney: Angus and Robertson, 1946.
Maenu'u, Leonard P. 'Traditional Farming in the Solomon Islands'. In *The Melanesian Environment: Papers Presented at the Ninth Waigani Seminar held at Port Moresby*, edited by John H. Winslow, 139–45. Canberra: Australian National University Press, 1977.
———. *Bib-Kami na Ano: Land and Land Problems in Kwara'ae*. Honiara: University of the South Pacific, Solomon Islands Centre, 1981.

Bibliography

Makis, Ephraim, and P. A. S. Dahanayake. *Foreign Aid in Papua New Guinea*. Canberra: National Centre for Development Studies, Australian National University, 1985.

Mamaloni, Solomon. 'Politics of Rural Development: A Political Structure for Rural Development'. In *The Road Out: Rural Development in the Solomon Islands*, edited by Stephen Oxenham, 79–84. Suva: University of the South Pacific, 1981

──────. 'Rural Development Then and Now: Legacy of Colonial Miscalculation'. In *The Road Out: Rural Development in the Solomon Islands*, edited by Stephen Oxenham, 1–4. Suva: Institute of Pacific Studies of University of the South Pacific, 1981.

Manner, I. H. I. 'Buma village, West Kwara'ae. Malaita, the Solomon Islands'. In *Agroforestry in the Pacific Islands: Systems for Sustainability*, edited by W. C. Clarke and R. R. Thaman, 43–7. Tokyo: United Nations University Press, 1993.

Mantel, Kurt. 'History of the International Science of Forestry'. *International Review of Forestry Research* 1 (1964), 2–27.

Marjoram, Tony, and Sue Fleming. 'Levers in the Solomon Islands'. *Raw Materials Report* 2, no. 1 (1982).

Marten, K. D. 'Silvics of Species: *Agathis macrophylla*'. Interim Report. Honiara: Forestry Dept, 1970. Cyclostyled.

──────. *Solomon Island Tropical Rainforest Ecology*. Silvicultural Research Report No. 5/2/79. Honiara: Forestry Dept, 1979.

──────. 'Forest History in the Solomon Islands'. Unpublished paper, 1997. Photocopy.

Marwick, J. G., comp. *The Adventures of John Renton*. Kirkwall: Kirkwall Press, 1935.

Mauler, J. von, and William Kesslitz. 'The Scientific Mission of SM *Albatros*, 1895–1898'. *Report from the K. K. Geographical Society in Vienna*. Vol. 1. Edited by A. B. E. von Böhmerscheim, translated by V. C. Wasem. Vienna: Lechner, 1899.

Methodist Church. *Reports of the Australasian Wesleyan Methodist Society*. Sydney: Methodist Church, 1905, 1907.

Miles, J. A. R. *Infectious Diseases in the Pre-European Pacific*. In Press, Dunedin: University of Otago.

Milner, Edward, 'The Chainsaw Massacre'. *Post Courier, Weekend Magazine*, Port Moresby, 13 May 1994, 33.

Monberg, Torben. *The Reactions of People of Bellona Island towards a Mining Project*. Copenhagen: International Work Group for Indigenous Affairs, 1976.

Morauta, Louise, John Pernatta, and William Heaney, eds. *Traditional Conservation in Papua New Guinea: Implications for To-day*. Port Moresby, Papua New Guinea: Institute of Applied Social and Economic Research, 1982

Naess, Arne. *Ecology, Community, and Lifestyle: Outline of an Ecosophy*. Translated and revised by David Rothenberg. Cambridge: Cambridge University Press, 1989.

Nectoux, François, and Yoichi Kuroda. *Timber from the South Seas: An Analysis of Japan's Tropical Timber Trade and its Environmental Impact*. Gland, Switzerland: World Wildlife Fund International, 1989.

Neil, P. E. 'Climber Problems in Solomon Islands Forestry'. *Commonwealth Forestry Review* 63, no. 1 (1984): 29–33.

Nelson, Paul J. *The World Bank and Non-Governmental Organizations: The Limits of Apolitical Development*. New York: St. Martin's Press, 1995.

Nerdrum, J. G. B. 'Indtryk og oplevelser under et 7 aars ophold paa Salomon-oerne'. *Norske Geog. Selskabs Aarbog* 13 (1901/2): 22–58.

Neumann, Roderick P. 'Forest Rights, Privileges and Prohibitions: Contextualising State Forestry Policy in Colonial Tanganyika'. *Environment and History* 3 (1997): 45–68.

New Zealand Herald (Auckland), 30 May 1992.

New Zealand Methodist Times (Christchurch), 6 Dec. 1924.

News Drum (Honiara), 1980.

Novar, Viscount. 'Inaugural Meeting of the Empire Forestry Association'. *Empire Forestry* 1 (1922): 5–6.

Nunn, Patrick D. 'Recent Environmental Changes on Pacific Islands'. *Geographical Journal* 156 (1990): 125–40.

──────. *Environmental Change in the Pacific Basin*. Chichester, England: Wiley, 1999.

Bibliography

Oliphant, J. N. 'The Development of more Intensive Use of Mixed Tropical Forest'. *Empire Forestry Review* 16 (1937): 29–37.
Olsen, M. F. and M. H. Turnbull. *Assessment of Growth Rates of Logged and Unlogged Forests in Solomon Islands*. National Forest Resources Inventory Project Working Paper 18. Honiara: AIDAB and Ministry of Natural Resources, 1993.
Open Door (Auckland), 1922.
Oxenham, Stephen, ed. *The Road Out: Rural Development in Solomon Islands*. Suva: Institute of Pacific Studies of University of the South Pacific, 1981.
Pacific Islands Monthly (Sydney), 1935–June 1988; (Suva), July 1988–1996.
Palmer, B. S. *The Interaction of Churches and State in the British Solomon Islands over the Designated Schools Scheme, 1964–1967*. Paper, n.d. Cyclostyled.
Papadakis, Elim. *Environmental Politics and Institutional Change*. Cambridge: Cambridge University Press, 1996.
Park, Geoff. 'The Polynesian Forest: Customs and Conservation of Biological Diversity'. In *Science of Pacific Island Peoples*. Vol. 2. Edited by John Morrison, Paul Geraghty and Linda Crowl, 132–53. University of the South Pacific: Suva, 1994.
Parker, Russell. *Maekera: The Life Story of Hereditary Chief Nathan Kera and the Saikile Community of Solomon Islands, as Told to Russell Parker*. Tenterfield, New South Wales: Solomonesia, 1994
Parkinson, R. 'Ethnography of Ontong Java and Tasman Islands with Remarks re: The Marqueen and Abgarris Islands'. *Pacific Studies* 9, no. 3 (July 1986): 1–31.
Peluso, Nancy Lee. *Rich Forests, Poor People: Resource Control and Resistance in Java*. Berkeley and Los Angeles: University of California Press, 1992.
Pelzer, Karl. *Pioneer Settlements in the Asiatic Tropics*. New York: Pacific Institute, 1945.
Pendleton, R. C. 'The Rain Shadow Effect on the Plant Formations of Guadalcanal'. *Ecological Monographs* 19, no. 1 (1949): 75–93.
Peters, Leon F., and William J. Neuenschwander. *Slash and Burn: Farming in the Third World Forest*. Moscow, Idaho: University of Idaho Press, 1988.
Peterson, William. *Population*. New York: Macmillan, 1961.
Pleydell, G. J. 'Solomon Islands Timber: Developing a new export industry'. *Progress* 83 (1969): 80–8.
_____. *Timbers of the British Solomon Islands*. London: United Africa Co., 1970.
Ponting, Clive. *A Green History of the World: The Environment and the Collapse of Great Civilizations*. New York: Penguin, 1993.
Potton, Craig. *Tongariro : A Sacred Gift*. Auckland: Lannsdowne, 1987.
Price Waterhouse, Economics Studies and Strategies Unit. *Forestry Taxation and Domestic Processing Study*. Final Draft Report for the Solomon Islands Government, Ministry of Finance, and the Ministry of Forests, Environment and Conservation. [Honiara]: Price Waterhouse, 1995.
Raabe, H. E. *Cannibal Nights: The Reminiscences of a Free Lance trader*. New York: Geoffrey Bles, Payson and Clark, 1927.
Radford D. A. and R. J. Blong. *Natural Disasters in Solomon Islands*. 2nd edn. Sydney: School of Earth Sciences, Macquarie University, 1992.
Radford, Kent H. 'The Ecologically Noble Savage'. *Cultural Survival Quarterly* 15, no. 1 (1991): 46–8.
Rainbow, R. G. and S. Teteha. *Shifting Cultivation in Solomon Islands*. Honiara: Curriculum Development Centre, 1983.
Rangarajan, M. 'Imperial agendas and India's forests: The Early History of Indian Forestry, 1800–1878'. *The Indian Economic and Social History Review* 31, no. 2 (1994): 147–167.
Raucaz, L. M. *In the Savage South Solomons: The Story of a Mission*. Lyons: Society for the Propagation of the Faith, 1928.
Reed, A. H. *The New Story of the Kauri*. Rev. edn. Wellington: A. H. and A. W. Reed, 1964.
Rence, Gordon. 'Land, Trustees and Timber'. *Pacific Perspective* 8, no. 2 (1979): 18–20.
Rentz, John. *Marines in the Central Solomons*. Washington, DC: United States Marine Corps, 1952.
Ribbe, Carl. *Zwei jahre unter den Kannibalen der Salomo-Inseln: Reiseerlebnisse und Schilderungen von Land und Leuten*. Dresden: Blasewicz, Druck und Verlag der Elbgau-Buchdruckerei, Hermann Beyer, 1903.

Bibliography

Rich, Bruce. *Mortgaging the Earth: The World Bank, Environmental Impoverishment and the Crisis of Development.* London: Earthscan, 1994.

Richardson, S. D. *The Role of Forest-based Industries in the Economic and Social Development of West Irian.* New York: United Nations Development Programme, 1968.

Riddell, R. *Foreign Aid Reconsidered.* Baltimore: The Johns Hopkins University Press, 1987.

Rivers, W. H. R. *Essays on the Depopulation of Melanesia.* Cambridge: Cambridge University, 1922.

Robin, Libby. *Defending the Little Desert: The Rise of Ecological Consciousness in Australia.* Melbourne: Melbourne University Press, 1998.

Roche, Michael, J. Dargavel. and J. Mills. 'Tracking the KTC from Kauri to Karri to Chatlee'. In *Australia's Ever-changing Forests 2*, edited by J. Dargavel and S. Feary,187–204. Canberra: The Australian Forest History Society and the Centre for Resource and Environmental Studies, 1993.

Roche, Michael. *History of New Zealand Forestry.* Wellington: New Zealand Forestry Corporation, 1990.

Ross, Harold M. 'Stone adzes from Malaita, Solomon Islands: An Ethnographic Contribution to Melanesian Archaeology'. *The Journal of the Polynesian Society* 79, no. 4 (Dec. 1970): 411–20.

———. *Baegu: Social and Ecological Organization in Malaita, Solomon Islands.* Urbana, Chicago: University of Illinois, 1973.

Roughan, John. 'Get a Handle on the Future by Knowing the Past'. Paper presented to the United Party, Honiara, 1996.

———. 'Solomon Islands Nongovernmental Organizations: Major Environment Actors'. *Contemporary Pacific* 9, no. 1 (Spring 1997): 157–66

Routley, R., and V. Routley. 'Destructive Forestry in Australia and Melanesia'. In *The Melanesian Environment*, edited by John Winslow, 374–97. Canberra: Australian National University, 1977.

Roux, J. C. 'Traditional Melanesian Agriculture in New Caledonia and Pre-contact Population Distribution'. In *Pacific Production Systems: Approaches to Economic Prehistory*, edited by D. E. Yen and J. M. J. Mummery, 161–173. Occasional Papers in Prehistory, No. 18. Canberra: Australian National University, 1990.

Rudel, Tom and Jill Roper. 'Regional patterns and Historical Trends in Tropical Deforestation, 1976–1990: A Qualitative Comparative Analysis'. *Ambio* 25, no. 3 (May 1996): 160–6.

Runte, A. '"Worthless lands" – Our National Parks: The Enigmatic Past and Uncertain Future of America's Scenic Wonderlands'. *American West* 10 (May 1973): 4–11.

Russell, T. 'The Culture of Marovo, British Solomon Islands'. *Journal of the Polynesian Society* 57 (1948): 306–29.

Salisbury, R. F. *From Stone to Steel: Economic Consequences of a Technological Change in New Guinea.* Melbourne: Melbourne University Press, 1962.

Scheffler, Harold W. *Choiseul Island Social Structure.* Berkeley and Los Angeles: University of California Press, 1965,

Schenk, John. *Forest Ecology and Biogeography in the Solomons.* National Forest Inventory Project Working Paper 15. Honiara: AIDAB and Ministry of Natural Resources, 1993.

Scherzer, Karl von. *Narrative of the circumnavigation of the globe by the Austrian frigate Novara, (Commodore B. von Wullerstorf-Urbair): undertaken by Order of the Imperial Government, in the years 1857, 1858 & 1859, under the immediate auspices of His I. and R. Highness the Archduke Ferdinand Maxmilian, Commander-in Chief of the Austrian Navy.* Vol. 2. London: Saunders, Otley, 1862.

Scheyvens, Regina. '"Engendering" Environmental Projects: The Case of Eco-timber Production in the Solomon Islands'. *Development and Practice* 8, no. 4 (Nov. 1998): 439–53.

Schoeffel, Penelope, Gerard Fitzgerald, and Alison Loveridge. *Forest Utilisation in the Solomon Islands: Social Aspects and Issues.* National Forest Resources Inventory Project Working Paper 19. Honiara: AIDAB and Ministry of Natural Resources, 1994.

Schumacher, E. F. *Small is Beautiful: A Study of Economics as if People Mattered.* London: Blond and Briggs, 1973.

Scobie, Paul. 'Angry Solomon Islanders want to Regain Control of their Forests'. *Habitat* 10 (Oct. 1982): 5–6.

———. 'Solomon Islands Rainforest – A Chance for Conservation'. *National Parks Journal* 31 (May 1987): 15–16.

Bibliography

Seijama v. Luna, 24 Feb. 1981. *The Solomon Islands Law Reports*. Honiara, 1982.
Selwyn, John Richardson. 'The Islands of the Western Pacific'. *Proceedings of the Royal Colonial Institute* 25 (1893/4): 361–92, 587–607.
Shakespeare, W. *The Tempest*. Edited by Northrop Frye. Baltimore, Maryland: 1959.
Shelton, H. M., J. H. Schottler, and G. Chaplin. *Cattle Under the Trees in the Solomon Islands*. Brisbane: Dept of Agriculture, University of Queensland, 1987.
Shepherd, C. Y. *Report on Agricultural Policy for Fiji and the Western Pacific High Commission Territories*. Noumea: Western Pacific High Commission, 1944.
Shield, E. D. *Plantation Opportunity Areas in the Solomon Islands*. National Forest Resource Inventory Project Working Paper 13. Honiara: AIDAB and Ministry of Natural Resources, 1992.
Shineberg, Dorothy. *They Came for Sandalwood*. Melbourne: Melbourne University Press, 1967.
Shlomowitz, Ralph. 'Mortality and Workers'. In *Labour in the South Pacific*, edited by Clive Moore, Jacqueline Leckie and Doug Munro, 124–7. Townsville: James Cook University, 1990.
Sirikolo, Myknee Qusa. *The Honiara Botanical Gardens*. Honiara: Ministry Of Natural Resources, 1992. Photocopy.
Smith, Harold Hamel, and Fred A. G. Pape. *Coconuts: The Consols of the East*. London: Tropical Life, 1912.
Snow, Deborah, and John Collee. 'The Rape of an Island Paradise'. *The Observer* (London), 29 Sept. 1996, 4–5.
Sol-Tree Nius (Honiara), 1996.
Solomon Islands College of Higher Education. *School of Natural Resources, Handbook, 1992*. Honiara: Solomon Islands College of Higher Education, 1992.
Solomon Islands Development Trust. *Summary Reports, 1990, 1991*. Honiara:, SIDT, 1990, 1991.
Solomon Islands Government [Including British Solomon Islands Protectorate]. Advisory Council of BSIP. *Minutes of Meeting of Advisory Council, 1951*. Suva: Western Pacific High Commission, 1952.
_____. Agricultural and Industrial Loans Board. *Annual Report of the Agricultural and Industrial Loans Board, 1973*. BSIP, 1974.
_____. *Agricultural Gazette*, BSIP. Vol. 1, no. 1 (1931–1932).
_____. Departments of Agriculture, Lands and Forestry. *Report of the Interdepartmental Committee of Shifting Cultivation and Soil Exhaustion in West Kwara'ae*. Honiara: Departments of Agriculture, Lands and Forestry, 1969.
_____. Department of Agriculture. See Webb, I. S.
_____. Office of the Prime Minister. *National Development Plan for the Solomons, 1975–1979*. Honiara: Office of the Prime Minister, 1975.
_____. Office of the Prime Minister. 'A National Forest and Timber Policy'. In *Forestry and Foreign Investment: Government of the Solomon Islands*. Honiara: Office of the Prime Minister, 1984.
_____. Office of the Prime Minister. Solomon Islands Development of the Forestry Sector and Related Industries/Resources, Policy Paper No. 4/94. Policy Evaluation Unit, Office of the Prime Minister, 19 Dec. 1994, copy in writer's possession.
_____. Foreign Investment Board. *Forestry and Forests Investments: Forest Resources of the Solomon Islands*. Foreign Investment Bulletin, 5. Honiara, 1984.
_____. Forestry Department/Division Annual Reports, 1952–1989.
_____. Forestry Department/Division. See Chaplin, G. E. and A. J. Neumann; Cheatle, R. J.; Gilbert, G. D.; Ifui, J. Anderson; Iputu, S.; Kranenburg, Bill; Marten, K. D.; Thorpe, Tim, and Ian Frazer; Trenaman, K. W.; Watt, G. R.
_____. Forestry Division, Ministry of Natural Resources. *A Review of Research Progress: 1981–1986*. Forestry Note No. 16 16/86. Honiara: Forestry Division, Ministry of Natural Resources, 1986.
_____. Forestry Division. *Forestry Policy Statement*. Honiara: Forestry Division, August 1989.
_____. Education Dept. *The Use of Forests*. Honiara: Curriculum Development Unit, 1984.
_____. *Forestry Policy*. White Paper: BSIP 12. Honiara: Dec. 1968.
_____. Legislative Assembly of BSIP. *Debates*. 1975.
_____. Legislative Council of BSIP. *Debates*. Honiara: Government Printing Office, 1968.

Bibliography

———. Legislative Council. Report of a Special Select Committee appointed by the Legislative Council. Legislative Council Paper No. 86 of 1968.
———. Ministry of Agriculture and Lands [Scott, George]. Land: Proposal for a Government Policy, Enclosure, Customary Land Records Bill, 1992, revised first draft, 29 Aug.1990. Honiara: Survey and Cartographic Division, Ministry of Agriculture and Lands, 1990.
———. Ministry of Finance. *Forestry Review 1995*. Honiara: Ministry of Finance, 1996.
———. Ministry of Natural Resources. *See* AIDAB and Ministry of Natural Resources.
———. *Newsletter*, BSIP. Honiara, 1957–1961.
———. Ombudsman's Office. *Report of the Ombudsman*, 1988, 1989, 1990, 1991.
———. Parliament of Solomon Islands. *Debates*. 1980–1990.
———. Planning Unit. *Sixth Development Plan 1971–1973*. BSIP. Hong Kong: Cathay Press, 1971
———. *Statistics to 31 March, 1909*. Sydney: Protectorate of British Solomon Islands, 1909.
———. *Statistical Bulletin: Solomon Islands 1986 Population Census Report 1: (supplement) Summary of Population by Socio-Economic Characteristics Final Results*. Honiara: Solomon Islands Statistics Office, 1988.
———. Ordinance. Lands and Titles (Amendment) Ordinance, 1964.
———. Ordinance. Forests and Timber Ordinance, Oct. 1972, no. 11 of 1972.
———. Ordinance. Forests and Timber (Amendment) Ordinance, 1977.
Solomon Islands New Drum (Honiara), 1977.
Solomon Nuis (Honiara),1995.
Solomon Star (Honiara), 1985–1997.
'Solomons Operation Works for Pacific'. *World Wood* 11, no. 1 (Jan. 1970): 3–5.
Solomons Tok Tok (Honiara), 1980–1990.
Solomons Voice (Honiara), 1991–1997.
Somerville, Boyle T. 'Ethnographical Notes on New Georgia, Solomon Islands'. *Journal of the Royal Anthropological Institute* 26 (1897): 357–413.
South Pacific Commission. *Pacific Islands Population*. Report prepared by the South Pacific Commission. Auckland: South Pacific Commission, 1994.
Southeast Asia Business Times (Kuala Lumpur, Malaysia), 26 Oct. 1996.
Southern Cross Log. Auckland: Melanesian Mission, 1896–1904; Sydney: Melanesian Mission, 1904–1913.
Sparkhawk, W. N. *Notes on Forest and Trees of the Central and Southwest Pacific Area*. [New York?]: US Dept. of Agriculture, Forest Service, 1945.
Spriggs, Matthew. 'Landscape, Land Use, and Political Transformation in Southern Melanesia'. In *Island Societies: Archaeological Approaches to Evolution and Transformation*, edited by Patrick Vinton Kirch, 6–19. Cambridge: Cambridge University Press, 1986.
———. 'Why Irrigation Matters in Pacific Prehistory'. In *Pacific Production Systems: Approaches to Economic Prehistory*, edited by D. E. Yen and J. M. J. Mummery, 174–189. Occasional Papers in Prehistory, No. 18. Canberra: Australian National University, 1990.
———. 'Pleistocene Agriculture in the Pacific: Why Not?' In *Sahul in Review*, edited by M. A. Smith, Matthew Spriggs and B. Fankhauser, 137–43. Canberra: Dept of Prehistory, Australian National University, 1993.
Star, Paul. 'From Acclimatisation to Preservation: Colonists and the Natural World in Southern New Zealand, 1860–1890'. Ph. D. diss., (draft). Dunedin: University of Otago, 1997.
Steen, H. *The U. S. Forestry Service*. Seattle: University of Washington, 1976.
Steensberg, Axel. *New Guinea Gardens: A Study of Husbandry with Parallels in Prehistoric Europe*. New York and London: Academic Press, 1988.
Steeves, Jeffrey S. 'Unbounded Politics in Solomon Islands Leadership and Party Alignments'. *Pacific Studies* 19, no. 1 (Mar. 1996): 115–38.
Steley, Dennis. 'Juapa Rane: The Seventh Day Adventist Mission in the Solomon Islands, 1914–1942'. Master's thesis, University of Auckland, 1983.
Stevenson, Janelle, and John R. Dobson. 'Palaeoenvironmental Evidence for Human Settlement of New Caledonia'. *Archaeology in Oceania* 30, no. 1 (Apr. 1995): 36–42.
Stoddart, Charles H. 'The Forests of the Solomon Islands'. *Journal of Forestry* 44, no. 12 (1946): 1041–3.

Bibliography

Stoddart, D. R. 'Biogeography of the Tropical Pacific'. *Pacific Science* 46, no. 2 (1992): 276–93.
Struben, Roy. *Coral and Colour of Gold*. London: Faber and Faber, 1961.
Sun (Honiara), 1982.
SWIFT-Netherlands Foundation. *Solomon Western Islands Fair Trade Newsletter* 1 (June 1996).
Sydney Herald, 1900.
Sydney Mail, 1912.
Sydney Morning Herald, 1913.
Tacconi, Luca, and Jeff Bennett. 'Biodiversity Conservation: The Process of Economic Assessment of a Protected Area in Vanuatu'. *Development and Change* 26 (1995): 89–110.
Tahuniara, Samson. '*Posi*: A Unique Yam Cultivation Practice in Guadalcanal'. In *Land Use and Agriculture: Science of Pacific Peoples*, Vol. 2, edited by John Morrison, Paul Geraghty and Linda Crowl, 181–90. Institute of Pacific Studies, University of the South Pacific, 1994.
Tausinga, J. D. 'Our Land, Our Choice: Development of North New Georgia'. In *Independence, Dependence, Interdependence: The First 10 Years of Solomon Islands Independence*, edited by Ron Crocombe and Esau Tuza, 55–66. Honiara: Institute of Pacific Studies, University of the South Pacific and the Solomon Islands College of Higher Education, 1992.
Taylor, Nancy M. *The New Zealand People at War: The Home Front*. Vol. 2. Wellington: Government Printer, 1986.
Tedder, Margaret M. 'Old Kusaghe'. *Journal of the Cultural Association of the Solomon Islands* 4 (1976): 41–95.
Tedder, Margaret. M. and J. L. O., 'Yam Cultivation on Guadalcanal'. n.d. [c. 1970?]. Cyclostyled.
Thaman, R. R. 'Ethnobotany of Pacific Island Coastal Plants'. In *Fauna, Flora, Food and Medicine: Science of Pacific Island Peoples*, Vol. 3, edited by John Morrison, Paul Geraghty and Linda Crowl, 147–84. Suva: Institute of Pacific Studies, The University of the South Pacific, 1994.
The Australasian Methodist Missionary Review (Sydney), 1907–1917.
The Australian (Canberra), 7 June, 21 Sept. 1995.
The Australian Women's Weekly (Sydney), 1988.
The Fiji Times (Suva), 1994.
The Planters' Gazette (Sydney), 1922.
The Straits Times (Singapore), 19 May 1994.
The Sun (Sydney), 1910–1911.
The Transformed Isle: Barbarism to Christianity: A Genuine Portrayal of Yesterday and To-day, the Story of Fifteen Years Among the Head-hunters of the Island of Vella Lavella. Australian Religious Film Society, 1920.
Theroux, Paul. *The Happy Isles of Oceania: Paddling the Pacific*. London: Penguin, 1992.
Third Division Historical Committee. *Pacific Pioneers: The Story of the Engineers of the New Zealand Expeditionary Force in the Pacific*. Dunedin: Coulls, Somerville and Wilkie, 1945.
Thompson, E. P. *Whigs and Hunters: The Origins of the Black Act*. Harmondsworth, England: Penguin, 1977.
Thorpe, Tim, and Ian Frazer. *Project Appraisal System: A System for Appraising Reforestation Projects on Customary Land*. Honiara: Forestry Division, Ministry of Natural Resources, 1992.
Thorpe, Tim. *Customary Land Reforestation Project: Draft Agreement, 20 May 1992*. Honiara: Forestry Division, 1992.
_____. *Experience with Portable Sawmills in the Solomon Islands*. Heads of Forestry Meeting, Western Samoa. Honiara: Forestry Division, Ministry of Natural Resources, 1992.
_____. 'The Effectiveness of Aid Delivery: A Comparative Study'. M. Phil. (draft), Massey University, Palmerston North, 1994.
Tietenberg, Tom. *Environmental and Natural Resource Economics*. Glenview, Illinois: Scott, Foresman and Co., 1988.
Tok Blong Pasifik (Victoria, British Columbia), Vol. 49, No. 2, June 1995.
Tokotok/Sun (Honiara), Oct. 1982
Townsend, M. M. 'Problems of Land Tenure on Malaita'. *Atoll Research Bulletin* 85, (1961): 27–32.
Trenaman, K. W. 'Forestry in the Solomon Islands since 1944'. Forestry Dept. Library, Munda, Solomon Islands, Jan. 1959. Cyclostyled.
_____. *Land for Forestry*. Forestry Department Technical Note 7/63, encl. Honiara: Department of Forestry, November 1963.

Bibliography

_____. *Timber Development Prospects in the Solomons – Based Mainly on the Findings of a Pilot Study by FAO/ESCAFE on Timber Trends and Prospects in the Asia-Pacific Regions.* Forestry Department Technical Note 1/60, Honiara: Department of Forestry, 1960.

_____. *The Forest Estate.* Forestry Department Technical Note, 1/62. Honiara: Department of Forestry, 1962.

_____. *Timber Development Prospects in the Solomons with Appendix of Summary of Asia and Oceania study by FAO.* Forestry Department Technical Note 3/63 (a supplement to Technical Note 1/60). Honiara: Department of Forestry, 1963.

Troup, R. S. *The Silviculture of Indian Trees.* 3 vols. Oxford: Clarendon Press, 1921.

_____. *Silvicultural Systems.* Oxford: Clarendon Press, 1928.

_____. *Forestry and State Control.* Oxford: Clarendon Press, 1938.

_____. *Colonial Forest Administration.* Oxford: Clarendon Press, 1940.

Tucker, Richard P. 'Managing Subsistence Use of the Forest: The Philippine Bureau of Forestry, 1904–60'. In *Changing Pacific Forests: Historical Perspectives on the Forest Economy of the Pacific Region*, edited by John Dargavel and Richard Tucker, 105–15. Durham, North Carolina: Forest History Society, 1992

United States Senate Committee on Banking and Currency. *Participation of the United States in the International Monetary Fund and International Bank for Reconstruction and Development.* 79th Cong., 1st sess., 1945. S. Rept. 452, part 2, 'Minority Views'.

Vasu, S., et al. 'Selection and Performance of Preservice and Inservice Candidates in Forestry'. *Pacific-Asian Education* 5, no. 2 (1993): 67–72.

Verguet, L. 'Arossi et ses Habitants'. *Revue d'Ethnologie* 4 (1885): 193 –232.

Wahl, O. *Local Timber Processing Study, Phase II, Solomon Islands.* Mission Report. Rome: FAO, 1979.

Walker, F. S. *The Forests of the British Solomon Islands Protectorate.* London: Crown Agents, 1948.

Wall, J. R. D. *Proposals for Land Development after Logging on North New Georgia, Solomon Islands.* Project Report 99. London: Overseas Development Administration, 1980.

Wall, Lindsay. 'Guadalcanal Stone Adzes'. *Journal of the Solomon Islands Museum Association* 2 (1974): 37–40.

_____. 'The Stone Carvings of Nggatokae'. *Journal of the Solomon Islands Museum Association* 2 (1974): 34–6.

Ward, Marion W. *Women and Employment in Solomon Islands.* Pacific 2010 Series. Canberra: National Centre for Development Studies, Australian National University,1995.

Ward, R. Gerard. 'The Pacific Bêche-de-Mer Trade with special Reference to Fiji'. In *Man in the Pacific Islands: Essays on Geographical Change in the Pacific Islands*, edited by R. Gerard Ward, 91–123. Oxford: Clarendon Press, 1972.

Watt, G. R. *A Note on the Benefits which could be Derived from Harvesting and Processing of Timber from the Existing Resource, from Different Areas of Plantation, and from 130,000 Acres of Plantation on Public land.* Research Report E 7. Honiara: Forestry Department, 1974.

_____. *Social Cost Benefit Analysis of Wood Processing Industries that could be Set up to Utilize Logs from the Replanting Programme.* Research Report E. 5. Honiara: Forestry Department, 1974.

Watt, G. R., and K. W. Trenaman. *Timber Planting Programme Projection 1974 to 1985 and Log Production Cost Estimates.* Forestry Department Technical Note 1/74. Honiara: Forestry Department, 1974..

_____. *Timber Planting Programme, 3; A review of Research Progress: 1981–1986.* Honiara: Forestry Department, 1986.

Webb, I. S. *The Effects of Tractor Logging on Some Soil Properties and on the Growth of Tree crops.* Auki, Solomon Islands: Department of Agriculture, 1974.

_____. *The Influence of Logging Operations on the Soils of Kolombangara: Preliminary Report.* Auki, Solomon Islands: Department of Agriculture, 1973.

Webster, John. *The Last Cruise of the Wanderer.* Sydney: Cunninghame, [1863].

Weisler, M. I. 'Henderson Island Prehistory: Colonization and Extinction on a Remote Polynesian island'. *Biological Journal of the Linnean Society* 56 (1995): 377–404.

Wendt, Neva. 'Environmental Problems in the South Pacific: The Regional Environment Programme Perspective'. In *Resources, Development and Politics in the Pacific Islands*, edited by Stephen Henningham and R. J. May, 185–94. Bathurst, New South Wales: Crawford House, 1992.

Bibliography

Westoby, Jack. 'The Role of Forest Industries in the Attack on Economic Underdevelopment'. In *The State of Food and Agriculture 1962*, edited by Food and Agriculture Organisation, 88–128. Rome: FAO, 1962.
———. *The Purpose of Forests: Follies of Development*. Oxford and New York: Basil Blackwell, 1987.
Whitmore, T. C. *Guide to the Forests of the British Solomon Islands*. London: Oxford, 1966.
———. 'Land Flora: Geography of Flowering Plants'. *Philosophical Transactions of Royal Society* 255 (1969): 549–66.
———. 'The Vegetation of the Solomon Islands'. *Philosophical Transactions of the Royal Society Bulletin* 255 (1969): 259–70.
———. *Change with Time and the Role of Cyclones in Tropical Rain Forest on Kolombangara, Solomon Islands*. Oxford: Commonwealth Forestry Institute, 1974.
———. *Tropical Rain Forests of the Far East*. Oxford: Clarendon Press, 1975, 1984.
———. *An Introduction to Tropical Rain Forests*. Oxford: Clarendon Press, 1990.
Wickler, S., and Matthew Spriggs. 'Pleistocene Occupation of the Solomon Islands, Melanesia'. *Antiquity* 62 (1988): 703–06.
Williams, Michael. *Americans and Their Forests: A Historical Geography*. New York: Cambridge University Press, 1989.
Wood, P. J., and G. R. Watt, *An Evaluation of British Aid to Assist Forest Development in the Solomon Islands, 1965–1980*. London: Overseas Development Administration, 1982.
Woodford, Charles Morris. *A Naturalist Among the Headhunters: Being an Account of Three Visits to the Solomon Islands in the Years 1886, 1887, and 1888*. London: G. Phillip, 1890.
World Bank. *Pacific Island Economies: Toward Efficient and sustainable Growth. Vol. 5: Solomon Islands: Country Economic Memorandum*. Report 11251-EAP, Country Dept. 111, East Asia and Pacific region. Washington DC: World Bank, 1993.
Worster, Donald. 'Transformations of the Earth: Toward an Agroecological Perspective in History'. *Journal of American History* 76, no. 4 (1990): 1087–1106.
Worster, Donald. *Nature's Economy: A History of Ecological Ideas*. 2nd edn. Cambridge: Cambridge University Press, 1994.
Yen, D. E. 'Agricultural Systems and Prehistory in the Solomon Islands'. In *Southeast Solomon Islands Cultural History: A Preliminary Survey*, edited by R. C. Green and M. M. Cresswell, 61–74. Royal Society of New Zealand, Bulletin 11. Wellington: The Royal Society, 1976.
———. 'The Domestication of Environment'. In *Foraging and Farming*, edited by David R. Harris and Gordon C. Hillman, 55–75. London: Unwin Hyman, 1989.
———. 'Environment, Agriculture and the Colonisation of the Pacific'. In *Pacific Production Systems: Approaches to Economic Prehistory*, edited by D. E. Yen and J. M. J. Mummery, 258–77. Canberra: Australian National University, 1990.
Young, Florence, S. H. *Pearls From the Pacific*. London: Marshall Brothers, 1925.
Zoloveke, Gideon. 'Traditional Ownership and Land Policy'. In *Land in Solomon Islands*, edited by Peter Larmour, 1–9. Suva: Institute of Pacific Studies, University of the South Pacific and Ministry of Agriculture and Lands, Solomon Islands, 1979.

Index

A

Abe, Christopher, 340, 349, 354, 358
Advisory Council, 151, 166
Agriculture (*see also* Gardening; Cash crops), 18, 172
  swidden, 20, 328 fig.
  Walker's views on, 140
Agriculture, Department of
  and bush fallow, 323
  and agroforestry, 324
  relations with Forestry Department, 154, 187, 323–4
Agroforesty, 35
  and Agriculture Department, 324
  and *taungaya* regime, 197
  project on Malaita, 262–3, 324
Aid (*see also* Australia; Britain; European Community; Japan; New Zealand; Non-government organisations; Taiwan)
  and American Peace Corps, 269
  and conservation, 344–5, 370
  and donors' concerns, 258, 275–6
  and North and South, 370
  and reforestation, 203, 246, 260
  and women, 262–3, 275
  dangers of, 265, 274–6
  donors, 2, 254–76
  from governments, 254–67
  from volunteers, 268–9
Alebua, Ezekiel, 241, 310
  and law, 249–50
  and moratorium, 249
  and Soltrust, 271
  Minister of Natural Resources, 341
Allan, Colin, 140, 161, 164

Allardyce Harbour, 32, 187, 193, 210, 265, 291
Allardyce Lumber Co., 169, 170, 180 , 210–13, 233
  and Code of Practice, 355
  and Forestry Division, 237, 281
  court cases, involvement in, 251, 278, 281
  criticism of draft legislation by, 250
  on New Georgia, 246–7, 281–2, 284
  on Vella Lavella, 282, 286–7
Alu, 66
  and Allardyce, 212, 246, 281
  commercial logging on, 210, 212
  Forest Area, 174
  logging by Monckton on, 65–70, 67–8 figs
  reforestation on, 257–8, 260
  regeneration of trees, 70
  State land on, 167
  timber control at, 237
  timber rights on, 89
Ambuaffer, Peter, 39–40;
America, 1, 163
Americans
  and Board of Economic Warfare, 114
  and earth-moving, 128
  and Maasina Rulu, 128–9
  and timber supplies, 113–14, 126–7
  construction battalions (Seabees), 117, 119 fig., 120, 122
  loggers, 231–2
  occupation of Solomon Islands, 116
Andresen, Edmond, 293, 314, 345
Angiki, Duran, 358
Anuta, 20, 33, 376
Aola, 59, 170;
Apa, Martin, 313, 315
Arboriculture (*see also* Trees; crops), 20, 22
Are'are, 129, 300, 304

Index

Area Committees
  and land rights, 181, 218
  replaced by area councils, 226
Area Councils, 226, 240
  Choiseul, 244
  Dorio, 243
  Guadalcanal, 226–7
  Santa Isabel, 293
  Vella Lavella, 283–4
  west Kwaio, 243.
Arosi district, 56, 307, 343
Aruligo, 230, 312,
Asian Development Bank, 203, 269
Assistance, development (*see* Aid)
Atori, 229, 235
Auki, 134, 229, 290, 304
  government station at, 55, 59
  firewood shortage at, 155
  Forestry Division at, 263, 267, 340

B

Baeanisia, Abraham, 269, 303
Baga (Mbava)
  and British Solomons Forestry Co., 171
  commercial logging on, 210
  State land on, 167
Baines, Graham, 269, 294
Bank, Central, 326
  and government, 340, 346, 357
  governor threatened, 356,
Bara, Anthony, 302–3
Barley, J. C., 87, 111
Barora, 218, 219, 221, 234, 256
Barter (*see* Trade, internal)
Bauro, 307, 310
Bea, Mary, 358
Beaufort Bay, 134, 151
Beliefs
  about ancestors, 28–30, 34, 231, 288, 335
  about cause of death, 97
  about sacred sites, 27–8, 50, 56, 287, 288, 311, 335
  and fear of darkness, 29
  and garden ritual, 33
  and magic, 19, 34
  and religious ritual, 34
  and sorcery, 96
  and spirits, 28–9, 54
Bellona
  geology of, 6
  World Heritage site on, 315
Bilney, Gordon, 352, 355
Bina, 181, 229, 243, 299

Boatbuilding, 71, 75, 82, 83
Boseto, Leslie, 284–5, 294
Botany (*see* Forests; Forestry Department/Division)
Bougainville, 72, 99
  and Australia, 345
  and land bridge, 7
  Buka in, 18
  military saw-millers on, 120
  resource conflicts of, 220
Boyers, James, 213, 246, 284, 293, 313
Britain (*see also* Colonial Office), 38
  aid from, 194, 200
  and Colonial Development and Welfare Act, 133, 150
  and colonial timber supplies, 136, 137, 361–2
  and Commonweath Fund for Technical co-operation, 269
  and forestry, 136–7
  and School of Forestry, Oxford University, 109, 137, 165
  and sustainability, 354–5
  and timber control, 237
  forestry specialists from, 191, 197, 201, 207, 352
  Imperial Forestry Institute of, 109
  Labour government of, 133
  Overseas Development Administration of, 203, 207, 237, 257, 258, 265, 269, 275, 299, 352, 354–5
  Royal Indian Engineering College of, 137
  reforestation grants from, 199, 202–3, 207
  soil study by, 194
  warships of, 132
British Solomon Islands Defence Force, 125
Buala, 267, 340
Burns, Philp and Co, 57, 58, 59, 76, 84, 102
  and land at Tadhimboko, 146–7
  interest in timber, 65
  mortagors, 69
Bush fallow (*see* Gardening), 20

C

Caldwell, J. T., 85–6
Campbell, F. M., 55, 56
Carmel, Anthony, 271
Cassells, Ross, 324–6, 365
Cattle
  and reforestation, 265
  introduction of, 49
  projects, 329
  under the trees project, 203, 265, 274

Index

Chan Wing Ltd, 354
Chinese
  builders, 58, 71, 75
  cooks at Vanikolo, 97, 99
  shipwrights at Mbatuna, 125
Chirovanga, 54, 289, 290
Choiseul Bay, 288, 289
Choiseul
  and churches, 290
  and Eagon, 233, 244–5, 252, 261, 288–90, 367
  and Forestry Division, 244–5
  and Island Resources Ltd, 233
  and Sagawa Development Co., 232
  and Sirovanga Association (VASA), 289
  and Taisol, 233
  availability of land on, 45
  defences of, 38
  Japanese on, 116
  Methodists on, 54
  people's perception of, 289
  premier of, 288–90
  subsistence value of forests of, 324–6
  timber stocks of, 244
Christian Fellowship Church, 218, 219–20, 247
Christianity, 38, 62
  and forest conservation, 335, 374
  and Maasina Rulu, 129
  extension of, 60, 291
  influence on settlement patterns, 56
Church of Melanesia, 291–3
  and social justice, 359
  on Santa Isabel, 291, 294
Churches (*see also* Missions and individual churches), 2
  and environmental issues, 272
  and small-scale logging, 367
  and social justice, 358–9
  of Solomon Islands Christian Association, 273
Colonial Office, 39, 66
  advice from forestry experts, 109, 110
  and immigration, 40, 46, 139
  and timber tax, 144–5
  and Trade Scheme, 127
  forest policy of, 111–12, 133, 138, 144, 183, 363
  and Vanikolo licence, 85, 86
  view on royalties, 87, 108–9, 144–5
Colonisation, 36, 40
  reasons for, 38–9
  and forest reserves, 189
Commerce
  missionary views of, 43–6
  post-war, 127

Community forestry, 277, 374
  and EU, 265
  and Forestry Division extension, 261, 263–4, 264 fig.
  and NGOs, 271, 274
  and Trenaman, 180–1
  and United Church, 285–6
Compagnie Caledonienne des Nouvelles Hebrides, 84, 86
Conservation
  and Dennis, 314
  and Melanesians, 32–3
  and Rainforest Information Centre, 248
  and Tausinga, 221, 224, 245–8
  in Europe, 136
  in India, 136
  in Marovo, 287
  in New World, 1
  in Vella Lavella, 282
  in wartime, 127
  not a concern, 86, 361–3
  of forests, 89–90, 335, 363
  utilitarian, 140
  views of, 370–3
Coote, Audley, 72, 87
Corrin, Jennifer, 302–3
Councils, local government, 125
  and management of forest reserves, 151, 154, 161–2, 364
  and timber revenues, 202
  ex-gratia payments to, 166
  of Eastern Islands, 212
  of Star Harbour, 307
  replacement of, 218
Court
  and debit tax, 359
  and right of appeal, 176
  and tax, 278, 297
  civil case in, 302
  High, 229, 307
  injunctions, 340
  law, 228, 247, 250–1, 282, 331
  native, 176
Cox, R, 149, 170
Crops
  cash, 45, 319–20, 329
  land for, 172, 187, 326–7, 328–31
Cultivation (*see* Gardening; Plantations)
Culture
  and ecology, 28, 336–7, 371–2
  in environment, 27–8, 373
Customary Land Appeal Court (CLAC), 240
  right of appeal to, 176, 181
  role in certification of timber rights, 182

Index

Cyclones, 11–16
  effects of, 14, 16,17, 204–5, 336
  Ida, on Santa Isabel, 176, 210, 211 figs, 291
  Namu, 249, 311, 320
  on Vanikolo, 98, 155

D

D'Urville, Dumont, 11, 91
Dala, 181, 304
Darwin, Charles, influence of, 41, 136
Davies, Harold, 143, 148
Decolonisation, 2, 254
Deforestation, 30, 136
Dennis, Geoff, 189, 268, 314
Dethridge, R. A., 125, 134, 143, 153
Development, 3
  and cattle projects, 329
  and environment, 269–74
  and Hyundai, 226
  and indigenous land tenure, 139
  and Integrated Forest Industries, 230–1
  and Kayuken Pacific, 242
  and local logging company, 358
  and loggers, 234
  and NGOs, 269–74, 365
  and United Church, 284–5
  Choiseulese view of, 289
  economic, 161
  European view of role in, 59–60
  government programme for, 147, 150, 175
  government's view of, 337, 365
  Maasina Rulu's view of, 128–9
  Malaitans' view of, 38–40
  missionaries' views, 43–6
  partners, 360
  pre-war concept of, 43
  of resources, 337
  role of timber in, 164, 170, 184
  Solomon Islanders' views of, 60–1, 174, 184, 337, 366
  Vella Lavellans' view of, 286
  view of, from Guadalcanal, 174
  view of, from Santa Isabel, 292
Devolution (*see* Government, Solomon Islands; Provinces)
Dillon, Peter, 84, 91
Disease (*see also* Malaria)
  Aids pandemic, 327
  beri-beri, 98l
  dysentery, 88
  introduced, 31, 32, 38
  on Vanikolo, 92, 97–8
Dodo Creek, 146, 204

E

Earthquakes, 4, 14, 155
Eco-timber (*see* Timber)
Eco-tourism, 72, 287, 288, 312–13
Economy (*see also* Plantation economy)
  and Great Depression, 69, 98, 75, 102, 109
  and international institutions, 340–1, 351, 360
  and structural adjustment, 340, 360
  monetary, 319–23, 340, 341, 346–7
  non-monetary, 323–8
Education, 43–6
  about soil protection, 273
  and Eagon, 289
  and missions, 54, 75, 76
  and NGOs, 269–74
  at Komuniboli rural training centre, 269
  at St Dominic's rural training centre, 273
  by extension services, 267
  formal, 336–7
  King George VI school, 359
  of forest personnel, 267–8
  planters' views of, 46
  technical, 178
  Solomon Islands College of Higher Education (SICHE), 263, 267–8, 359
Eke, Abraham, 304–5, 342
El Niño Southern Oscillation, 31
Enoghae, 220, 222, 259, 367
Environment
  and churches, 284–6, 290
  and development, 285, 337
  forest, 27–30
  regional awareness of, 249
  theology of, 294, 335
Eto, Silas, 218
European Community (European Union)
  aid for training, 311
  and forestry sector review, 258
  finance for reforestation, 257–8, 261, 274–5
  project on Santa Cruz, 316
  Sustainable Forest project on Santa Isabel, 265, 294, 367
Europeans, 1
  and natural resources, 41, 46–7, 140
  cultural perceptions of, 56
  employees at Vanikolo, 96–7, 99, 105, 155–6, 162
  in saw-milling ventures, 178–81
  views of role in landscape transformation, 59–61
Evans, W. (Bill). R., 188, 213
Explorers, Spanish, 20, 30

Index

F

Fairley, Rigby Co. Ltd (*see also* San Cristoval Estates Ltd), 85
Fatutaka, 33
Fauna, 6, 18, 19, 30, 33, 34
    extinctions, 30, 34
    introduced, 49–50
Fiji
    cassava from, 56
    Fiji Kauri Timber Company, 134
    floristic province, 7
    forestry training in, 150
    Koster's Curse from, 69
    labour trade in, 37, 38, 40
    military force from, 120
    missionaries from, 50
    Nadarivatu in, 109
    Native Land Trust Board of, 139
    timber of, 64, 108, 145
    workers from, 99
Filose, F., 97, 107–8
Fire, and heath, 32
    in clearing, 47–8, 196
    in gardening, 20, 21, 22, 25
    on grasslands, 23, 31
Firewood, 48
    demand for, 64, 108, 154–5
    in Queen Elizabeth II National Park,188–9
    on Malaita, 263
Fish, 35, 46, 85, 228, 321, 341
Fletcher, Murchison, 109, 110
Flora, 7, 18
    introduced, 22, 38, 50, 56
Florida Islands (*see* Nggela)
Food crops (*see also* Root crops; Tree crops)
    for pigs, 33
    greens, 22
    introductions, 22, 38, 56
Footaboory, Benjamin, 39–40
Foreign Investment Board, 225, 240–1, 293, 306
Forest Areas
    component of Forest Estate, 165
    declaration of, 172–5
    resistence to, 174, 184, 364
Forest Reserves, 138–9, 141, 150–1
    and vacant land for, 183
    component of Forest Estate, 165
    expansion of, ceases, 185
    government rationale for,150, 170
    replaced by Controlled Forests, 175
Forestry Department/Division (*see also* Aid; and specific donor countries), 2, 221
    advisers' reports, 275
    and Agriculture Department, 153, 187, 323–4
    and Alebua government, 245
    and allowable cut, 238–9, 253
    and Australian aid, 265–7
    and Axiom's licence, 293
    and colonial timber companies, 209–18
    and Eagon, 233
    and education, 267–8, 292
    and Finance Department, 348–9, 356
    and Forest Policy Statement (1989), 249–50, 324
    and Forest Resources Information System (FRIS) and inventory, 8, 266, 274, 297, 312, 341–2, 349–50
    and freehold land, 165
    and Hyundai, 225–8, 282–3
    and independence, 207–8
    and indigenous saw-milling, 178–81
    and Kayuken, 229–30, 242–3, 300–1, 303
    and Kenilorea government, 242
    and leased land, 150, 161, 165
    and levy for reforestation, 201–2, 261, 279–80, 342
    and log prices, 238, 242
    and Malaita province, 242–3, 298–302
    and Mamaloni governments, 250–3, 299, 313–14, 348–58, 366, 368
    and market trends, 198, 199
    and Marving Bros, 302, 304, 313–14
    and milling, 224, 281, 289, 299, 300
    and moratorium, 222, 229, 232, 239, 242–3, 248–9, 250–1, 305, 306
    and non-timber products, 324, 336
    and North New Georgia Corporation Act, 246–7
    and production, 174, 179, 186–91
    and proposed reserves, 150–1
    and protection, 150, 165, 175, 184, 186–91
    and Qoloni's reports, 243, 302
    and reforestation, 174, 175, 180, 191–9, 204, 205, 206, 239, 265–6
        by European Community, 258
        of customary land, 243, 261
    and regeneration, 157, 158, 161, 181, 191, 256
    and Silvania, 287–8
    and uncertain law, 241, 243, 249
    and White Paper (1968), 175, 197, 202
    and White Paper (1989), 249–50
    annual reports of, 350
    assumptions about logging by, 224–5
    at Poitete, 203, 206
    botanical studies of, 201, 268
    budget for, 279
    commercial unit of, 267, 342

Index

devolution to provinces of, 250–3, 279–80, 299, 316–18
enforcement by, 281, 290
entomological research by, 204, 257
establishment of, 109, 135, 140, 150–2, 154
extension work of, 179–81, 261, 263–4, 264 fig., 267, 324, 364, 368
herbarium, 201, 268
kauri management plan, 158, 161, 191
licensed quota, 350
loss of autonomy by, 206
management unit, 158
negotiations with Lever's, 190–1
nursery techniques, 206
policy advice for, 249
political pressure on, 180–1, 198, 242–3, 248–9, 250, 293, 301
relationship with Minister, 348–9
role of, in direct dealing with loggers, 182, 184, 185, 237
Select Committee on Forest Policy (1981), 239
Select Committee on Forest Policy (Forest Policy Review, 1974), 181–2, 198–9, 203, 218
silviculture research of, 140, 152, 189, 200–1
site studies by, 194
source of funding for, 274–5, 368
species research by, 152, 191–3, 199, 257
staffing, 109, 152, 158, 206, 237, 242, 243, 249, 268
statutory body for, 181, 206–7
surveys by, 239, 242, 249
Timber Control Unit, 237–9, 241–2, 250, 253, 266–7, 290, 340, 342, 347, 350–1, 354–6, 360, 366
Timber Marketing Board, 238
trees on Kolombangara, 259–60
utilisation research by, 200
view of customary land by, 190–1, 324;
work of, 186–208
Forestry, plantations (*see also* Reforestation; Forestry Department/Division, reforestation), 2, 10
and *Empire Forestry*, 137
and Forest Estate, 163, 165–79, 181, 186, 190, 260, 352, 363, 369
and KFPL, 259–60; and purchase of timber-cutting rights, 175–7
and Santa Cruz, 261
and Troup, 137
and Whitmore's research, 188, 190
area for, 165, 184–5, 191, 366
areas planted, 193, 197–8, 203, 205, 207, 225 fig., 256

Cattle under the trees project, 203
ecologically sustainable, 187–8, 190, 207, 369
economically sustainable, 2, 158, 186–8, 190, 207, 255, 260, 265, 369
extraction rates from, 197–8, 254–6, 351 fig.
first plantings in, 193–4
funding of, 199, 201–3, 207
in Indian subcontinent, 136–7, 183
interest by CDC in, 203, 258–60
Japanese interest in, 197, 202–3
land for, 138–9, 162, 165–72, 183
on New Georgia, 245–6
pests of, 204–6, 207, 257
planting target for, 97, 203, 207, 254–6, 261
scale of, 190
silvicultural methods for, 196–7, 205, 207, 257
species for, 198–9, 205–6, 256–7
stocking levels of, 207, 256–7
value of, 257, 352–4
Forests (*see also* Tree species; Forestry Department/Division), 1
adaptation, cyclones, 13–16, 12 figs, 336
and direct dealing by owners, 182, 184
and spirits, 29
and World War Two, 128, 141
botany of, 141, 201
changing ecology of, 34, 49, 187–8, 190, 328
changing perceptions of, 229–331, 333–8
clearing of, 2, 19–21, 26, 37, 47–8, 50–4, 56, 60, 129, 326–8, 336
communities, 8 fig.
composition of, 13, 328
Conservator of (Chief Forestry Officer; Commissioner of Forests)
and political interference, 293
and provinces, 252
legal action by, 228
role in logging agreements, 182, 240, 251, 282–3
Controlled, 175, 184
dynamics of, 11–13
expected cut from, 197; human transformation of, 19, 31, 34, 327–8, 369–70
meaning of, 27–30, 335
of Malesia, 375
on atolls, 63
policy
of Colonial Office towards, 112–13, 133, 138, 144, 183, 363
of government towards, 86–90, 135, 181–2
protected, 90

Index

regeneration of, 66, 70, 127
resources of, 3, 19, 323, 369
sacred sites in, 28; stocks of, 135, 166, 244
subsistence value of, 138, 323, 369
succession in, 12, 23
surveys of, 138–41, 152, 153
types, 4–17, 9 fig., 24, 26, 34, 46, 129
value of Walker's report for, 141
virgin, 46–7, 56
zonation within, 10
Fote, 262, 304
France, 43, 85, 212
 interests of, at Vanikolo, 63, 86, 87, 111
Freeman, Paul, 230, 232, 306, 310

G

Gardening, 1, 19, 35
 and erosion, 30, 32, 323
 and shortened fallow, 187, 327, 328 fig., 377
 area required for, 24, 362–3
 effects of, 30, 34, 187, 243
 intensification of, 24, 25, 273, 376–7
 land suitable for, 23
 Logie's view of, 151
 methods, 19–21, 23, 35, 273
 ritual, 33
 site selection for, 26
 technology of, 22–5, 35, 37–8
 Walker's views of, 140, 323
Gardens, 27
 and timber survey, 138
 botanical, 56, 334 fig.
 clearing for, 128–9, 151, 243
 introduced crops, 38
 land for, 172
 leguminous planting in, 273
 mission, 54
 source of subsistence, 33, 138, 319–20
 source of power, 33–4
 sup sup, 273
Gaviro, Sam
 and customary land, 261
 and devolution, 250
 and Eagon, 244–5
 and education, 267, 292
 and Kayuken, 243
 and Marving Bros, 313
 and Ministry of Natural Resources, 368
 and Solmac, 228
 and Timber Control Unit, 238
 and staffing levels, 238, 242
 on 'hit list', 303

political pressure on, 293
powers of, 243, 284
Gilbert Islands, 39, 150
Gizo, 59, 267, 340
Gizo Island
 Forest Area of, 174
 Lever's logging on, 169, 205
Glassa, son of Ware, 66, 67
Goldie, Helena, 50, 53
Goldie, John F., 45, 50, 53
 and local timber, 75
 plantation interests, 46
 view of Solomon Islanders of, 45–6
Goodsir, W. E., 134–5
Gorai, chief, 24
Governing Council, and local participation in logging, 178–9
Government, Protectorate (see also Forestry Department/Division), 2, 36, 41, 42, 43, 60, 84
 and forestry expertise in, 146
 and funding of Forestry Department, 150
 and independence, 182
 and KTC, 157–9
 and loan for timber industry, 153
 and logging on customary land, 146
 and Malaita, 128–9, 132, 154, 181
 and role of Solomon Islanders in forestry, 161–2, 178–81
 and self-government, 133, 178
 and subsistence economy, 138, 140, 323–4
 and taxes, 43, 55
 and Tenaru Timbers, 143–7, 150
 and timber royalties, 109, 157–8
 and tree ownership, 88–9, 139
 and VKTC, 94, 107–9
 at Tulagi, 56–60
 claims against, 72
 conflict of interest within, 145
 development policies of, 39–40, 43, 132, 139, 150, 161, 175, 182–4
 High Commissioner, 109, 139
 influence on settlement patterns of, 55–6
 land policies of, 88, 139, 146–7, 162, 166, 182–4, 363
 landscape transformation by, 56–61
 logging postponed by, 134–5
 need for timber by, 143
 pacification by, 41, 43, 61
 path building by, 55
 post-war reconstruction, 127
 role of district officer in, 95, 98
 Sixth Development Plan of, 175, 197
 Solomon Islanders' suspicions of, 166
 timber control by, 66, 109

Index

timber policies of, 66–7, 86–90, 112, 138–40, 162, 175–7, 182–5, 197–8, 363
wartime requisition of logs by, 114
wartime requisition of saw-mills by, 125
Government, Solomon Islands (*see also* Forestry Department/Division; Forest Legislation; Land, State),185
  allegations of bribery within, 310, 345, 354
  and aid donors, 341, 350, 352, 354–5
  and Australia, 344, 348, 350–1, 354–6
  and Constituency Development Fund, 349, 356
  and control of forested land, 313–15, 316
  and education for forestry personnel, 268
  and Finance Ministry, 346–9, 366
  and FRIS, 341–2, 349–50
  and Hyundai, 226–8
  and Integrated Forest Industries, 230–2
  and Kayuken, 229–30
  and Lever's, 224–5
  and milling industry, 299–300, 305, 310, 342–4, 346, 365
  and moratorium, 242–3, 248–9, 250
  and NGOs, 269–74, 314
  and political parties, 248, 310, 340–1, 345, 359
  and provincial government, 278–9, 299, 316–18, 348, 359, 368
  and public service, 340, 341, 348–9, 354, 359
  and reform of timber industry, 341–5, 347
  and sale of forestry assets, 352, 354
  and Select Committee on Forest Policy (1981), 238–9
  and Silvania, 287–8, 344
  and Solmac, 228–9, 305
  and timber control, 238, 250, 253, 340, 347, 349, 351
  and timber revenues, 319–23, 341–6, 348–9, 366
  and White Paper (1989), 249–50
  conflict of interest within, 228, 302, 304, 306, 341, 346, 357
  financial management of, 340, 341–2, 348–9, 356–7, 359, 366
  Forest Act (1984) of, 239–40
  interest in KFPL, 259–60
  Leadership Code Commission of, 358
  provincial funding by, 277–80
  record on legislation, 250–3
  revenue, 230, 238, 253, 280, 287, 322–3
Grantham, High Commissioner, 135, 139, 364
Grasslands
  and fertility, 24
  Thermeda australis, 31, 50, 189
  Imperata cylindrica, 31, 50
Guadalcanal, 152
  and Berjaya, 308, 311–13
  and Cape Esperance, 228, 229
  and extension work, 264
  and Forest Development Co., 229
  and Foxwood, 178, 312
  and FRIS, 312
  and Hyundai, 225–8, 312
  and Kayuken Pacific, 229–30
  and loggers, 134, 241, 311, 312, 313
  and Maasina Rulu, 128
  and Sagalu Exim, 228
  and Solmac, 228, 305
  and Taisol, 299
  and timber processing, 311, 312
  armies on, 115–19
  and Tadhimboko lands, 170
  cyclones on, 311, 320
  earthquakes, 14
  eco-tourism in, 312
  floods in, 311, 312
  Forest Areas of, 174
  forest survey of, 133
  forestry plantings on, 152, 189
  former settlements on, 32
  grasslands of, 31
  irrigation on, 25
  JICA's interest in, 203
  lack of harbour in north, 148
  Lever's land on, 143
  Melaneian Mission on, 44
  military logging and saw-mills, 118–26, 119–23 figs.
  mission land acquisition on, 42
  plantations on, 71
  policy review for, 239
  population of, 24, 30
  Protection of Historic Places Ordinance of, 311
  provincial government of, 311, 313
  Queen Elizabeth II park on, 188
  rural training centre on, 269, 312
  saw-milling at, 71, 143–9
  site of capital, Honiara, on, 154–5, 188
  south coast of, 14, 312
  Spaniards on, 20, 30
  squatters on, 188–9, 377
  Tenaru Timbers on, 143–9
  timber stocks of, 135, 166, 229, 230, 312
  vele man of, 29

Index

H

Head-hunters, 41
    effects of, 29, 31, 32
    of New Georgia Islands, 37
Hadley, Chris, 152, 191, 192
Halavo, 123, 124
Haling brothers, 155–7
Hauhui, 71, 147, 300
Haununu, 307, 308
Hilly, Francis Billy, 258, 343 fig.
    and Australia, 341, 344–5, 371
    and Berjaya, 310
    and forest policy, 341–4
    and levy, 280
    and local processing, 310
    and logging industry, 341–2, 344
    and Marving Bros, 314
    and National Coalition Party , 280, 287, 340
    and Silvania, 287–8
    and structural adjustment, 341
    government of, 340, 345
Hingava, chief, 50, 61
Holt-MacCrimmon, Lucy, 71, 108
Hombu Hombu, 125, 126 fig.
Hong Kong, 212, 224
Honiara, 143, 189, 286, 292, 293, 300, 308
    and Forestry Division, 263
    beggars in, 327
    cyclone at, 311
    demonstration plot in, 273
    land sale in, 341
    Suny Tong's murder in, 302–3;
Horticulture (*see* Gardening)
Houenipwela, R. N., 340, 347, 356
Hunter-gatherers, 18, 19, 35

I

Ilu, 122, 146
Immigration
    Asian labour, 40, 46
    recommended, 139
    skilled workers, 178, 179 fig.
Indonesia, 1, 170, 224
International Monetary Fund, 340
    and land tenure, 378
    and Ulufa'alu government, 360
    structural adjustment program of, 341

J

Jamakana, Frank, 219, 247
Japan, 83
    aid from, 258
    and earth-moving, 128
    and *sogo shosha*, 1
    and World War Two, 113, 115–16, 120, 127, 130
    Ataka from, 213
    log importer of, 315
    loggers from, 232
    logs to, 162, 212, 224
    ships of, 102
    timber demand of, 163, 164, 170
    traders from, 85, 99
Japan International Co-operation Agency (JICA), 203
Japanese Overseas Afforestation Association (JOAA), 197, 202–3, 260, 368
Jericho, 220, 222
Jones, Fred, 95, 112, 113, 134, 135
Jones, Walter, 212, 292, 315

K

Kane, R, 86, 109
Kari, Hilda, 312, 314, 360
Kauri Timber Company (*See also* Vanikoro Kauri Timber Company), 92, 94, 155–61, 162
    and Australian market, 156–7, 160–1
    and labour, 155–6
    and log quality, 159–60, 159–60 figs
    and plant of, 156, 158; and Tevai, 158
    and timber taxation, 145, 157–8
    and Trenaman, 157–8
    contractors, 155
    extraction methods of, 158–9, 363
    logging agreement of, 157
    losses of, 157, 161
    production by, 155, 157, 158, 162
    road construction by, 156
Kausimae, David (Nelson Kausimae Nanu), 180, 191, 192
Kavusu, David Livingston, 169
Kazukuru, 246–7, 251, 281
Keating, Paul, 344, 371
Kelesi, Marioano, 180
Kemakeza, Allan, 314, 348, 349, 354
Kengava, Clement, 244, 288–90, 317
Kenika, Ben, 232, 241

Index

Kenilorea, Peter, 182, 219, 228, 243
   and constituents, 248
   and Kayuken, 230, 275
   and moratorium, 242
   and Western Province, 245
Kera, R. T., 169, 207, 208, 238
Kes, Eric, 355, 358
Kirakira, 28, 232, 267, 303, 306, 340
Kokenggolo
   bomb damage, 130–1
   Methodist Mission at, 45, 50, 52–4, 75
Kokota, 272, 294, 355
Kolombangara
   and Cattle under the trees project, 203, 274, 323
   and Commonweath Development Corporation, 203
   and Kwan How Yuan (KHY) Pty Ltd, 170
   and Lever's negotiations, 190–1
   and Mamaloni government, 259–60; 278–9, 367
   and royalties, 234
   and timber rights, 182
   Forest Area, 174
   forest survey of, 152
   forest types, 10, 32
   hospital at, 83
   irrigation on, 25
   JOAA trial plots on, 202–3
   landowners of, 259–60, 279, 367
   Lever's lands on, 166
   Lever's logging on, 205, 213–18
   logs from, 170
   National Forestry Training Institute on, 263
   population of, 24
   rainfall on, 214
   rural training centre on, 273
   saw-milling at, 149, 170
   timber on, 64
   Whitmore's research on, 188, 190
Kolombangara Forest Products Ltd (KFPL), 259–60, 366–7, 374
   and natural forest, 286
   area planted by, 260
   exports of, 260
   outgrower interest in, 263, 280
Komuniboli, 269, 312
Kong Ming Khoo, 300, 301–3, 307
Koraga, 219, 220, 247
Korovou, 55, 59
Kuki, Sosimo, 269, 312
Kukum, 117, 118, 120, 126, 128, 152
Kwainarara, Enele, 311
Kwaio 302
   east, 230
   west, 229, 242, 243, 260, 302, 303, 307

Kwara'ae, 21, 29, 35, 243, 272
   botanical taxonomy of, 27, 201
   west, 298, 304

L

Labour
   and closed districts, 66
   and land clearing, 40 fig.
   convict, 125, 154
   disputes, 96–8, 155–7
   cost of recruiting of, 96
   indentured, 46, 95, 155
   skilled, 95–6, 156, 171–2, 178
   punishment of, 98
   recruiters, 85
   supply, 41, 43, 98
   trade, 31, 37, 38–40
   wages of, 65, 95
Lalani, Keleto, 302–4
Land (*see also* Land tenure)
   acres of, under coconut palms, 48–9
   alienated, return of, 181–2
   amount of, alienated, 41, 42
   and Certificates of Occupation, 42, 89, 90, 146, 164, 166, 167
   and Committee of Inquiry into Lands and Mining (1976), 181
   and forestry reserves, 138–41, 150–1
   and Lands Commission, 88, 111, 140, 164, 166
   and Special Lands Commissioner, 164
   British attitudes to, 39, 138–41, 150–1
   claims to, 40, 42, 72–3, 84, 86, 88, 138, 146–7
   concerns of Forest Policy Review (1974), 181–2
   consciousness growing, 147
   Crown (*see* Land, State)
   customary (Native), 150
      adjudicating timber claims on, 182
      and cash cropping, 327, 329–40
      compensation for rights' grant, 164
      logging of, 174, 241, 256, 364–5
      reforestation of, 203, 243, 261
      status of, 161
   deemed vacant, for forestry, 150–1, 164, 165–7, 363
   Europeans settlers' views of, 46–7
   freehold, 42, 89, 167
   Government (*see* Land, State)
   indigenous attitudes to, 39, 139, 166, 364, 377–8
   leasehold, 42, 71, 89, 90, 161, 167
   occupied, 39

Index

policy and Trenaman, 166
Public (*see* Land, State)
purchase, 39, 42, 56
registration of, 166, 172
settlement policy, 172, 176–7, 182
State, 90, 168
    appropriation for forestry, 138–9, 141
    forestry reserves on, 150, 164
    policies towards, 259–60, 313–14, 278–9, 366–7
    provincial claims on, 278–9
tenure
    and forest survey, 138, 140
    and lineage claims, 26, 28, 42
    conversion of, 166
    model of, 26 fig.
    ownership, 41–2, 363
    timber rights, 328–32
vacant, 39, 42, 151
Walker's view of, 139, 150
Land Trust Board, 139–40, 141, 164, 166
Laperouse, explorer, 63, 91
Lapli, John, 315
Lauru (*see* Santa Isabel)
Lawrence (Paeu) River, 98, 104
Legislation
    forest, 2, 89, 90, 141, 156, 185
    Amendment, 1972, 176, 185
    and Lands Commission, 151, 161–2, 186
    and Minister's role, 226
    and North New Georgia Corporation Act, 246–7, 287
    draft, 151, 161
    Forests and Timber Act (Amendment), 1977, 182,185, 218, 228, 232, 237, 238, 367
    Forests and Timber Act (Amendment), 1984, 229–30, 239–41, 243
    Forests and Timber Ordinance, 1969, 175, 177, 184, 237, 238
    Forests Ordinance, 1960, 165–6, 184–5, 187
    Forests Resources and Timber Utilisation Act (Amendment), 1990, 251–2
    Forests Resources and Timber Utilisation Act (Amendment), 1991, 252
    inadequacies of, 225, 232, 233, 238, 243, 251
    King's Regulation of 1913, 87, 185
    King's Regulation of 1922, 87, 164, 167, 185
land
    forestry implications of, 169–70
    Land Regulations, 41

Lands and Titles Amendment Ordinance, 1972, 176
Lands and Titles Amendment, 1977, 182
Lands and Titles Ordinance, 1959, 164, 165, 166
Waste Land Regulations, 41–2, 88, 146, 164, 166, 363
Legislative Assembly
    and licences, 180
    and New Georgia, 177
    and State land, 182
    Select Committee on Forest Policy of (1974), 181
Legislative Council, rejection of Forest Areas, 175
Lepping, George, 281
Lepping, Margaret, 233
Leslie, A., 181, 198, 199, 206
Lever's Pacific Plantations Pty Ltd
    and logs, 64
    and rents, 150
    and Santa Cruz, 85
    Gavutu headquarters of, 58
    lands of, 42, 143, 144, 146–7, 150, 166
    sale of assets, 259
    saw-mill on Russell Islands, 148
Lever's Pacific Timber Ltd, 170, 171, 213–24, 234
    and Earthmovers, 246
    and Enoghae incident, 220–2, 367
    and Mamaloni, 224
    and New Georgia, 213, 246
    and timber processing, 190–1, 197, 246
    and workforce welfare, 222–3 figs, 225
    at Vona Vona, 234
    departure of, 222–3, 245, 246
    export licence of, 220
    introduction of skilled workers by, 178
    investment by, 224
    logging methods of, 205, 214–17 figs., 245
    on Gizo, 213
    on Kolombangara, 213–18
    prices paid for logs by, 169, 175
    timber research by, 200
Lever's Solomons Ltd, 213, 220
    on Pavuvu, 313
    sale by, 313
Life expectancy (*see* Population)
Loggers, 2
    and direct dealing with, 2, 169–70, 175, 177, 181–2, 184
    and reforestation, 191, 260–1, 366
    and transfer pricing, 157, 162, 236–7, 244
    Asian, 2, 162
        business culture of, 225, 235, 295
        views of Solomon Islanders of, 365–6

Index

joint ventures with, 241, 306
   problems of, 171–2, 295, 304, 313, 367
Logging
   and churches, 272, 286
   and erosion, 90, 188
   and flooding, 311
   and NGOs, 272, 294, 300, 304, 306
   and siltation, 213, 288, 307
   and soils, 187–8, 229, 231, 256, 282, 287
   and water, 230, 231, 282, 288
   extraction rates, 2, 239, 341, 345
   location of, 76, 104
   methods, 69, 76–81, 98, 101–13, 125, 197, 227, 230, 231–2, 256, 282, 355
   on customary land, 140, 190, 241, 256
   practices condemned, 110, 127, 287
   selective, 144, 256
   small-scale (*see also* Timber; Saw-milling)
      and foreign loggers, 365, 367–8
      and government, 153–4, 177–81, 271, 285, 294, 304, 365
      and Soltrust, 271, 285, 294, 295, 304, 316
      and United Church, 285–6
      on Pavuvu, 314
      on Santa Isabel, 294–5
   unsustainable, 2, 255, 344, 365
   volume recovery rate in, 201
Logging companies (*see* Appendix 7; *see also* Allardyce Lumber Co; KTC; Lever's Pacific Timbers; Tenaru Timbers; VKTC), 237
   Axiom, 292–5
   Berjaya Group (Cayman) Ltd, 308–13, 345
   British Solomons Forestry Co. (Nanpo Ringo Kaisha Ltd), 169, 170, 210
   Cape Esperance, 228–9, 305
   Costigan Bros., 212
   Dalsol Ltd, 246, 299, 311, 312
   Dethridge Timber Syndicate, 134–5, 143
   Eagon Resources Development Company, 233, 244–5, 261, 289–90, 324–5, 356, 367
   Earthmovers Solomons Ltd, 213, 246, 295, 297, 312
   Eastern Enterprises, 212, 292–3, 295, 297, 315
   Fletcher Holdings, 142
   Forest Development Co., 229
   Foxwood, 178, 220, 292
   Golden Springs, 247–8, 278, 287, 304, 313
   Goodwill Company Ltd, 288
   Goodwill Industries, 343
   Howell Enterprises, 230
   Hyundai Timber Co. Ltd, 225–8, 282–3, 302, 312
   Integrated Forest Industries, 230, 246, 299, 307–8, 312, 354
   Isabel Timber Company, 272, 293–5, 297, 355–6
   Island Logging, 304–5
   Island Resources Ltd, 233
   Kalena Timber Company, 210, 213, 238, 278, 284, 287, 358
   Kayuken Pacific, 228–30, 242–3, 260, 275, 298–303, 304, 307–8
   Kololeana, 343
   Kumpulan Emas Berhad, 295, 354
   Mala Timber Development Company, 71
   Marving Bros, 288, 302, 304, 313–14
   Mears, G. E. C., and Ireton, D. A., 170, 210
   Mega Corporation, 343
   Monarch Leasing Company, 307
   Neilsen Pty Ltd, 141
   Odalisk (SI) Ltd, 282
   Pacific Timber Company, 166, 291
   Pacific Timbers, 292
   Rural Industries Ltd, 232, 305, 307
   Sagalu Exim Pty Ltd, 228
   Sagawa Development Co., 232
   Shortland Development Co., 210
   Silvania Products, 287–8, 293, 344–5
   Sollumber, 306–7
   Solmac Construction and Timber Co., 228, 305
   Somma Ltd, 343, 346, 358
   South Pacific Timber Corporation, 246
   Star Harbour Timber Co., 307, 310
   Taisol Investment Corporation, 233, 260, 298–302, 304
   Waibona Logging and Milling Co, 301–4
Logging industry (*see also* SIFIA; Timber industries), 1
   and export duties, 66, 145, 157–8, 322, 342, 346–7
   and reforms, 342
   and saw-milling licences, 342–4, 346
   and stockpiling logs, 342
   and timber control, 267
   and training of Solomon Islanders, 178
   Code of Practice for, 355
   duty remissions for, 346
   economic importance of, 2, 65, 175–6, 299, 322–3, 341, 342
   employment in, 212, 229, 289
   exports by, 171 fig., 351 fig.
   on Alu, 65–8
   wartime interest by, 133–5
   workforce of, 178
Logging licences
   and Conservator of Forests, 182, 240–1, 243

Index

and Davies' proposal, 146
and Dethridge's plan, 153
conditions of, 86
demands for, 249
for Alu, 66, 69
for Vanikolo, 85, 86
moratorium on, 109, 227, 242–3, 248–9, 251
procedures to obtain, 226–7, 240–1
    incorrect, 225–4, 245–6, 293, 306–7, 313–14, 331–2
refusal of, 233, 293
Logie, John, 150, 152–4, 188, 364
Lucas, W.H, 46, 65, 162
Luialamo, George, 302, 345, 353
    and Waibona, 300, 303–4, 313–14
Lulei, Denis, 315, 345,
Lungga river, 119, 188, 293, 118, 143, 311
Lupa, 219, 220, 221

M

Maasina Rulu, 128–9, 132, 364
    and community development, 147
    and government support for enterprise, 154
    and lands commission, 140
    political aims of, 128–9
    view of colonial government by, 129, 134, 135, 141, 142
    view of development by, 128–9
MacCrimmon, Neil, 70, 106, 147
    leases of, 71, 89
Magusaiai, 34, 66.
Makambo, 57, 58, 59, 65, 124;
Makili, Lawrence, 272, 294
Makini, Jules, 295–6
Makira
    and Berjaya, 309–10, 313
    and business licence, 305–6
    and Extension Service, 263
    and forest survey, 152
    and Howell Enterprises, 230
    and Integrated Forest Industries, 230–2, 231 fig., 246, 299, 305, 306–8
    and Kayuken, 301, 307–8
    and loggers, 241
    and moratorium, 305, 306
    and Roman Catholic development, 269
    and Rural Industries Ltd, 232, 305
    bribery alleged on, 307
    colonial pacification of, 41
    disputes on, 307–8, 313
    former settlements on, 32
    Hagaparua Sawmilling Association on, 232
    human sacrifice on, 33
    Maasina Rulu on, 128
    minerals on, 72
    mission land acquisition on, 42
    population decline on, 38, 56
    provincial government of, 305–8, 310
    resettlement on, 305
    timber stocks of, 135, 230, 306
    whalers at, 63
    wrecks at, 63, 231
Mala Development Company, 71, 106
Malaita
    and A. E. Enterprises, 304–5
    and cash cropping, 331
    and Golden Springs, 304
    and Kayuken Pacific, 230, 242–3, 298–304
    and labour trade, 38
    and Malayta Company, 70
    and SIDT, 300
    and Sollumber, 306
    and Solmac, 228
    and Soltrust, 304
    and Taisol, 298–302, 304
    and Tenaru Timbers, 147
    and Waibona, 300–2, 304
    and World Wildlife Fund, 272
    application to log on, 134
    artificial islands of, 34
    clearing for towns and gardens on, 128
    commercial enterprise encouraged, 154
    cost of recruiting from, 96
    cyclone on, 320
    defences on, 38
    Development Council of, 180
    disputes on, 304, 328
    forest survey on, 133
    human resources of, 39
    increasing population of, 187, 243
    Maasina Rulu of, 128–9, 132
    mission land acquisition on, 42
    moratorium on, 242–3, 248–9, 298
    provincial government of, 242–3
    recommended cut for, 299
    reforestation on, 229, 261, 262, 263
    relocation in, 128
    saw-milling on, 71, 106, 178, 180–1, 229, 235, 300
    soil degradation on, 187, 323
    subsistence needs of, 261, 263
    taro disease on, 56
    timber stocks of, 135, 229, 299
    UN Development Advisory Team at, 229, 299
    West Kwaio Producers Cooperative Association Ltd of, 304

Index

Malaitans
  attitudes to land, 39–40
  at Vanikolo, 95, 96, 98, 99
  colonial pacification of, 41
  development sought by, 39–40, 178–81
  on Alu, 69; on Guadalcanal, 189
  petitions of, 39–40
  qualities of, 48
Malaria
  aetiology of, 56
  and clearing of forest, 56, 108
  impact on settlement size of, 24
  on Vanikolo, 92, 97
  vector of, 29
Malaysia, 1, 163, 212, 224, 344, 345
Malayta Company, 45, 70
Malesia, 6, 7, 8
Mamaloni, Solomon, 220, 230, 258, 272 fig., 276, 290, 314, 359–60, 378
  and aid to forestry sector, 350–6, 366
  and Australia, 344–5, 347, 352, 354–6, 358
  and devolution, 249–51, 253, 299, 316–17
  and forest policy, 238–9, 348–50
  and Hyundai, 225
  and Kayuken, 301, 303, 307
  and KFPL, 259–60, 367
  and Lever's 224
  and levy, 279–80
  and logging industry, 232, 233, 340–1, 346–7
  and moratorium, 292, 346
  and Pavuvu conflict, 314–15, 358, 367
  and public service, 348–50
  and Salaka's dismissal, 227
  and structural adjustment, 340, 346
  deportation by, 306
  return of State land by, 278–9
Maomatekwa, Premier of Malaita, 242, 298
Marau, 54, 312
Marovo, 272, 343, 344
  and mission logging, 76
  faunal classificatory system of, 27
  forest survey of, 152
  land owners' donations, 83
  Seventh Day Adventist Mission in, 45–6, 54
Marten, Ken, 187, 200, 257
Mataniko river, 118, 128;
Mbanga, 50, 54, 75
Mbatuna, 42, 76, 77, 125, 143, 153
Mbuma, 54, 83, 125, 143, 229, 243,
Media
  and criticism of misappropriation, 356
  global, 315
  local, 237, 271, 272 fig., 350, 344–5, 356–8, 359
  regional, 344, 345, 352, 357–8
Medicine
  European, 43
  from trees, 268
  Melanesian, 19, 23,
Melanesia
  gardening in, 20
  name, 11
  rainforests of, 7
Melanesian Mission, 37, 42, 44, 56
  Southern Cross of, 64
  offer of trees, 84
  and Filose, 108
Melanesians (*see also* Solomon Islanders), 1, 3, 35
  ancestral, 19, 30
  and betel nut, 23
  and botanical classification, 8
  and conservation, 32–3
  and control of erosion, 21
  and forest, 25, 27, 335
  at Vanikolo, 96–9, 105
  food production by, 26
  gardens of, 24
  hunting by, 31, 34
  landscape transformation by, 30, 32
  population of, 31
  subsistence ethic of, 127
Melbourne, 92, 94, 104, 155
Men
  attributes of, 29, 30, 335
  dealing in land, 331
  work of, 19, 20, 22, 150, 331
Mendaña, Alvaro, 62–3
Methodist Mission, 3, 42
  and paths, 54–5
  establishment of, 50, 54–5, 51–3 figs.
  saw-mill of, 75–6
  and commerce, 45–6
Minchin, Devon, 210, 212
Mining, 138, 141, 159, 356
Missionaries, Christian (*see also* Melanesian Mission; Methodist Mission; Queensland Kanaka Mission; Roman Catholic; Seventh Day Adventist Mission; South Sea Evangelical Mission), 3, 36, 40
  and education, 43–6
  and population relocation, 56
  and saw-milling, 73, 75–84
  attitudes to development, 43–6
  clearing of land by, 50–4
  land acquired by, 42
  role of, 41

Index

Mole, 289, 290
Monckton, Eric, 65–9, 87, 89, 125
Monckton, Minna, 69
Mongga, 149, 170,
Mono
    whalers at, 63
    land claims on, 72
    conquerers from, 212
Morea, James, 307–8
Mortality (*see* Population)
Moses, Gideon, 311, 313
Mt Austin, 122, 128, 154, 188, 189, 195
Munda, 169
    and IHDP, 285
    and War, 128, 130–1 figs
    Forestry Department at 201
    Methodist Mission at, 50, 51–2 figs.

N

Naqu, A. L., 152, 191, 192
Nature (*see also* Culture)
    unpredictable, 28
    wild, 27, 371–3
Ndai, logging of, 304–5.
Ndekurana, 219, 220;
Ndene (*see* Santa Cruz)
Ndovele, 284, 286, 355
New Caledonia
    flora, 7, 8
    landscape transformation of, 30, 375
    sandalwood from, 64
    timber exports to, 178
New Georgia, 38
    Allied control of, 116
    and Allardyce, 251, 281–4
    and land settlement, 172
    and World Wildlife Fund, 272
    application to log on, 134
    concession declined, 171
    Forest Area on, 174
    forest survey of, 133
    irrigation on, 25
    landscape transformation of, 128
    logging at Viru, 210, 213
    Methodist Mission at, 45
    north, 218–21, 245–8
    population of, 24, 245
    purchase of timber-cutting rights on, 176, 182
    reforestation on, 257–8
    resistence to land sales, 169
    Saikile Development Company on, 213, 234, 284
    state land on, 167
    taro disease on, 56
    timber rights on, 191, 218, 219, 227, 251
New Guinea (*see* Papua New Guinea)
New Hebrides (*see* Vanuatu)
New Zealand
    aid to forestry, 203, 242, 261–3
    and agroforestry, 262–3, 275, 324
    kauri exports of, 92, 94
    landscape transformation of, 375, 376
    loggers from, 92, 95, 99
    Melanesian Mission from, 43
    military loggers and millers from, 116–25, 120–3 figs.
    strategic vulnerability of, 38
    timber requirements of, 133
    wartime timber restrictions of, 116
Ngalimbiu, 126, 144;
Nggatokae
    logging near beach of, 143
    mortars from, 19
Nggela, 124
    and land bridge, 7, 18
    log purchase from, 65
    military logging on, 123
    mission land acquisition on, 42
    sale of firewood from, 154
Nila, 54, 69
Noel, O. C., Resident Commissioner, 139, 140
Non-government organisations (NGOs) 2, 269–74, 306, 357, 365, 367
    Appropriate Technology for Community and Environment, 273
    Development Services Exchange, 271
    Foundation for the South Pacific, 269
    Friends of the Earth, 221
    Greenpeace, 265, 272–3, 294, 314, 357
    Iumi Tugetha Holdings Ltd, 271–2
    Maruia Society, 287
    Nature Conservancy Division, 271–2
    NSW Rainforest Information Centre, 248
    Save the Children, 314
    Solomon Islands Development Trust, 269–71, 272 fig., 286, 292, 300, 303, 314, 335
    Soltrust, 271, 285, 294–5, 304, 316, 370
    World Vision, 269
    World Wide Fund for Nature, 294
    World Wildlife Fund, 272
Nori, Andrew, 230, 342
Noro, 124, 212,
North New Georgia Corporation, 218–21, 234
    Act, 245, 246
    conflicts within, 219, 247;
Nusuzonga, 50, 130

Index

O

O'Keefe, J. B., 149, 170
Ombudsman (*see* Qoloni, Isaac)
Onepusu, 43, 180,
Orodani, Francis, 314, 345
Osi Lagoon, 134, 135,
Osifelo, Fred, 181

P

Pacific Forum, 344, 351
Pacific Islands Company, 41, 64
Paeu
    government station at, 14, 95
    Laperouse at, 63
    timber company at, 92, 93 figs., 96 fig., 102, 105, 108, 111, 113,162
Paia, H., 247, 251
Papua New Guinea (*see also* Bougainville), 1
    and land bridge, 7
    and World War Two, 115, 120, 121
    Bismarck Islands in, 7
    cyclones in, 13
    flora of, 7–8
    forestry in, 317, 351
    human migration from, 19
    landscape transformation of, 375
    montane forest of, 10
    rainforest of, 6
    South Pacific Forum meeting in, 351
Parks
    and land reservation, 190
    Mt Austin, 154
    on Kolombangara, 189
    Queen Elizabeth II National, 188
Paths
    and Maasina Rulu, 129
    government attitude towards, 55–6
    mission encouragement of, 54–5, 55 fig.
    pre-European, 55
Pests
    Amblypeta cocopharga, 204
    Anopheles mosquito, 29, 30, 56
    beetle (*Brontispa froggatti*), 49
    caterpillar (*Hyblaea puera*), 204
    fungus, 257
    insect, 22
    Koster's Curse (*Clidemia hirta*), 69
    Merremia peltata and spp., 204, 235, 257, 265
    moth (*Tirathaba rufivena*), 49
    Operculina riedeliana, 204
    Oxymagis horni, 204
    shoot borer (*Hysipyla robusta*), 204, 257
    taro blight, 56
    teredo worm (*Teredo navilis*), 64
    ticks, 49
    weed tree (*Brousnnettia papirifera*), 189
Philippines, 1, 163, 224
Phillips, F. B., Judge, 86, 87, 164
Piaito, William, 191, 192
Pigs, 21, 30, 33, 64
Pijin, 95
Plantation economy, 2, 36–61
    and subsistence economy, 323
    establishment of, 39, 43
    labourers, 43, 46
    wage structure, 46
Plantations, coconut, 41, 48
    government's, 2, 56
    missions', 44–5, 50
    war damage to, 127, 130–2
Planters, European, 3, 40
    and Great Depression, 69, 154
    and logging, 65–6
    clearing of land by, 47–8, 49 fig., 60, 64, 336, 362
    cost of clearing by, 66
    views of Solomon Islanders, 46
Poitete, 259, 263
Political parties, 248
    People's Protection Party, 174
    National Coalition Party, 310, 340–2
    People's Alliance Party, 340
    SI National Unity and Reconciliation Party, 340, 345, 359
Polynesians, 30, 35
Population(*see also* Settlement)
    and dysentery epidemic, 88
    and European settlers, 40, 41
    and Maasina Rulu, 128
    and timber needs, 138, 155
    controls on, 35
    decline of, 25, 26, 37, 56, 66, 86, 91, 138, 361–2
    density of, 165, 327, 362
    extinction predicted of, 41, 85, 86
    growth rate of, 326
    human, 24, 35
    infant mortality of, 35
    life expectancy of, 35
    needs of, 326, 366
    of bush, 39
    on north Guadalcanal, 30, 31
    on Vanikolo, 91
    pressure of, 24–5, 197, 374–5

Index

Solomon Islander, 40
stable, 26
youthful, 335
Prehistory, 1, 18, 19.
Provinces (*see also* Government, Solomon Islands, devolution)
    and business licences, 278, 306
    and control of forestry, 251–3, 316, 317–18, 367
    and forestry levy, 279–80
    and loggers
        in eastern and central islands, 299–318
        in western islands, 277–99
    and national government, 278–80
    and Provincial Government Act, 212, 306
    attempt to remove, 359
Pu Veterei, chief, 22

Q

Qoloni, Isaac
    and Eagon, 245
    and Hagaparua, 306
    and Hyundai, 227
    and Kayuken 243, 302
    and parliament, 252
    and Taisol, 299
    court action by, 307
Queensland, 37, 38, 39–40

R

Ramoli, Ben, chief, 89, 111
Ramoni, Premier of Makira, 230, 310
Ratu, Nelson, 314, 317
Reef Islands
    arboriculture on, 22
    food crops of, 22
    recruiters at, 85
Reefs, damage to, 30, 281, 307
Reforestation,(*see also* Forestry Department/ Division, reforestation) 66, 90, 174, 180
    and loggers, 258–61, 289–90
    at Vanikolo, 110–11
    techniques of, 187, 196–7, 205, 255
    costs of, 197, 260, 261
    on customary land, 180, 203, 243, 354
    on Malaita, 229, 243
Relocation (*see* Settlements)

Rennell Island
    and FRIS, 315
    mineral on, 72
    timber stocks on, 230, 315
Renton, John, 20, 22
Reserves (*see* Parks)
Ringgi Cove, 216–17, 220
Riti, Philemon, 284
Roads (*see also* Paths)
    at Vanikolo, 156
    for development, 170
    in wartime, 117–18
Rohinari, 54, 300
Roman Catholic Church, 269
    and logging, 290
    and social justice, 358–9
    and *sup sup* garden, 273
    in Shortland Islands, 286
    St Dominic's rural training centre of, 273
    women's programmes of, 290
Roman Catholic Mission, 42, 44, 54
    and pit saw, 83
    and relocation, 56
    saw-milling of, 83, 125, 170
Root crops, 18, 19–22
    cassava (*Manihot* spp.), 56
    Cyrtosperma, 25
    pana (*Dioscoria esculenta*), 21, 27
    sweet potato, *kumara* (*Ipomoea batatas*), 38, 56
    taro (*Colocasia esculenta*), 21, 22, 23, 25, 27
    yams (*Dioscoria alata*), 21, 22, 23
Roughan, John, 269, 300, 314
Rove, 152, 268, 354
Roviana Lagoon
    and leaf supply, 75
    Lapita fragment at, 30
    Methodist Mission at, 45, 51
    soils of, 50
Russell Islands
    and Greenpeace, 272–3, 314, 357
    and Mamaloni government, 314–15
    and Marving Bros, 313–15
    and provincial government, 314
    and Soltrust, 271
    land claims on, 313, 315
    Lever's land on, 313
    Lever's milling on, 148
    logs from, 64
    military milling on, 123
    resettlement on, 313–14
    timber rights on, 220
    timber stocks of, 213, 220

Index

S

Sa'a, 37, 29, 71
Salaka, Peter, 213, 226–7, 228–9, 239
Samoa
    and labour trade, 37
    floristic province of, 7
    missionaries from, 50
San Cristobal (*see* Makira)
San Cristoval Estates Ltd, 85, 86, 92
    and agreement with government, 108–9
    financial problems of, 106
San Jorge Island
    claim to land on, 72–3
    vegetation of, 31
Santa Cruz (Ndene)
    and fear of sorcery, 96
    and kauri management plan, 158, 191
    and KTC, 158
    application to log on, 134, 135
    arboriculture on, 22
    kauri on, 113, 135
    logging option sought on, 113
    recruiters at, 85
    reforestation on, 261, 256, 257–8
Santa Cruz district (*see also* Temotu)
    colonial pacification of, 41
    crops on, 23
    cyclone damage in, 14
    French interests in, 84
    irrigation in, 25
    labour from, 95, 98, 99
    population decline in, 38
Santa Isabel
    and Axiom, 292–5
    and Church of Melanesia, 291, 294
    and cyclone Ida, 176, 199, 202, 211 fig.
    and Earthmovers, 295
    and Eastern Development Enterprises, 297
    and Greenpeace, 272, 294
    and Isabel Timber Company, 293–5, 287, 355–6
    and Kumpulan Emas Berhad, 295, 297
    and timber leases, 86
    and Pacific Timber Company, 291
    and Soltrust, 294
    and timber revenue, 202
    and World Wide Fund for Nature, 294
    and World Wildlife Fund, 272
    Development Authority of, 291, 294
    development education on, 292
    environmental sensitivity on, 297
    Forest Area on, 174
    forest survey of, 133
    land purchased on, 167
    land rights on, 291
    licenced cut on, 297
    local logging proposition for, 153
    mission land acquisition on, 42
    Mothers' Union on, 294
    population of, 291
    reforestation project on, 265–6, 274
    soil damage on, 187, 291
    Spanish at, 62
    Sustainable Forest Project on, 265
Santo (Espiritu Santo), sandalwood trade, 64
Savo, 124
    forests of, 31
    volcano on, 31
Saw-milling (*see also* Tenaru Timbers; Timber industries)
    and A. E. Enterprises, 304–5
    and Atasi, 180, 229, 235
    and British Solomon Islands Timber Co., 149, 180
    and Burns Philp, 65
    and Eagon, 289
    and Hagaparua Sawmilling Association, 232, 306
    and Kayuken, 299–300
    and Rafea and Kwaleunga Sawmill Co-operative, 181
    and Soltrust, 271, 294
    and Taisol, 299
    and United Church, 285
    and West Kwara'ae Sawmill Co., 178
    at Bina, 299–300
    at Faisi, 66
    at Mamara, 71
    at Mbatuna, 76–83, 81 fig., 125, 143, 153, 177
    at Mbuma, 143, 177
    at Su'u, 71
    at Tenavatu, 148
    by British Solomon Timber Company, 149, 177
    by Cox, 170, 171
    by Foxwood, 178
    by Kwan How Yuan, 170
    by Maasina Sawmill Ltd, 243
    by O'Keefe, 149, 170, 171, 177
    by Vuragare Association, 234
    economics of, 271, 361
    expatriate, 177–8
    legislation for, 177
    local, Forestry Department's view, 177–81, 184
    local, Logie's view, 153–4
    Methodist mission, 54, 75
    military, 117–25, 123 fig.

Index

missions', 362
number undertaking, 237
on Guadalcanal, 311
on Vanikolo, 106
problems with,148–9
production from, 149, 177
rationalisation of, 299–300, 365
sales of logs for, 170, 177
Sorenson's plans for, 72
wartime production from, 141
with portable *wokabaot* mills, 181, 271, 294
Saw-pit
    at Mbuma, 83–4
    at Roviana, 75, 75 fig.
Seghe, 124, 125,
Sesepe, 123, 124
Settlement
    human, 1, 18–9
    location of, 30
    patterns of, 35
    spread of, 30
Settlements, 27, 374
    and logging, 92
    clearing for, 93
    consolidation of, 56
    in mountains, 32
    influence of Christianity on, 38, 56, 76
    lack of, on coast, 41
    mobility of, 33
    relocation of, 56, 329–40, 376–7
    size of, 24, 33
Settlers
    and development, 43
    and growth of towns, 154–5
    and Maasina Rulu, 128–9
    European, 2, 40
    first, 18, 30
    landscape transformation by, 30, 56–61
    views of, 41, 60–1
Seventh Day Adventist Church, 218
    and Komuniboli, 269
    at Ndovele, 286
Seventh Day Adventist Mission, 42, 45–6, 50
    adherants of, 169
    and commerce, 46
    and log supply, 83, 143
    and gardens, 54
    and villages, 50, 55
    dietary restrictions of, 46
    hospital, 83; housing, 73 fig.
    saw-milling by, 76–83,125, 143, 153, 177
Ships
    Anastasia, 63
    Astrolabe, 63
    Belama, 154
    Bouselle, 63
    Houto, 92
    James Cook, 103
    Malaita, 76
    Melanesian, 178
    Rob Roy, 64
    Southern Cross, 64, 85, 95
    Swift, 286
    Titus, 51
Shortland Island (*see* Alu)
Shortland Islands (*see also* Alu), 38
    and Allardyce, 281
    and commercial logging, 210
    and sago, 23
    clay pots from, 19
    depopulation of, 66, 86
    logging team at, 67 fig., 69
    society, 24
    timber control at, 267
Sikaiana, 63, 64
Singapore, 224
Slash-and-burn (*see* Gardening)
Smith, Adrian, 358–9
Société Française des Nouvelles Hebrides, 84, 85
Societies (*see also* Beliefs; Trade), 37
    and leadership, 24, 30, 33–4, 37, 176–7, 218–19, 332–3
    characteristics of, 35
    Melanesian, 225
    size of, 35
    values of, 332–5
Soga, chief, 291
Soils
    alkaline ultramafic, 10
    and logging, 187, 188 fig., 205
    and nutrient cycle, 16–17, 16 fig., 21–2
    and sedimentation of reefs, 30, 307
    erosion of, 6, 17, 21, 24, 30, 32, 47, 162, 323
    fertility of, 19, 21, 23, 25, 31, 32, 47, 187, 265, 273
    impact of cyclones on, 14, 16
    instability of, 4
    leaching of, 17
    on New Georgia, 50, 245
    protection of, 150, 165, 187, 273
    study of, 194
Solomon Islanders (*see also* Melanesians; individual Solomon islands), 2
    and direct dealing with loggers, 169–70, 175, 177, 181–2, 184, 296 fig., 363–4, 365
    and European landscape transformation, 60–1
    and Forest Areas, 174, 364
    and land, 166, 169, 183
    and missions, 60–1, 362

and perception of forest resource, 184, 328–38, 373–4
aptitude for clearing land of, 48
as logging workers, 95–8, 155–6
botanical taxonomy of, 27
environmental knowledge of, 30, 35, 336–7
in forestry matters, 161–2
needs and wants of, 76, 84, 234, 284–5, 329–33, 335–8, 365–6
suspicions of government of, 166
timber control by, 238, 284
view of independence by, 333
views of Australia by, 345
Solomon Islands, 1
annexation of, 38
colonial pacification of, 41, 56
fauna of, 6, 19
flora of, 6, 7, 8
geography and geology of, 4, 6
income, 2
paths in, 54–6
population of, 40, 326
rainfall of, 4, 31
seasonal winds of, 102
sovereignty of, 41–2
Solomon Islands Forest Industries Association (SIFIA)
and Code of Practice, 355
and complaints, 358
and duty on logs, 342
and FRIS, 349
and government, 348
and local processing, 341
and restructuring, 344
court action by, 278
Solomon Islands Labour Corps, 125
and milled timber, 153
Sorenson, Peter Neils, 72, 87
South America, 1, 8.
South Pacific Regional Environmental Programme, 249, 268–9
South Sea Evangelical Church, 229
and environmental issues, 272
South Sea Evangelical Mission, 42, 43–5
South-east Asia, 2
flora of, 6, 7, 8
gardening in, 24
human migration from, 19
timber processing in, 164
transfer-pricing in, 236–7
Standard Logging Agreement, 239–41, 244, 355–6
Star Harbour, 232, 306, 313,
Structures
European, 49, 57–9 figs, 153, 154, 222

government, 56–9, 125, 142
indigenous, 73, 147, 153–4
mission, 50, 51–3 figs, 54, 55 fig., 73–4, 73–4 figs, 75
wartime, 117–24
Su'u, 55, 71, 106, 147;
Subsistence economy (*see also* Gardening; Economy, non-monetary), 323–4, 377
value of, 138, 140, 324–6, 366
Suny Tong (Suny Wun Sai Tong), 233, 260, 299, 303
Supa, N., 293, 354
Swabey, C., 157, 191
Sykes, R. A., 109, 110

T

Taiwan, 233, 273, 299, 300
Tambea, 230, 234,
Tan Sri Dato Tan Chee Yioun, 308, 309 fig., 310
Tangarare, 54, 170
Tara, Tarcisius, 357–8
Tausinga, Job Dudley, 219, 220–1, 224, 304
and North New Georgia Corporation Act, 245–6
resource control, 245, 247
support for Kengava, 289
views on conservation, 245–8
Technology, 1
and logging, 95–6, 98, 101–3, 156, 158, 162, 179, fig., 188 fig.
chain-saws, 147, 156, 171
draft animals, 69
firearms, 37–8
for ply-making, 157
machines, 50, 101, 103, 147–8
metal-based, 2, 37–8, 63
rolling stock, 101, 103, 104, 105, 147
stone, 20, 25, 35, 37
tools, 37, 47, 48
weapons, 37
Temotu (*see also* Santa Cruz district)
and Allardyce, 212
and Eastern Development Enterprises, 315–16
and EC, 316
Development Authority of, 316
and Soltrust, 316
and central government, 316
Tenaru Timbers Pty Ltd, 143–50
and government control, 150, 161
and timber duty, 144–5, 157
Tenavatu, 147, 148
Tenema, 89, 105

Index

Tetepare
   and logging, 288
   state land on, 167
   timber stocks on, 65
Tevai, 92, 105
   and government control of, 143–4
   and kauri management plan, 158
   application to log on, 134, 135
   French interests in, 84–5
   lease of, 71, 110–11, 158
Thomson, B. R., 200, 257
Tikopia, 305, 376
   canarium introduced, 22
   landscape transformation of, 30–1, 136
   pig-extirpation on, 33
   settlement of, 30
Timber (*see also*, Firewood; Saw-milling)
   American publicity about, 133–4
   consumption, 148 fig.
   domestic demand for, 2, 76, 84, 153–5, 177–8
   economic importance of, 202
   for export, 64, 106, 150, 178
   for liquid containers, 94
   for plywood, 94, 178
   for reconstruction, 125, 126 fig., 127
   for shipbuilding, 62–3, 94
   imported, 56, 73, 84, 143, 148 fig,
   industries (*see also* SIFIA, Saw-milling)
      and reforestation, 198
      bribery in, 249, 345, 354
      eco-timber prices, 285, 316
      eco-timber, 265, 271, 272 fig., 274, 294, 295, 314, 316, 341
      government support for, 170
      potential of, 198
      pre-World War Two, 62–90
      processing of, 157, 178, 190–1, 197, 198–9, 299, 310, 341
      role in development of, 164
   indigenous names of, 138
   leases, 66–7, 71, 85, 86
   local, 2, 75
   markets for, 1
   milled
      indigenous demand for, 147, 153–4, 177–81
      quality of, 148–9, 177
   needs of the people surveyed, 138
   overseas demand for, 135, 163
   prices of, 76, 83, 143, 226, 285
   production of, 106, 125, 148 fig., 149 fig., 177–8
   products, 64, 76, 82–3 figs., 83
   qualities of, 2
   royalties
      and Allardyce, 212
      and export duty, 157–8, 201–2
      and Integrated Forest Industries, 231
      and KTC, 145
      and reforestation levy, 202
      and VKTC, 107–9
      calculated on superfeet, 109
      calculated on trees felled, 107–9
      compared to duty on timber, 145
      disputes over, 219, 232
      on Shortland Island, 66–7, 69
      on Vanikolo, 87, 89
   treated, 177
   wartime demand for, 116–27
Tina river, 293, 169
Tonga, 7
Tools (*see* Technology)
Toshio Hashimoto, 232, 246, 299
Trade (*see also* Commerce)
   barter, 24
   domestic, 19, 35
   goods, 37, 63
   in nuts, 34
   in vegetables, 34
   overseas, 37, 41, 63–4;
Traders, 40, 64, 84
Traditions, oral, 27
Treasury Island (*see* Mono)
Tree species (*see also* Trees; crops)
   Afzelia bijuga' (sic.), 64
   Agathis alba, 161
   Agathis australis, 84, 91
   Agathis macrophylla (kauri), 7, 9, 84–6, 91–114, 100 fig., 156, 193, 256
   Agathis robusta, 193
   Agathis spp., 194
   Albizia spp., 11
   andila, 48
   Araucaria cunninghamii, 161, 193
   *Araucaria hunsteinii,*160, 193
   Araucaria spp., 194
   Banyans, 48
   Brian Boru, 94
   Bruguira spp., 10
   Calophyllum inophyllum, 33
   Calophyllum kajewski, 11, 32, 83, 84, 125, 193
   Calophyllum spp. (koila, koilo), 10, 66, 67, 70, 84, 94, 101, 119, 123, 143, 147, 212, 213, 272, 285
   Campnosperma brevipetiolata, (karamati or ketekete) 97, 9,10, 11, 13, 32, 156, 193, 194, 204, 204–5, 210–1, 256, 257
   Casuarina equisetifolia, 33, 64
   Casuarina papuana, 9, 10
   Cedrela australis, 194

Index

Cedrela mexicana (pencil cedar), 194 fig.
Cedrela odorata, 193, 256
Chryosophyllum roxburghii, 123
climax species, 11, 13
Dillenia salomonensis, 11, 13, 32, 210, 213
Dillenia spp., 7, 237
Dipterocarpaceae, 7
Endospermum medullosum, 11, 94
Eucalyptus deglupta, 194, 195 fig., 202, 204, 245, 256, 257, 260, 265
Eucalyptus marginata (jarrah), 156
eucalyptus, 48
Gmelina arborea, 202, 245, 256, 257, 259, 260, 263, 280
Gmelina molluccana, 11, 193, 263
Gonystylus macrophyllum, 161
hardwoods, 11, 48
Hibiscus tilacus, 9, 33
Intsia bijuga, 94, 237, 281
lauan (meranti), 163
Macaranga similis and spp., 11
Maesopsis eminii, 193, 194
Mangifera solomonensis, 119, 127, 144, 148, 161
mangroves, 56
Maranthes corymbosa, 11
Mastixiodendron smithii, 123
megamega, 21
Metroxglon solomononesis, 9, 10, 23
Ochrosia spp., 33
Palquium spp., 126
Pandanus, 9
Paraserianthes falcataria, 9
Parashorea malaanonan, 161
Parinari salomonesis, 11
Pemphis acidula, 73
pioneer species, 11, 13
Pipturus spp., 33
Pometia pinnata, 23, 71, 119, 126, 144, 147, 148, 212
Pterocarpus indicus, 94, 285
Rhizophora spp, 10
Santalum spp., (sandalwood), 64, *Schizomeria serrata*, 11, 13
Schleinitzia novaguineensis, 33
shade-tolerant species, 11
Shorea marcroptera, 160
softwoods, 11
Sweitenia macrophylla, 193, 204, 256, 257, 262
Tectona grandis, 256
Terminalia brassii, 7, 9,10, 67, 69, 70, 134, 194, 203, 210, 213, 245, 256

Terminalia calamansanai, 9, 11, 193, 194, 204, 256–7
Terminalia ivorensis, 193
Terminalia spp. 126, 193
Terminalia superba, 193
tui, 21
Vitax cofassus, 7,23, 31, 123, 220, 285
Xanthostemon spp., 73
Xylocarpus granatum, 94
Tree, food, 19, 20
   apple (*Spondias dulcis* and *S. dulcis*), 22
   bananas (*Musa* spp.), 22
   betel nut (*Areca catchu*), 23
   breadfruit (*Artocarpus altilis*), 22, 23
   canarium (*Canarium* spp.),18, 21, 22–3
   coconut (*Cocos nucifera*), 22, 63
   domestication of, 22
   Malay apple (*Eugenia malaccensis*), 22
   Oceanic lychee (*Pometia pinnata*), 23
   pandanus (*Pandanus* spp.), 23
   Polynesian chestnut (*Inocarpus fagiferus*), 23
   pomelo (*Citrus grandis*), 22
   sago (*Metroxylon solomonense*), 23
   to'oma (*Terminalia solomonensis*), 22
Trees (*see also* Forests; Saw-milling)
   felling of, 20–1, 25–6, 47–9
   planting of, 33
   stocking rates, 47, 65
   dimensions, 56
   price paid for, 65, 69, 71
   ownership of, 88–9, 127, 139, 261, 290.
Trenaman, Keith W., 157, 162, 186, 189 fig.
   and aid, 199, 203
   and customary land, 180, 190
   and forest protection, 187, 190
   and functions of department, 179–80, 206
   and Lever's, 190–1, 213
   and reforestation, 196–7, 199, 203
   and timber processing, 197, 202
   establishment of Forest Estate by, 165–77, 183–5, 191, 364, 369, 374
   introduction of levy by, 202
Tropa, A., 230, 239
Tropical Forest Action Plan, 352, 378
Troup, R. S., 109, 133, 137, 165
Tua, Ben, 158
Tua, chief, 86, 111
Tuhanuku, Joses, 310, 314, 342, 345, 358
Tulagi, 95, 154
   and War, 113, 115, 116, 123
   landscape transformation at, 56, 57–60, 57–9 figs.
Tuti, Dudley, 291, 293

Index

U

Ughele, 74, 284
Ulufa'alu, Bart, 359–60, 378
Unilever, 220, 221
United Church (*see also* Methodist Mission), 284
   and small-scale logging, 285–6
   and societal needs, 284–5
   and SWIFT, 285–6, 294
   development programme of, 285
United Nations Organisation, 163
   and decolonisation, 364
   Development Advisory Team, 229, 245, 261
   FAO
      advice on log exports, 181, 199
      advice on policy, 249
      advice on timber processing, 198
      advice on timber varieties, 197, 198
      and effects of logging, 256
      draft forest legislation, 250
      and prediction of timber demand, 164
      warning, 236
   UNICEF, 273
United States of America (*see* America)
Unusu, Arthur, 278, 280, 287, 345
Utupua, 85–6

V

Valuables, customary, 19, 33.
Vangunu
   and Silvania Products, 287, 344
   commercial logging on, 210, 213
   forest survey of, 152
   forest types on, 10
   SDA mission at, 77
Vanikolo, 14, 105
   and population 85, 86, 87, 92, 111, 362–3
   area of, 92
   canarium of, 22
   climate of, 92, 98
   closure of logged areas on, 110
   distribution of rents on, 112
   dysentery epidemic on, 88
   forest species of, 7
   forest survey of, 109
   Forestry Department on, 152, 191, 192 fig.
   French interests at, 84–5, 88
   Japanese shell-fishers at, 85
   kauri management plan for, 158, 191
   Lands Commission at, 88, 111
   Laperouse shipwreck at, 63
   leases for, 85–6, 111–12
   local council of, 161
   regeneration on, 110–11, 191, 256
   shipping to, 102, 113–14
   terrain of, 98
   timber licence conditions on, 86–8
   timber rights on, 87
   timber stocks on, 94, 104
Vanikoro Kauri Timber Company (VKTC; *see also* Kauri Timber Co), 71, 135
   and diseases of workforce, 97–8
   and government, 94–5, 98, 107–9
   and Kauri Timber Company, 113
   and reforestation, 110–1
   and rights to unused land, 112
   and royalties 107–9
   employees of, 71, 95, 96–7, 98, 99, 100–2 figs, 102, 104 fig., 106, 113
   establishment of, 91–2
   extraction pattern of, 104–5, 363
   firewood for, 108, 154
   kauri sent to Malaita by, 71, 106
   liquidation of, 113
   log exports by, 106, 107 fig.
   and logging contractors, 113
   logging methods of, 98, 101–3, 110
   logs requisitioned from, 113
   performance by, 105–6, 113–14
   rolling stock of, 50, 101, 103, 103 fig., 104, 105–6
   saw-mill of, 92, 106
   timber exports by, 106, 150
   wartime losses of, 113
   working conditions of, 96–7
Vanuatu, 233
   floristic province of, 7
   geology of, 7
   landscape transformation of, 30
   sandalwood of, 64
   timber exports to, 178
Vegetation (*see also* Trees), 10
   Cordyline terminalis, 23
   heath (*Gleichenia* spp.), 31
   herbaceous, 11
   identification of, 8
   Lycopodium cernum, 31
   strand, 47
   succession of, 23
Vella Lavella
   and Hyundai, 282, 284
   and Odalisk, 282
   application to log, 134
   Goldie's plantation on, 46
   land disputes on, 282

logging damage on, 282
military logging on, 117–18 figs
people's perception of, 286
Villages (see Settlements)
Viru, 48, 292
    and Eagon, 290, 352, 356, 367
    Forestry Department at, 171, 193, 204, 210, 257–8
    logging of, 288
Visale, 54, 74, 228;
Volcanoes, 4, 6, 31
Vona Vona, 220, 234
Vouza, Jacob, 166
Vura, 169, 228;
Vuragare Development Association, 220, 234

W

W. R. Carpenter
    interest in timber, 142
    shippers and merchants, 57, 58, 71, 84, 102
Wainoni Bay, 54, 74, 170,
Walker, F. S., 31, 137, 201, 362
    forest survey by, 110, 133, 138–41, 147
Ware, chief, 66–7
Water supply
    and Controlled Forests, 175, 184, 187
    and logging, 90
    on Choiseul, 290
    protection of, 150, 165
Weapons (see Technology)
West, the (of World) 2
    botanical classification, 8
Western Province
    and Choiseul, 244–5
    and environmental consciousness, 269
    and KFPL, 259–60
    and logging companies, 278
    and revenue, 278
    and Silvania, 287, 345
    and levy, 280
    log exports from, 278
    log extraction in, 280
    log value of, 280
    provincial assembly of, 220
    State land in, 278–9
Western Solomons
    Allied offensive in, 116
    cassava introduced to, 56
    Cheyne at, 64
    mission land acquisition in, 42
    timber stocks of, 135, 187

Wheatley, Norman, 50, 83
Whitmore, T. C.
    and human disturbance of forest, 32
    botanical research of, 7, 11, 13, 190, 201
    forest reserve recommended by, 188
Wickham, Frank (the first), 50, 51
Wickham, Frank, 267
Women
    work of, 19, 20, 22
    and Forestry Department/Division, 150, 268
    and development projects, 262–3, 275
    and *sup sup* garden, 273
    empowerment of, 290
    Mothers' Union, 294
    and logging, 331
Wong, James, 233, 292
Woodford, Charles Morris, 40–2
    and development, 87
    and forestry, 89–90
    and French, 84
    and Japanese, 85
    and kauri, 85
    and logging on Shortland Island, 66, 70
    and plantation economy, 89
    botanical interests of, 56, 59
    and coconut industry, 59
    deportation of Sorenson by, 72
    opinion of Solomons' timbers of, 64
    purchase of Tulagi by, 56
    refusal of land claim by, 72–3
    support for logging by, 86
    view of Solomon Islanders of, 40–1, 361–2
World Bank
    and logging, 256, 351
    and Ulufa'alu government, 360
    and utilisation of tropical timbers, 164
    funding sought from, 203
    policy advice from, 249
    pressure on government, 341
World War Two, 2, 90, 113–4, 116–41
    Allied forces in Protectorate during, 116, 127
    demand for timber in, 116–27, 153
    environmental effects of, 116, 130–2 figs., 336, 362
    war damage compensation from, 127

Y

Young, Florence, 44, 45.

www.ingramcontent.com/pod-product-compliance
Lightning Source LLC
Chambersburg PA
CBHW021812300426
44114CB00009BA/144